HOW TO USE THIS BOOK

- In this book, you will find access to thousands of carefully constructed questions.
- The questions are organised into sets so that you can determine progress towards learning objectives.
- The Question Sets are structured and categorised by topic and subtopic to help you plan using Learning by Questions.
- Each Question Set is set out in this book with all its questions, answers and diagrams or illustrations to allow you to check suitability for your class.
- Once you have found the perfect Question Set you'll be able to find it on our website at www.lbq.org using the Quick Search References.

Everything in the book is online at www.lbq.org

LbQ can be delivered in these ways:

Self-Paced Questioning

Set a task for your pupils – they each receive questions on their devices, respond at their own pace and receive instant feedback.

Ad Hoc Questioning

Pose questions on the spur of the moment – the pupils respond on their devices.

Teach

Turn any question into a slide, ideal for modelling concepts on your classroom display.

Self-Paced Questioning

Lbq.org is built on tens of 1000s of questions. Questions are grouped into carefully scaffolded Question Sets to provide structured support for learning and to help pinpoint problems

When you click the 'Start' button, you're opening the door for your class to start a differentiated learning journey.

START BUTTON

Pupils connect with a simple code and start receiving questions straight away.

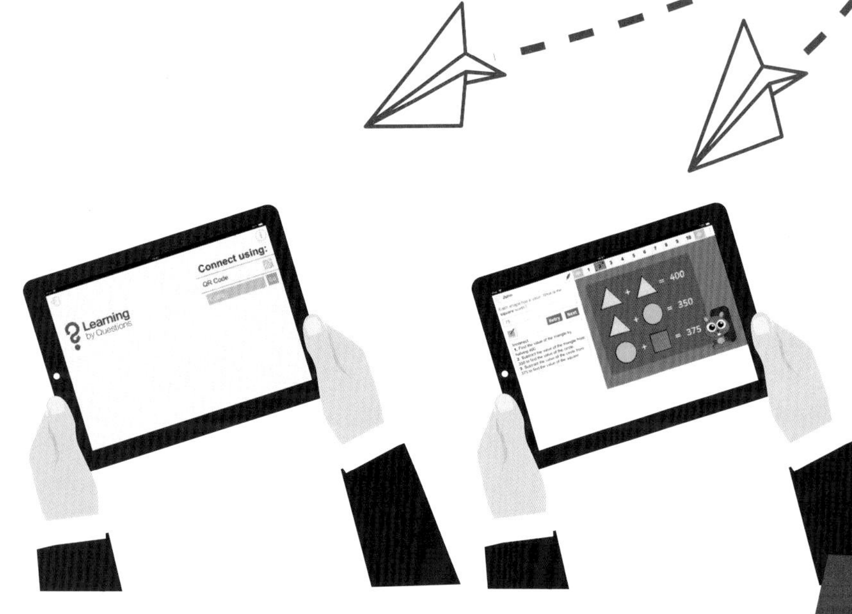

*After the initial free trial, a subscription is required for Self-Paced and Ad Hoc tasks. Teach mode remains free.

When your pupils hit a challenging question,
you'll know about it.

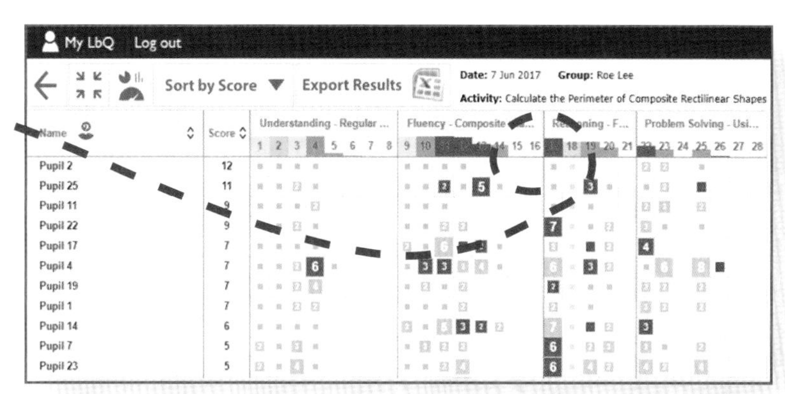

Drill down to see every answer and be right on top of every misconception.

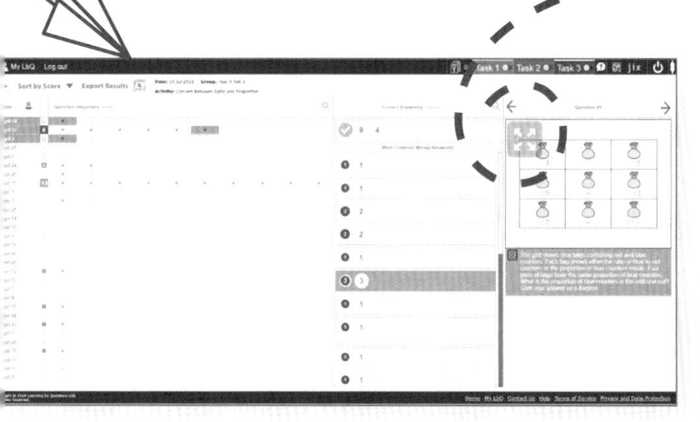

Pause, intervene, explore, explain and model using 'Teach' mode.

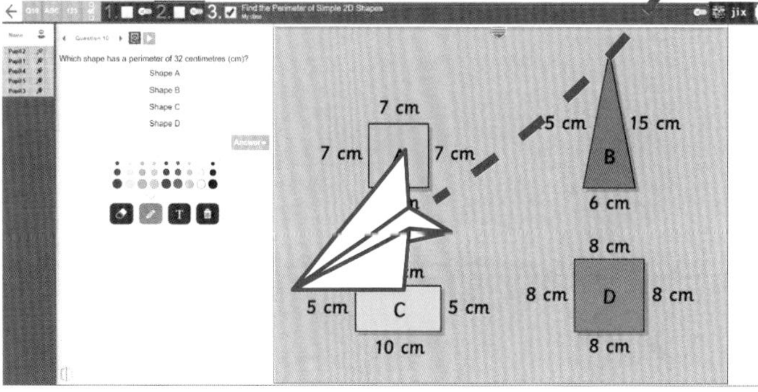

Try the question again with Ad Hoc questioning.

Everything in this book is online at www.lbq.org

SE THE BOOK TO PICK THE RIGHT TASKS FOR YOUR CLASS AND
E READY FOR INTERVENTION OPPORTUNITIES.

Ad Hoc Questioning

Have your pupils got tablets, laptops, Chromebooks, PCs? Lbq.org makes it fast, easy and super-productive to engage your whole class.

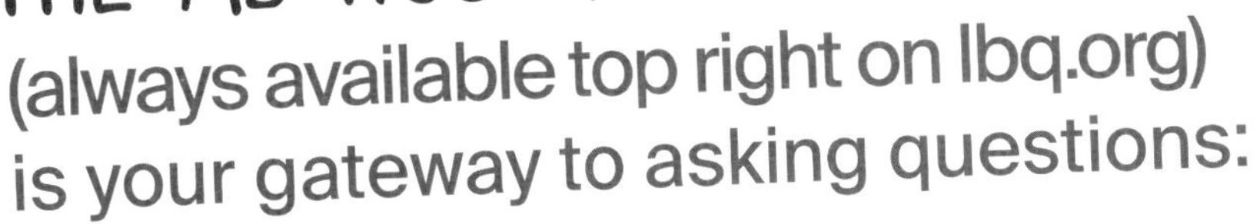

(always available top right on lbq.org) is your gateway to asking questions:

at any time,
of everyone,
about anything,
even during a Self-Paced task.

Are your class struggling with a challenging question?

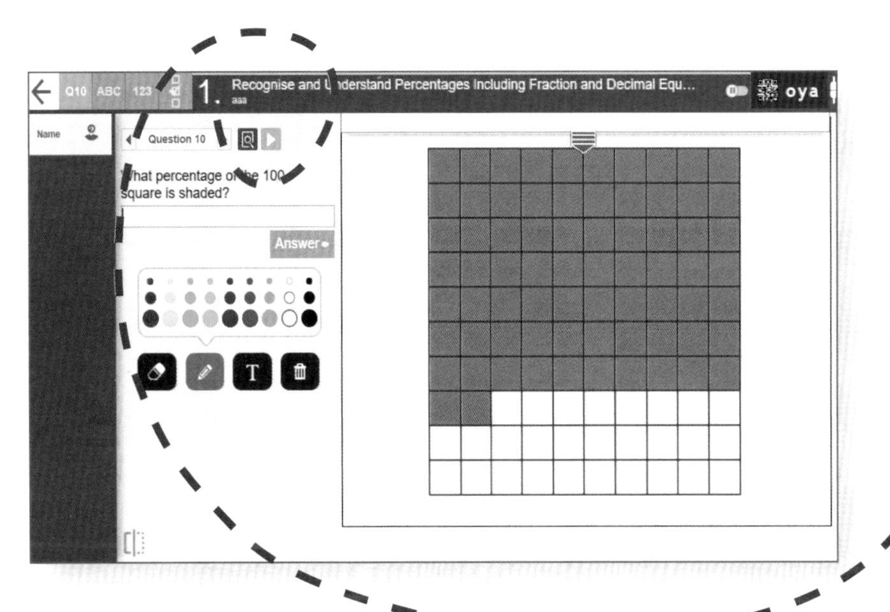

Use the green button to send the question to your class and give everyone another chance to answer.

AD HOC QUESTIONING IS **FREE** FOR 60 DAYS!

Register FREE at lbq.org

Vant to build on an existing lbq.org question?

Annotate to explain, model, modify and extend...

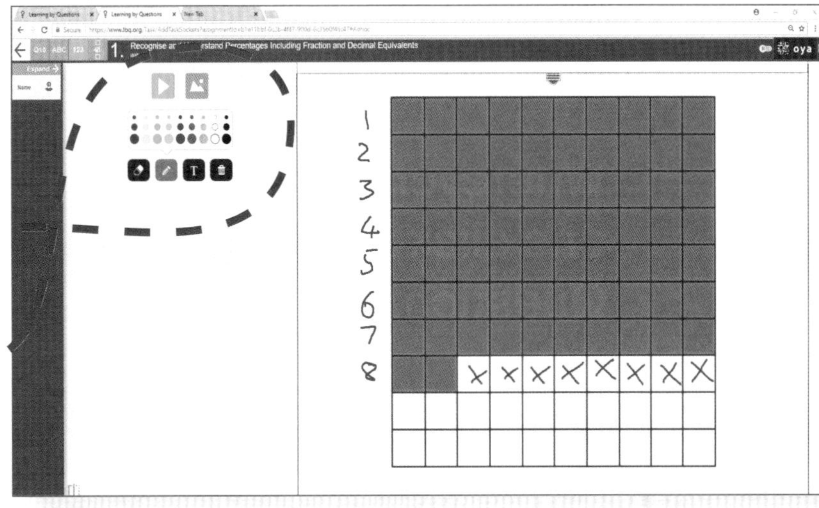

...and use Ad Hoc to send as a new question to your class.

Make your own questions on the fly.

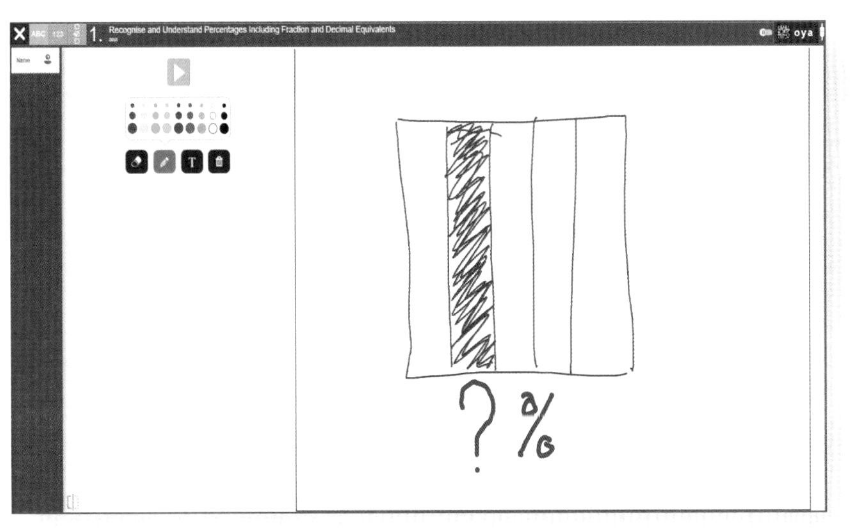

Use our teach tools to create the right question at the right time.

Write a question, draw a question or just ask a question.

Forget 'hands up'. With Ad Hoc questioning everyone answers, every question, every time!

USE THE BOOK TO CHOOSE, ORGANISE AND REHEARSE QUESTIONS TO TEACH IN YOUR CLASSES.

Teach

If you don't have pupil devices, you can still turn any question into a whole-class teaching resource on your classroom display.

Wherever you see the 'Teach' icon, you're one click away from turning a question or Question Set into a teaching resource.

TEACH ICON

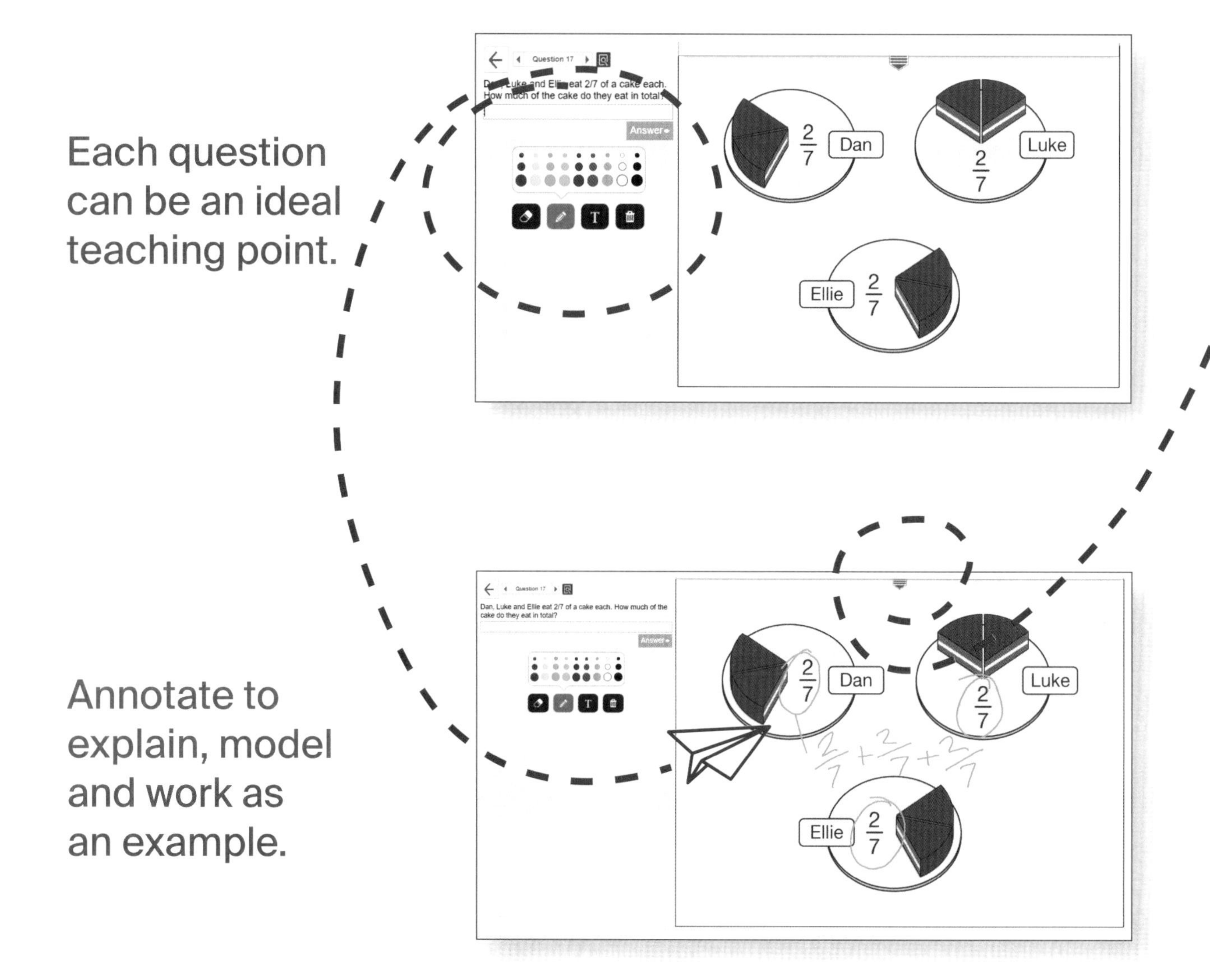

Each question can be an ideal teaching point.

Annotate to explain, model and work as an example.

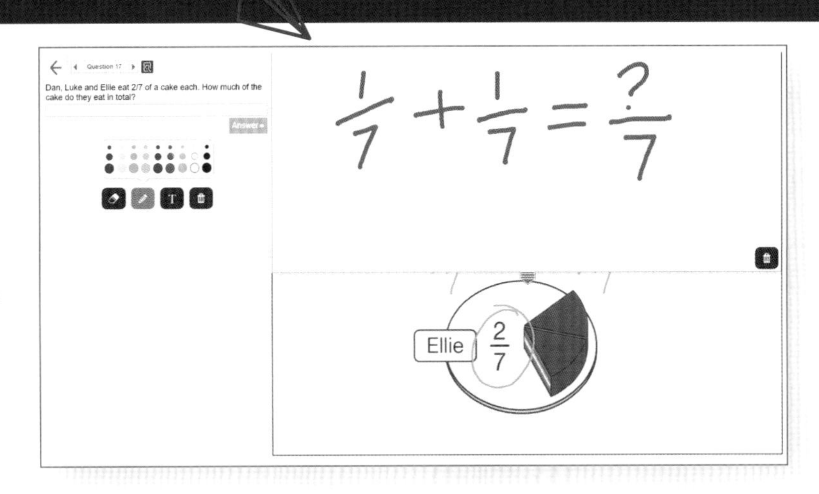

Use the pull-down pad to construct your own questions.

Work through multiple questions like a slide show.

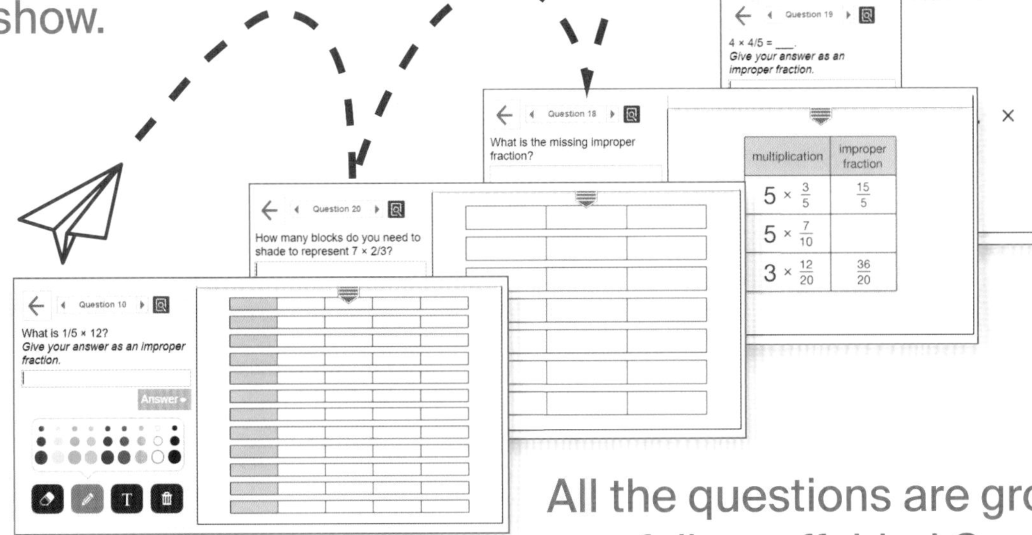

All the questions are grouped into carefully scaffolded Question Sets and provide great support for progression from understanding to problem solving.

TEACH MODE IS **FREE** TO ALL REGISTERED USERS

Register FREE at lbq.org

USE THE BOOK TO HELP FIND, ORGANISE AND REHEARSE QUESTIONS TO INCLUDE IN YOUR LESSONS.

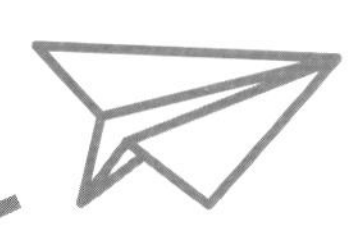

Tailored Instant Feedback with Every Question

Feedback corrects misconceptions and reinforces learning to accelerate progress.

LbQ's digital platform provides pupils and teachers with instant feedback.

The feedback provided is designed to reflect the kind of specific feedback a teacher would provide, given the time to do so.

Our feedback gives specific guidance and pupils are given the opportunity to retry the question.

Feedback is high quality and consistent across all Question Sets, relating directly to learning points taken from the National Curriculum.

Confidence and resilience building is developed through feedback. Encouraging students to try again takes away the 'fear of failure'.

STEPPED GUIDANCE

We don't just give the correct answer. We give feedback that provides short steps to reach the right answer. This is effective in its simplicity, providing guidance on the action needed to find the correct answer.

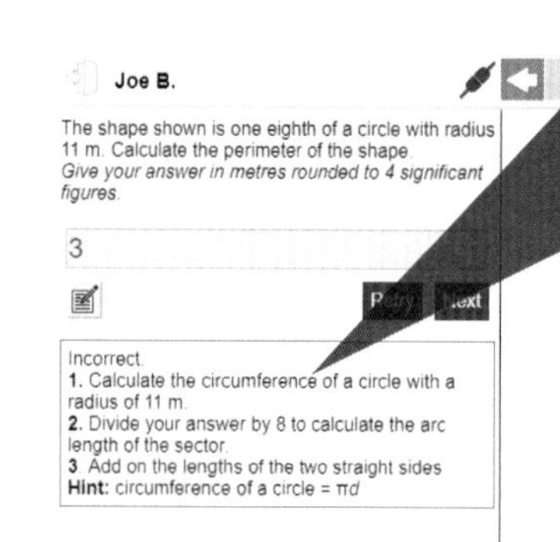

Incorrect.
1. Calculate the circumference of a circle with a radius of 11 m.
2. Divide your answer by 8 to calculate the arc length of the sector.
3. Add on the lengths of the two straight sides.
Hint: circumference of a circle $=\pi d$

11 m

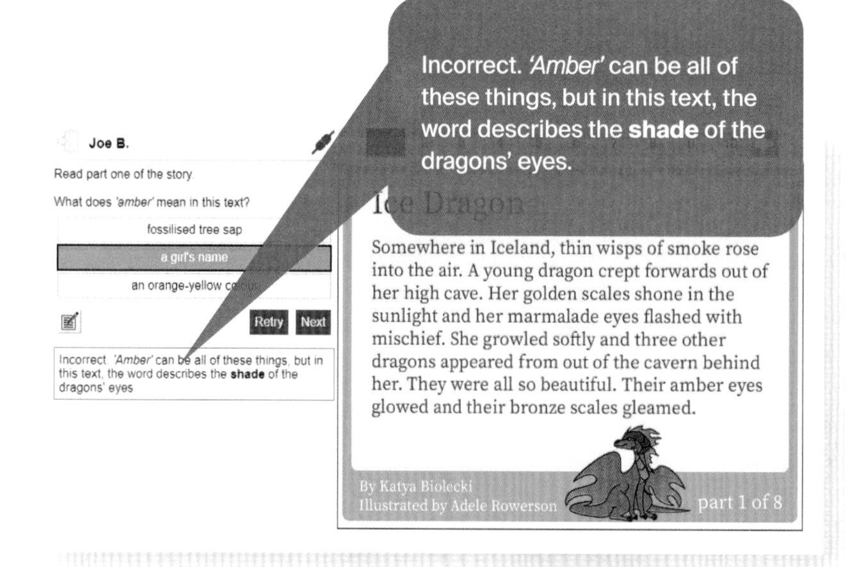

Incorrect. *'Amber'* can be all of these things, but in this text, the word describes the **shade** of the dragons' eyes.

COMMON MISCONCEPTIONS

Common misconceptions are addressed in feedback to encourage pupils to find the correct solution themselves.

REINFORCING DETAIL

Correct-answer feedback provides learners with reinforcing detail that boosts confidence and produces opportunities for deeper learning experiences.

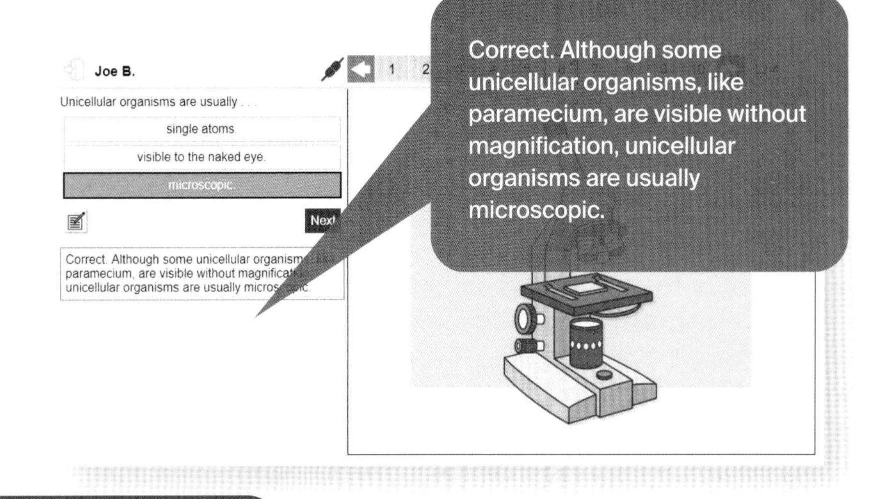

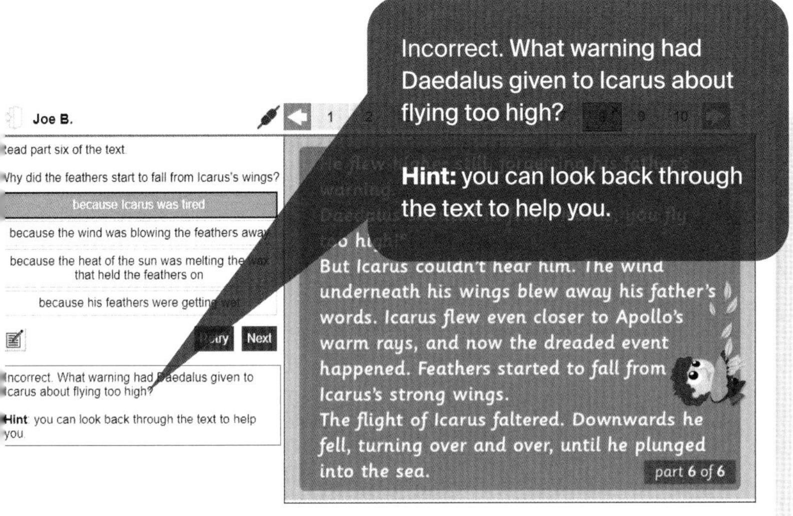

GROWTH MINDSET ENCOURAGEMENT

Feedback that provides information on how to correctly answer the question next time encourages a growth mindset by emboldening pupils to have another go.

SUPPLEMENTARY INFORMATION

Supplementary information in feedback for correct answers compounds understanding and nurtures curiosity.

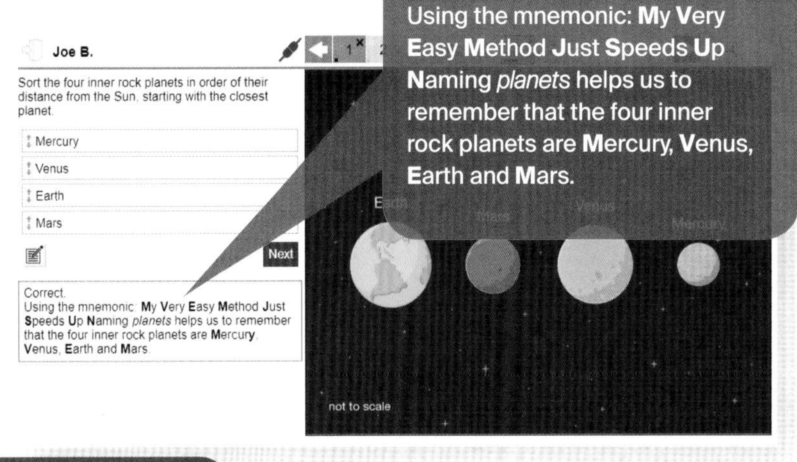

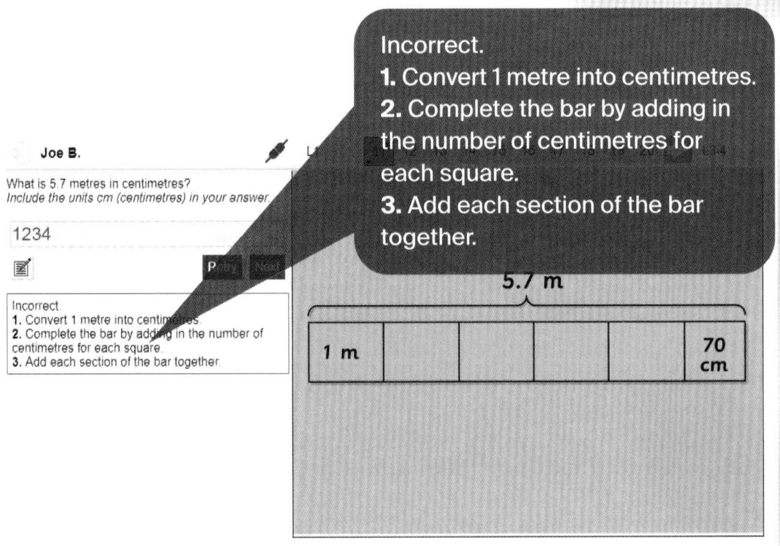

SUPPORTING IMAGERY

Feedback that explains how to use the image provided allows less-secure learners to use pictorial representations rather than abstract understanding.

Mastery Across the Subjects

Learning by Questions (LbQ) is a classroom app filled with curriculum-aligned Question Sets and immediate pupil feedback to supercharge learning. LbQ reduces teacher workloads with automatic marking and real-time lesson reports to verify learning as it happens. LbQ covers KS2 maths, English and science, plus KS3 maths, English, chemistry, physics and biology.

Teachers use LbQ's live data to pinpoint where intervention is needed and why. Pupils use devices such as tablets, to progress through scaffolded Question Sets at their own pace and receive immediate feedback as they answer. Teachers receive live analysis and results are automatically recorded to support assessment and planning and to save time.

Mastery Approach

The mastery approach informs the creation of all of our Question Sets. At LbQ, we believe that mastering a topic or concept is vital. Each pupil's journey to mastery may be different to their fellow students' journeys, so it is important to us that scaffolding or guiding of learning is built into our Question Sets.

LbQ embraces digital tools to make the mastery approach to learning more achievable than it has ever been before.

Our mastery Question Sets are specifically designed to progress pupils only when they demonstrate a grasp of prerequisites, helping them to master concepts, build a growth mindset and take ownership of their learning. Equally, if a student demonstrates mastery level understanding, progression rules will allow the pupil to move on after a set amount of correct answers, pushing higher ability pupils at pace.

Mathematics

Mastery

Maths mastery Question Sets take pupils on a learning journey through four different cognitive levels of a topic: understanding, fluency, reasoning and problem solving. To aid pupil progression, every question in every level provides contextualised feedback and addresses key misconceptions. Maths mastery Question Sets provide students with exposure to problem solving and reasoning questions, but not before previous skills and knowledge have been mastered.

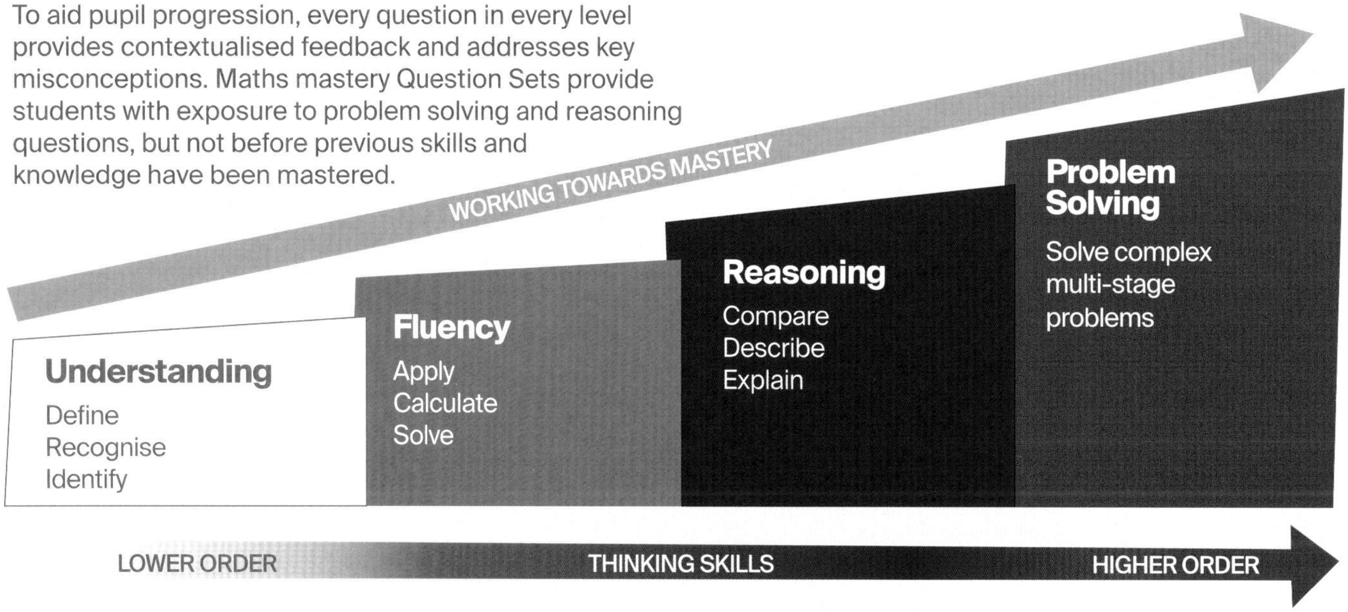

WORKING TOWARDS MASTERY

Understanding
Define
Recognise
Identify

Fluency
Apply
Calculate
Solve

Reasoning
Compare
Describe
Explain

Problem Solving
Solve complex multi-stage problems

LOWER ORDER — THINKING SKILLS — HIGHER ORDER

Practice

There are elements within the curriculum that require a different approach to learning. Practice Question Sets in maths drill one or two levels of cognition, where pupils recall and apply knowledge rapidly. Such Question Sets are perfect for practising times tables, number bonds and other basic recall of operations needed to progress in maths.

English

Guided Reading

This collection of Question Sets in English focuses on guided reading for KS2 and on understanding text for KS3. They assist in developing the reading skills of retrieval, inference and understanding vocabulary through exposure to a range of classic and original texts.

WORKING TOWARDS MASTERY

Defining
Pre-Read Question Sets
What does...mean?
What do you expect this text to be about?

Recalling
Retrieval Question Sets
How many...?
Which character...?
When did... take place?

Inferring
Inference Question Sets
Why is/does/did...?
Which statement best describes the...?

Analysing
Language Question Sets
What do you think the author meant by using the phrase...?
Why do you think the author uses the word...?
Structure and Features Question Sets
What genre is this text?
Which paragraph explains...?
What is the purpose of...?

LOWER ORDER — THINKING SKILLS — HIGHER ORDER

Mastery

LbQ's English mastery Question Sets encourage students to understand the nuances of grammar and terminology before moving on to the application of rules. English mastery Question Sets push students to face common misconceptions head-on and provide explorative feedback that encourages deeper thinking and understanding.

Practice

Practice drill questions of spellings and word classifications are an invaluable part of learning. LbQ has a range of practice Question Sets in English, linked directly to the requirements of the National Curriculum, including spelling lists associated with each year group.

Science

Mastery

Science mastery Question Sets enable pupils to progress through a carefully stepped sequence of learning. Blocks of questions are structured to guide learners as they move from knowledge, through understanding and application to analysis. Real-world applications are woven through the Question Sets, making learning relevant and relatable.

WORKING TOWARDS MASTERY

Knowing
I can recognise...
I can name...

Understanding
I can use a model to describe...
I understand how...

Applying
I can apply my learning to...
I can suggest ways to...

Analysing/ Evaluating
I can predict...
I can evaluate claims that...

LOWER ORDER — THINKING SKILLS — HIGHER ORDER

Practice

LbQ's science practice Question Sets cover areas of the curriculum where repetition of a learning point is essential to ensure that it is thoroughly embedded. They often focus on a skill or area of the curriculum where a weaker grasp of knowledge may become a barrier to future learning, e.g. energy transfer terminology.

Investigation

Question Sets support the skills of planning an investigation and analysing results. Pupils will be given opportunities to plan investigations and review data. In KS3, students are provided with further Question Sets that present a mix of exam-style questions and a wider application of the data.

Subscriptions

FULL ACCOUNT

Start a 60 day no obligation trial of the FULL Account and view affordable pricing at www.lbq.org/trylbq

FULL Account subscription includes:
- Connecting pupil devices,
- instant pupil feedback,
- automatic marking,
- real-time lesson analysis,
- lesson results saved.

Use LbQ books to plan and quickly locate Question Sets at www.lbq.org/books

FREE ACCOUNT

Register at www.lbq.org/trylbq

FREE Account includes:
Tens of 1,000s of questions with learning feedback for use by teachers on a classroom display.

Teach

Other Titles in the Series

Learning by Questions

PRIMARY KS2	SECONDARY KS3
Maths Year 3 Mathematics Primary Question Sets Year 4 Mathematics Primary Question Sets Year 5 Mathematics Primary Question Sets Year 6 Mathematics Primary Question Sets	**Maths** Year 7 Mathematics Secondary Question Sets Year 8 Mathematics Secondary Question Sets Year 9 Mathematics Secondary Question Sets
Science Years 3&4 Primary Science Question Sets Years 5&6 Primary Science Question Sets	**Biology** Years 7–9 Biology Secondary Question Sets
English Years 3&4 English Primary Question Sets Years 5&6 English Primary Question Sets	**Chemistry** Years 7–9 Chemistry Secondary Question Sets
	Physics Years 7–9 Physics Secondary Question Sets
	English Years 7–9 English Secondary Question Sets

Quick Search Reference Guide
All Question Sets, all subjects, all years

US Series

US Math
Grade 5 Mathematics Question Sets
Grade 6 Mathematics Question Sets
Grade 7 Mathematics Question Sets
Grade 8 Mathematics Question Sets

See www.lbq.org/books for title availability

Understanding a Question Set

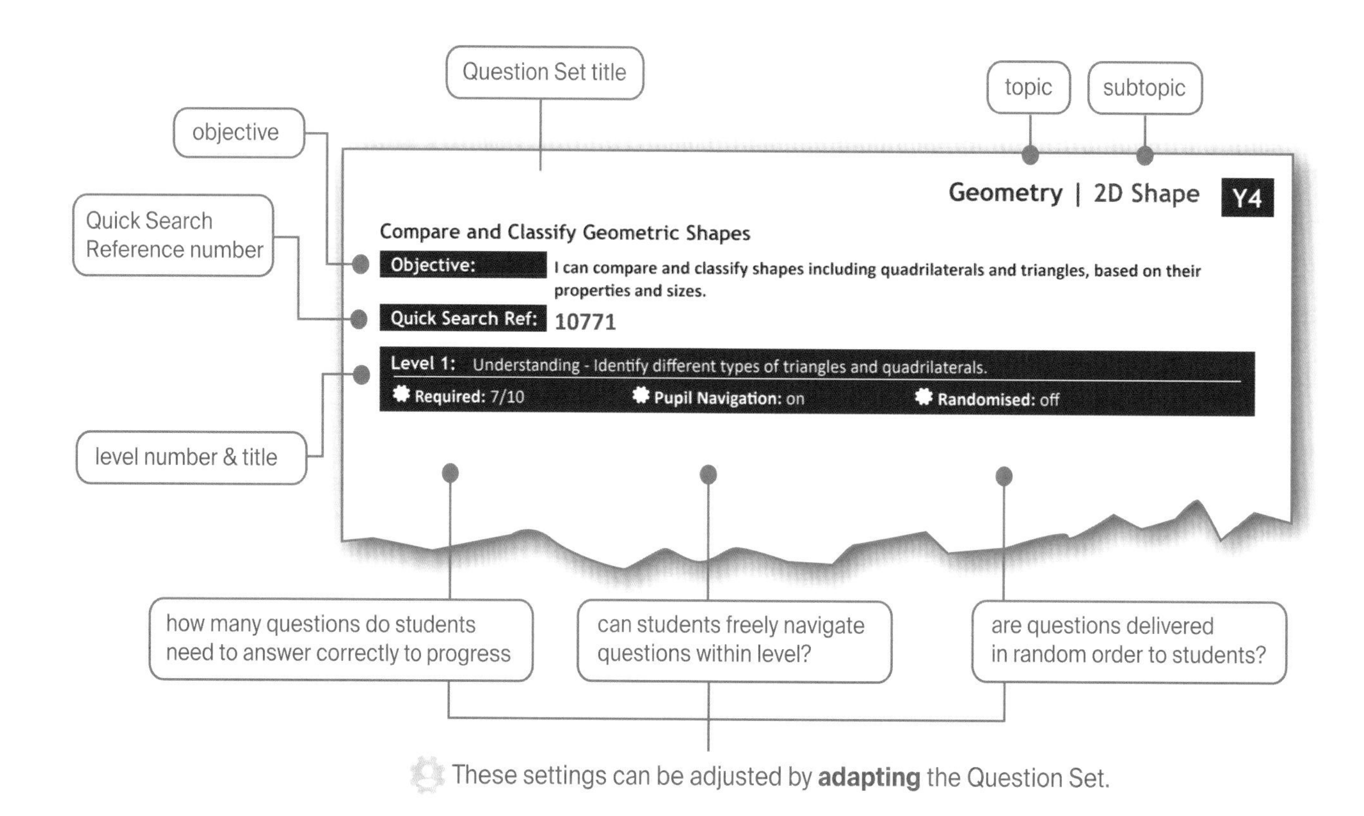

These settings can be adjusted by **adapting** the Question Set.

Understanding a question

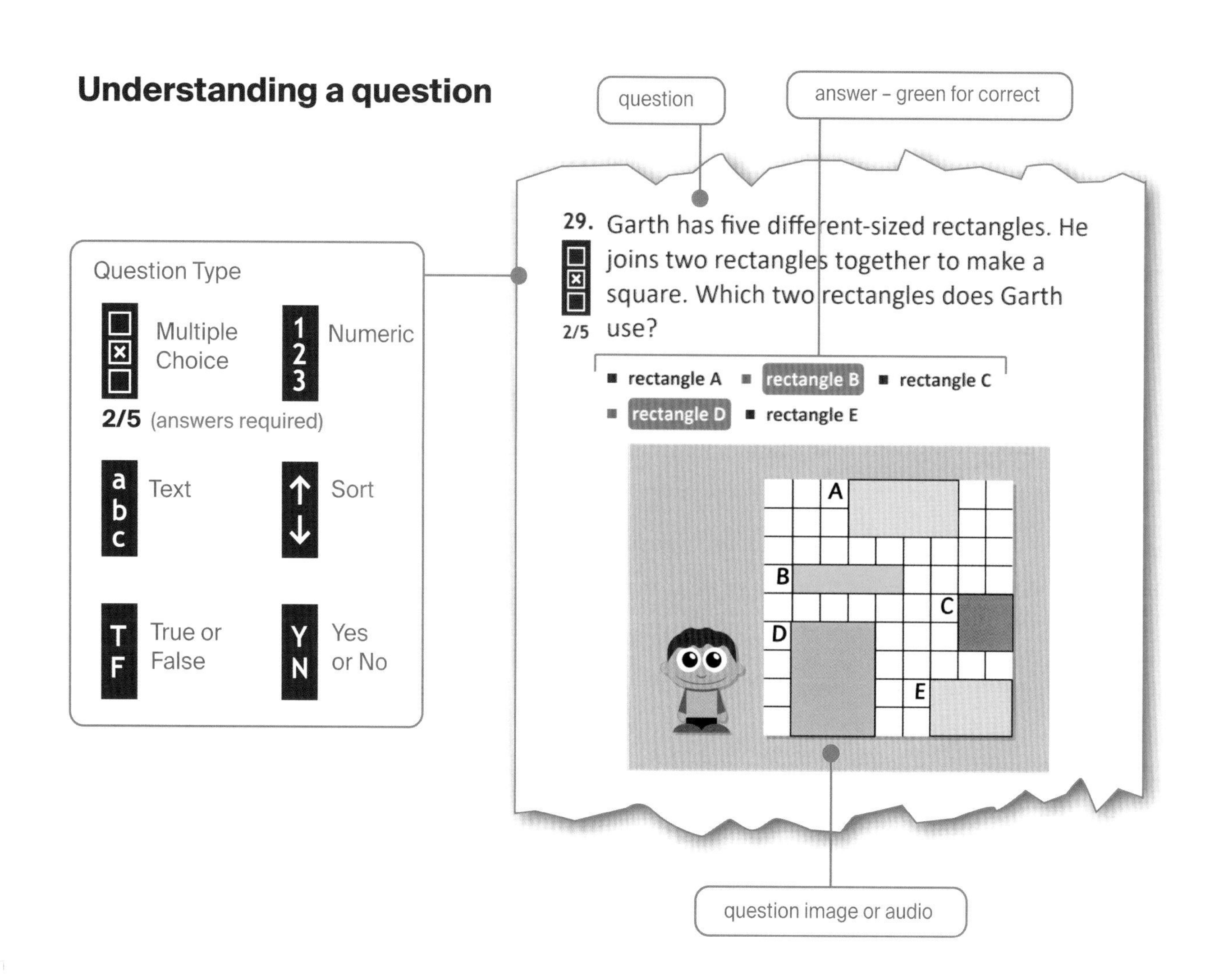

Finding a Question Set From This Book on the LbQ Platform

The **year, topic** and **subtopic** classifications used in this book relate directly to those used on the LbQ platform. The fastest way to find a specific question set on lbq.org is via the **Quick Search Reference Number** (e.g. 10771).

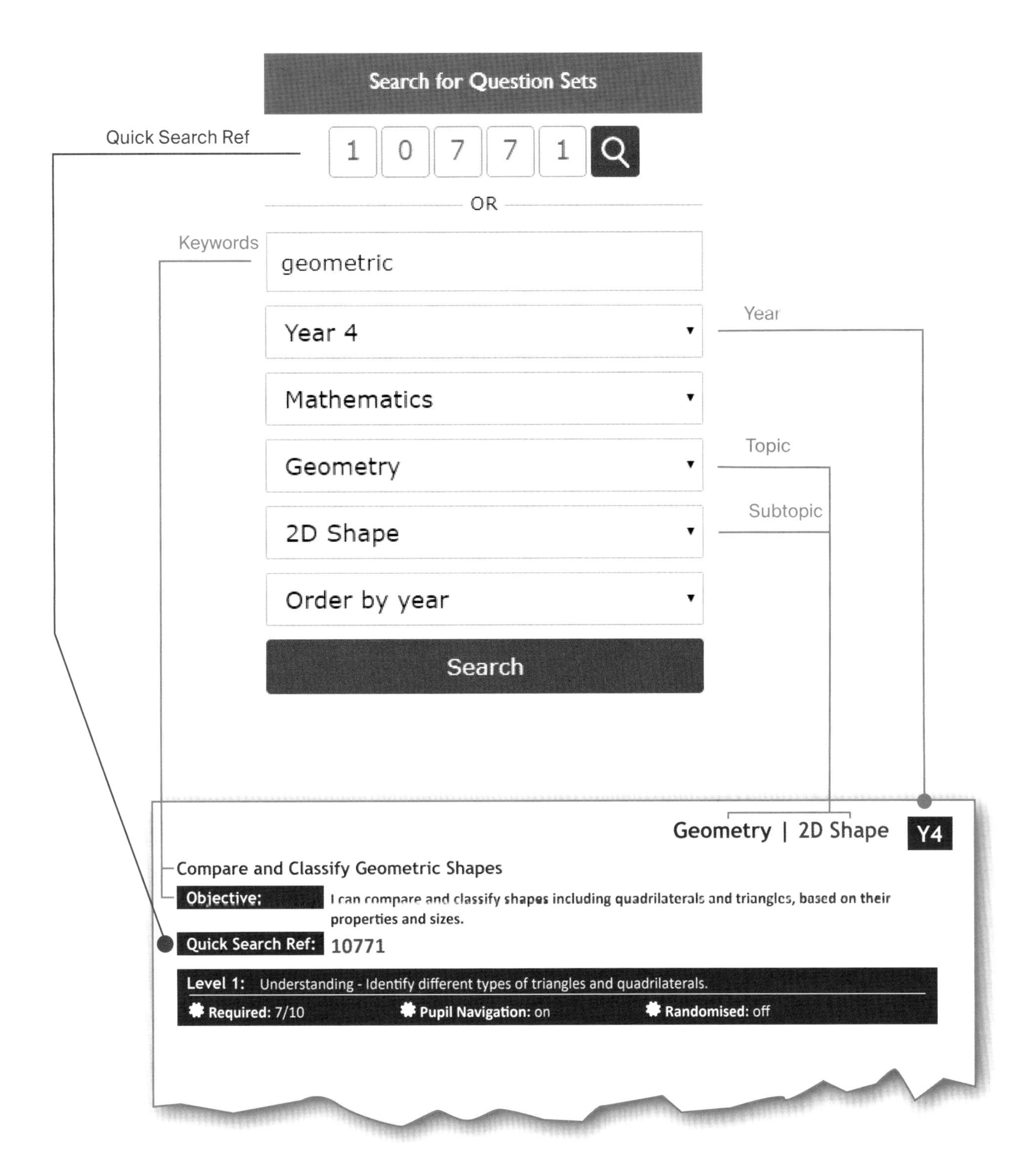

Note: The Question Sets detailed in this book are correct at time of compilation (December 2018) and correspond directly to the Question Sets as published on www.lbq.org.

Owing to the nature of www.lbq.org we will from time to time extend, update or modify the Question Sets published there, which will give rise to discrepancies between this book and the online resources.

Topic Directory Y9

QS Ref:

Geometry | Similarity and Congruence

Geometry | Pythagoras and Trigonometry

Statistics

Statistics | Scatter Graphs

Statistics | Averages and Range

Statistics | Frequency Polygons

Probability

Probability | Probability Scale

Probability | Probability of Single Events

Probability | Probability of Combined Events

Probability | Sets and Venn Diagrams

Ratio and Proportion

Ratio

Use Scale Diagrams and Maps

Objective: I can use scale factors for diagrams and maps.

Quick Search Ref: 10380

Level 1: Understanding: Finding actual lengths using scales.

Required: 7/10 Pupil Navigation: on Randomised: off

1. A scale on a map tells you . . .

1/4

- the actual distance in kilometres from one town to another.
- what different symbols on the map mean.
- the orientation of the map.
- the ratio between distances on the map and actual distances.

2. A scale on a drawing, plan or map can be given in words or represented by a ratio. A plan has a scale where 1 centimetre represents 2 metres. What is the ratio of measurements on the plan to real-life measurements?

1/6

- 1 : 2
- 200 : 1
- 2 : 100
- 2 : 1
- 1 : 200
- 100 : 2

3. Reagan draws a television with a scale where 1 centimetre on the drawing represents 5 centimetres on the television. What is the actual height of the television?
Include the units cm (centimetres) in your answer.

- 50 centimetres
- 10 centimetres
- 50 cm
- 50
- 10 cm
- 10

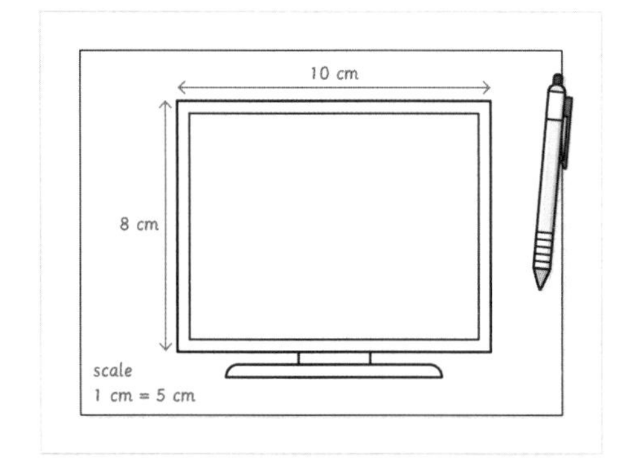

4. A floor plan of a classroom is drawn with a scale of 1 centimetre to 2 metres. What is the actual length of the classroom?
Include the units m (metres) in your answer.

- 16 m
- 16 metres
- 16
- 8 centimetres
- 8
- 8 cm

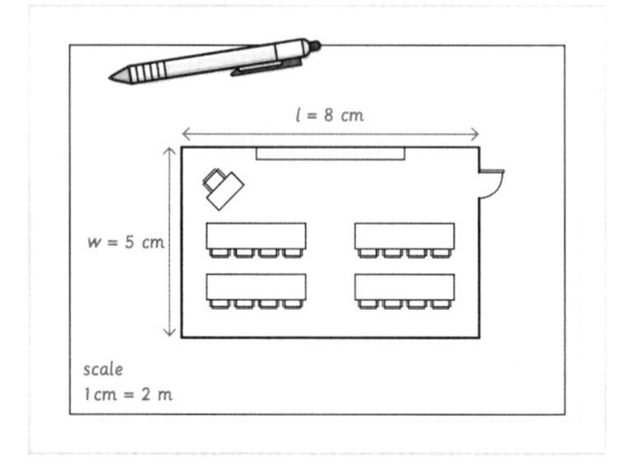

5. Emily draws a car with a scale of 1 : 20. What is the length of the car in metres?
Give your answer as a decimal and include the units m (metres).

- 22.5
- 450
- 22.5 cm
- 4.5 metres
- 4.5 m
- 22.5 centimetres
- 4.5
- 450 centimetres
- 450 cm

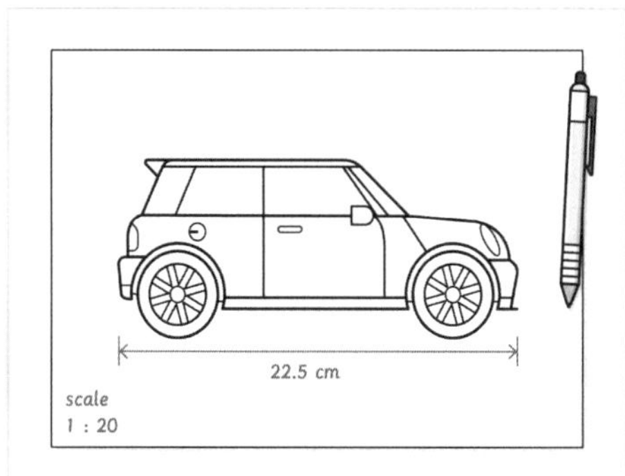

Level 1 *continued*

6. A map has a scale of 1 cm to 500 m.
What is the actual distance in kilometres from Skipwick to Malton?
Give your answer as a decimal and include the units km (kilometres).

- **3.25 kilometres** ■ 6.5 centimetres ■ 3250 metres
- 3.25 ■ 3,250 metres ■ **3.25 km** ■ 3250 ■ 6.5 cm
- 3,250 m ■ 6.5 ■ 3,250 ■ 3250 m

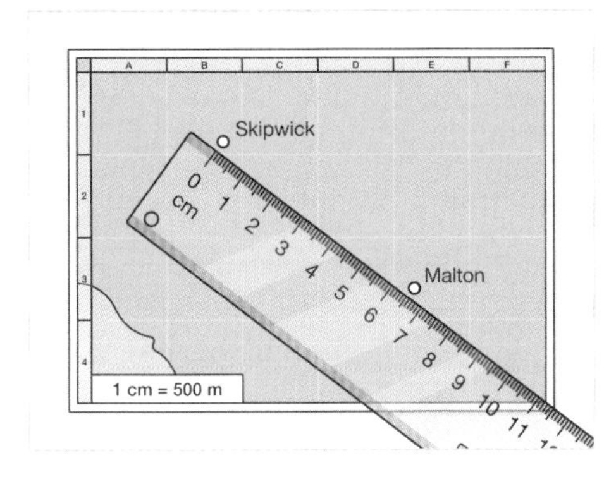

7. Halima draws a building with a scale of 1 : 250.
What is the actual width of the building in metres?
Include the units m (metres) in your answer.

- 24 cm ■ 6,000 cm ■ 24 ■ **60 m** ■ **60 metres**
- 6000 ■ 24 centimetres ■ 6000 cm
- 6000 centimetres ■ 60 ■ 6,000
- 6,000 centimetres

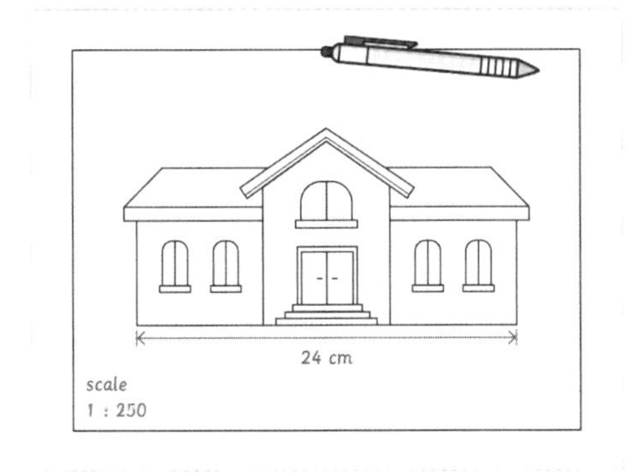

8. A design for a chair is drawn with a scale of 1 cm to 4 cm.
What is the actual height of the chair?
Include the units cm (centimetres) in your answer.

- 28 centimetres ■ **112 cm** ■ 112
- **112 centimetres** ■ 28 cm ■ 28

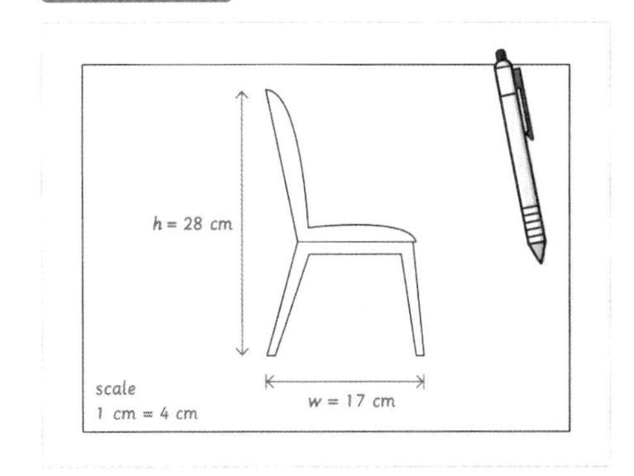

9. Jake draws a plan of his garden with a scale of 1 : 50.
What is the length of his garden in metres?
Include the unit in your answer.

- **12 m** ■ 1,200 centimetres ■ 24 ■ 1200 ■ 12
- **12 metres** ■ 24 centimetres ■ 1200 cm ■ 24 cm
- 1,200 ■ 1,200 cm ■ 1200 centimetres

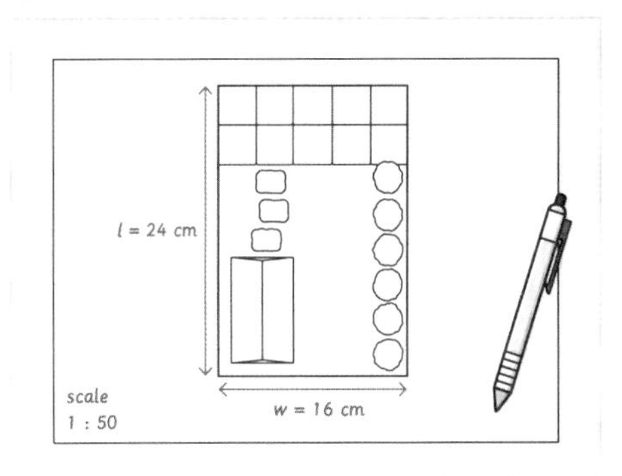

10. Alice draws a plan of her bedroom with a scale of 1 cm to 25 cm.
What is the width of her bedroom in metres?
Include the units m (metres) in your answer.

- **4 m** ■ 400 centimetres ■ 16 cm ■ **4 metres**
- 400 ■ 4 ■ 16 ■ 400 cm ■ 16 centimetres

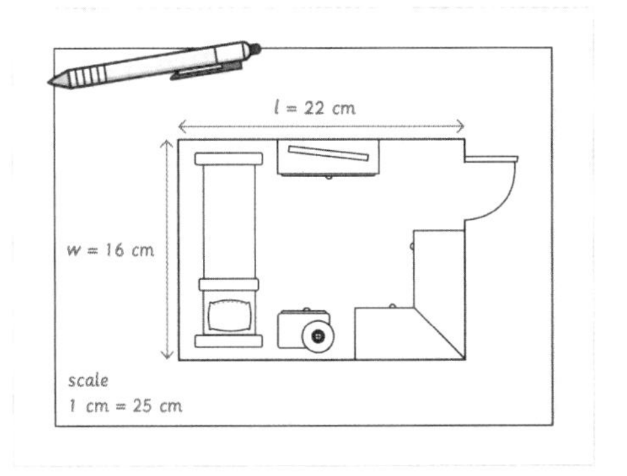

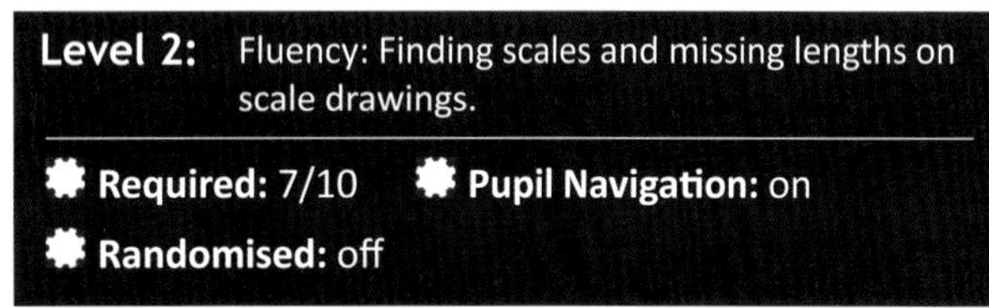

11. A scale drawing is made of a room that is 5 metres long by 3 metres wide. How many centimetres does 1 centimetre on the drawing represent?
Include the units cm (centimetres) in your answer.

- **20 cm**
- **20 centimetres**
- **20**

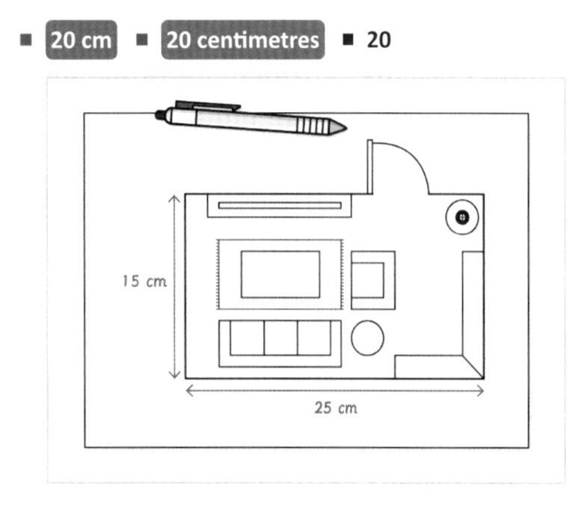

12. Duncan draws a 28 metre by 50 metre swimming pool. What scale has Duncan used?
Give your answer as a simplified ratio in the form 1:n, where 1 represents the distance on the drawing.

- **1:200**
- **1:2**
- **25:5000**
- **25:5,000**

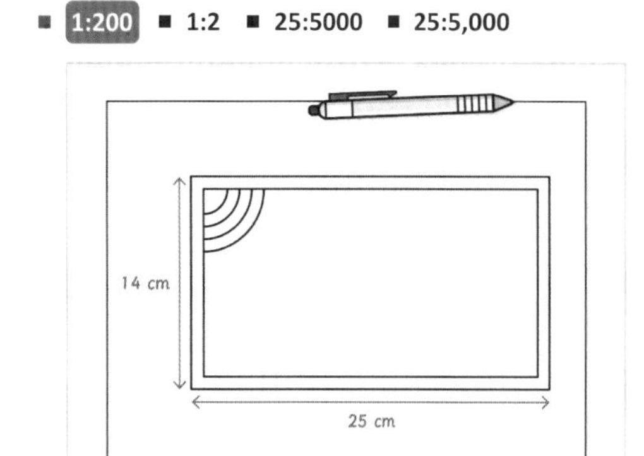

13. Corey draws a 1.8 metre-long shed using a scale of 1 centimetre to 5 centimetres. What is the missing length on the drawing?
Include the unit centimetres (cm) in your answer.

- **36 cm**
- **36 centimetres**
- **36**

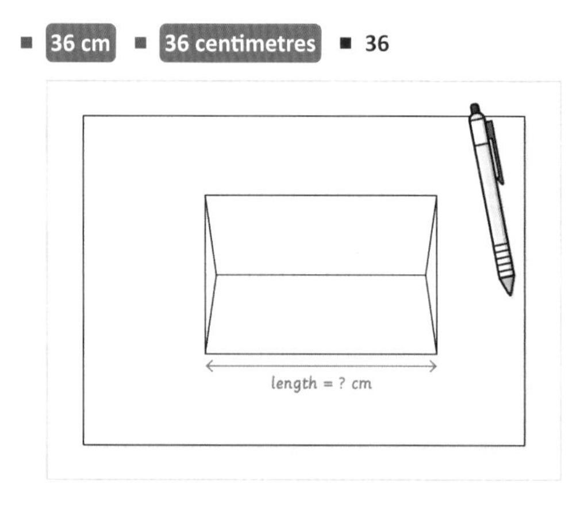

14. Erin draws a 1.2 metre-wide cupboard with a scale of 1 : 4. What is the missing width on her drawing?
Include the units cm (centimetres) in your answer.

- **30 cm**
- **30 centimetres**
- **30**

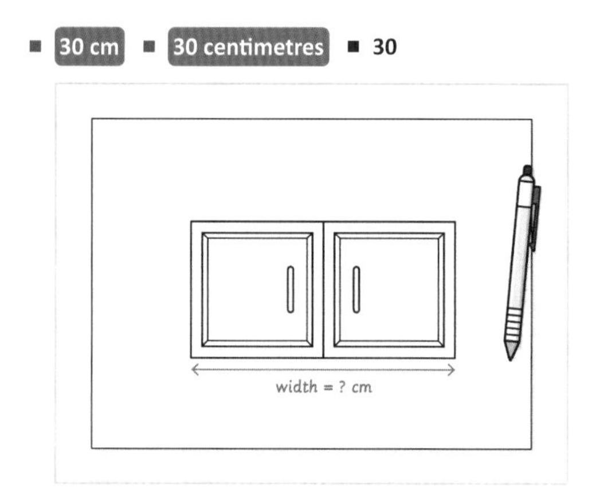

15. Tori draws a plan of a 15 metre by 12 metre classroom. What scale has Tori used?
Give your answer as a simplified ratio in the form 1:n, where 1 represents the distance on the drawing.

- **30:1,500**
- **1:0.5**
- **1:50**
- **30:1500**
- **1:1/2**

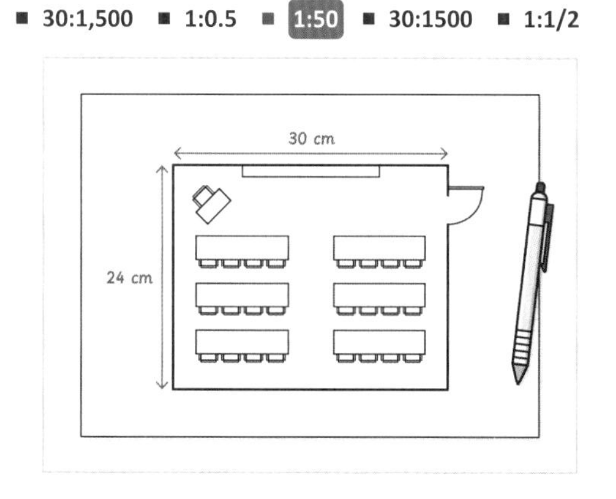

Level 2 *continued*

16. A 294 metre-long boat is drawn at a scale of 1 centimetres to 5 metres.
How long should the drawing be in centimetres?
Give your answer as a decimal and include the units cm (centimetres).

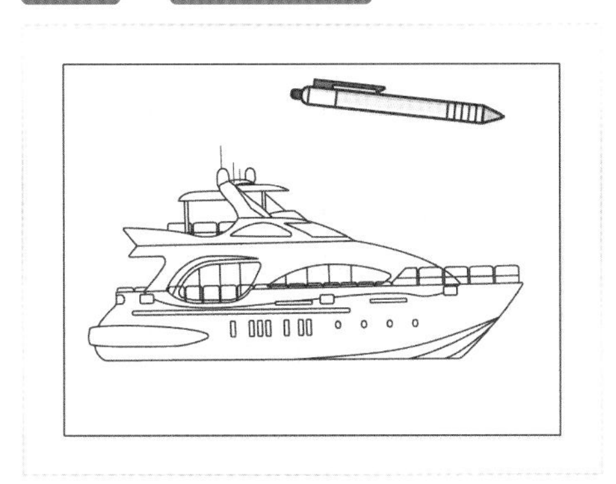

17. A map of an area with a length of 30 kilometres and a width of 20 kilometres is drawn using a scale of 1 : 50,000. What is the missing length on the map?
Include the units cm (centimetres) in your answer.

- 60 cm
- 60 centimetres
- 60

scale = 1 : 50,000
length = ? cm

18. A square play area with sides of 15 m is drawn using a scale of 1:200. How long should each side of the square be on the drawing?
Include the units cm (centimetres) in your answer.

- 7.5 cm
- 7.5 centimetres
- 7.5

19. Frankie draws a 180 centimetre-long table with a scale of 1 : 5. How long should his drawing be in centimetres?
Include the units cm (centimetres) in your answer.

- 36 centimetres
- 36 cm
- 36

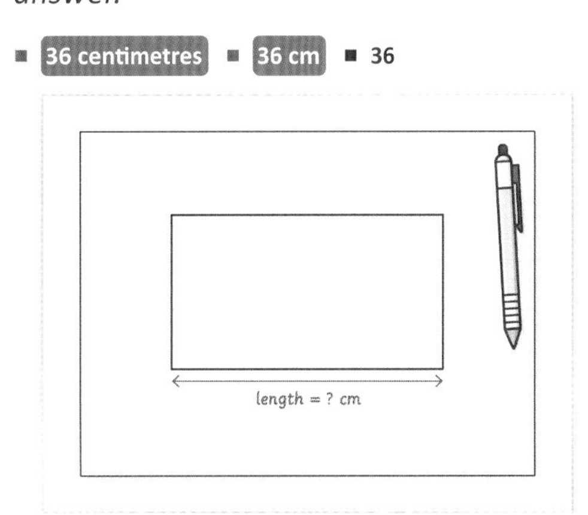

20. A church that is 170 metres tall is drawn at a scale of 1 : 200. What should the height of the drawing be in centimetres?
Include the units cm (centimetres) in your answer.

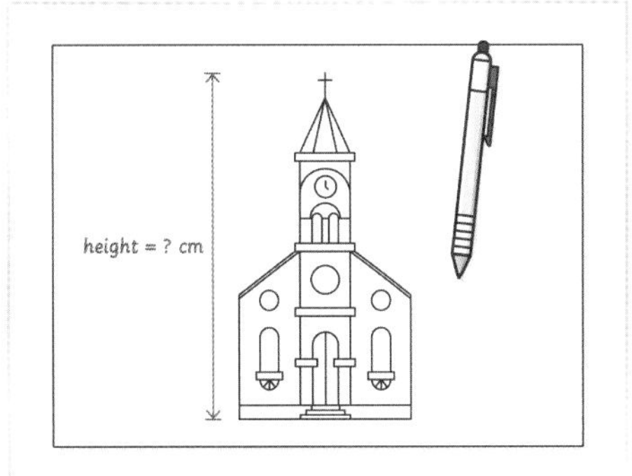

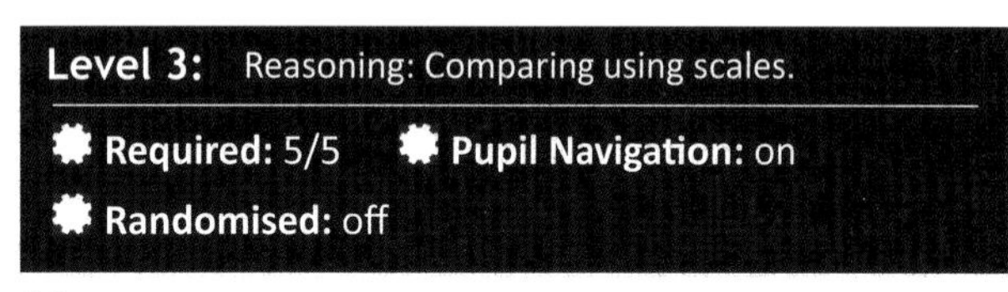

21. A drawing is made of a garden using a scale of 1 : 50.
What is the ratio of the **area** of the drawing to the area of the actual garden?
Give your answer in the form 1:*n*.

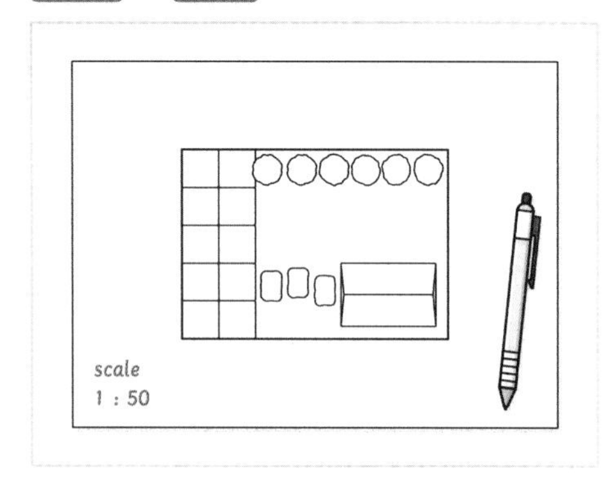

22. Jack is making a scale drawing of a football pitch that is 120 metres by 75 metres. Jack is drawing it on an A4 piece of paper that is 297 mm by 210 mm.
What would be a sensible scale for Jack to use? Explain your answer.

- Open question, no set answer

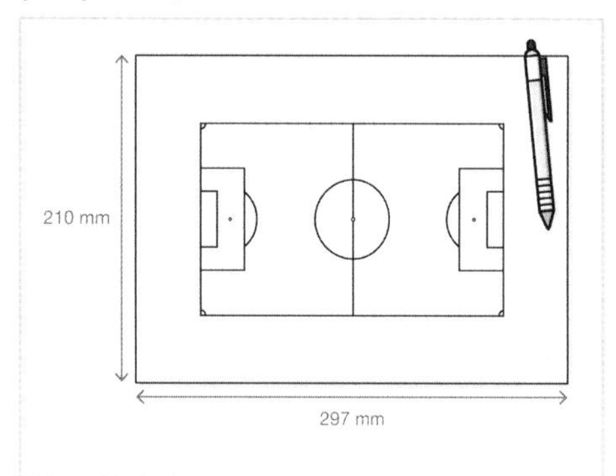

23. The man and the tree are drawn to the same scale. Use the average height of a man to estimate the height of the tree.

1/3

- 1–5 metres
- 5–10 metres
- 10–15 metres

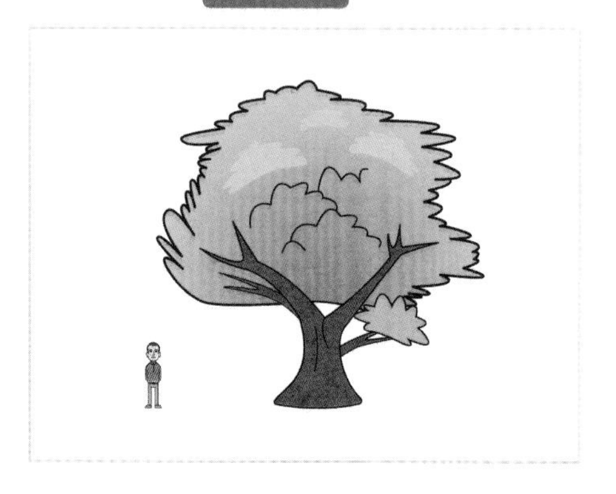

24. The same hiking trail is drawn on three different maps that each have a different scale. Sort the maps according to how long the drawing of the trail is on each map, from shortest to longest.

- map C 1 cm on the map represents ¼ km.
- map A scale = 1 : 24,000
- map B 100 metres is represented by 5 mm on the map.

25. A plan of a desk is drawn to a scale of 1 : 5. If the desk is 180 cm longer than the drawing, how long is the drawing?
Include the units cm (centimetres) in your answer.

- 45 centimetres
- 36 cm
- 45
- 45 cm
- 36 centimetres
- 36

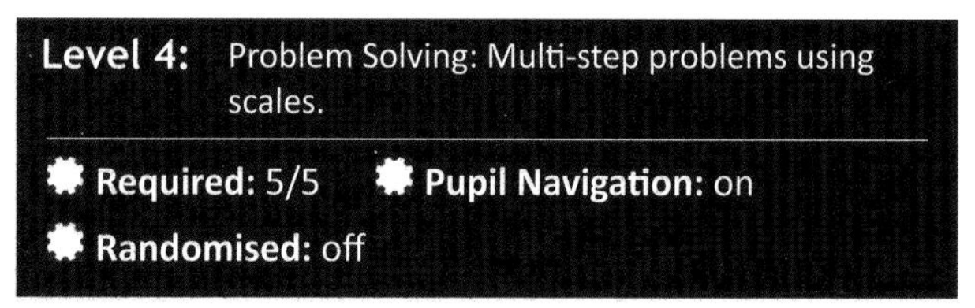

26. A rectangular room is drawn using a scale of 1 : 50. The drawing of the room is 15 centimetres long and 12 centimetres wide.
Carpet for the room costs £9.50 per square metre.
What is the cost of carpeting the room?
Include the £ sign in your answer.

- 427.5
- £427.50
- £427.5
- 427.50

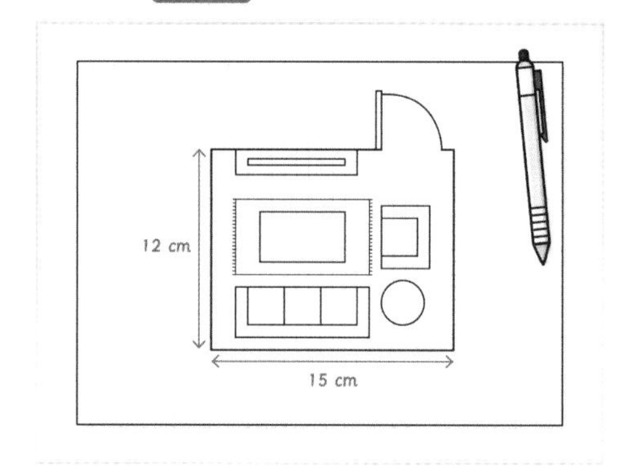

Level 4 *continued*

27. Yasmin is travelling from Caldwell to Stanmouth. If Yasmin drives at an average speed of 80 km/h, how many hours will the journey take her?
Give your answer as a decimal and don't include units.

- 2.5

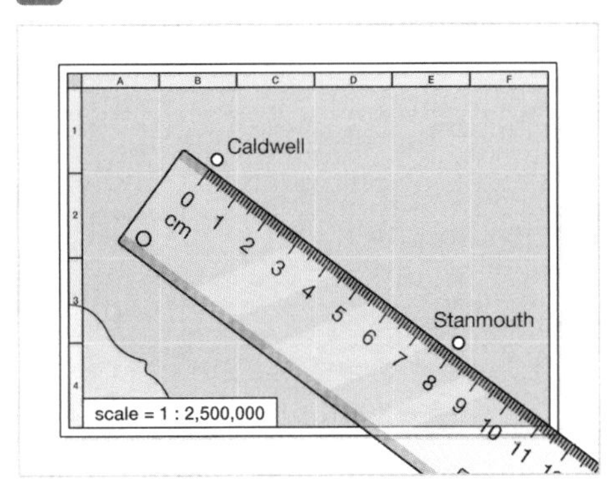

28. Here is a plan of the ground floor of a house. What is the area of the lounge in metres squared?
Don't include the unit in your answer.

- 56

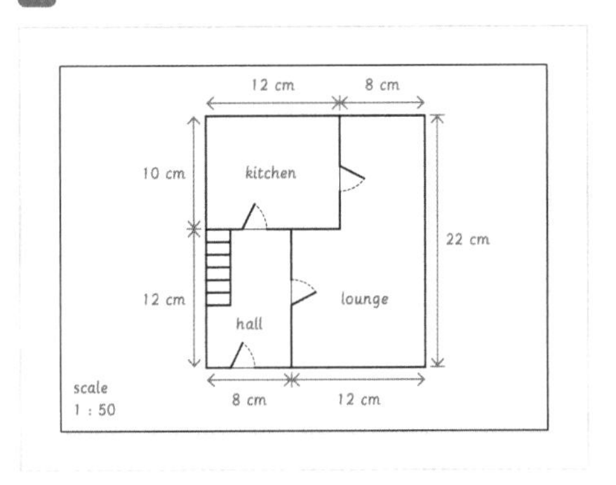

29. Tom wants to tile a wall in his kitchen. Each tile is 20 cm by 10 cm and a pack of 10 tiles costs £5.95. How much will it cost to tile the wall?
Include the £ sign in your answer.

- £119 ■ £119.00 ■ 119.00 ■ 119

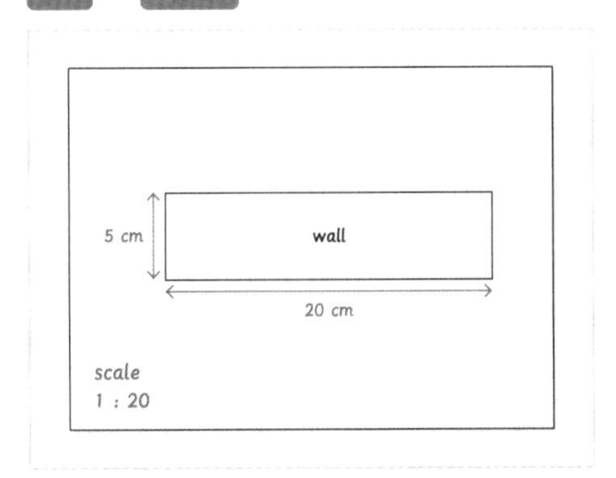

30. Sophie and Rabia are making a scale drawing of their school hall.
Sophie uses a scale of 1 : 40 and Rabia uses a scale of 1 : 50. Sophie's drawing is 10 centimetres longer than Rabia's. How long is the actual hall in metres?
Include the units m (metres) in your answer.

- 20 metres ■ 20 m ■ 20

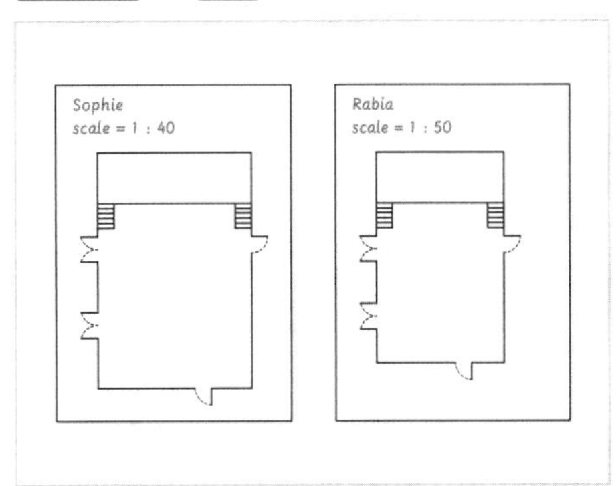

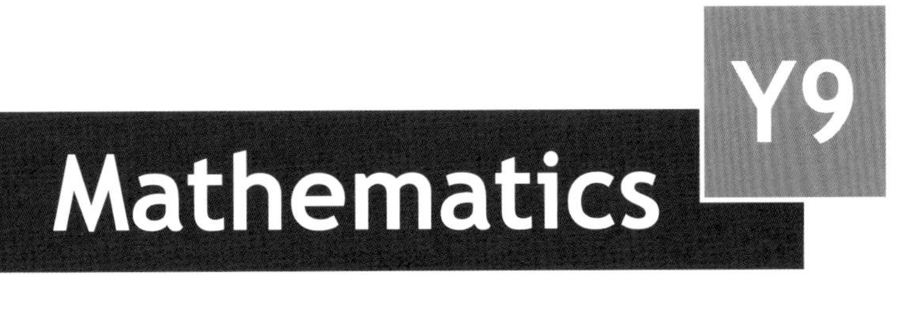

Properties of Number

Powers and Roots

Interpret and Compare Numbers in Standard Form

Objective: I can interpret and compare numbers in standard form.

Quick Search Ref: 10670

Level 1: Understanding: Interpret numbers in standard form.

Required: 7/10 **Pupil Navigation:** on **Randomised:** off

1. Write 10^5 as an ordinary number.

abc ■ 100,000 ■ 100000 ■ 1000000 ■ 1,000,000

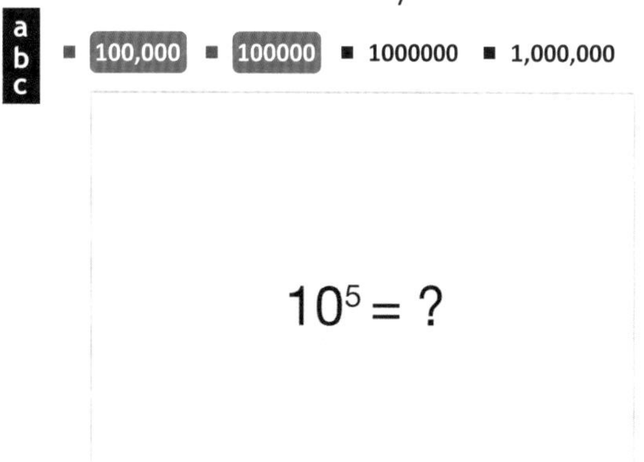

2. Write 2×10^4 as an ordinary number.

abc ■ 20,000 ■ 20000 ■ 0.0002

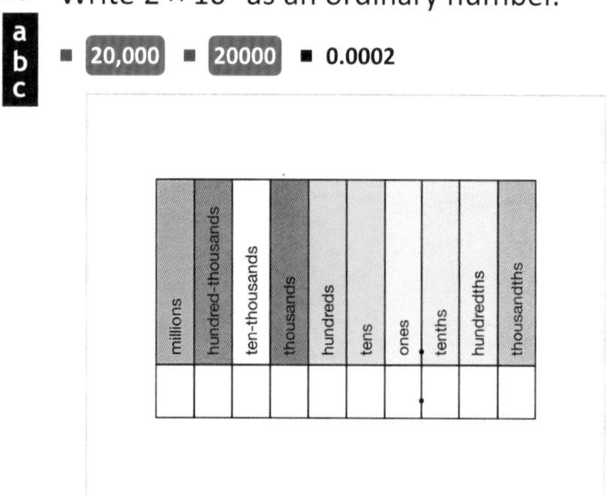

3. 380,000 can be written in the form $A \times 10^5$. Give the value of A to one decimal place.

abc ■ 3.8 ■ 38.0 ■ 38

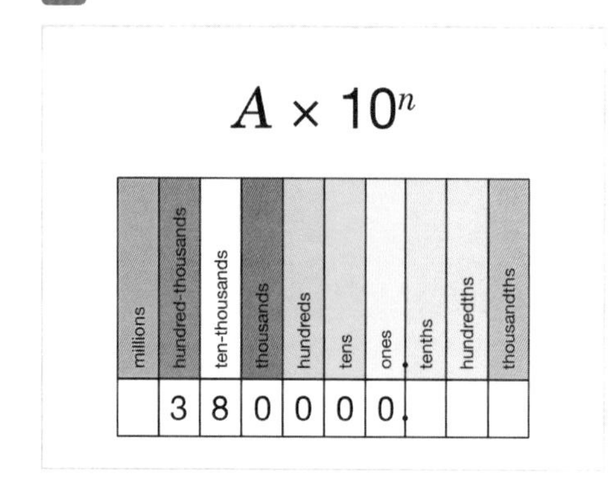

4. 36,200,000 can be written in the form $A \times 10^n$. Give the value of n.

abc ■ 7

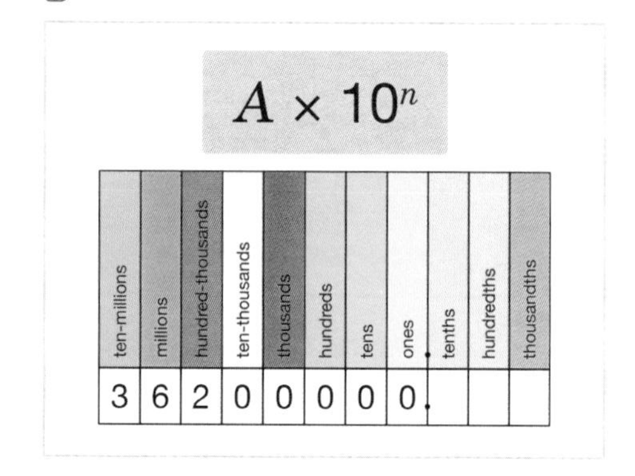

5. Write 10^{-4} as an ordinary number.

abc ■ 0.0001 ■ 0.00001

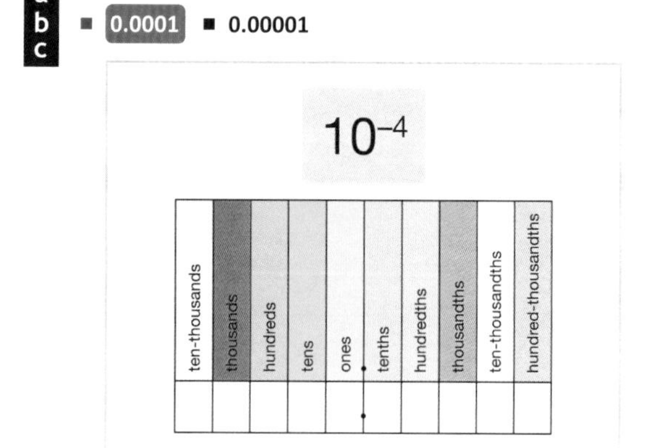

6. Write 3×10^{-5} as an ordinary number.

abc ■ 0.00003 ■ 300000 ■ 300,000

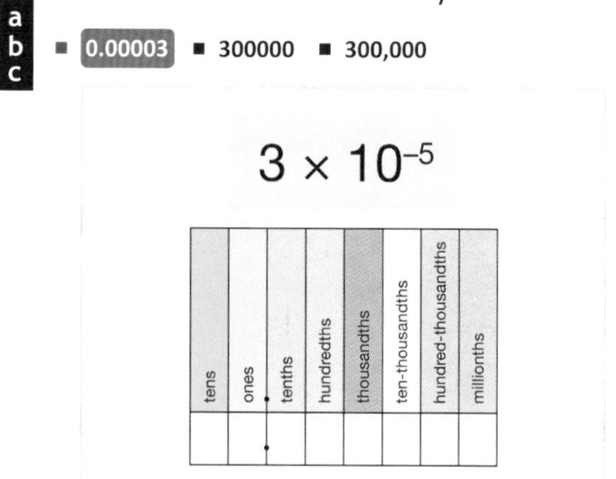

Level 1 *continued*

7. 0.00049 can be written in standard form in the form $A \times 10^n$. Give the value of n.

■ -4 ■ 4

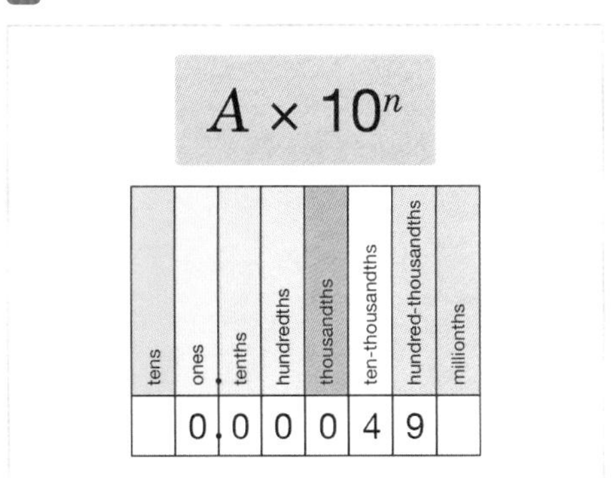

8. Write 5.09×10^6 as an ordinary number.

■ 5090000 ■ 5,090,000 ■ 0.00000509 ■ 509000000 ■ 509,000,000

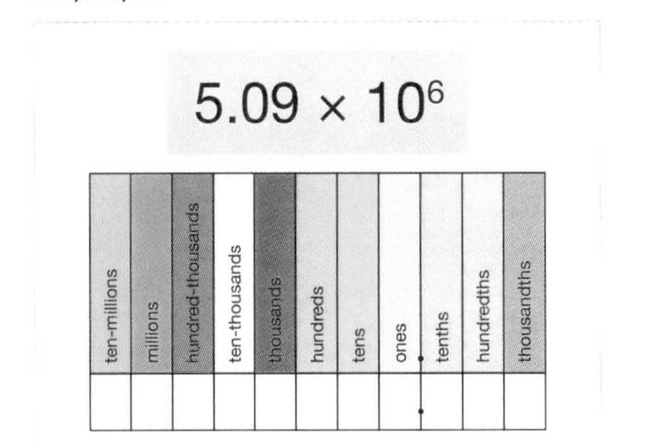

9. Write 7.09×10^{-6} as an ordinary number.

■ 7090000 ■ 0.00000709 ■ 7,090,000

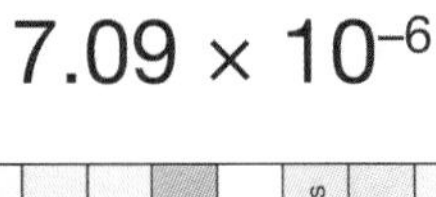

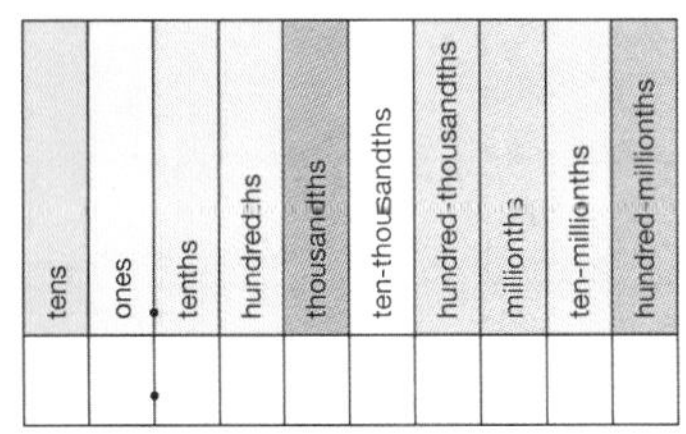

10. 0.00408 can be written in the form $A \times 10^n$. Give the value of n.

■ -3 ■ 3

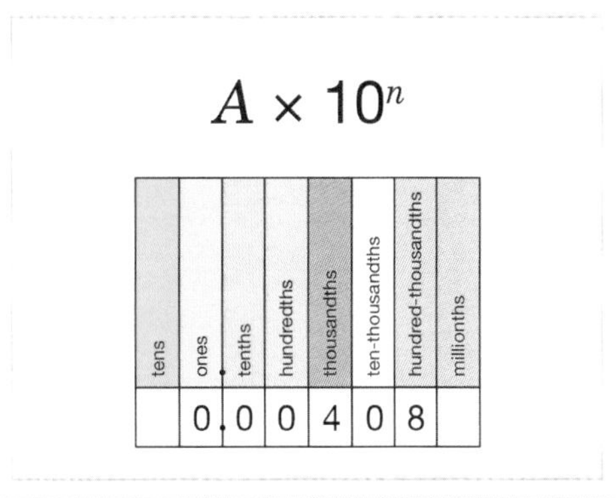

Level 2: Fluency: Compare numbers in standard form.

Required: 7/10 **Pupil Navigation:** on

Randomised: off

11. Select the number that is greater than 3.8×10^4.

1/4

■ 3.777×10^3 ■ 5.87×10^2 ■ 1.2×10^5 ■ 7×10^0

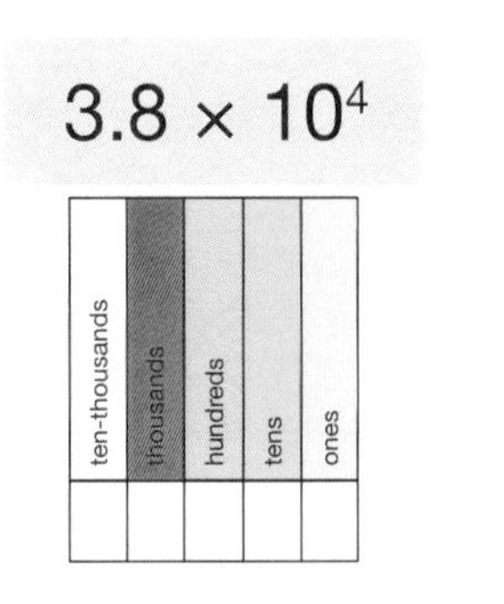

12. Select the number that is less than 7.1×10^{-6}.

1/4

■ 5.4×10^{-5} ■ 4.1×10^{-4} ■ 3×10^{-3} ■ 9.76×10^{-7}

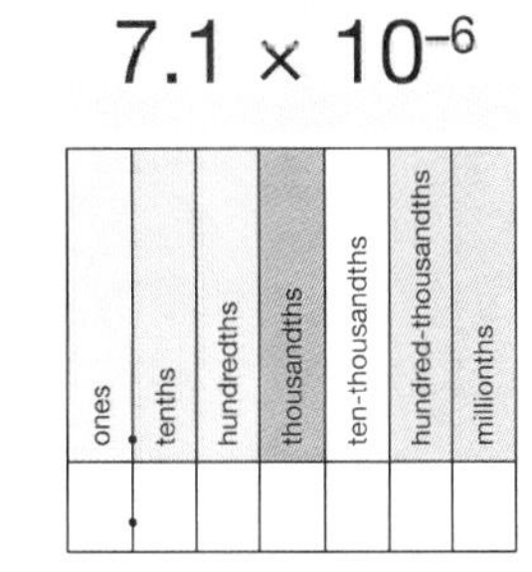

Level 2 *continued*

13. Sort the these numbers by size, from smallest to largest.

- 5.87×10^0
- 4.3×10^1
- 3×10^2
- 1×10^3

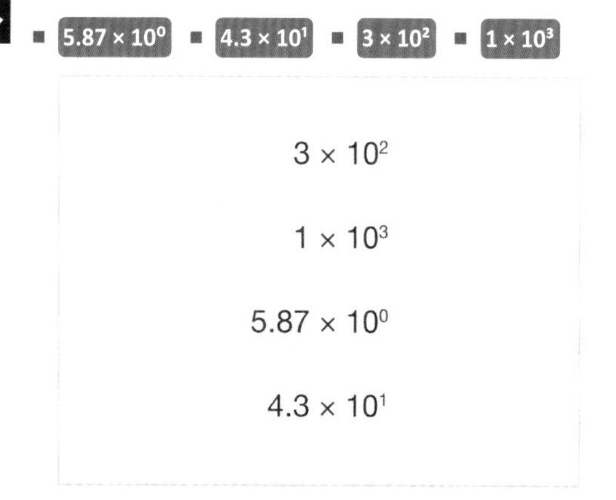

14. Select the number that is less than 8×10^{23}.

1/4

- 1.04×10^{24}
- 2×10^{26}
- 1.1×10^{25}
- 9.999×10^{22}

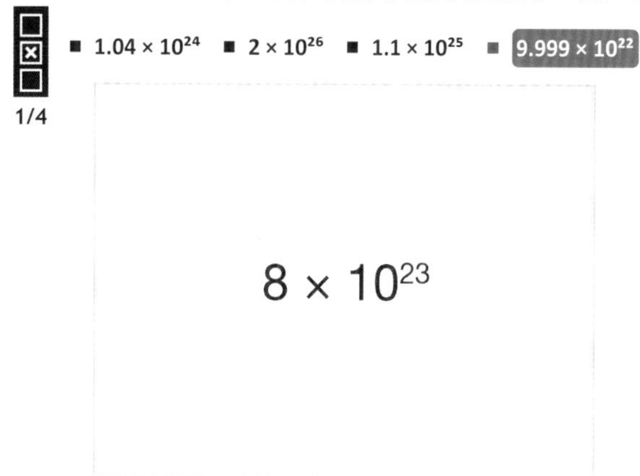

15. Select the number that is greater than 5.2×10^{-9}.

1/4

- 2×10^{-8}
- 8×10^{-12}
- 6.7×10^{-10}
- 7.3×10^{-11}

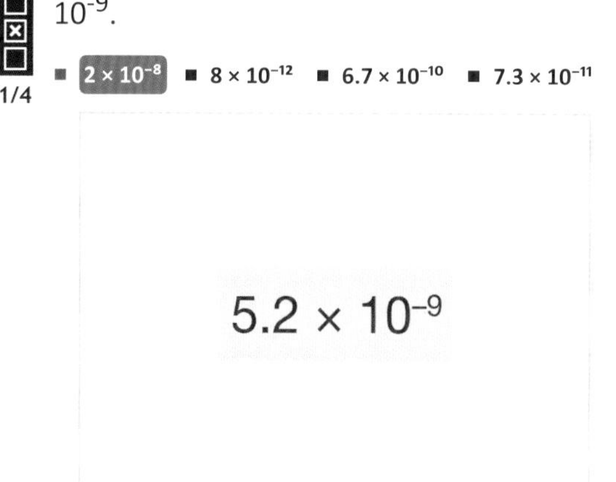

16. Sort the these numbers by size, from largest to smallest.

- 7.3×10^8
- 3×10^1
- 2.3×10^1
- 2.4×10^{-3}

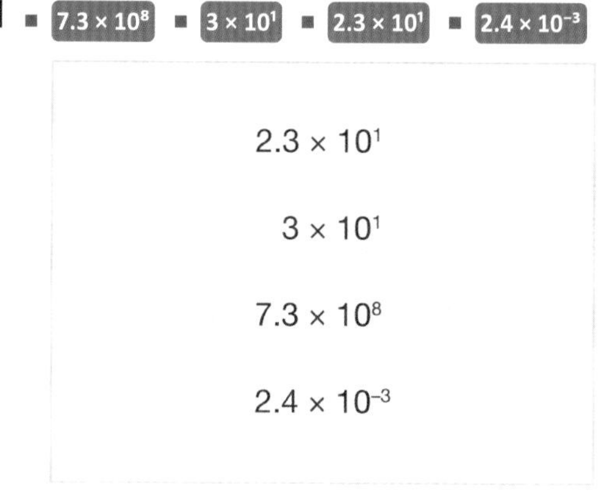

17. Select the number that is less than 6.09×10^{11}.

1/4

- 4.09×10^{13}
- 5.1×10^{11}
- 5.02×10^{12}
- 1×10^{14}

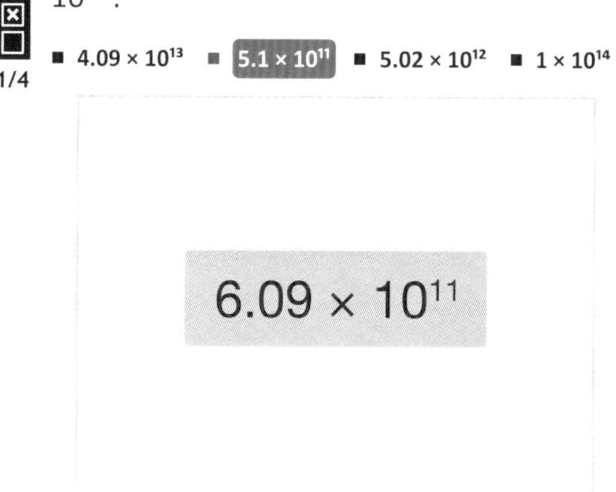

18. Select the number that is less than 8.1×10^{-4}.

1/4

- 1×10^{-3}
- 9×10^{-4}
- 9×10^{-5}
- 9.3×10^{-4}

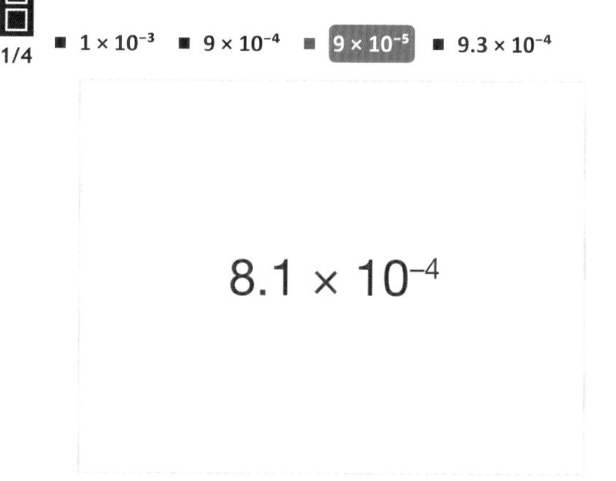

Level 2 continued

19. Sort the these numbers by size, from smallest to largest.

- 9.8×10^{-11}
- 8.3×10^{-3}
- 2.4×10^{0}
- 7×10^{6}

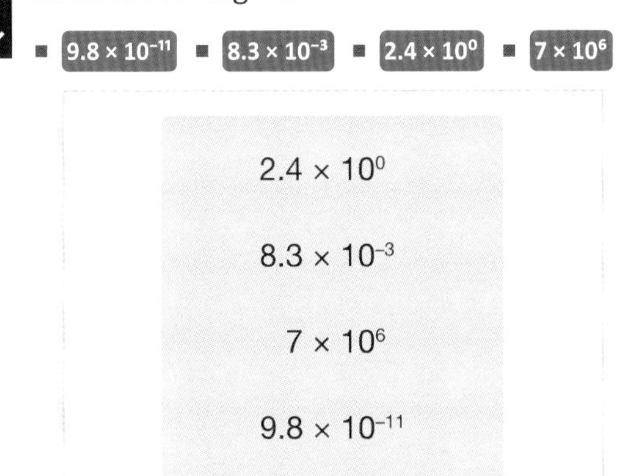

20. Select the number that is greater than 4.2×10^{21}.

1/4

- 2×10^{22}
- 7.7×10^{20}
- 4.19×10^{21}
- 9×10^{19}

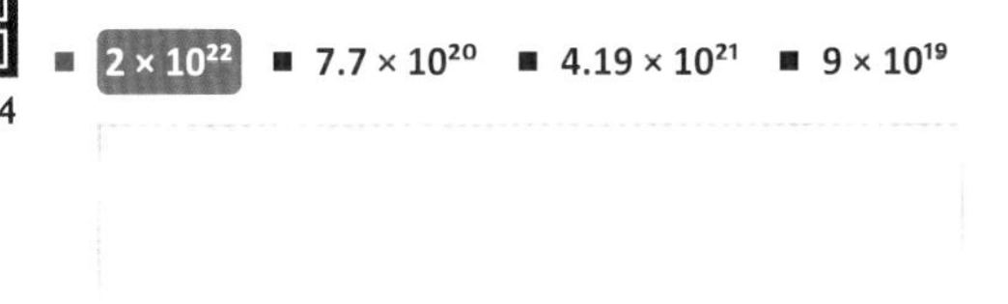

Level 3: Reasoning: Make inferences when comparing standard form numbers.

Required: 5/5 **Pupil Navigation:** on

Randomised: off

21. Select the numbers that are correctly written in standard form.

3/6

- 2.6×10^{1}
- 4.2×10^{0}
- 480
- 0.9×10^{4}
- 3.7×10^{10}
- -8×10^{7}

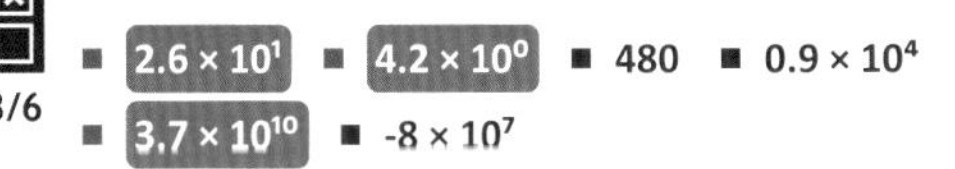

22. Steven says that 14×10^{18} is written in standard form. Is he correct? Explain your answer.

- Open question, no set answer

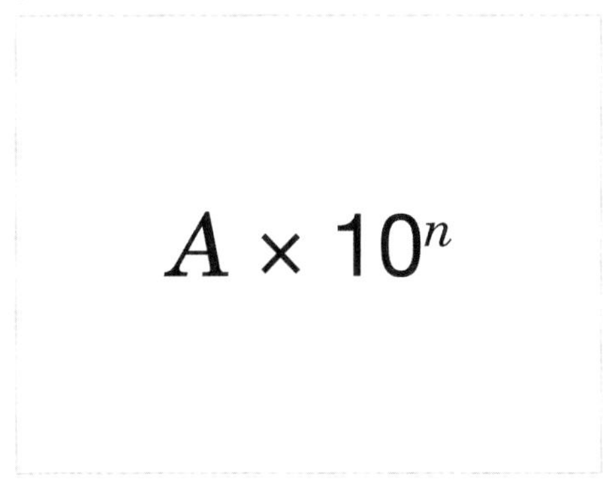

23. Tallia says that 3×10^{4} is larger than 6.7×10^{3}. Is she correct? Explain your answer.

- Open question, no set answer

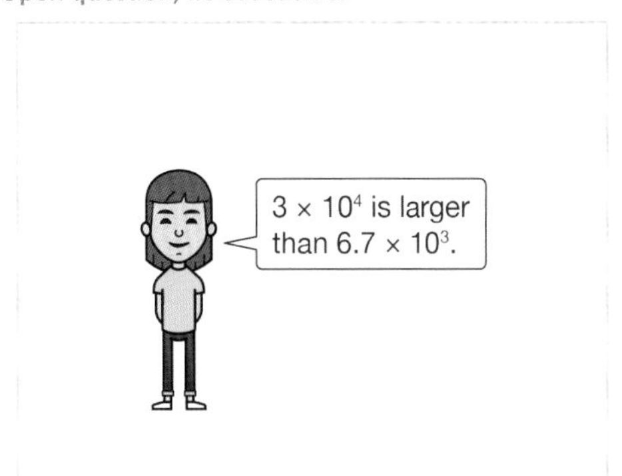

24. Maria says that $38 \times 10^{3} = 3.8 \times 10^{4}$. Is she correct? Explain your answer.

- Open question, no set answer

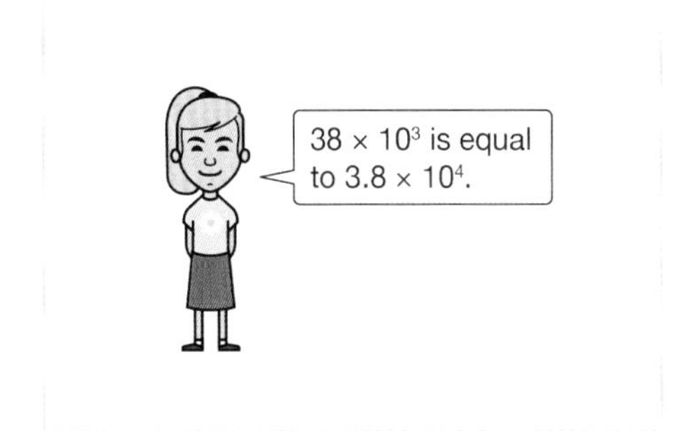

Level 3 *continued*

25. Select the number that is not equal to 5.3×10^4.

1/4

- 53×10^3
- 530×10^2
- $5,300 \times 10^5$
- $53,000 \times 10^0$

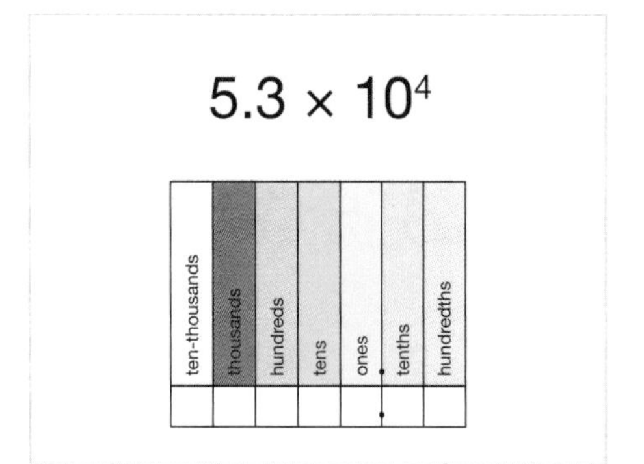

Level 4: Problem Solving: Solve multi-step problems involving comparing standard form numbers.

Required: 5/5 **Pupil Navigation:** on

Randomised: off

26. Sort the these numbers by size, from smallest to largest.

- 0.03×10^6
- 32,000
- 8×10^4
- 1.4×10^5

27. Sort the these numbers by size, from largest to smallest.

- 120×10^5
- 14×10^3
- 0.0016×10^6
- 180

28. The circumference of Earth is 4.0075×10^9 centimetres. How many kilometres is this? *Give your answer as an ordinary number. Include the units in your answer.*

- 40075 km
- 40,075 kilometres
- 40,075
- 40075 kilometres
- 40,075 km
- 40075

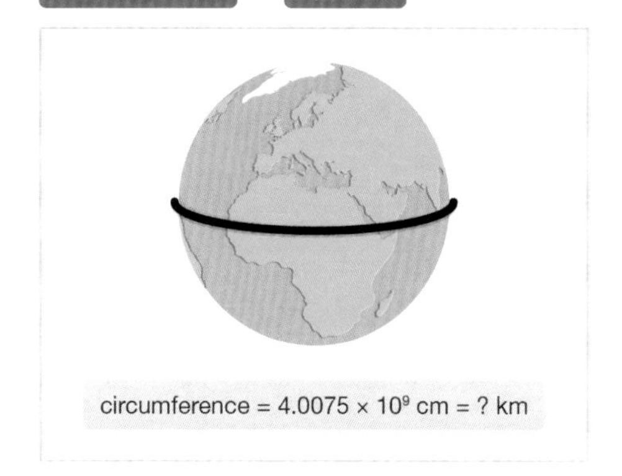

29. Sort these lengths by size, from shortest to longest.

- 0.8×10^0 kilometres
- 3.2×10^4 metres
- 4.2×10^8 millimetres
- 6×10^7 centimetres

30. The answer to the following calculation can be written in the form $A \times 10^n$. What is the value of n?

$3 \times 10^2 \times 4 \times 10^3$

- 6
- 5

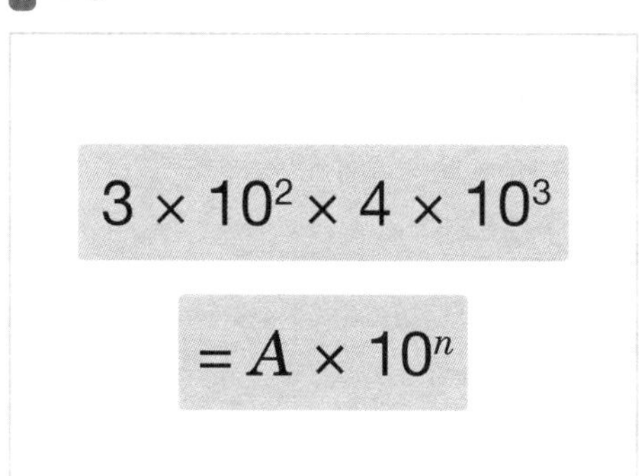

Mathematics Y9

Algebra

Sequences
Graphs
Straight Line Graphs
Quadratic Graphs

Recognise and Use Geometric Sequences

Objective:	I can recognise geometric sequences and find missing terms.
Quick Search Ref:	10569

Level 1: Understanding: Recognise a geometric sequence and find missing terms.

Required: 7/10 **Pupil Navigation:** on **Randomised:** off

1. Which of these is a geometric sequence?

1/4

- 2, 4, 6, 8, 10
- 10, 13, 16, 19, 22
- **2, 4, 8, 16, 32**
- 5, 9, 13, 17, 21

2. Which of these geometric sequences has a common ratio of × 10?

1/4

- 10, 20, 40, 80, 160
- 10, 30, 90, 270, 810
- **1, 10, 100, 1000, 10000**
- 1/10, 1/30, 1/90, 1/270, 1/810

3. Which of these is a geometric sequence?

1/4

- 250, 500, 750, 1000, 1250
- **250, 125, 62.5, 31.25, 15.625**
- 35, 42, 49, 56, 63
- 750, 700, 650, 600, 550

4. What is the next term in this sequence?

- **96**
- 51
- 72

3, 6, 12, 24, 48, ?

5. What is the next term in this sequence?

- **60,000**
- **60000**
- –6,066
- –6066

6, –60, 600, –6000, ?

6. What is the next term in this sequence?

- **1/112**
- 112
- 1/63

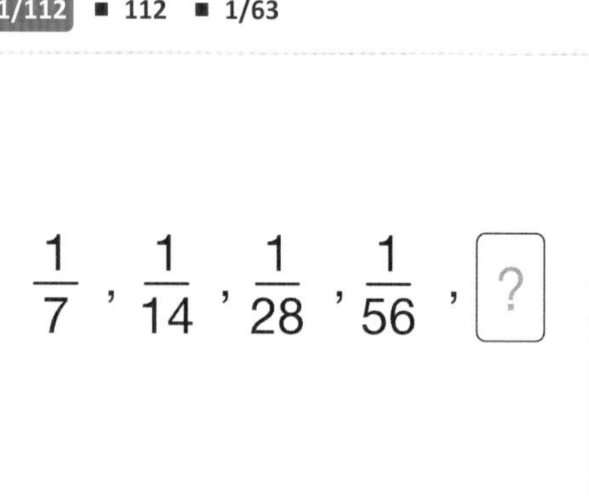

$\frac{1}{7}, \frac{1}{14}, \frac{1}{28}, \frac{1}{56}, ?$

7. What is the missing term in this sequence?

- **54x**
- 22x
- 30x

$2x, 6x, 18x, ?, 162x$

8. What is the next term in this sequence?

- **15y**
- 0y
- 0
- –90y

$240y, 120y, 60y, 30y, ?$

Level 1 *continued*

9. What is the missing term in this sequence?

a b c

- **1/27** (answer)
- 1/15
- 27

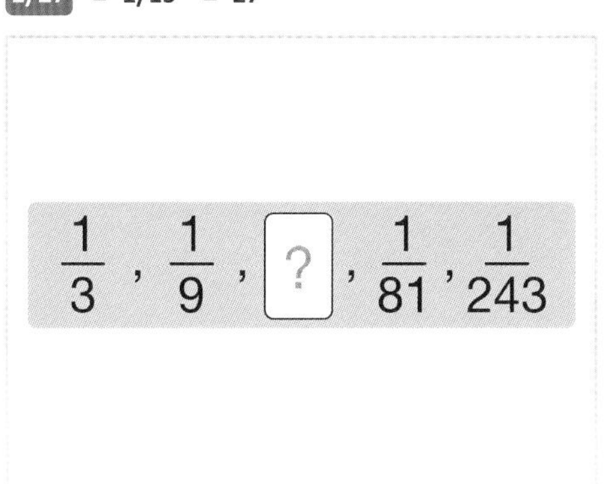

10. Which of these is a geometric sequence?

- 700, 750, 800, 850, 900
- 85, 75, 65, 55, 45
- 99, 108, 117, 126, 135
- **4.25, 8.5, 17, 34, 68** (answer)

1/4

Level 2: Fluency: Make inferences within geometric sequences.

Required: 9/10 **Pupil Navigation:** on

Randomised: off

11. What is the common ratio for this sequence?

1/6

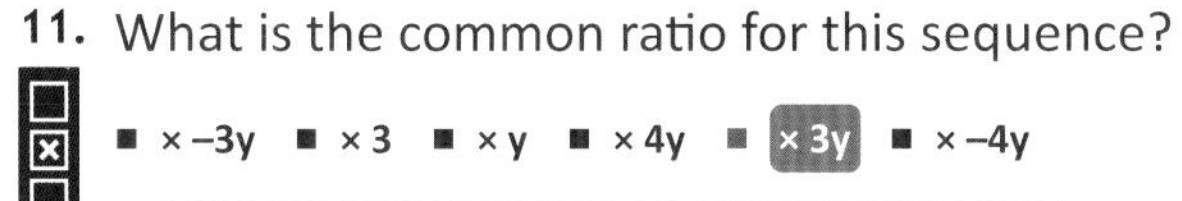

- × −3y
- × 3
- × y
- × 4y
- **× 3y** (answer)
- × −4y

$2, 6y, 18y^2, 54y^3, 162y^4$

12. The 3rd, 4th and 5th terms of a sequence are 7, 14, 28. What is the 1st term in this sequence?

1 2 3

- **1.75** (answer)
- 3.5

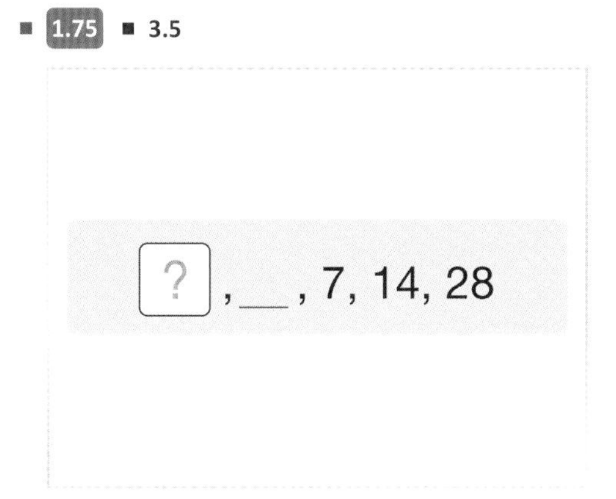

13. The 3rd, 4th and 5th terms of a sequence are 3, −6, 12. What is the 8th term in this sequence?

1 2 3

- **-96** (answer)
- -24
- 96
- 48

__, __, 3, −6, 12, __, __, ?

14. The 5th, 6th and 7th terms of a sequence are 2/5, 4/15, 8/45. What is the 9th term in this sequence?

a b c

- **32/405** (answer)
- 16/135

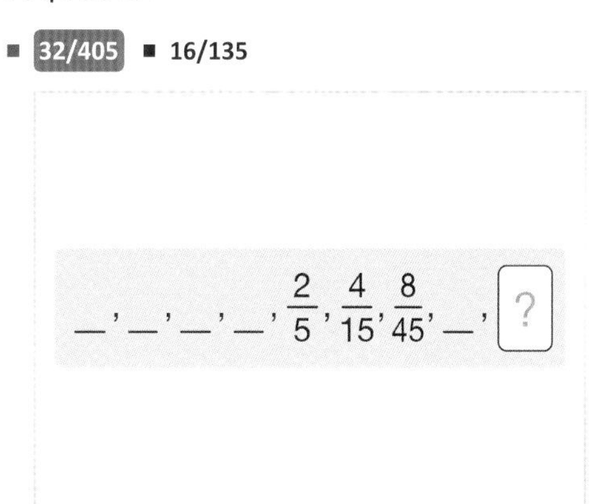

15. The first four terms of a sequence are *a*, *ar*, ar^2, ar^3. What is the 10th term in this sequence?

1/6

- ar^8
- a^9r
- **ar^9** (answer)
- $a^{10}r$
- ar^{10}
- a^8r

a, ar, ar^2, ar^3, __, __, __, __, __, ?

Level 2 *continued*

16. What is the next term in this sequence?

- 14,641 ■ 14641

1, 11, 121, 1331, ?

17. What is the 5th term in this sequence?

- -1/32 ■ 1/32 ■ 1/16

$\frac{-1}{2}, \frac{1}{4}, \frac{-1}{8}$, ___, ?

18. The 3rd, 4th and 5th terms of a sequence are 8, –8, 8. What is the 8th term in this sequence?

1 2 3

- -8 ■ 8

__, __, 8, –8, 8, __, __, ?

19. The 3rd, 4th and 5th terms of a sequence are 2, 6, 18. What is the 8th term in this sequence?

a b c

- 1458 ■ 486 ■ 162 ■ 1,458

__, __, 2, 6, 18, __, __, ?

20. The 5th, 6th and 7th terms of a sequence are 3/5, 6/25, 12/125. What is the 8th term in this sequence?

a b c

- 24/625

__, __, __, __, $\frac{3}{5}, \frac{6}{25}, \frac{12}{125}$, ?

Level 3: Reasoning: Find missing terms.

Required: 5/5 **Pupil Navigation:** on

Randomised: off

21. The first four terms of a sequence are 3, 6, 12, 24. What is the value of the 6th term divided by the 5th term?

1 2 3

- 2 ■ 1.2

3, 6, 12, 24, __, __

Level 3 *continued*

22. The first term of a geometric sequence is 2 and the common ratio is –1. What is the 100th term in this sequence?

■ -2 ■ 2

23. Gemma says that the sixth term of this sequence is 6. Is she correct? Explain your answer.

- Open question, no set answer

24. The first four terms of a sequence are 5, 10, 20, 40. What is the value of the 8th term divided by the 6th term?

■ 4 ■ 2

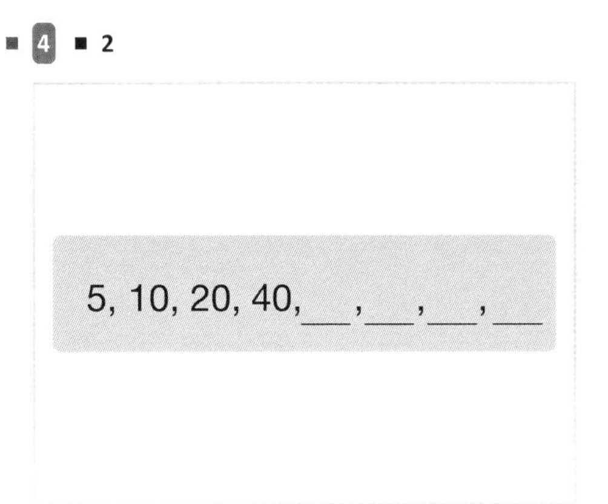

25. The table shows the number of bacteria present in a sample each minute for 5 minutes. Calculate the number of bacteria present after 10 minutes.

■ 2048 ■ 2,048 ■ 128

time (minutes)	number of bacteria present
0	2
1	4
2	8
3	16
4	32
5	64

Level 4: Problem Solving: Solve multi-step problems with geometric sequences.

Required: 5/5 **Pupil Navigation:** on
Randomised: off

26. The 1st and 3rd terms of a geometric sequence are 5 and 45 respectively. All of the terms in the sequence are positive. What is the 4th term in this sequence?

■ 135 ■ 405

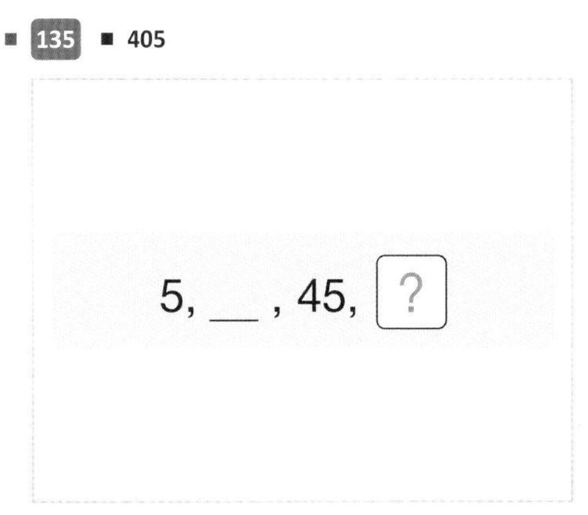

27. The 1st, 3rd and 4th terms of a sequence are 2, 18 and 54 respectively. For this sequence, what is the value of the following calculation?
(102nd term ÷ 100th term) × 2nd term

■ 54

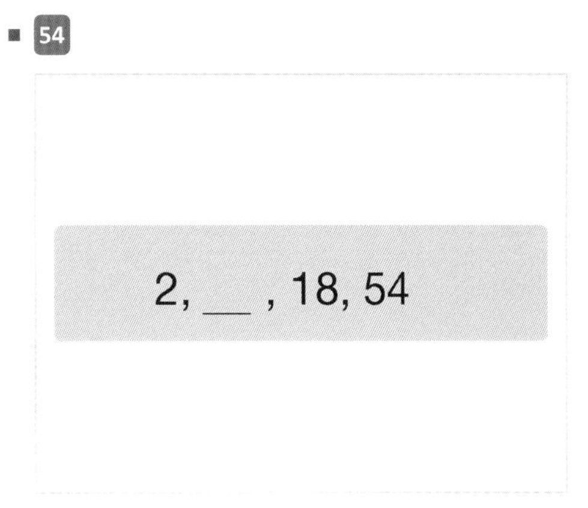

Level 4 *continued*

28. A bank account pays 2% compound interest per year. How many years will it take for an investment of £300 to be worth £331.22?
Hint: after one year, the investment is worth £306.

- 5
- 4

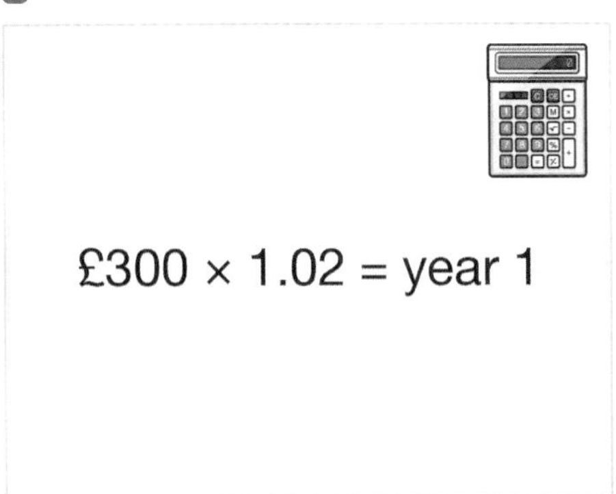

29. The 1st and 4th terms of a geometric sequence are 75 and 0.6 respectively. What is the sum of the first five terms?
Give your answer to 2 decimal places.

- 93.72

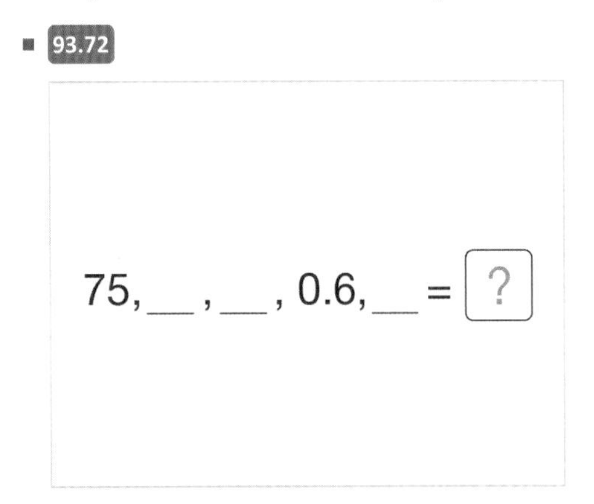

30. The height of square A is 64 cm. The height of square B is half the height of square A. This pattern continues through to square E. What is the perimeter of the shape which consists of squares A to E?
Give your answer in cm (centimetres) and include the units in your answer.

- 124 cm
- 376 cm
- 376 centimetres
- 376
- 124
- 124 centimetres

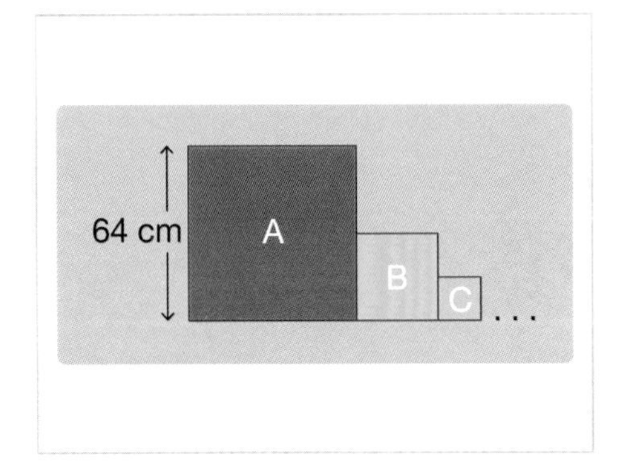

Work with Coordinates in All Four Quadrants

Objective: I can plot and read coordinates in all four quadrants.

Quick Search Ref: 10446

Level 1: Understanding: Plotting and reading coordinates in all four quadrants.

Required: 7/10 **Pupil Navigation:** on **Randomised:** off

1. What is the x-coordinate of this pair of coordinates?
(3, 4)

- **3**
- 4

(3, 4)

2. What are the coordinates of the origin?
Use a comma to separate the coordinates and enclose them in brackets.

- **(0, 0)**
- 0, 0
- (0 0)

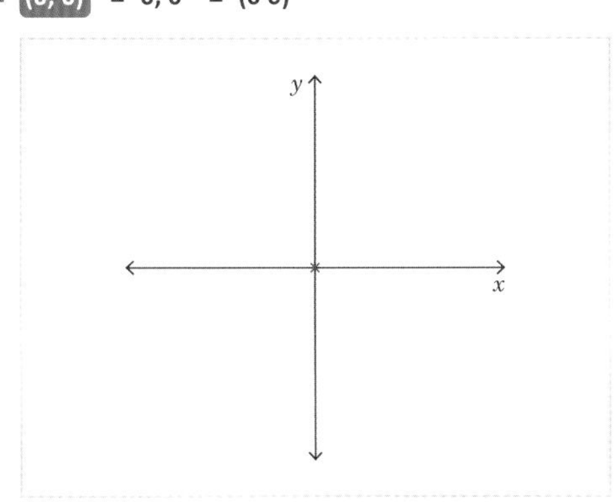

3. What is the y-coordinate of this pair of coordinates?
(–2, –3)

- **-3**
- -2

(–2, –3)

4. Arrange the quadrants to match the description of their coordinates.

- quadrant D
- quadrant A
- quadrant C
- quadrant B

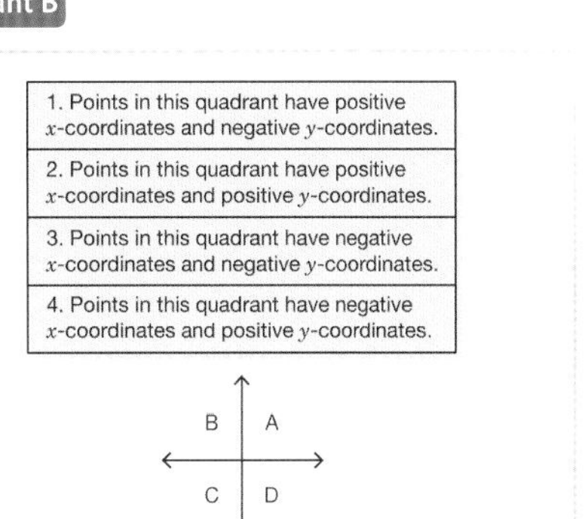

5. What are the coordinates of point A?
Use a comma to separate the coordinates and enclose them in brackets.

- (-3 2)
- **(-3, 2)**
- (2, -3)
- -3, 2

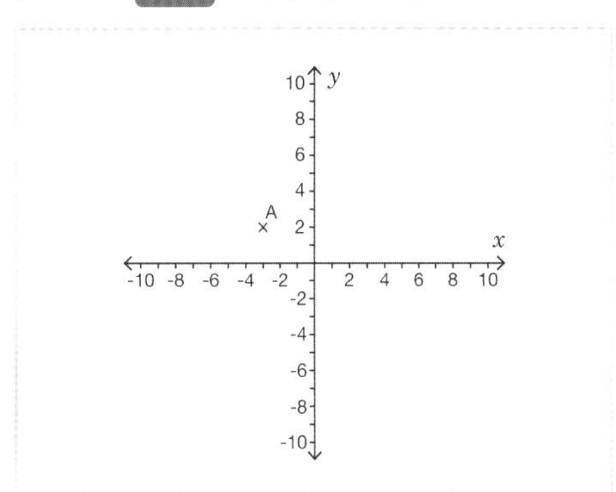

6. What are the coordinates of point A?
Use a comma to separate the coordinates and enclose them in brackets.

- (-6, 0)
- **(0, -6)**
- (0 -6)
- 0, -6

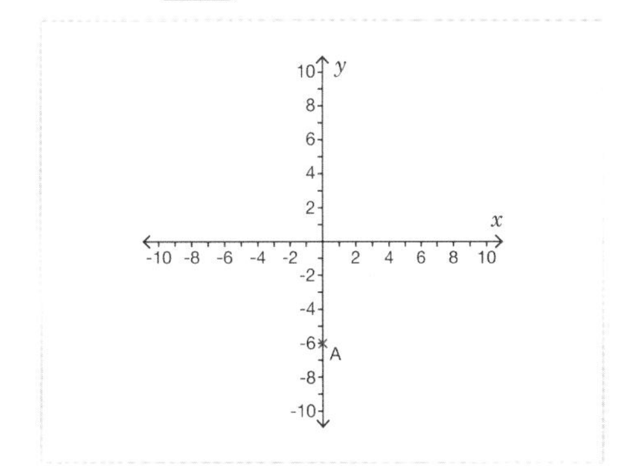

Level 1 *continued*

7. Which three pairs of coordinates are in the shaded region?

3/6

- (4, 3)
- (0, −6)
- **(2, 4)**
- (−4, 5)
- **(−2, −4)**
- **(−3, 5)**

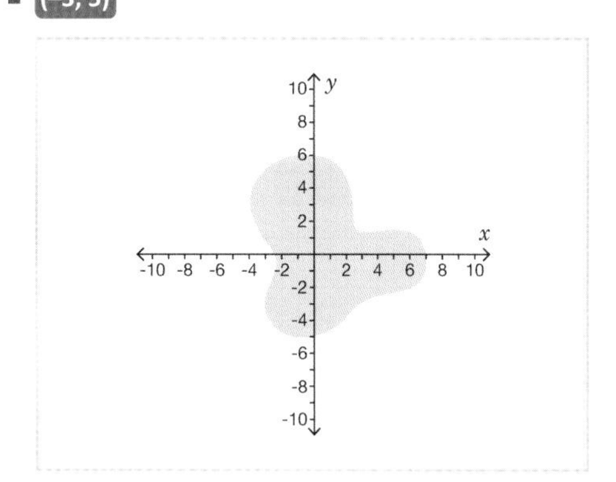

8. Which point has coordinates (−5, −2)?

1/4

- A
- **B**
- C
- D

9. What is the *y*-coordinate of point A?

- **4**
- -2

10. Which **three** pairs of coordinates are in the shaded region?

3/6

- **(−4, 4)**
- **(5, −3)**
- **(−5, 0)**
- (0, 0)
- (−6, 6)
- (−2, 0)

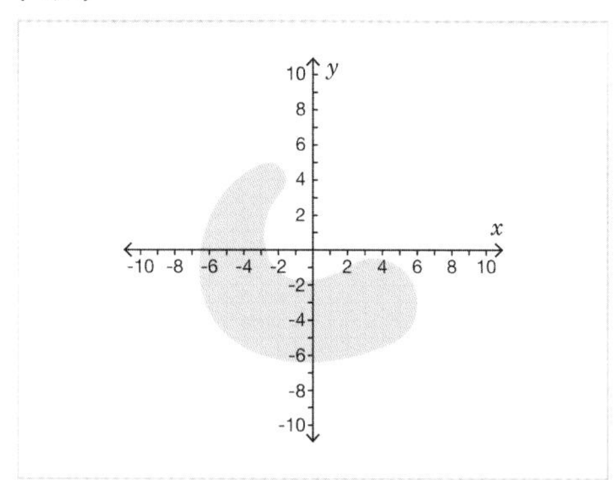

Level 2: Fluency: Making inferences about coordinates in all four quadrants.

Required: 7/10 **Pupil Navigation:** on

Randomised: off

11. The diagram shows points A, B and C. What coordinates would make a square?

- **(1, -6)**
- 1, -6
- (1 -6)
- (-6, 1)

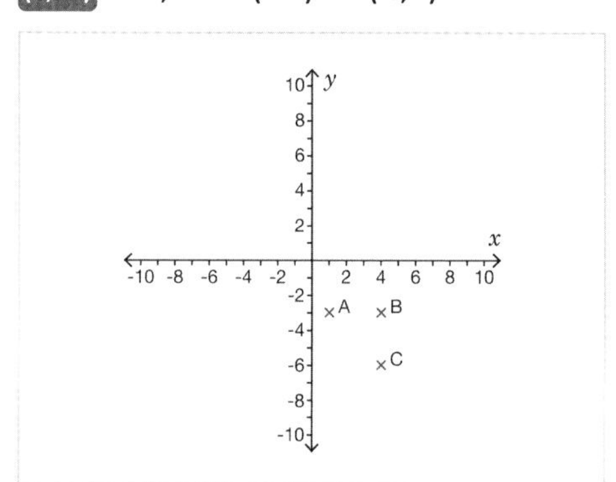

12. Select all of the points which lie on the line *y* = −2.

4/7

- **(0, −2)**
- (−2, 0)
- **(−2, -2)**
- **(2, −2)**
- The origin
- (−2, 2)
- **(8, −2)**

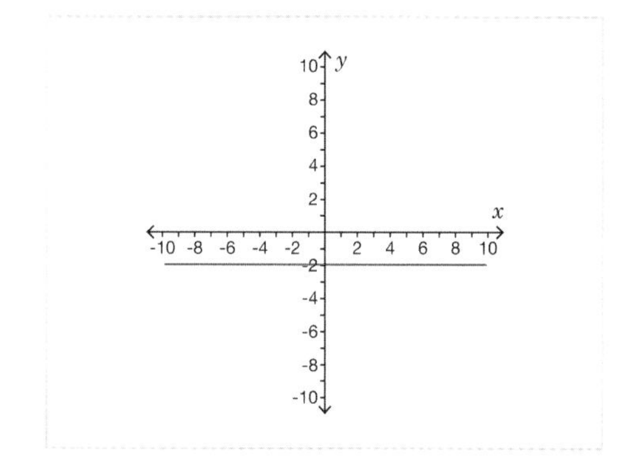

Level 2 *continued*

13. What are the coordinates of the midpoint of point A and point B?

abc

- (-2, -7) ■ (-4, -14) ■ -2, -7 ■ (-2 -7) ■ (-7, -2) ■ (-5, -7)

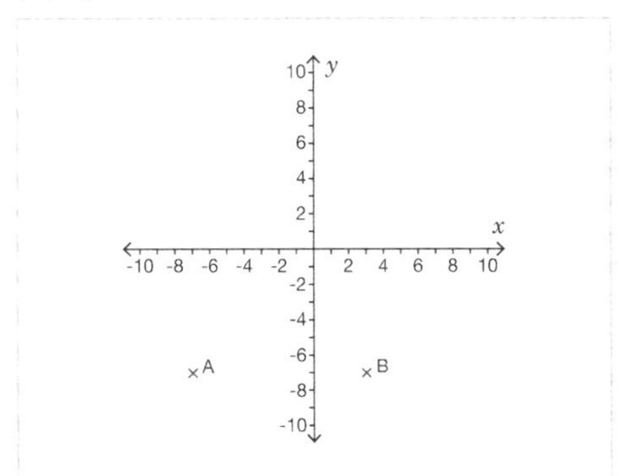

14. Select all of the points which lie on the y-axis.

2/6

- (0, 17) ■ (–10, 0) ■ (1, 0) ■ (0, 1) ■ (–4, 0)

■ (17, 0)

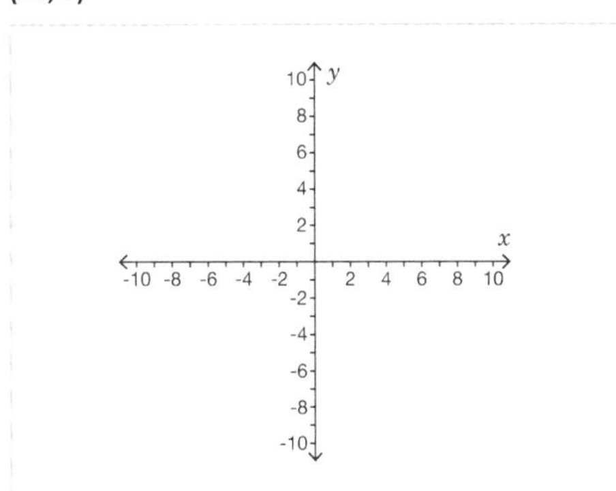

15. The rectangle ABCD is moved two units to the right and three units up. What are the new coordinates of point D?

- (-5 -3) ■ (-5, -3) ■ (-3, -5) ■ (-9, -3) ■ -5, -3

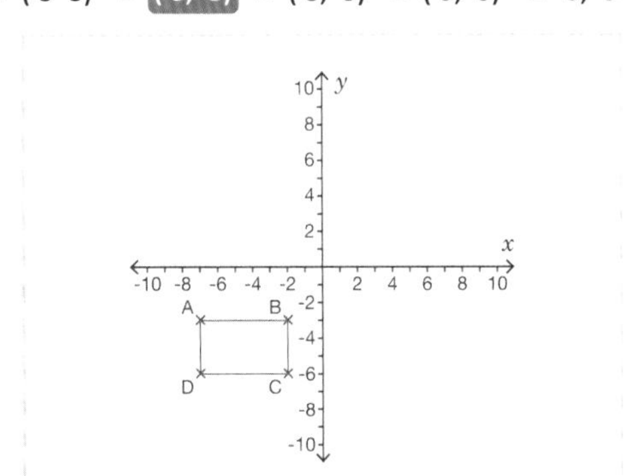

16. Point A is at (5, 7) and point B is at (5, –2). What are the coordinates of the midpoint of A and B?

- (5, 2.5) ■ (5, 4.5) ■ 5, 2.5 ■ (5 2.5) ■ (2.5, 5) ■ (10, 5)

17. The diagram shows points A, B and C. To make a kite, what are the coordinates of point D?

3/7 *Select three possible answers.*

- (–6, 7) ■ (–6, 4) ■ (8, –6) ■ (–6, 8) ■ (7, –6) ■ (–6, 10) ■ (–6, 9)

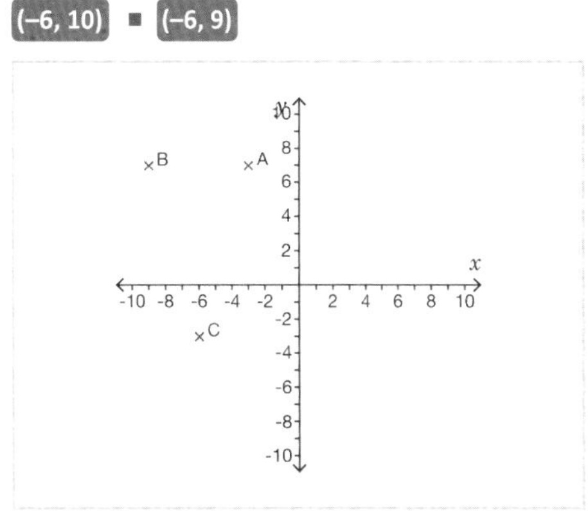

18. Select all of the points which lie on the *x*-axis.

4/7

- (1, 0) ■ (0, 1) ■ (–4, 0) ■ (0, –4) ■ the origin ■ (17, 0) ■ (0, 17)

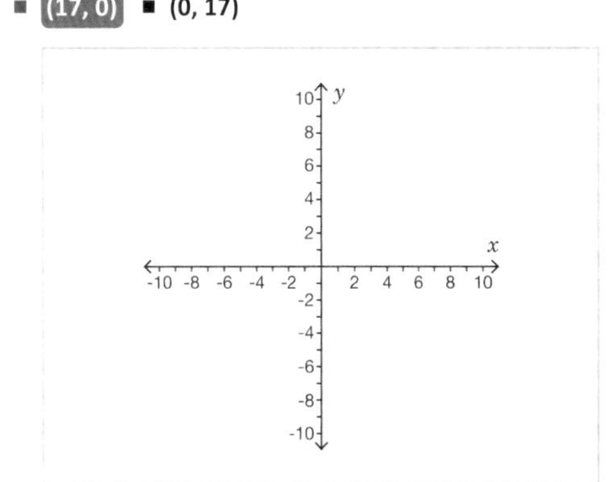

19. Point A is at (2, 3) and point B is at (8, 9). What are the coordinates of the midpoint of A and B?

abc

- (5 6) ■ (10, 12) ■ (5, 6) ■ (6, 5) ■ 5, 6

20. Select all of the points which lie on the line *x* = 3.

3/7

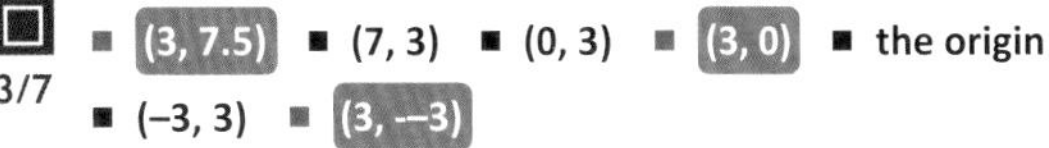

- (3, 7.5) ■ (7, 3) ■ (0, 3) ■ (3, 0) ■ the origin ■ (–3, 3) ■ (3, –3)

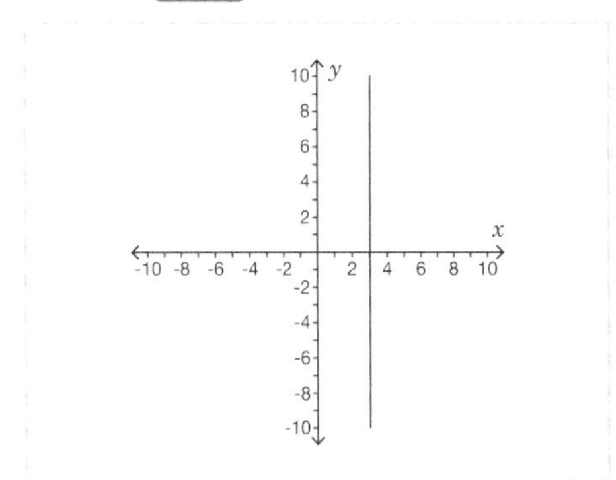

Level 3: Reasoning within coordinates in all four quadrants.

Required: 5/5 **Pupil Navigation:** on
Randomised: off

21. Iain says that the coordinates of point A are (–4, –2). Is he correct? Explain your answer.

- Open question, no set answer

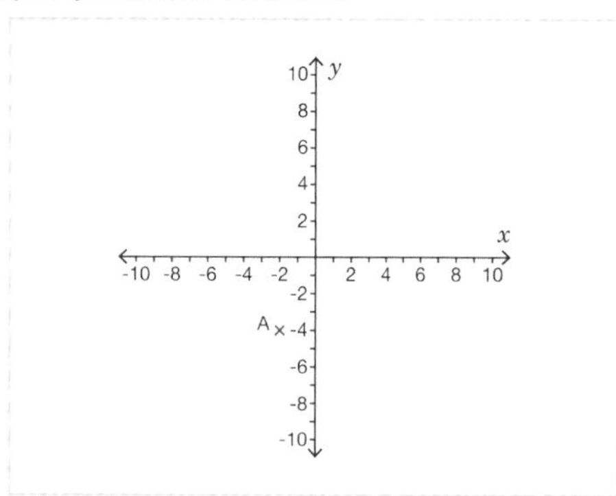

22. The triangle ABC is reflected in the *x*-axis. The coordinates of point C are (3, 4). What are the coordinates of the image of point C?

- (3, -4)
- (-4, 3)
- 3, -4
- (-3, 4)
- (3 -4)

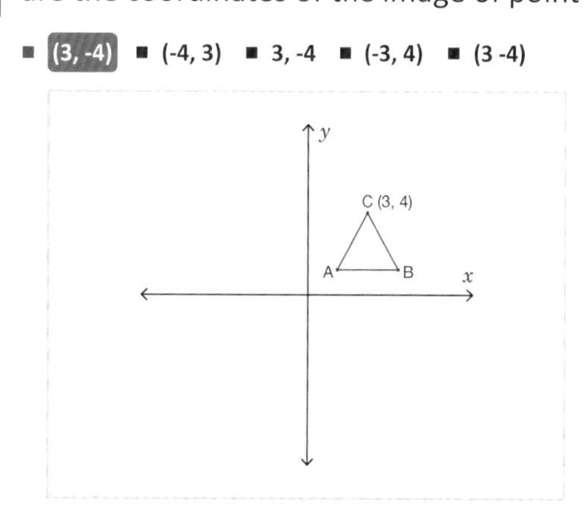

23. The diagram shows the line AB. Is it possible to draw points C and D such that ABCD is a square with a vertex in each quadrant? Explain your answer.

- Open question, no set answer

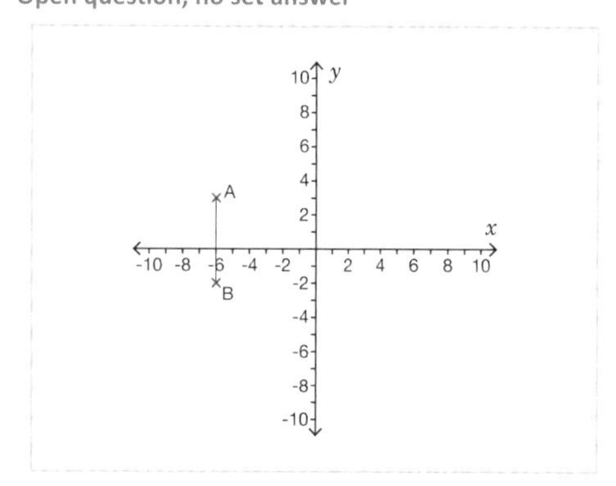

24. The diagram shows a line segment AB with point M being the midpoint. The coordinates of point B are (2, 10) and the coordinates of point M are (0, –3). What are the coordinates of point A?

- (-2, -16)
- (-16, -2)
- -2, -16
- (-2, -10)
- (-2 -16)

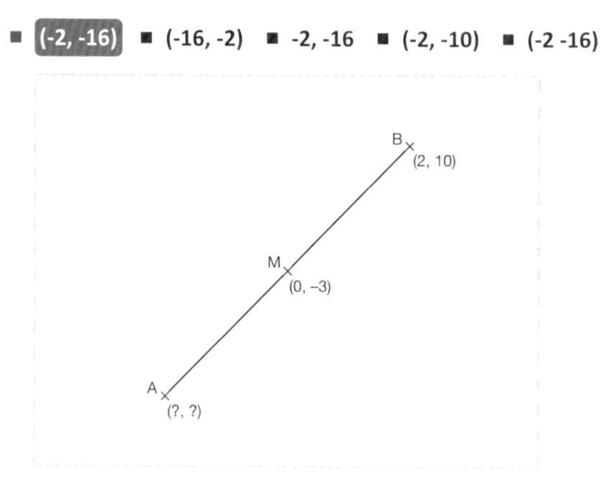

25. Govinda says that the point with coordinates (0, 8) lies on the *x*-axis. Is he correct? Explain your answer.

- Open question, no set answer

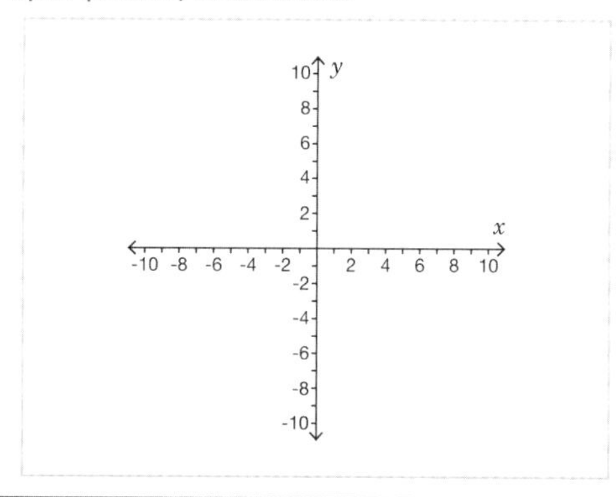

Level 4: Problem Solving: With coordinates in all four quadrants.

Required: 5/5 **Pupil Navigation:** on
Randomised: off

26. The diagram shows three points. To make a parallelogram a new point can be plotted at (3, 2) or (9, 2). What other coordinates would make a parallelogram?

- (7, 8)
- 7, 8
- (7 8)
- (8, 7)

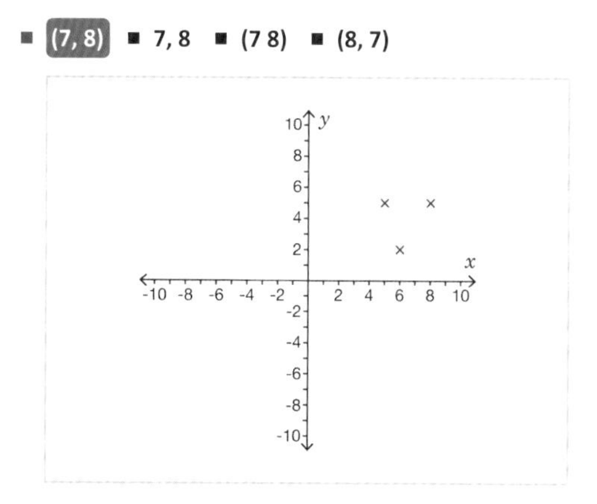

Level 4 *continued*

27. The two triangles in the diagram are identical. What are the coordinates of point D?

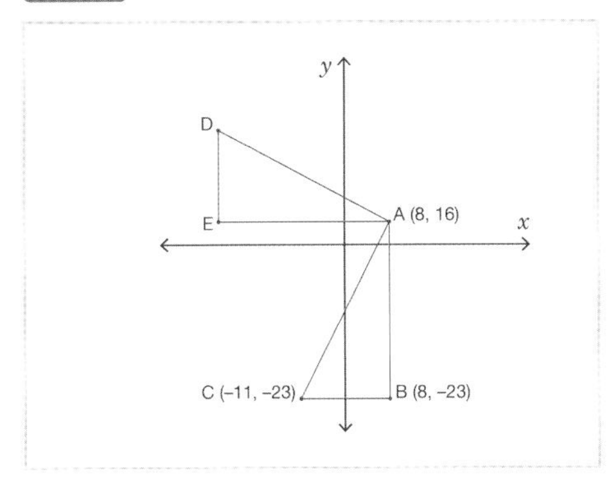

28. Point F has coordinates (x, y). x and y are both integers in the following:
$3 < x \leq 7$ and $-2 \leq y < 3$.
How many different possibilities are there for point F?

- **20**
- **16**
- **25**
- **30**

29. Use the following clues to find the coordinates of where the treasure is buried:

The treasure lies in the quadrant where x values and y values are both negative.
It does not lie on the axes.
It lies on the line $y = x$.
The x-coordinate is in the interval $-2 < x \leq 10$.
The x and y-coordinates are both integers.

- **(0,0)**
- **(-1,-1)**
- **-1,-1**
- **(-1-1)**

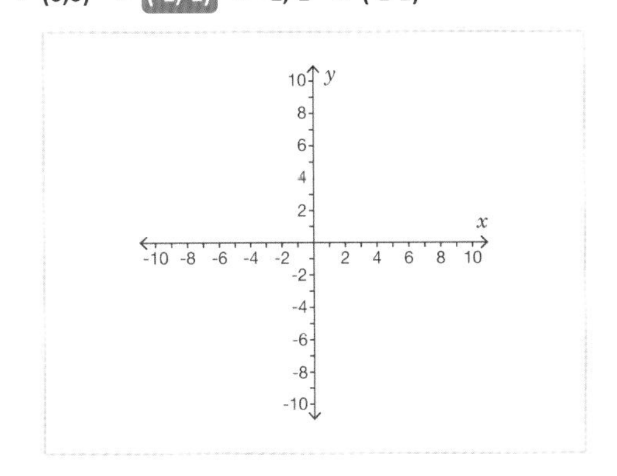

30. Katrina is thinking of a point with coordinates (x, y). She says that:
The product of x and y is 6.
$x > y$
The difference of x and y is 1.
$x < 3$
What are the coordinates of Katrina's point?

- **(-3, -2)**
- **(-2 -3)**
- **(-2, -3)**
- **(3, 2)**
- **-2, -3**
- **(2, 3)**

Find a Gradient and Y-Intercept of a Line from its Equation

Objective: I can find the gradient and the y-intercept of a straight line. calculate and interpret gradients and intercepts of graphs of such linear equations numerically, graphically and algebraically.

Quick Search Ref: 10504

Level 1: Understanding: Finding the gradient and y-intercept of a line when its equation is in the form y = mx + c.

Required: 7/10 **Pupil Navigation:** on **Randomised:** off

1. What is the gradient of the line shown in the diagram?

- **2**
- 3

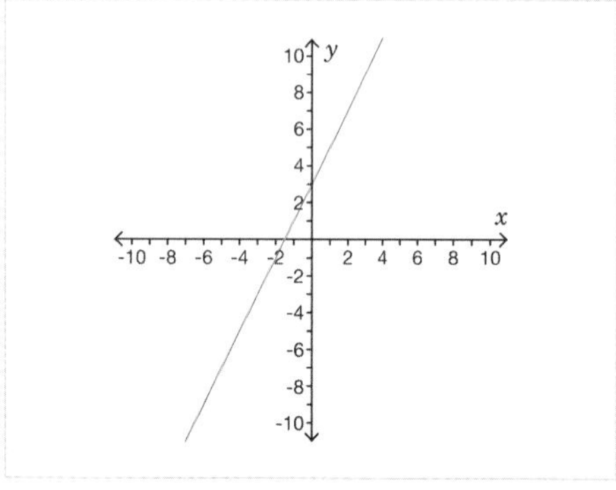

2. What is the y-intercept of the line $y = 2x + 5$?

- **5**
- 2

$y = mx + c$

m = gradient c = y-intercept

$y = 2x + 5$

3. What is the gradient of the line $y = 2x + 5$?

- **2**
- 5

$y = mx + c$

m = gradient c = y-intercept

$y = 2x + 5$

4. What is the y-intercept of the line $y = 5x – 4$?

- **-4**
- 5
- 4

$y = mx + c$

m = gradient c = y-intercept

$y = 5x - 4$

5. What is the gradient of the line $y = x + 7$?

- 0
- **1**
- 7

$y = mx + c$

m = gradient c = y-intercept

$y = x + 7$

6. Select the two properties of the line $y = 5x – 3$.

2/6

- gradient of 3
- y-intercept of 5
- **gradient of 5**
- **y-intercept of –3.**
- gradient of –3
- y-intercept of 3

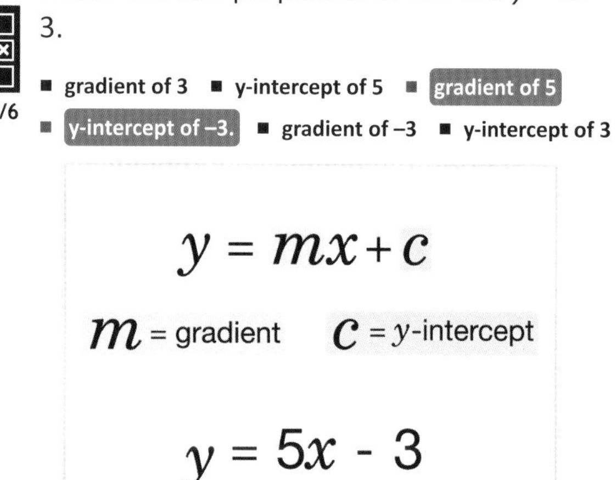

Level 1 *continued*

7. What is the y-intercept of the line $y = 4x$?

- 4
- 0
- 1
- The line does not cross the y-axis.

$$y = mx + c$$

m = gradient c = y-intercept

$$y = 4x$$

8. Select the two properties of the line $y = 3x - 7$.

2/6

- gradient of −7
- y-intercept of 3
- gradient of 7
- y-intercept of −7
- gradient of 3
- y-intercept of 7

$$y = mx + c$$

m = gradient c = y-intercept

$$y = 3x - 7$$

9. What is the gradient of the line $y = 3x - 4$?

- 3
- -4

$$y = mx + c$$

m = gradient c = y-intercept

$$y = 3x - 4$$

10. What is the y-intercept of the line $y = 2x - 5$?

- -5
- 2
- 5

$$y = mx + c$$

m = gradient c = y-intercept

$$y = 2x - 5$$

Level 2: Fluency: Using fractions and rearranging equations.

Required: 7/10 **Pupil Navigation:** on

Randomised: off

11. What is the gradient of the line $y = x/2 + 5$?

- 1/2
- 0.5
- 2
- .5

$$y = \frac{x}{2} + 5$$

12. What is the y-intercept of the line $y = 3x - ¼$?
Give your answer as a fraction.

- -1/4

$$y = 3x - \frac{1}{4}$$

Level 2 *continued*

13. What is the y-intercept of the line $y + 9 = 2x$?

■ -9 ■ 9

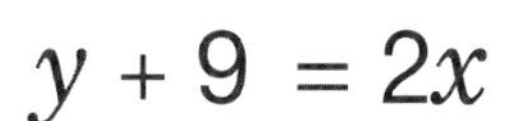

14. What is the gradient of the line $y - 7x = 5$?

■ 7 ■ -7

$$y - 7x = 5$$

15. What is the y-intercept of the line $2y = 8x - 5$?

Give your answer as an improper fraction.

■ -5/2 ■ -5 ■ -5/1 ■ 5/2 ■ -2.5

$$2y = 8x - 5$$

16. What is the gradient of the line $4x + 2y - 9 = 0$?

■ -2 ■ -4 ■ 4

$$4x + 2y - 9 = 0$$

17. What is the gradient of the line $y/2 - 3x = 2$?

■ 6 ■ 3 ■ -3 ■ -6

$$\frac{y}{2} - 3x = 2$$

18. What is the gradient of the line $y + 2x = 5$?

■ -2 ■ 2

$$y + 2x = 5$$

19. What is the y-intercept of the line $3y = 2x - 15$?

■ -5 ■ -15

$$3y = 2x - 15$$

Level 2 *continued*

20. What is the gradient of the line $3y - 6x + 7 = 0$?

1 2 3

■ -6 ■ **2** ■ 6

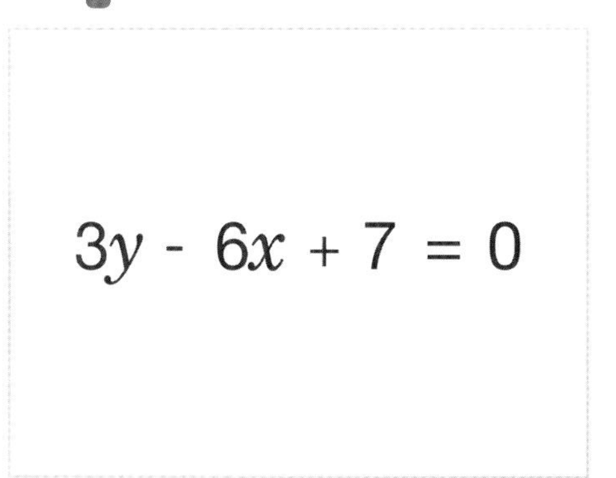

Level 3: Reasoning: Making inferences using y = mx + c.

Required: 5/5 **Pupil Navigation:** on

Randomised: off

21. Which line passes through the origin?

1/3

■ 4y = x + 8 ■ **3y + 2x = 0** ■ 7 + y = 5x

22. Leo says that the following lines are parallel:
$y - 3x = 1$ and $2y = 6x + 5$
Is Leo correct? Explain your answer.

a b c

- Open question, no set answer

23. What is the equation of the line shown in the diagram?

1/4

■ **y = 2x – 5** ■ y = 5x – 2 ■ y = 2x + 5 ■ y = 5x + 2

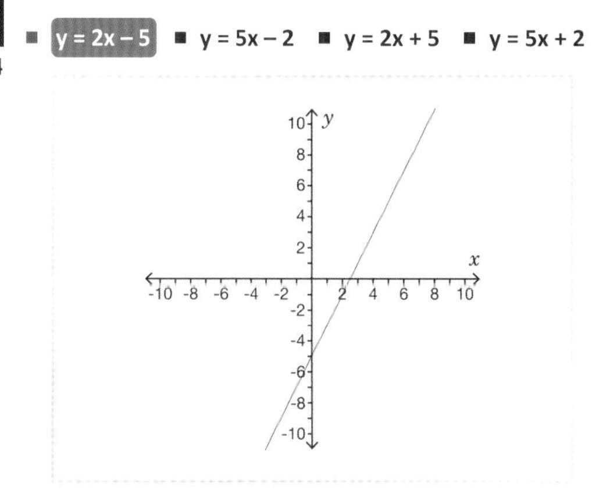

24. Phoebe has completed her homework on straight line graphs, but she has answered one question incorrectly. What is the correct answer to the question she got wrong?

1 2 3

■ **6**

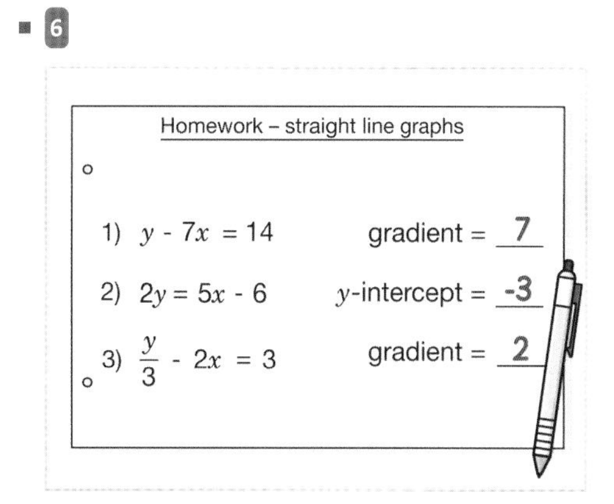

25. Which two lines have an identical gradient and y-intercept?

2/4

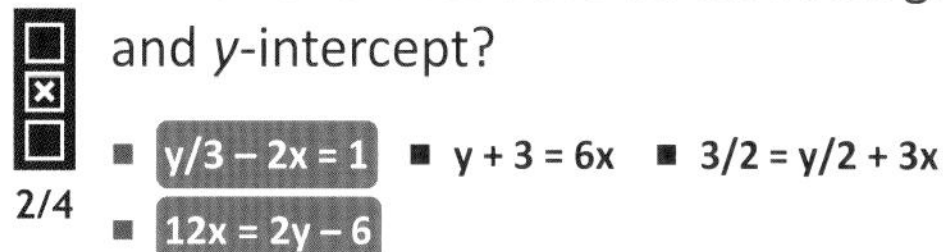

■ **y/3 – 2x = 1** ■ y + 3 = 6x ■ 3/2 = y/2 + 3x
■ **12x = 2y – 6**

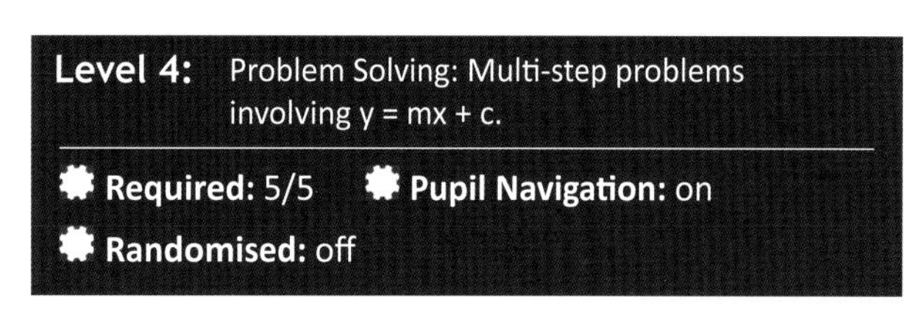

Level 4: Problem Solving: Multi-step problems involving y = mx + c.

Required: 5/5 **Pupil Navigation:** on

Randomised: off

26. Which is the correct diagram of the line $2y + 3x = 6$?

1/4

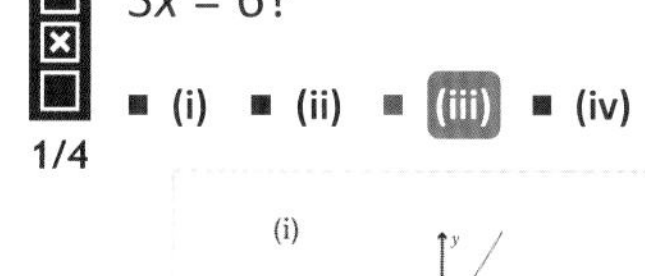

■ (i) ■ (ii) ■ **(iii)** ■ (iv)

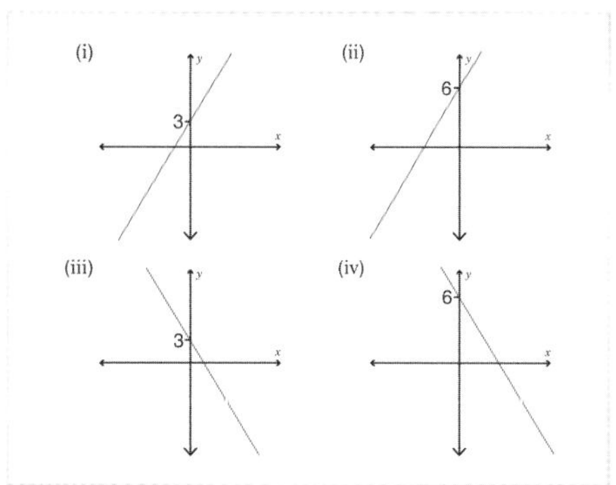

27. A straight line has a gradient of 3 and passes through the point (2, 7). What is the equation of the line in the form $y = mx + c$?

a b c

■ **y = 3x + 1**

Level 4 *continued*

28. The lines $y + 3x = 10$ and $2y + 8 = x$ meet at the point (4, –2). The two lines together with the y-axis make a triangle. What is the area of the triangle?
Don't include the units in your answer.

- 28
- 56

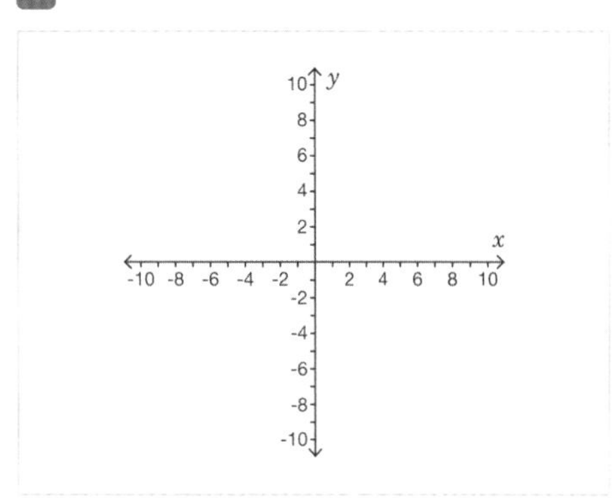

29. Give the equation of the line shown in the form $y = mx + c$.

- y = 1x - 3
- y = x - 3

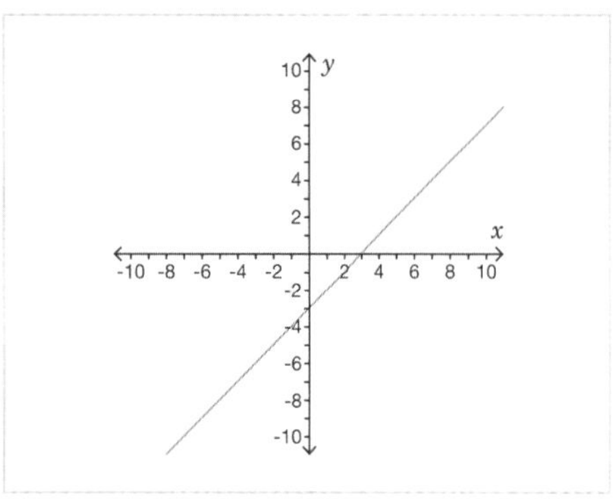

30. A straight line has a gradient of 2 and passes through the point (4, 5). What is the equation of the line in the form $y = mx + c$?

- y = 2x - 3

Find Gradients of Straight Lines

Objective: I can calculate the gradient of a straight line.

Quick Search Ref: 10487

Level 1: Understanding: Finding changes in x and y-coordinates and calculating simple gradients.

Required: 7/10 **Pupil Navigation:** on **Randomised:** off

1. The gradient of a straight line is a measure of . . .

1/3

- how long the line is.
- where it crosses the y-axis.
- how steep the line is.

2. How do you calculate the gradient of a line?

1/4

- Multiply the change in x-coordinates by the change in y-coordinates.
- Divide the change in x-coordinates by the change in y-coordinates.
- Divide the change in y-coordinates by the change in x-coordinates.
- Add the change in x-coordinates to the change in y-coordinates.

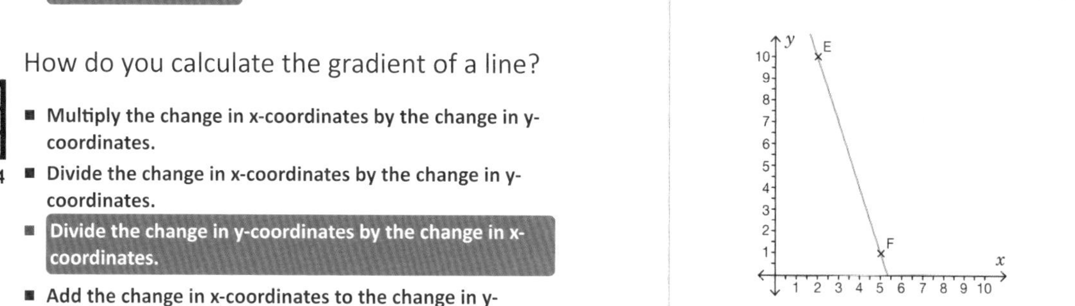

3. What is the change in y-coordinates from point A to point B?

- 6
- 7
- 2

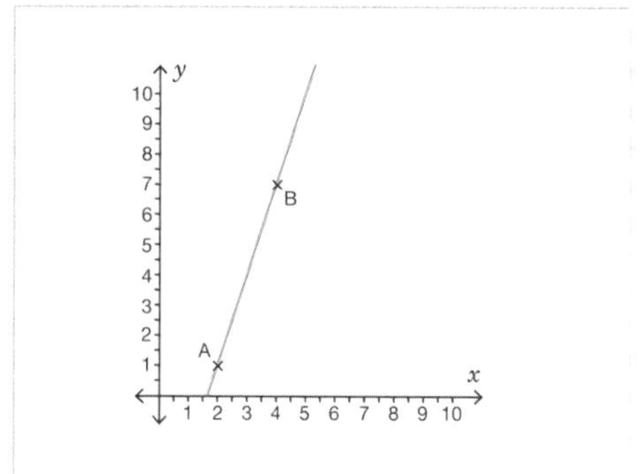

4. What is the change in x-coordinates from point R to point S?

- 8
- 9
- 2

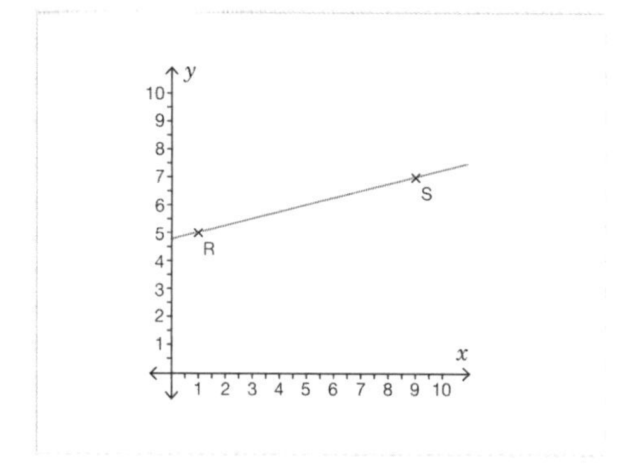

5. What is the change in y-coordinates from point E to point F?

- -10
- -9
- 9
- 3

6. What is the gradient of the line shown in the diagram?

- 2
- 0.5
- 4/2
- 1/2

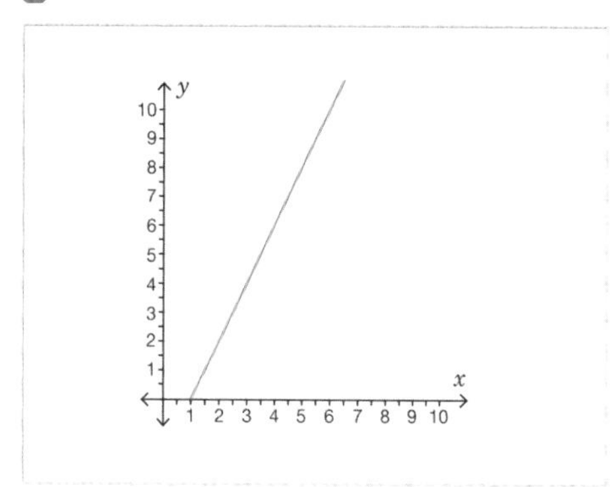

7. What is the gradient of the line shown in the diagram?

- 1
- 4/4

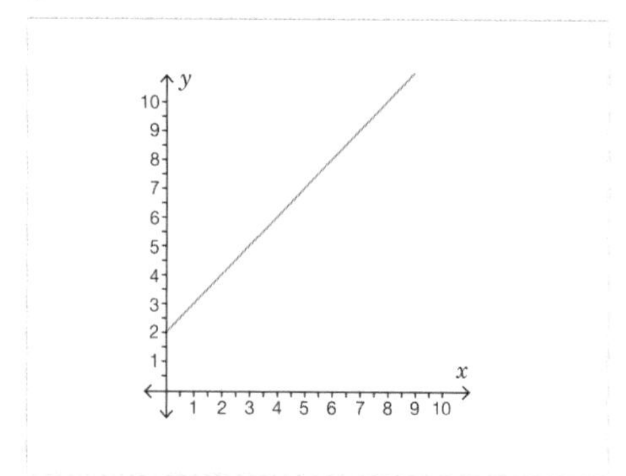

Level 1 *continued*

8. What is the change in x-coordinates from point U to point V?

1 2 3

■ 6 ■ **5** ■ 1

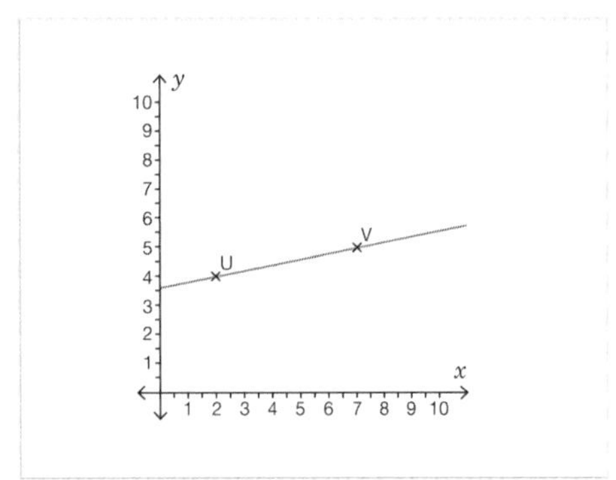

9. What is the change in y-coordinates from point C to point D?

1 2 3

■ **-8** ■ 8 ■ -9 ■ 4

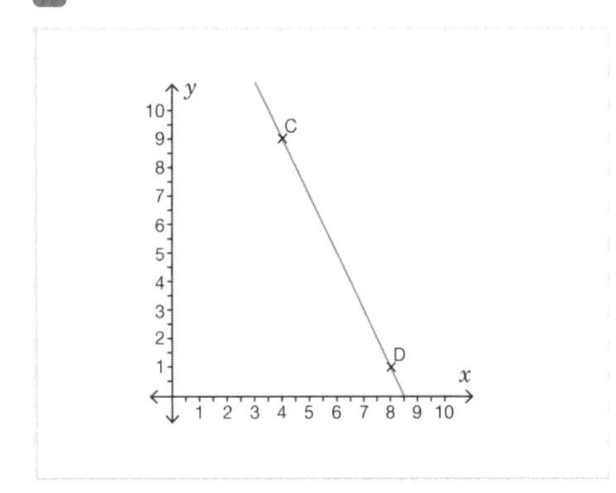

10. What is the gradient of the line shown in the diagram?

a b c

■ **3** ■ 1/3 ■ 6/2

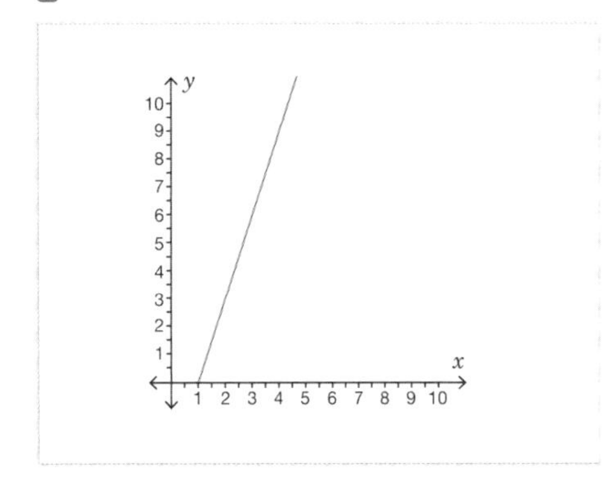

Level 2: Fluency: Calculating negative and fractional gradients.

Required: 7/10 **Pupil Navigation:** on

Randomised: off

11. What is the gradient of the line shown in the diagram?

a b c

■ **-3** ■ -1/3 ■ 3 ■ -2/6

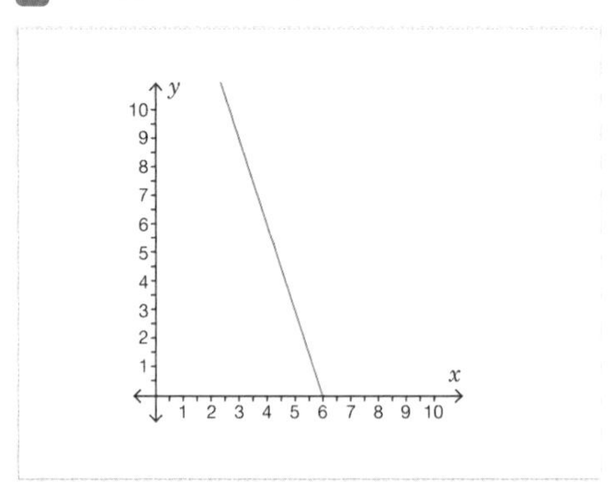

12. What is the gradient of the line shown in the diagram?

a b c

■ **0.5** ■ **1/2** ■ 2 ■ .5 ■ 2/4

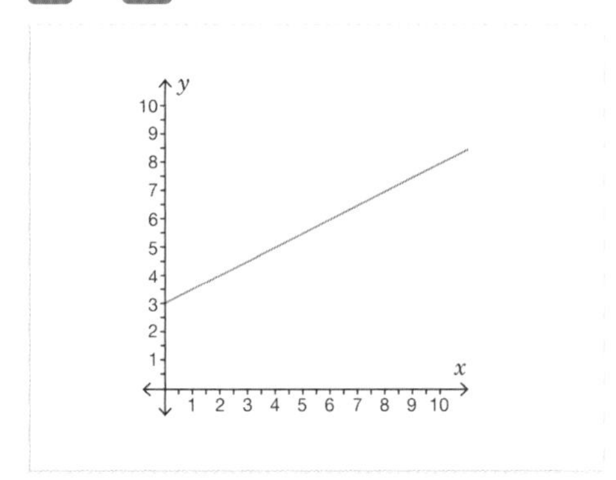

13. What is the gradient of the line that passes through points (1, 2) and (5, 10)?

a b c

■ ½ ■ **2** ■ 0.5

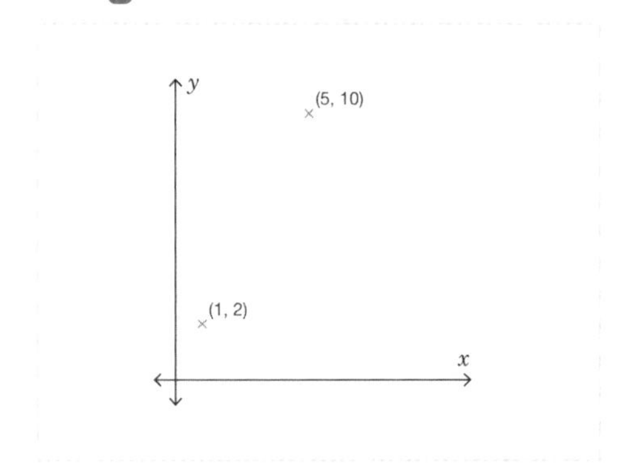

Level 2 *continued*

14. What is the gradient of the line shown in the diagram?
Give your answer as a fraction in its simplest form.

■ 2/3 ■ 1 1/2 ■ 3/2 ■ 6/9

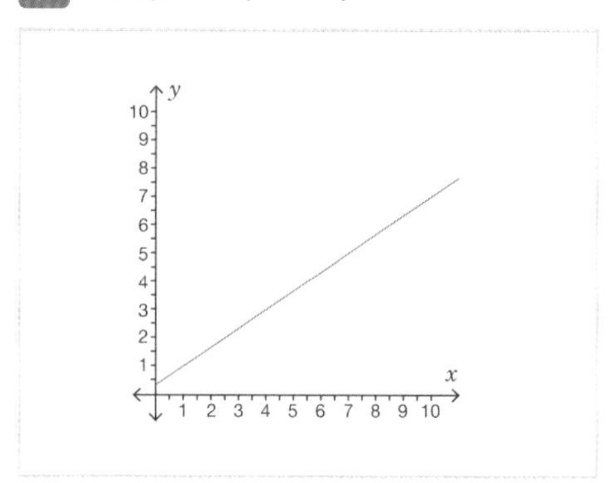

15. What is the gradient of the line that passes through points (1, 10) and (3, 1)?
Give your answer as a decimal.

■ -4.5 ■ 4.5 ■ -9/2 ■ 9/2

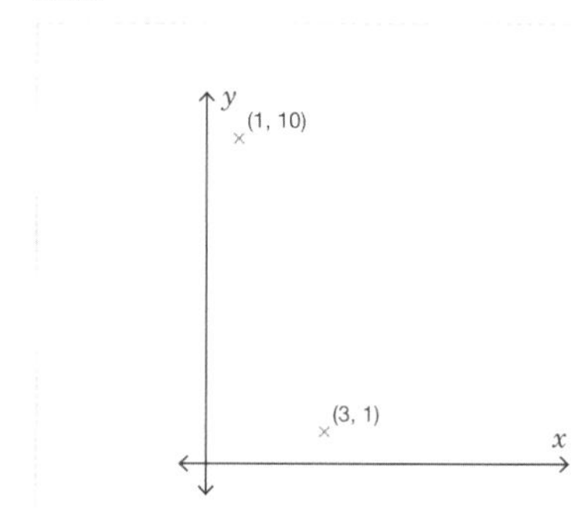

16. Fleur cycles up a straight road that climbs a height of 36 metres over a horizontal distance of 400 metres.
What is the gradient of the hill?
Give your answer as a fraction in its simplest form

■ 9/100 ■ 0.09

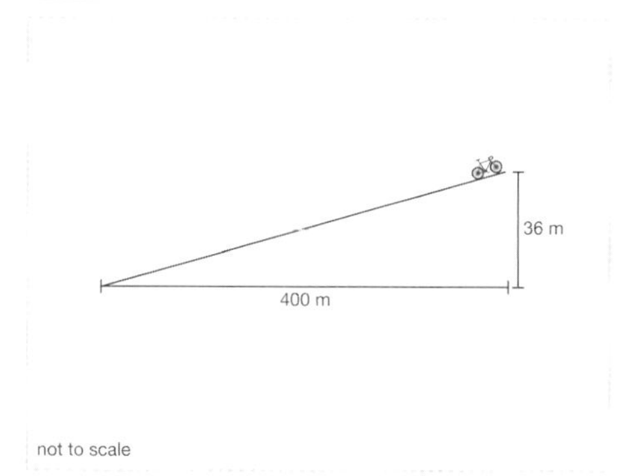

not to scale

17. A ladder is placed against a wall to reach a window 4 metres above ground. The bottom of the ladder must be placed 0.5 metres from the wall.
What is the gradient of the slope of the ladder?

■ 8 ■ 2

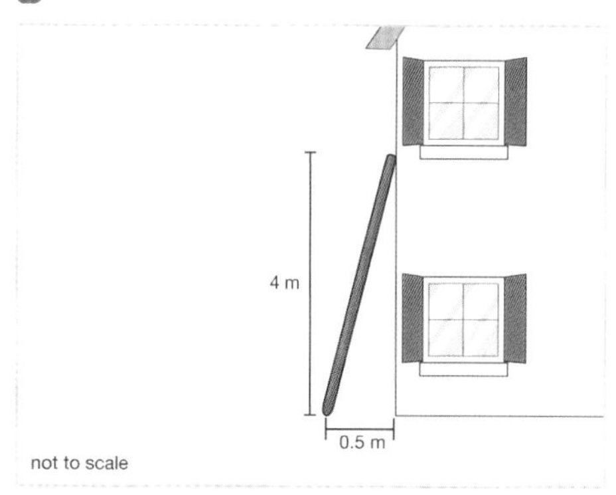

18. What is the gradient of the line shown in the diagram?

■ -1/4 ■ -4 ■ 2 ■ -0.25 ■ -2 ■ 4

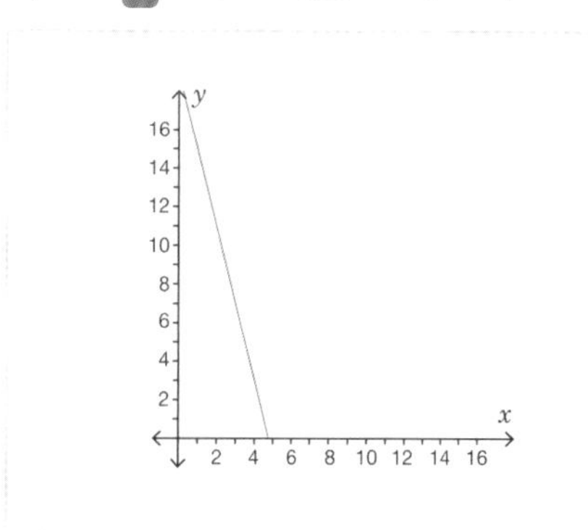

19. What is the gradient of the line that passes through points (2, 5) and (8, 8)?
Give your answer as a fraction in its simplest form.

■ 2 ■ 1/2 ■ 0.5 ■ 3/6

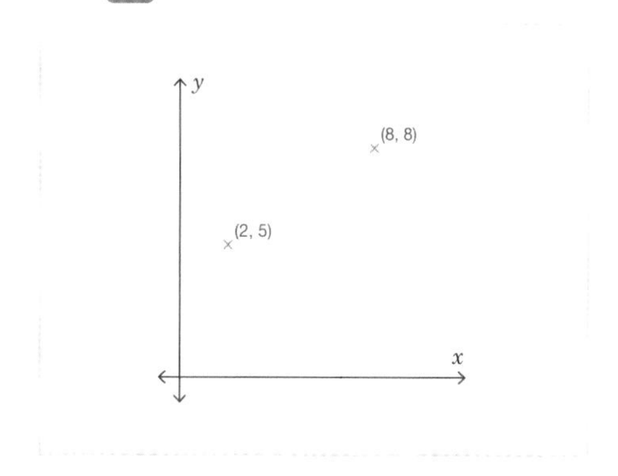

Level 2 *continued*

20. What is the gradient of the line shown in the diagram?
Give your answer as a fraction in its simplest form.

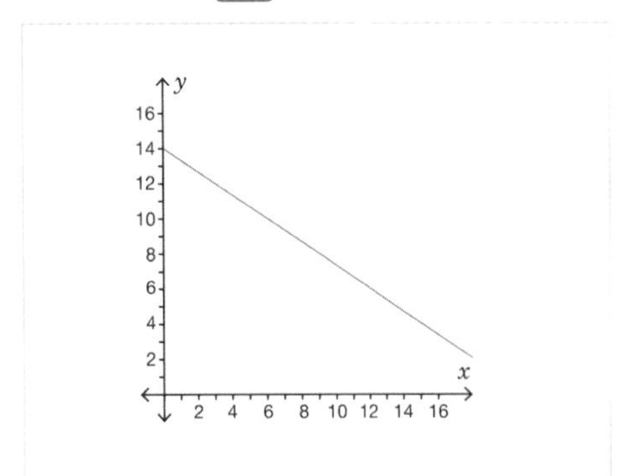

Level 3: Reasoning: Properties of gradients.

Required: 5/5 **Pupil Navigation:** on
Randomised: off

21. Hamza says the gradient of the line shown is 6. Is Hamza correct? Explain your answer.

- Open question, no set answer

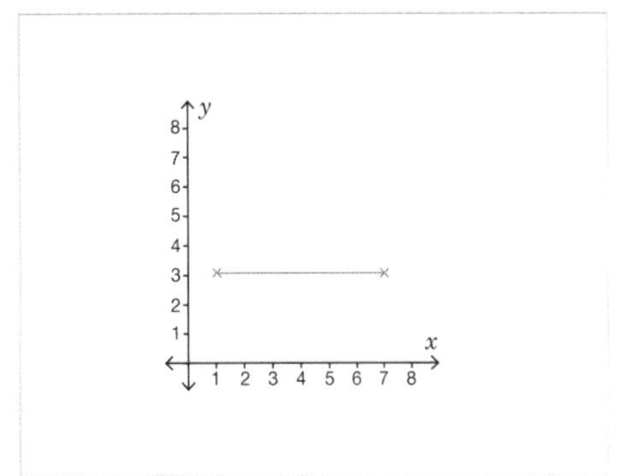

22. Sort the diagrams in the same order as the following gradients:
3/2, −1/3, 3, 2/3

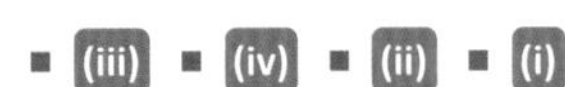

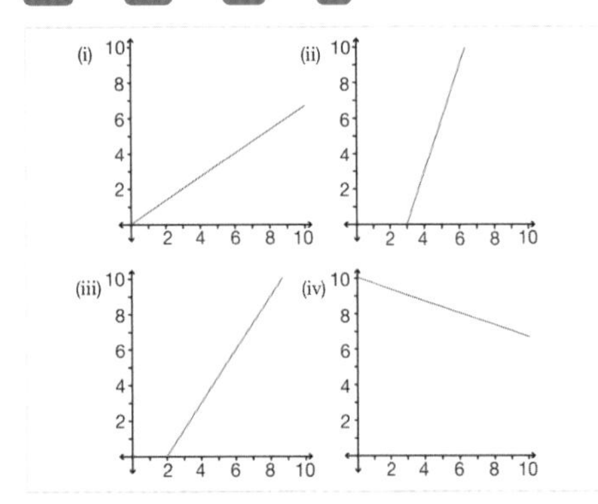

23. Joshua says that parallel lines have the same gradient. Is Joshua correct? Explain your answer.

- Open question, no set answer

24. Reuben, Suban and Teresa are calculating the gradient of the line in the graph. Reuben calculates the gradient of line AB, Suban calculates the gradient of line BC and Teresa calculates the gradient of line AC.
Which is the easiest method?

1/3

- calculating the gradient of line AB
- calculating the gradient of line BC
- calculating the gradient of line AC

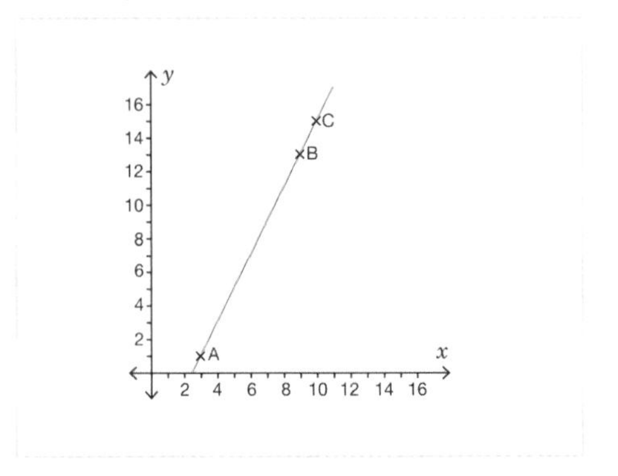

25. Tori says you can't calculate the gradient of the line shown in the diagram. Is Tori correct? Explain your answer.

- Open question, no set answer

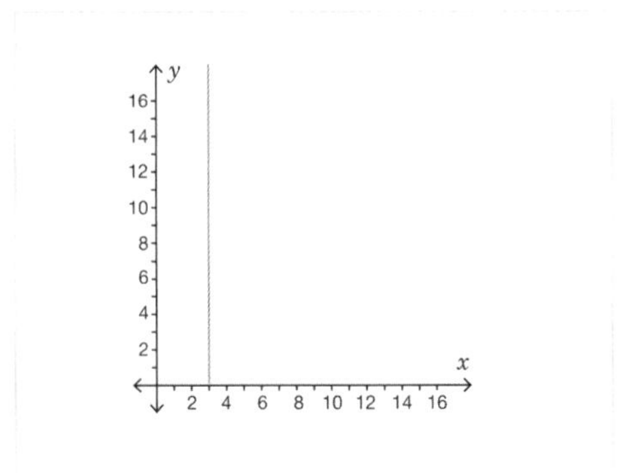

Level 4: Problem Solving: Multi-step problems involving gradients.

Required: 5/5 **Pupil Navigation:** on

Randomised: off

26. The following points are plotted on a coordinate grid to make a triangle.
A = (1, 1), B = (2, 6) and C = (5, 5).
Sort the lines in ascending order (smallest first) according to their gradient.

- BC
- AC
- AB

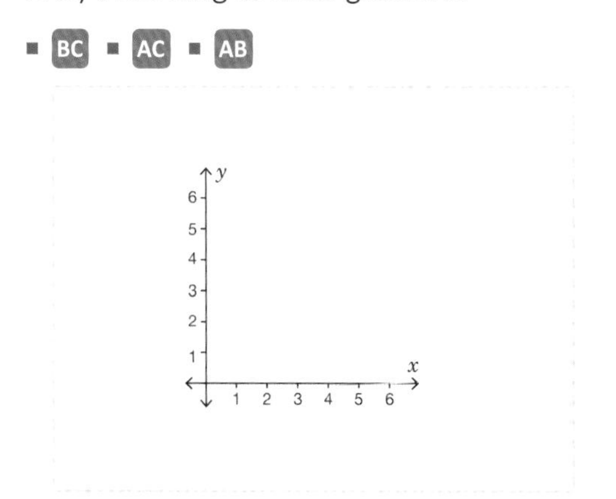

27. What is the gradient of the line $y = 3x + 2$?

- 3

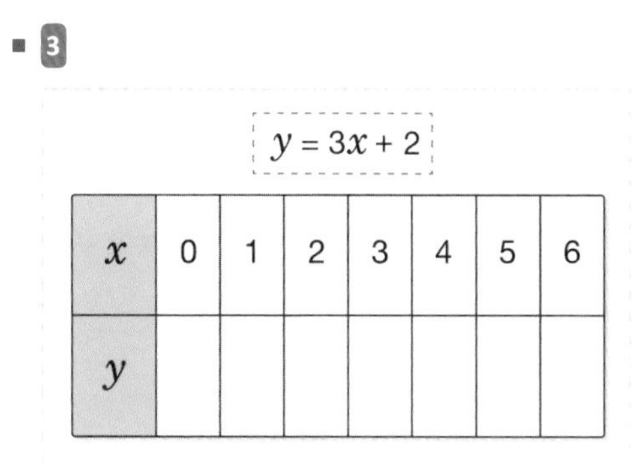

$y = 3x + 2$

x	0	1	2	3	4	5	6
y							

28. A fractional gradient can be converted into a percentage. Kyle makes a bicycle ramp with a gradient of 30%. If the ramp is 240 cm high, how long is the base of the ramp?
Give your answer in metres and include the units.

- 4000 m
- 8 metres
- 8 m
- 8
- 800 cm

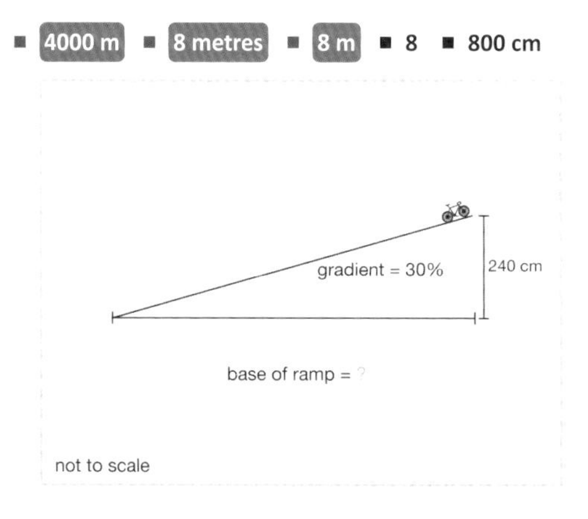

29. What is the gradient of the line $y = 7x + 1$?

- 7

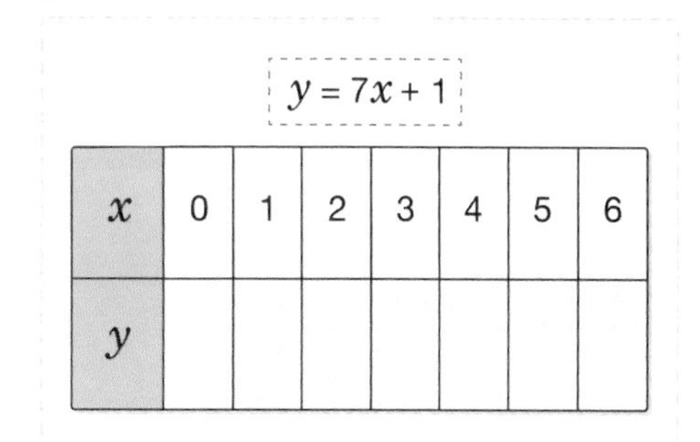

$y = 7x + 1$

x	0	1	2	3	4	5	6
y							

30. The line passing through the points (2, 4) and (11, k) has a gradient of two-thirds.
What is the value of k?

- 10

Find the Equation of a Line from its Graph

Objective: I can find the equation of a line from its graph.

Quick Search Ref: 10525

Level 1: Understanding: Find equations of lines from graphs.

Required: 7/10 **Pupil Navigation:** on **Randomised:** off

1. What is the gradient of this graph?

123 ■ 3 ■ -5

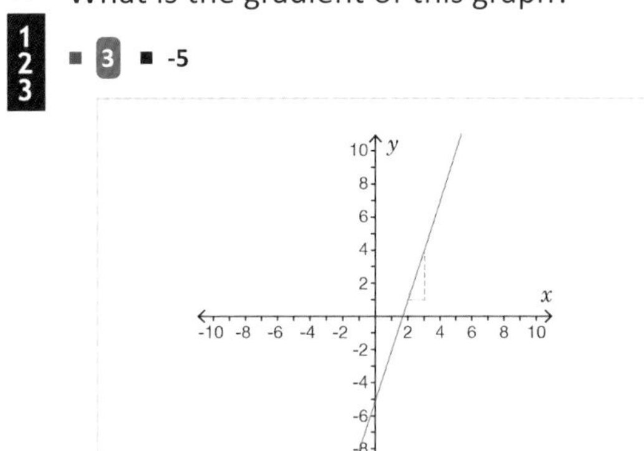

2. What is the y-intercept of this graph?

123 ■ -3 ■ 3 ■ 2

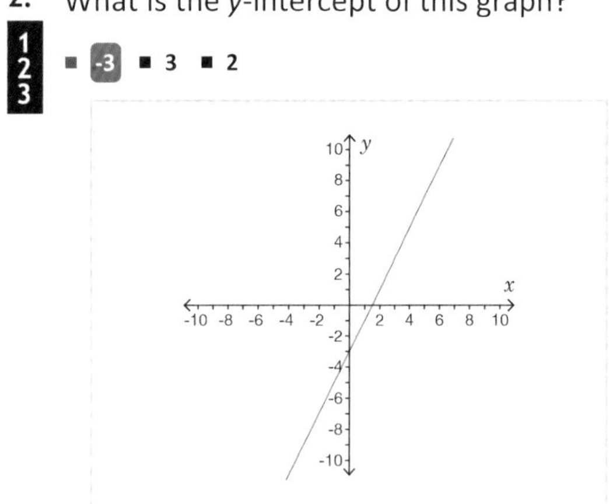

3. What is the equation of this line?

1/4

■ y = 2x + 4 ■ y = 4x + 2 ■ y = x + 4 ■ y = 4x + 1

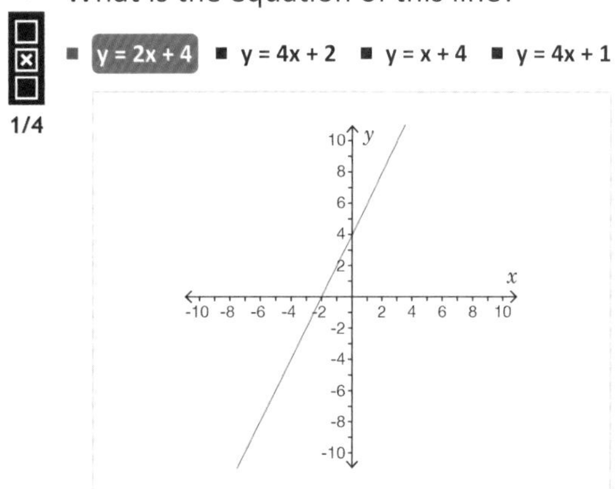

4. Which graph shows the line $y = 2x - 3$?

1/4

■ (i) ■ (ii) ■ (iii) ■ (iv)

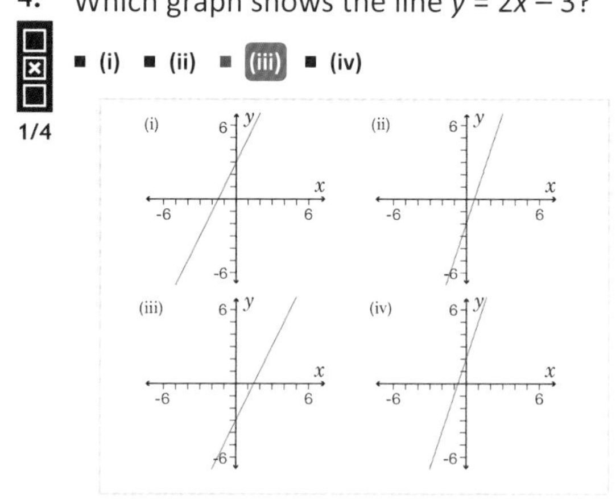

5. What is the equation of this line?

Give your answer in the form $y = mx + c$

abc ■ y = 4x + 1 ■ y = 1x + 4 ■ y = x + 4

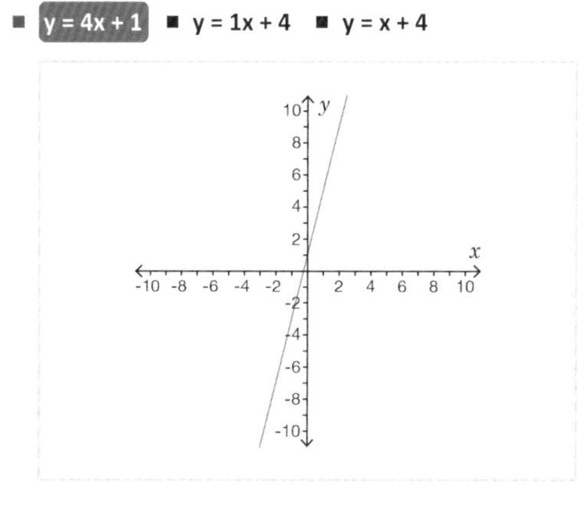

6. What is the equation of this line?

Give your answer in the form $y = mx + c$

abc ■ y = 3x - 5 ■ y = 3x + 5 ■ y = -5x + 3

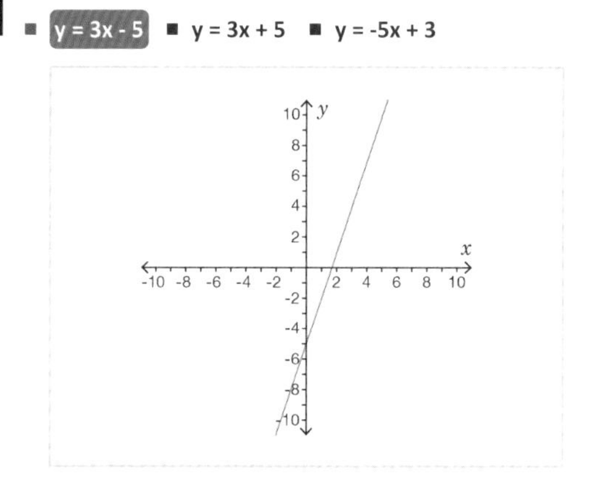

Level 1 *continued*

7. What is the equation of this line?

Give your answer in the form y = mx + c

- y = 1x + 3

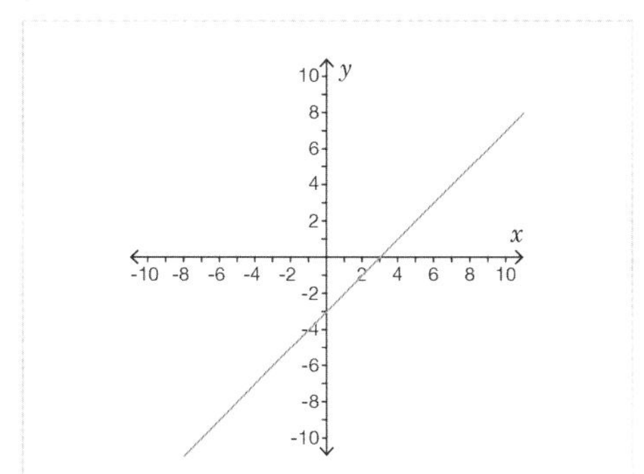

8. Which graph shows the line $y = 4x - 1$?

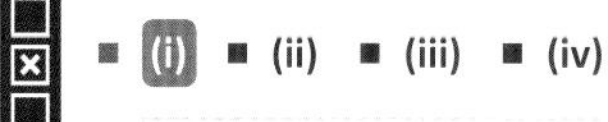

1/4

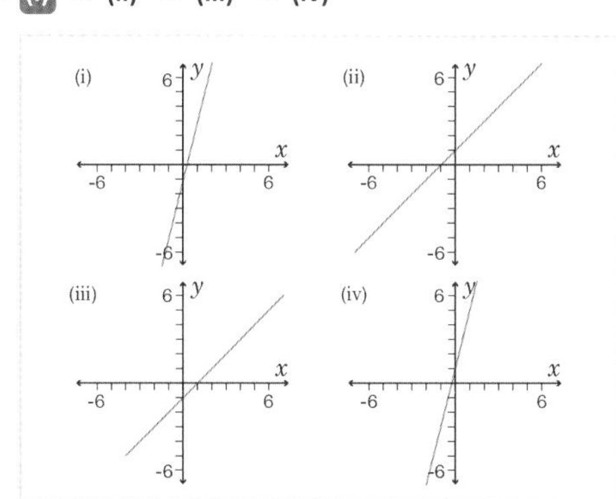

9. What is the equation of this line?

Give your answer in the form y = mx + c

- y = 5x + 1
- y = x + 5
- y = 1x + 5

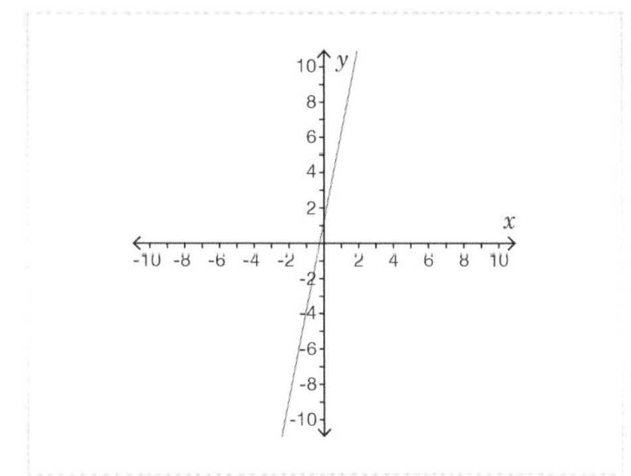

10. What is the equation of this line?

Give your answer in the form y = mx + c

- y = 3x - 4
- y = 3x + 4
- y = -4x + 3

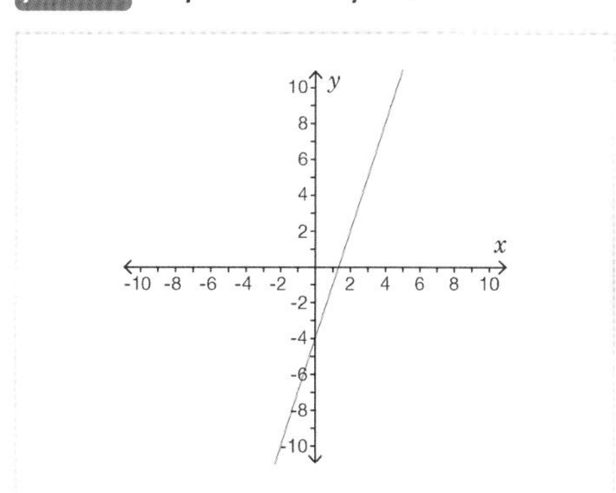

Level 2: Fluency: Use negative and fractional gradients and interpret graphs with partial information.

Required: 7/10 **Pupil Navigation:** on

Randomised: off

11. What is the equation of this line?

Give your answer in the form y = mx + c

1/6

- y = 2x - 3
- y = -3x + 2

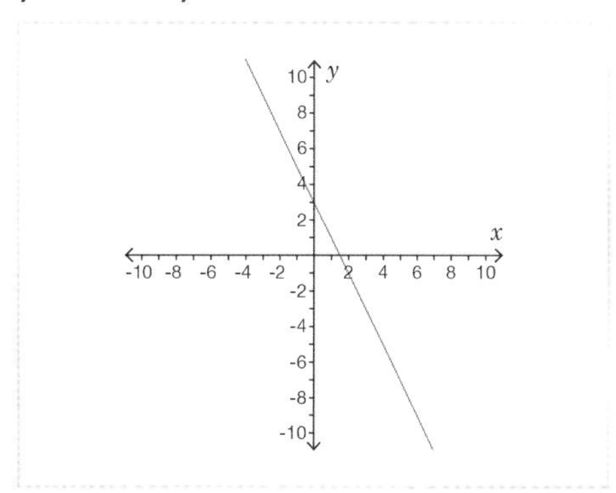

12. What is the equation of this line?

Give your answer in the form y = mx + c

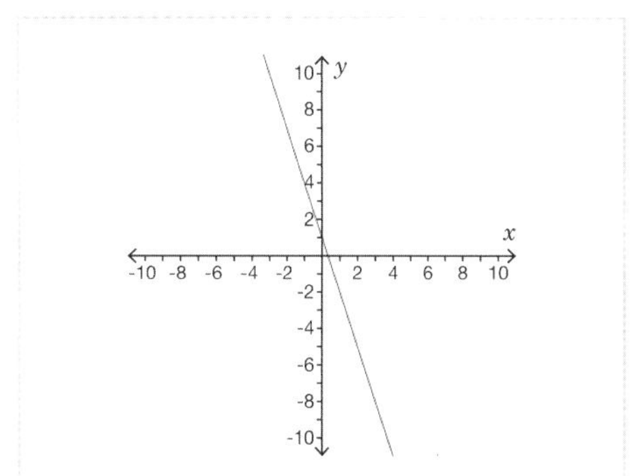

Level 2 continued

13. What is the equation of this line?
Give your answer in the form y = mx + c

■ y = 3x + 1/2 ■ y = 3x + 0.5 ■ y = 3x + 1

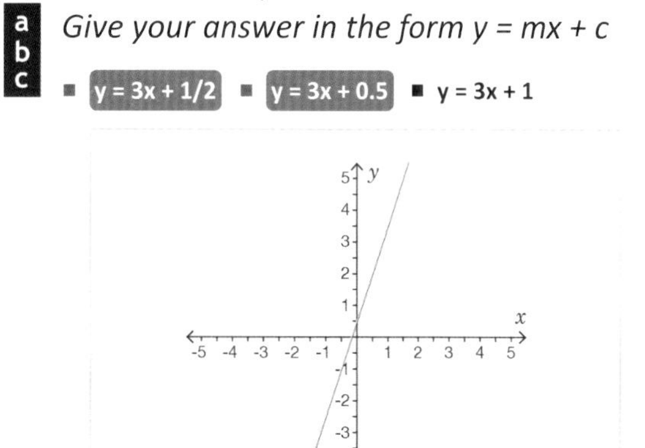

14. What is the equation of this line?
Give your answer in the form y = mx + c

1/4

■ (i) ■ (ii) ■ (iii) ■ (iv)

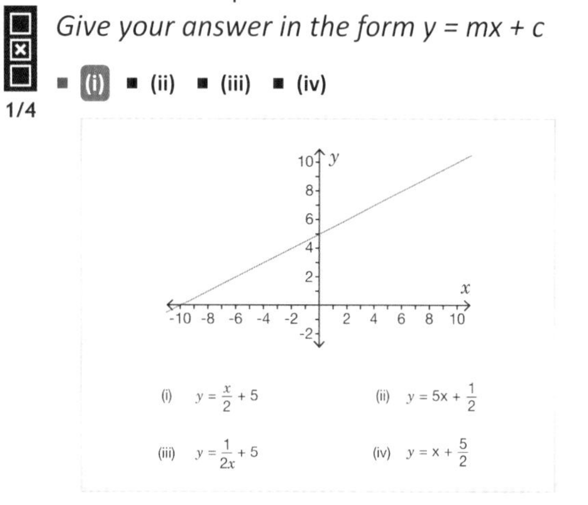

15. If the equation of this line was written in the form *y* = *mx* + *c*, what would be the values of *m* and *c*?

2/7

■ m = 3 ■ c = 2 ■ m = ⅓ ■ c = ½ ■ m = 2 ■ c = −2 ■ m = ½

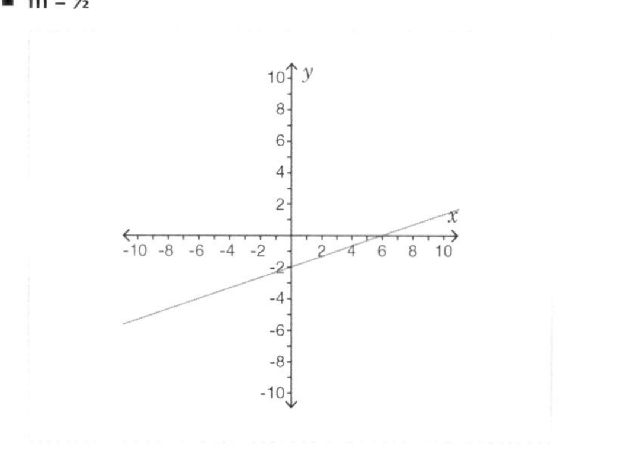

16. What is the equation of this line?
Give your answer in the form y = mx + c

■ y = 3x - 12

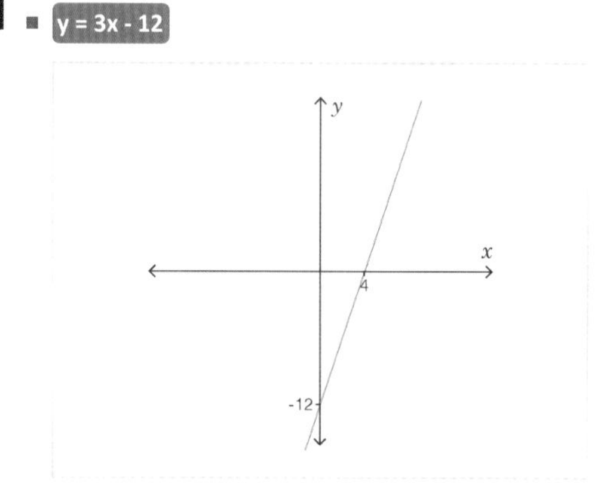

17. What is the equation of this line?
Give your answer in the form y = mx + c

■ y = -2x + 5 ■ y = 2x + 5

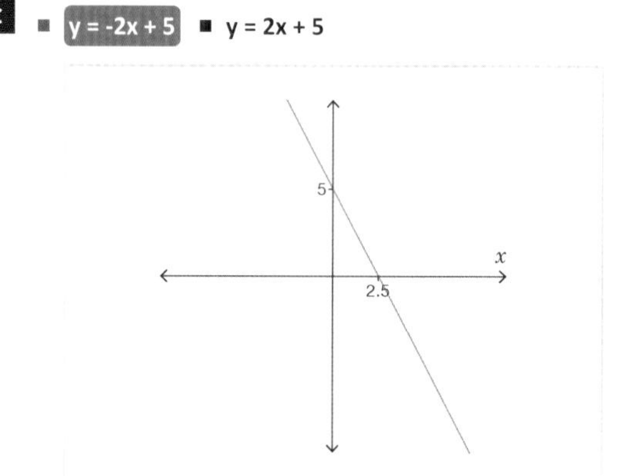

18. What is the equation of this line?
Give your answer in the form y = mx + c

■ y = -2x - 5 ■ y = 2x - 5

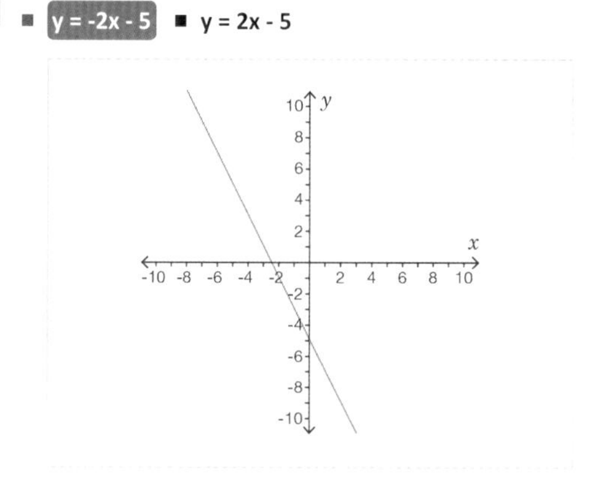

Level 2 *continued*

19. If the equation of this line was written in the form $y = mx + c$, what would be the values of m and c?

2/7

- m = ⅓
- c = ¼
- m = 4
- **c = 3**
- **m = ¼**
- c = ⅓
- m = 3

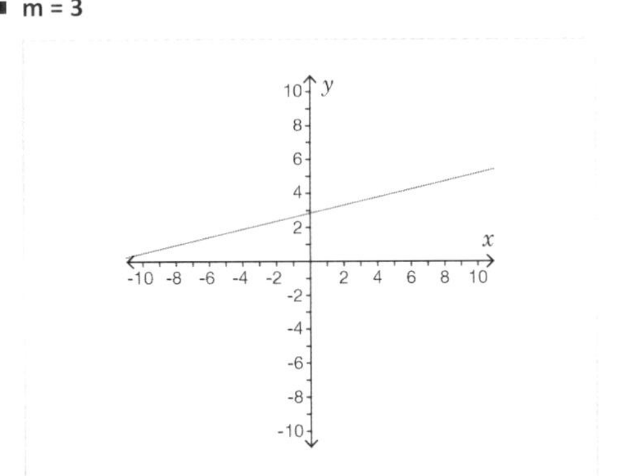

20. What is the equation of this line?
Give your answer in the form $y = mx + c$

- **y = 4x - 8**

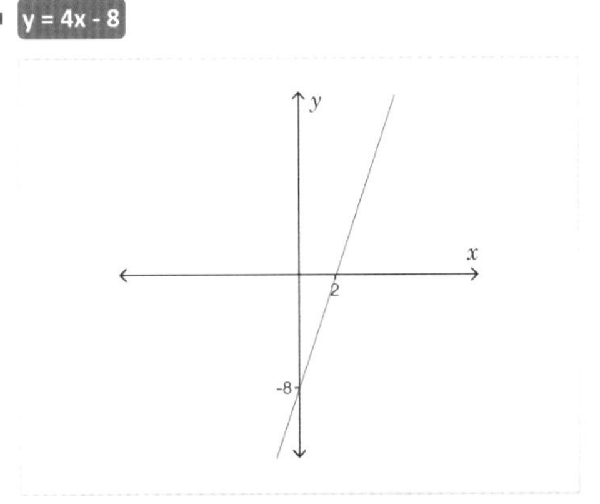

Level 3: Reasoning: Make inferences and address misconceptions with linear graphs.

Required: 5/5 **Pupil Navigation:** on
Randomised: off

21. Juan says the equation of this line is $y = 2x + 3$. Is Juan correct? Explain your answer.

- Open question, no set answer

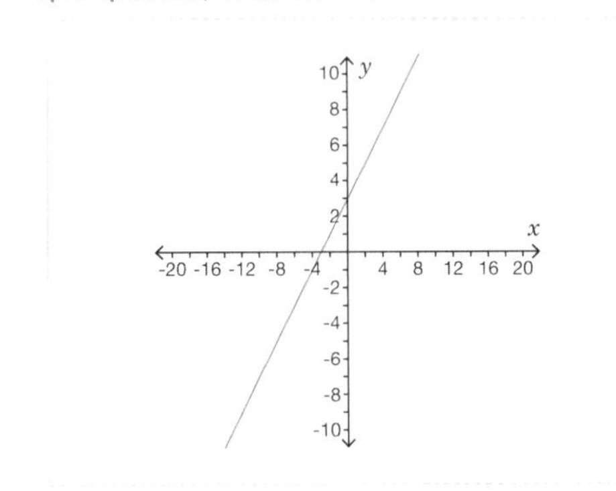

22. This graph shows the amount in pence that a mobile phone company charges for calls depending on the length of the call in minutes.
Select the equation for the total cost, C, in terms of the number of minutes, m.

1/4

- C = 7m + 2
- C = 5m + 2
- C = 7m + 5
- **C = 2m + 5**

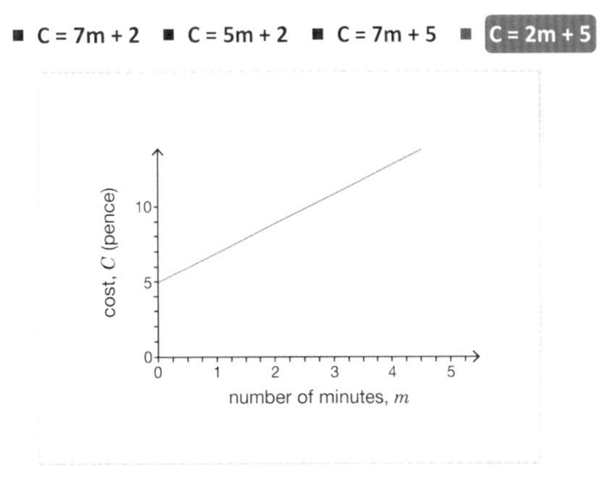

23. Sort the equations of these lines into the same order as the graphs. Put the equation of graph (i) at the top.

- **y = 2x – 1**
- **y = 3x – 1**
- **y = 3x + 1**
- **y = x + 2**

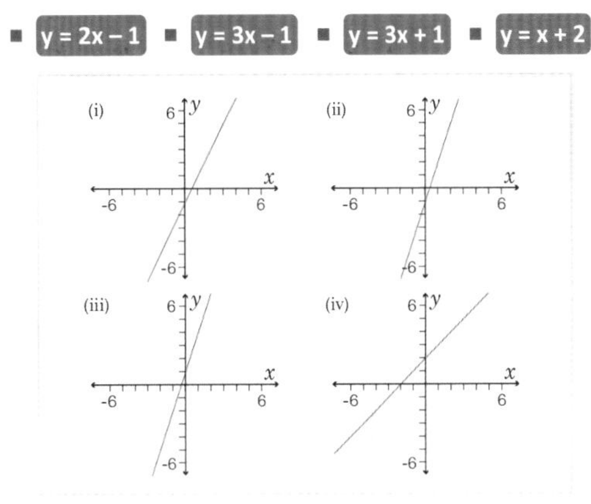

24. The line $y = 3x – 1$ is reflected in the x-axis.
What is the equation of the reflected line?

- **y = -3x + 1**
- y = -3x - 1
- y = 3x + 1

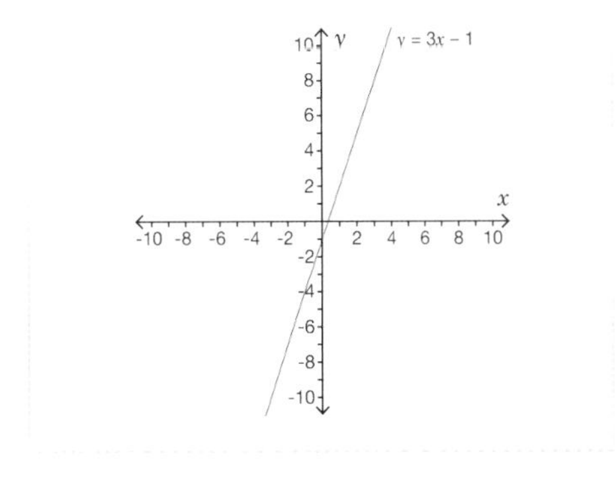

Level 3 *continued*

25. A line passes through the points (12, 40) and (15, 52). Where does this line intercept the *y*-axis?

- -8

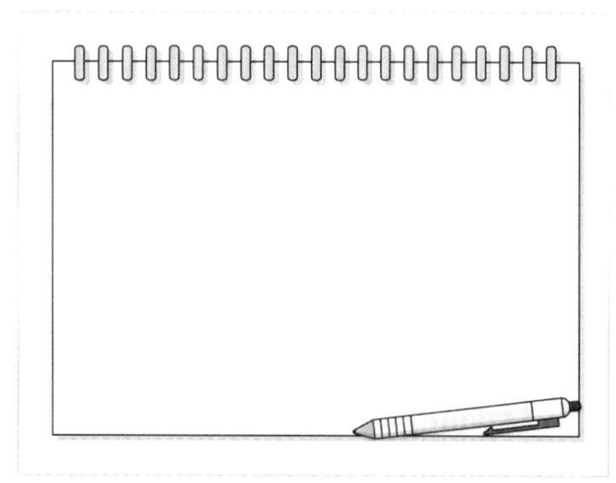

Level 4: Problem Solving: Solve multi-step problems to find equations of graphs.

Required: 5/5 **Pupil Navigation:** on
Randomised: off

26. Line 1 has an equation of $y = 4x - 10$. What is the equation of line 2?
Give your answer in the form $y = mx + c$.

- y = -2x + 6

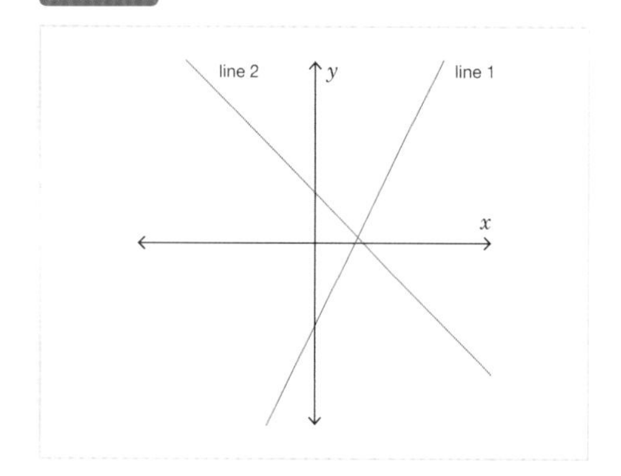

27. What is the equation of the straight line that passes through the points (2, 1) and (5, 10)?
Give your answer in the form $y = mx + c$

- y = 3x - 5

28. This graph shows the amount in pounds an electrician charges for work depending on time taken.
Give an equation for the total cost, *C*, in terms of the number of hours taken, *h*.

- C = 30 + 25 * h
- C = 25 x h + 30
- C = 30 + 25h
- C = 25 * h + 30
- C = 25h + 30
- C = 30 + 25 x h

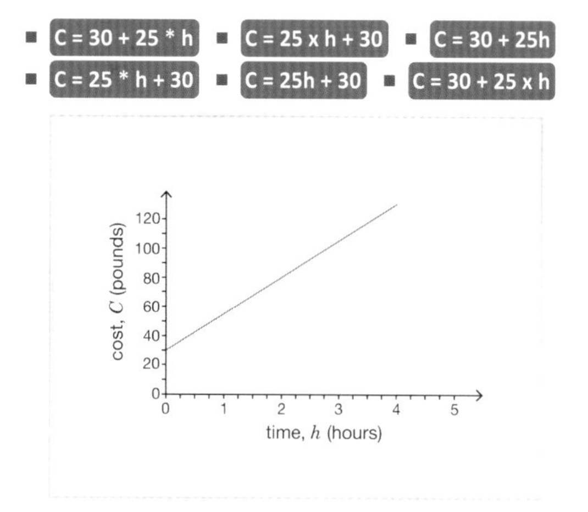

29. What is the equation of the straight line that passes through the points (1, 1) and (6, −9)?
Give your answer in the form $y = mx + c$.

- y = -2x + 3

30. This graph shows the amount, in pence, that a taxi company charges depending on the distance of the journey. Give an equation for the cost, in pence, of a journey, *C*, in terms of the number of miles travelled, *d*.

- C = 160 x d + 120
- C = 120 + 160d
- C = 120 + 160 * d
- C = 160d + 120
- C = 160 * d + 120
- C = 120 + 160 x d

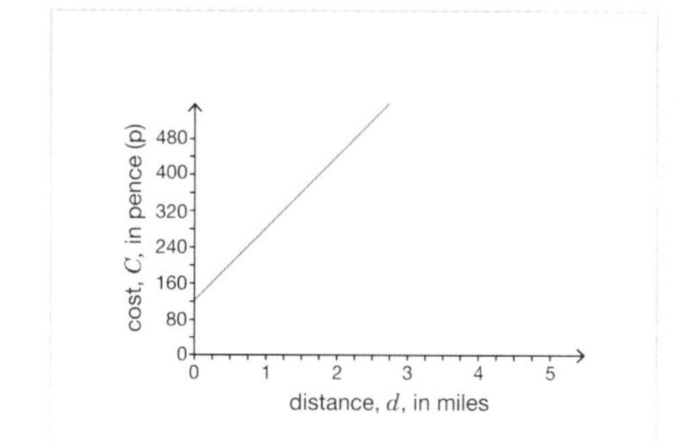

Interpret and Plot Straight Line Graphs

Objective: I can interpret and plot straight line graphs.

Quick Search Ref: 10470

Level 1: Understanding: Identifying coordinates and recognising equations of horizontal and vertical straight lines.

Required: 7/10 **Pupil Navigation:** on **Randomised:** off

1. Why does this straight line have the equation $x = 3$?

1/4

- The line goes through the x-axis at 3.
- The line goes through the y-axis at 3.
- **The x-coordinate for every point on the line is 3.**
- The y-coordinate for every point on the line is 3.

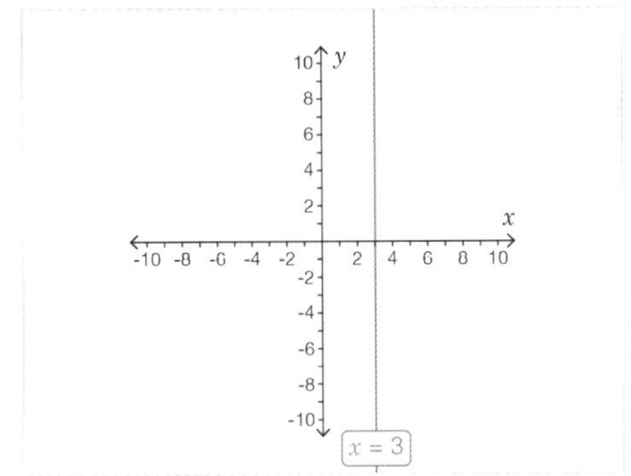

2. Which of the following points lie on the line $x = 3$?

1/4

- (5, 3)
- (−3, 3)
- **(3, 8)**
- (0, 3)

3. Why does the straight line have the equation $y = -2$?

1/4

- The line goes through the y-axis at −2.
- The line goes through the x-axis at −2.
- The x-coordinate for every point on the line is −2.
- **The y-coordinate for every point on the line is −2.**

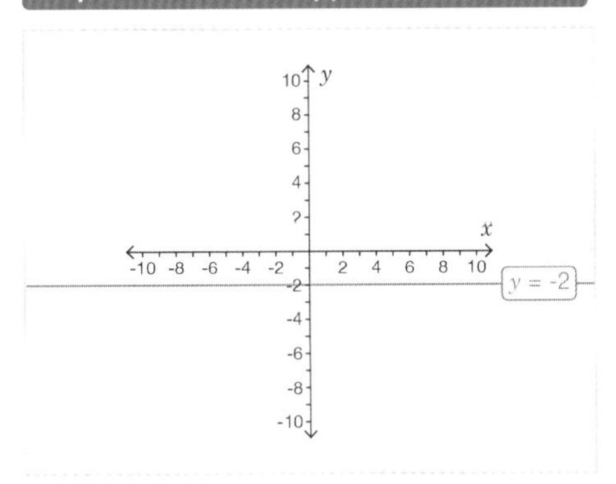

4. Which of these lie on the line $y = -2$?

1/4

-  **(2, −2)**
- (−2, 3)
- (−2, 0)
- (−2, 2)

5. What is the equation of the line shown in the diagram?

- **y = 4**
- y = 4x
- **4 = y**
- x = 4

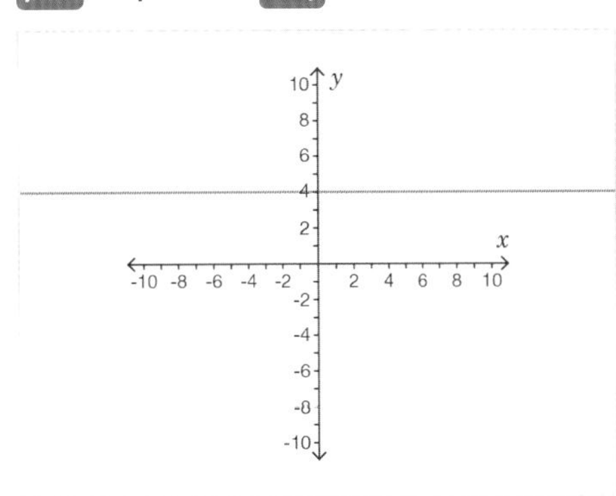

6. Which of the following points lie on the line $x = -1$?

1/4

- (0, 0)
- **(−1, 3)**
- (1, −1)
- (0, −1)

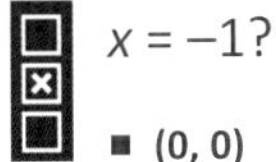

7. What is the equation of the straight line shown in this diagram?

- 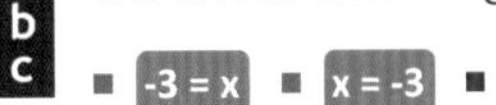 **-3 = x**
- **x = -3**
- y = -3

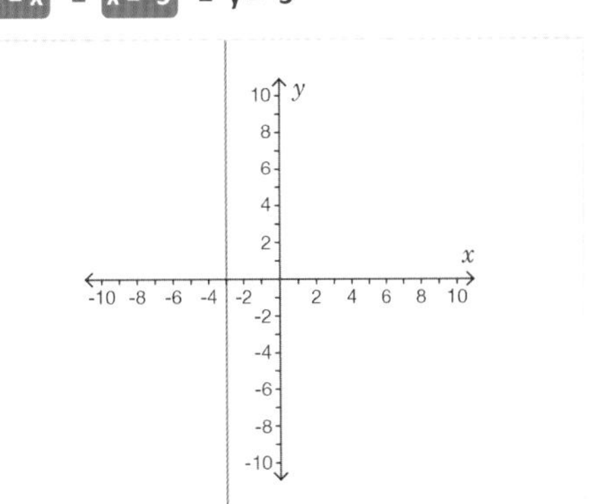

8. What is the equation of the straight line shown in this diagram?

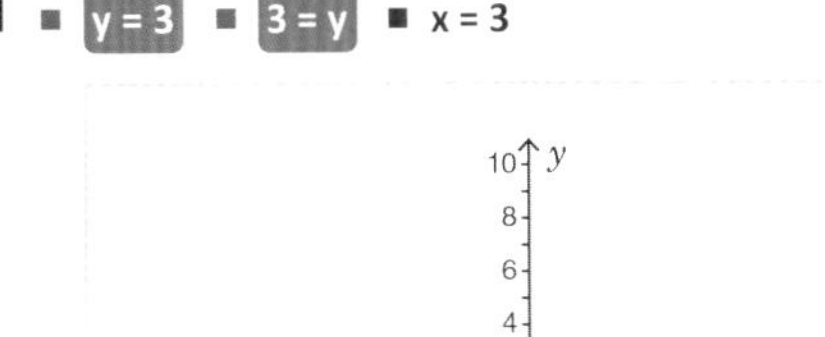

- **y = 3**
- **3 = y**
- x = 3

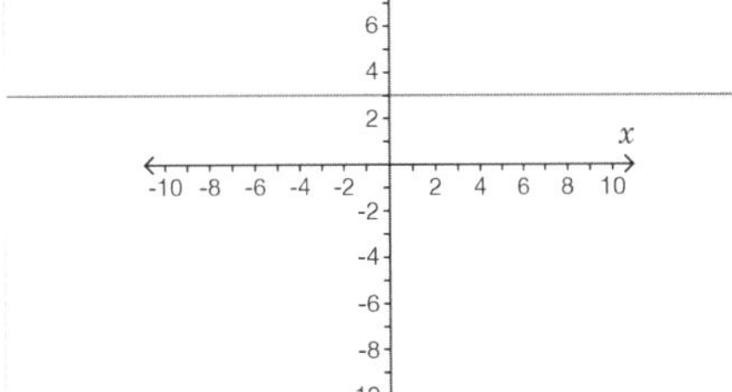

Level 1 *continued*

9. Which of the following points lie on the line $x = 2$?

1/4

- **(2, 0)**
- (0, –2)
- (1, 2)
- (–1, –2)

10. Which of the following points lie on the line $y = 2$?

1/4

- (2, 0)
- (0, –2)
- **(1, 2)**
- (–1, –2)

Level 2: Fluency: Finding coordinates and recognising equations of diagonal straight line graphs.

Required: 7/10 **Pupil Navigation:** on
Randomised: off

11. Which of the following points lie on the line $y = x$?

2/6

- (–2, 2)
- (3, –3)
- (1, 2)
- **(1, 1)**
- (4, 5)
- **(0, 0)**

12. What is the equation of the line shown in the diagram?

- y = 0
- **x = y**
- **y = x**
- y = 1
- x = 1

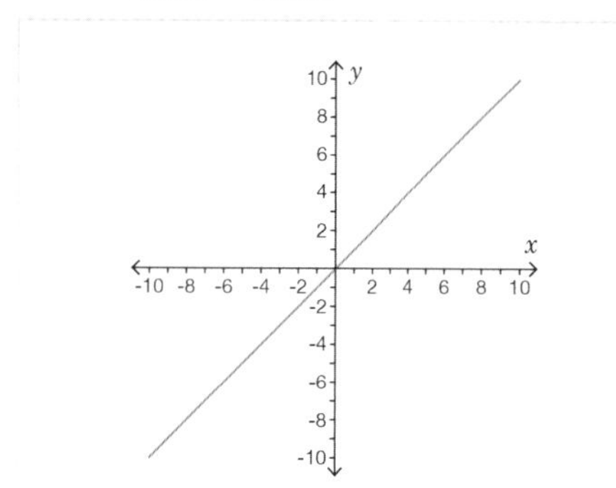

13. Which of the following points lie on the line $y = -x$?

2/6

- **(–2, 2)**
- (1, 2)
- (1, 1)
- (–4, –4)
- **(3, –3)**
- (8, –9)

14. Which of the following points lie on the line $x + y = 4$?

2/6

- (–2, 2)
- **(1, 3)**
- (4, 4)
- (5, 1)
- **(5, –1)**
- (1, 4)

15. What is the equation of the straight line shown in the diagram?
Give your answer in the form x + y = ?

- y + x = 2
- **x + y = 2**
- y = 2 - x
- y = -x + 2

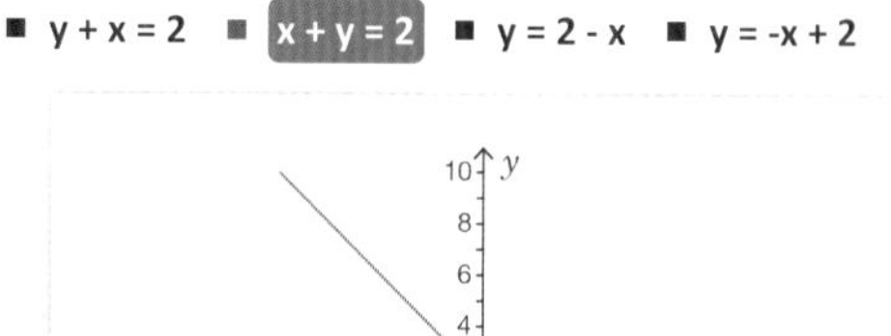

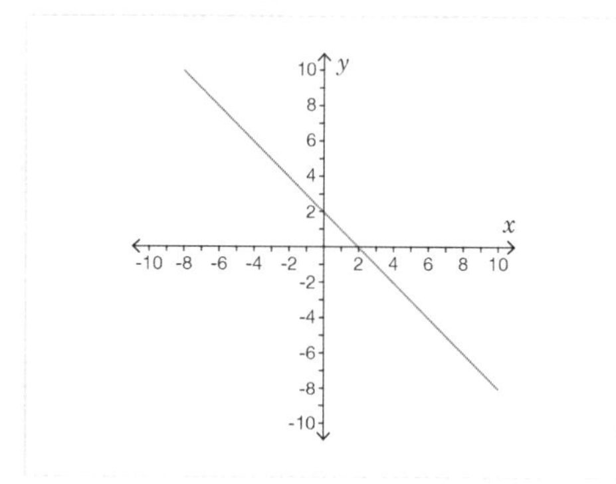

16. What is the equation of the line which joins the three points in the diagram?
Give your answer in the form y = ?

- -x = y
- y = x
- **y = -x**
- y + x = 0
- -y = x
- x = -y
- x + y = 0

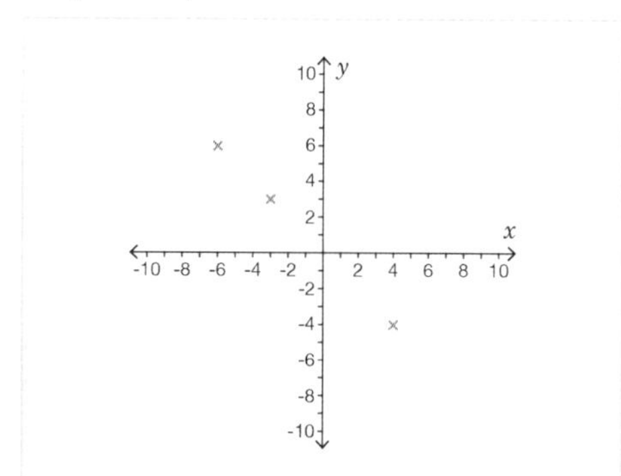

17. Point P lies on the line $y = -x$. What is the missing coordinate for point P?

- **3**
- -3

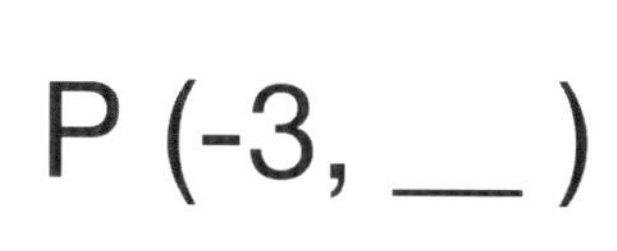

18. Which of the following points lie on the line $x + y = -2$?

3/6

- (6, –4)
- (0, 2)
- **(–6, 4)**
- **(4, –6)**
- (2, –2)
- **(–4, 2)**

Level 2 *continued*

19. Which of the following points lie on the line $x + y = 5$?

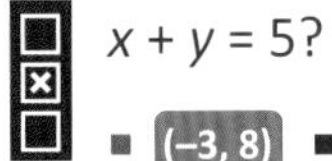

2/5

- (–3, 8)
- (5, 5)
- (5, –1)
- (–1, 5)
- (2, 3)

20. Why does this line have the equation $y = -x$?

1/4

- The y-coordinate is equal to the x-coordinate.
- The y-coordinate is equal to the negative of the x-coordinate.
- The x-coordinate is equal to 0.
- The y-coordinate is equal to 0.

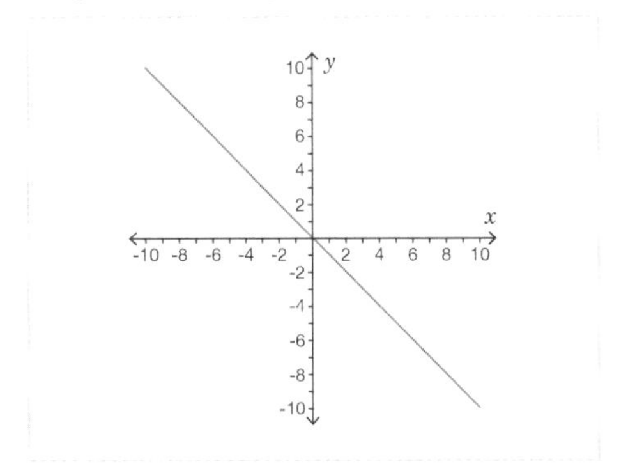

Level 3: Reasoning: With horizontal, vertical and diagonal line graphs.

Required: 5/5 **Pupil Navigation:** on

Randomised: off

21. The diagram shows the table of values for a straight line graph. What is the missing value?

- 

x	y
-4	7
0	3
2	?
5	-2

22. Elena says that the equation of this line is $x =$ –2. Elena is incorrect.
Explain why the equation cannot be $x = -2$ and give the correct equation.

- Open question, no set answer

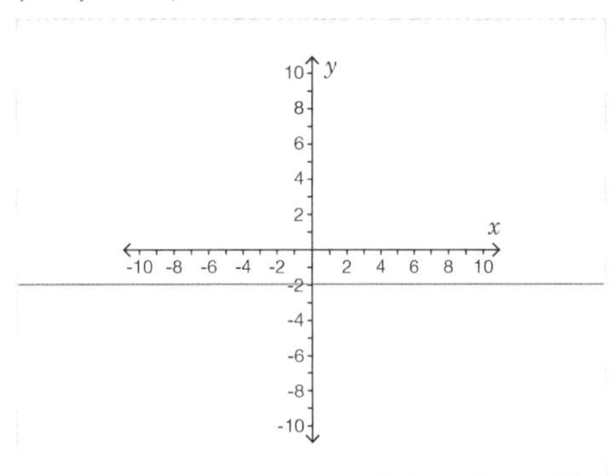

23. Milo says that the x-axis can also be called x = 0. Is he correct? Explain your answer.

- Open question, no set answer

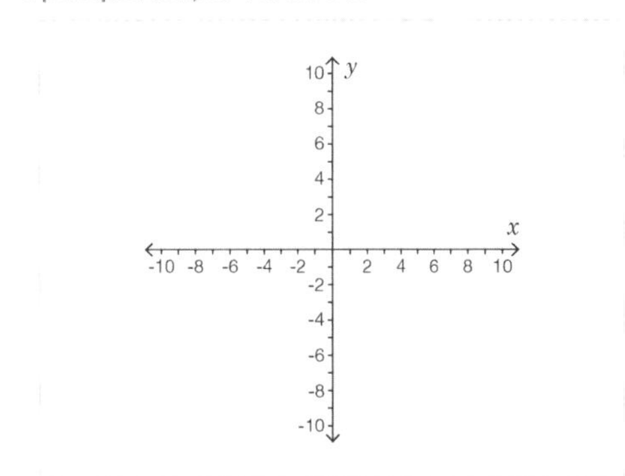

24. Which **two** of the following represent the line shown in the diagram?

2/6

- y = 0
- x = 0
- x-axis
- y-axis
- y = x
- y = –x

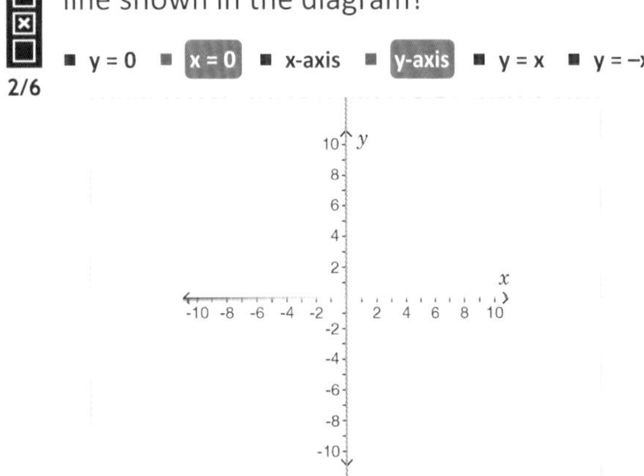

25. Raj says that the point with coordinates (–3, –1) lies on the line $x + y = -4$. Is he correct? Explain your answer.

- Open question, no set answer

Level 4: Problem Solving: With straight line graphs.

Required: 5/5 **Pupil Navigation:** on
Randomised: off

26. Archie draws the lines of $x = 2$, $x = 6$, $y = 1$ and $y = 5$ and notices that this produces a square. What are the coordinates of the centre of the square?

- **(4, 3)** ▪ **4, 3** ▪ **(4 3)**

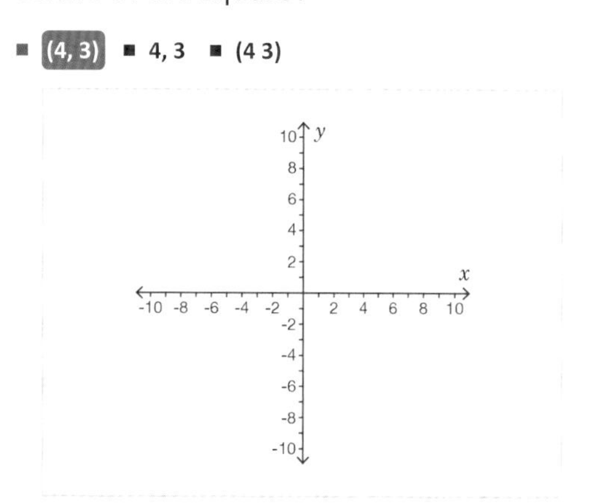

27. Siobhan and Patrick are talking about the same point. Siobhan says that the point lies on the line $x + y = -4$. Patrick says that the point lies on the line $y = x$. What are the coordinates of this point?

- **(-2, -2)** ▪ **(-4, -4)** ▪ **-2, -2** ▪ **(-2 -2)**

28. Select all of the lines that the point (0, 4) lies on.

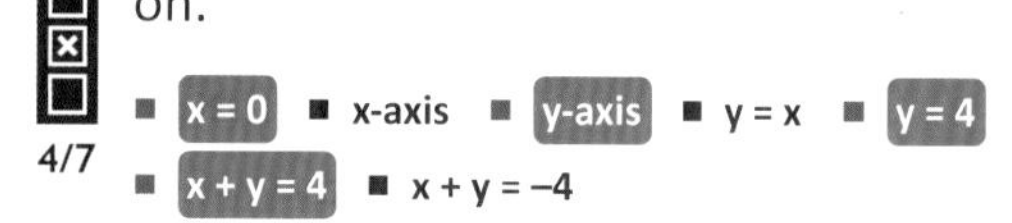

4/7

- **x = 0** ▪ **x-axis** ▪ **y-axis** ▪ **y = x** ▪ **y = 4**
- **x + y = 4** ▪ **x + y = −4**

29. I start at the point (3, −7). My position is reflected in the line $y = -x$. This position is then reflected in the line $y = 0$. What are the coordinates of my final position?

- **(7 3)** ▪ **(-7, -3)** ▪ **(7, 3)** ▪ **(11, 7)** ▪ **(3, -1)** ▪ **7, 3**

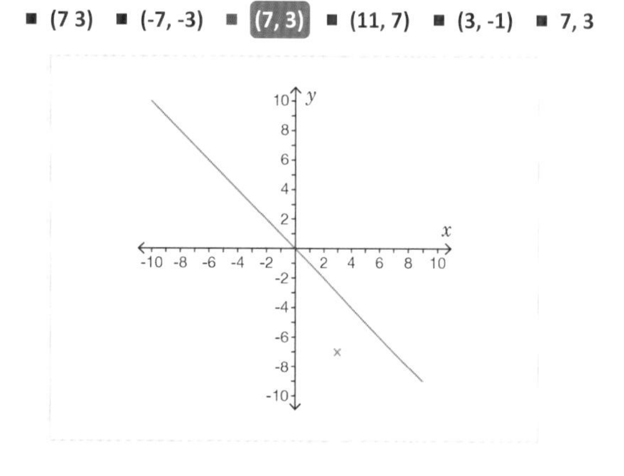

30. What is the area of the triangle bound by the following three lines?

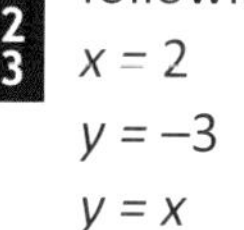

$x = 2$

$y = -3$

$y = x$

Do not include the units in your answer.

- **12.5** ▪ **25**

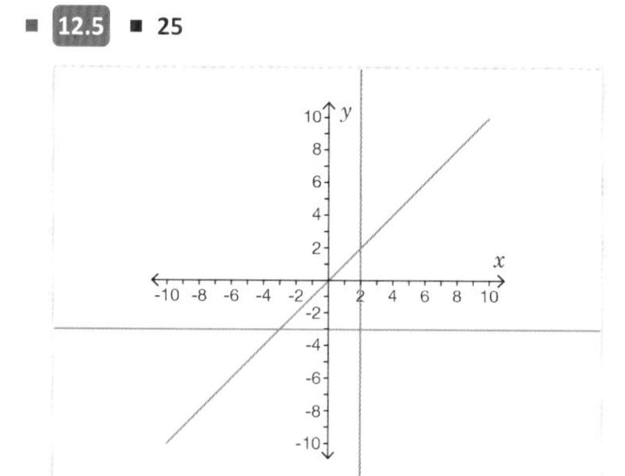

Make y the Subject of a Straight Line Equation

Objective: I can rearrange a straight line equation to make y the subject.

Quick Search Ref: 10522

Level 1: Understanding: Make y the subject of straight line equations (one step).

Required: 6/8 **Pupil Navigation:** on **Randomised:** off

1. Make *y* the subject of this equation:

$y + 3 = 2x$

Give your answer in the form y = . . .

- **y = -3 + 2x**
- **y = 2x - 3**
- y = 3 + 2x
- y = 2x + 3

$$y + 3 = 2x$$
$$-3 \qquad -3$$

2. Make *y* the subject of this equation:

$2y = 6x + 8$

Give your answer in the form y = . . .

- **y = 4 + 3x**
- **y = 3x + 4**
- y = 3x + 8

$$2y = 6x + 8$$
$$\div 2 \qquad \div 2$$

3. Make *y* the subject of this equation:

$y/2 = 4x - 1$

Give your answer in the form y = . . .

- y = 8x - 1
- **y = 8x - 2**
- **y = -2 + 8x**
- 2y = 8x - 2

$$\frac{y}{2} = 4x - 1$$
$$\times 2 \qquad \times 2$$

4. Make *y* the subject of this equation:

$3x = 7 + y$

Give your answer in the form y = . . .

- **y = -7 + 3x**
- **y = 3x - 7**
- y = 3x + 7
- y = 7 + 3x

$$3x = 7 + y$$
$$-7 \qquad -7$$

5. Make *y* the subject of this equation:

$y - 2x - 5 = 0$

Give your answer in the form y = . . .

- **y = 5 + 2x**
- **y = 2x + 5**
- y = -2x - 5

$$y - 2x - 5 = 0$$
$$+2x + 5 \qquad +2x + 5$$

6. Make *y* the subject of this equation:

$5 \quad y \quad 2x = 0$

Give your answer in the form y = . . .

- **y = -2x + 5**
- **y = 5 - 2x**
- -y = 2x - 5

$$5 - y - 2x = 0$$
$$+y \qquad +y$$

Level 1 continued

7. Make y the subject of this equation:
$y/3 = 3x - 2$
Give your answer in the form y = . . .

- y = -6 + 9x
- y = 9x - 6
- y = 9x - 2
- 3y = 9x - 6

$$\frac{y}{3} = 3x - 2$$
$\times 3 \quad \times 3$

8. Make y the subject of this equation:
$3y = 12x + 21$
Give your answer in the form y = . . .

- y = 7 + 4x
- y = 4x + 7
- y = 4x + 21

$$3y = 12x + 21$$
$\div 3 \quad \div 3$

Level 2: Fluency: Rearrange the equation of a straight line to make y the subject (multiple steps).

Required: 6/8 **Pupil Navigation:** on
Randomised: off

9. Rearrange the equation $2y - 6 = 4x$ to make y the subject.
Give your answer in the form y = . . .

- y = 3 + 2x
- y = 2x + 3
- y = 2x - 3
- y = 2x + 6

$$2y - 6 = 4x$$

10. Rearrange the equation $4y - 12x = 8$ to make y the subject.
Give your answer in the form y = . . .

- y = 3x + 2
- y = 2 + 3x
- y = 3x + 8
- y = -3x + 2

$$4y - 12x = 8$$

11. Rearrange the equation $y/3 + 2x = 4$ to make y the subject.
Give your answer in the form y = . . .

- y = -6x + 12
- y = 12 - 6x
- y = 6x + 12
- y = -6x + 4

$$\frac{y}{3} + 2x = 4$$

12. Rearrange the equation $6 - y = 7 - 3x$ to make y the subject.
Give your answer in the form y = . . .

- y = -1 + 3x
- y = -3x + 13
- y = 3x - 1
- y = 3x - 7 + 6
- y = 6 + 3x - 7

$$6 - y = 7 - 3x$$

Level 2 *continued*

13. Rearrange the equation 10 = $y/2 - 4x$ to make y the subject.
Give your answer in the form y = . . .

- y = 8x + 10
- y = 20 + 8x
- y = 8x + 20
- y = -8x + 20

$$10 = \frac{y}{2} - 4x$$

14. Rearrange the equation $2x + 8 = 4 - y$ to make y the subject.
Give your answer in the form y = . . .

- y = 2x + 12
- y = -2x - 4
- y = -4 - 2x
- y = 4 - 2x - 8
- y = -2x - 8 + 4

$$2x + 8 = 4 - y$$

15. Rearrange the equation $5x = y/3 - 13$ to make y the subject.
Give your answer in the form y = . . .

- y = 39 + 15x
- y = 15x + 39
- y = 15x - 39
- y = 15x + 13

$$5x = \frac{y}{3} - 13$$

16. Rearrange the equation $3y - 15x = 12$ to make y the subject.
Give your answer in the form y = . . .

- y = 5x + 4
- y = 4 + 5x
- y = 5x + 12
- y = -5x + 4

$$3y - 15x = 12$$

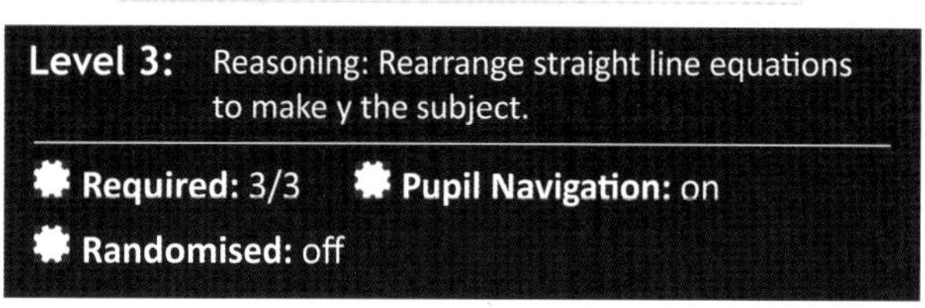
Level 3: Reasoning: Rearrange straight line equations to make y the subject.

Required: 3/3 **Pupil Navigation:** on
Randomised: off

17. George has completed his homework but he has answered one question incorrectly. What is the correct answer to the question he got wrong?
Give your answer in the form y = . . .

- y = 9 - 2x
- y = -2x +9

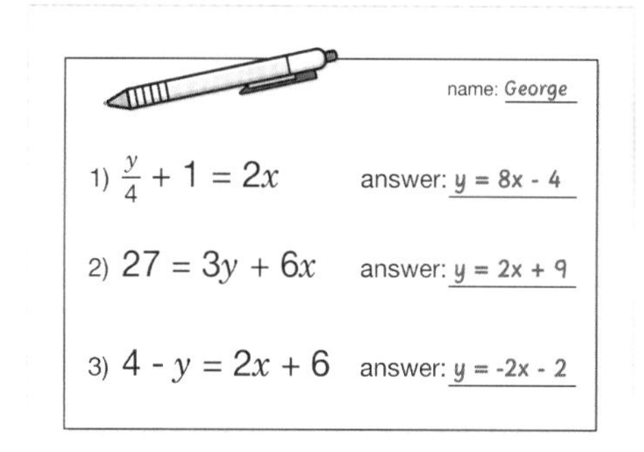

18. Which equation does not simplify to $y = 3x + 2$?

1/4

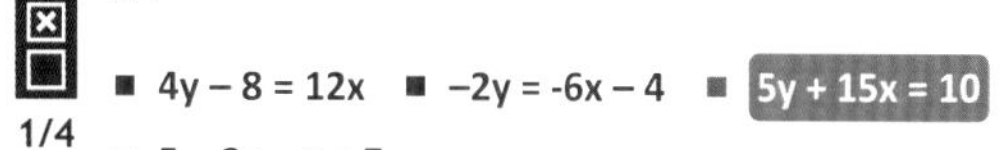
- 4y – 8 = 12x
- –2y = -6x – 4
- 5y + 15x = 10
- 5 = 3x – y + 7

19. Elise says that the graph of $2y = 8x - 10$ is identical to the graph of $7 = 4x - y + 2$. Is she correct? Explain your answer.

- Open question, no set answer

Level 4: Problem solving: Rearrange straight line equations to make y the subject.

Required: 3/3 **Pupil Navigation:** on
Randomised: off

20. What is the x-coordinate of the point where the lines $2y = 8x + 6$ and $5y = 10x + 45$ intersect?

- 3

21. What is the area of the triangle bound by the line $3(y + 2x - 4) = 6$ and the x- and y-axes in centimetres squared?
Don't include units in your answer.

- 9
- 18

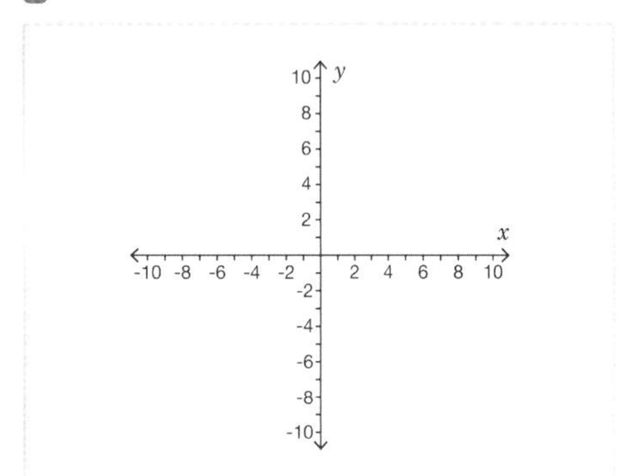

22. What are the coordinates of the point where the lines $3y - 15x = 21$ and $6 = 4x + 15 - y$ intersect?
Give your answer in the form (x, y).

- (2 17)
- (2, 17)
- 2, 17
- (17, 2)

Recognise and Produce Graphs of Linear Functions

Objective: I can recognise and produce line graphs.

Quick Search Ref: 10485

Level 1: Understanding: Completing tables of values with positive integer coefficients of x.

Required: 7/10 **Pupil Navigation:** on **Randomised:** off

1. What are the coordinates of point A?

- 2, 7
- **(2, 7)**
- (7, 2)
- (2 7)

point A

x	0	1	2	3
y	3	5	7	9

2. The diagram shows the table of values for the line $y = x + 3$. What is the missing value?

- **4**

$y = x + 3$

x	0	1	2	3
y	3	?	5	6

3. The diagram shows the table of values for the line $y = 2x – 1$. What is the missing value?

- **3**

$y = 2x - 1$

x	0	1	2	3
y	–1	1	?	5

4. What is the missing value in the table for the line $y = 4x + 1$?

- **13**

$y = 4x + 1$

x	0	1	2	3
y				?

5. The diagram shows the table of values for the line $y = 6x + 5$. What is the missing value?

- **-7**
- -17

$y = 6x + 5$

x	–2	–1	0	1	2
y	?	–1	5	11	17

6. What is the missing value in the table for the line $y = 3x – 8$?

- **-14**
- 2
- -2

$y = 3x - 8$

x	–2	–1	0	1	2
y	?				

7. For the line of $y = 7x – 9$, what is the value of y when $x = 8$?

- **47**

Level 1 *continued*

8. For the line of $y = 3x + 4$, what is the value of y when $x = 5$?

▪ 19

9. The diagram shows the table of values for the line $y = 2x + 2$. What is the missing value?

▪ 0 ▪ -4

$y = 2x + 2$

x	–2	–1	0	1	2
y	–2	?	2	4	6

10. What is the missing value in the table for the line $y = 11x – 3$?

▪ -14 ▪ -8

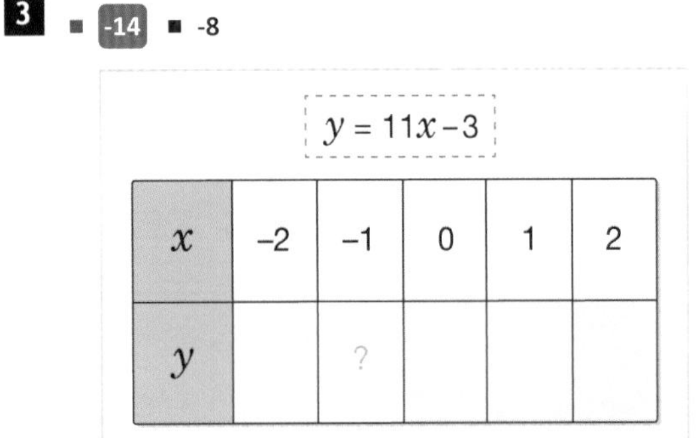

$y = 11x – 3$

x	–2	–1	0	1	2
y		?			

Level 2: Fluency: Plotting points with positive, negative and fractional coefficients of x.

Required: 7/10 **Pupil Navigation:** on

Randomised: off

11. If $x = 1$, which point lies on the line $y = 3x – 4$?

1/4

▪ A ▪ B ▪ C ▪ D

12. Which line has the equation $y = 4x + 3$.

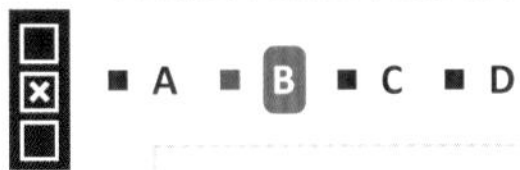

▪ A ▪ B ▪ C ▪ D

1/4

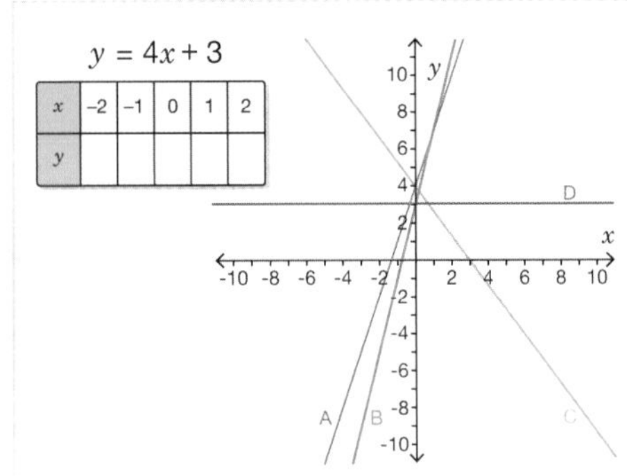

13. What is the missing value in the table for the line $y = –2x + 4$?

▪ 6 ▪ 2

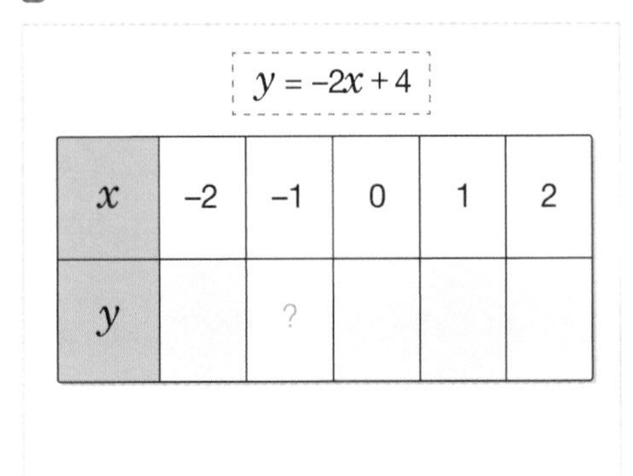

$y = –2x + 4$

x	–2	–1	0	1	2
y		?			

14. If $x = 2$, which point lies on the line $y = –3x – 2$?

▪ A ▪ B ▪ C ▪ D

1/4

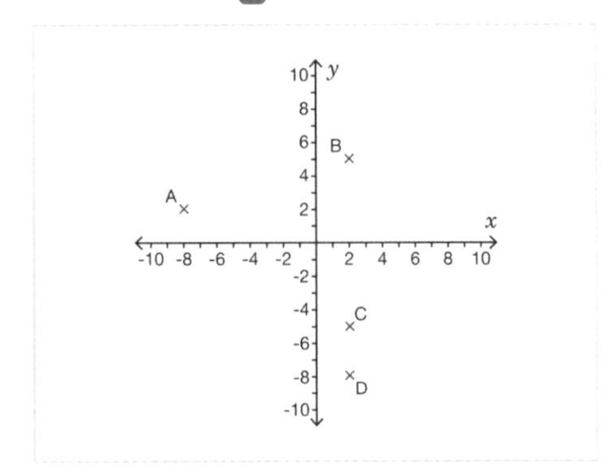

Level 2 *continued*

15. What is the missing value in the table for the line $y = \frac{1}{2}x + 1$?

1 2 3

■ 1.5 ■ 2

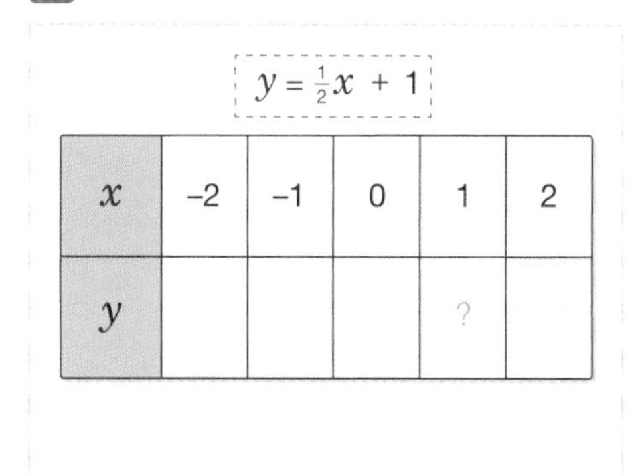

$y = \frac{1}{2}x + 1$

x	-2	-1	0	1	2
y				?	

16. Which line has the equation $y = -4x + 2$.

1/4

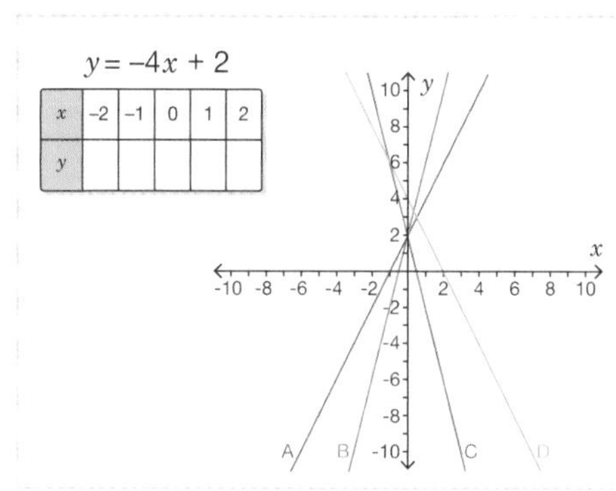

17. Which line has the equation $y = \frac{1}{4}x - 2$.

■ A ■ B ■ C ■ D

1/4

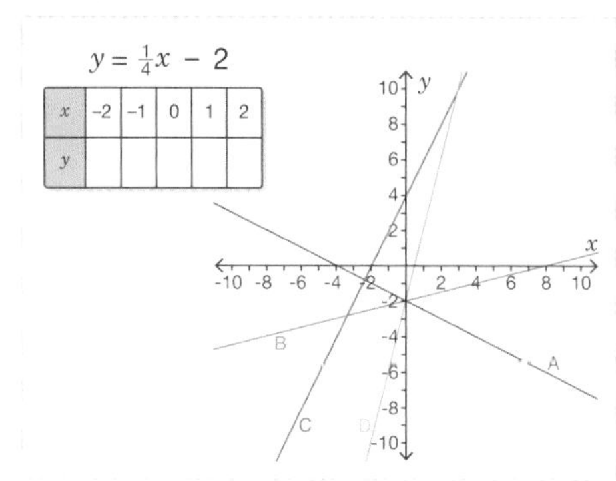

18. If $x = 2$, which point lies on the line $y = \frac{1}{3}x + 2$?

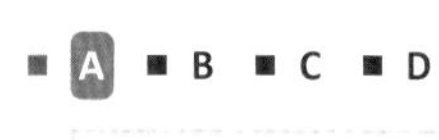

1/4

5 y
4
3
2
1
x
-5 -4 -3 -2 -1
1 2 3 4 5
-1
-2
-3
-4
-5
A
B
C
D

19. Which line has the equation $y = 6x - 5$.

1/4

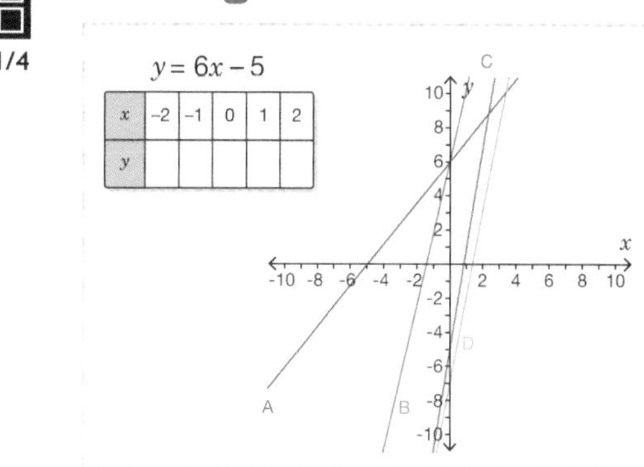

20. If $x = -2$. which point lies on the line $y = 5x + 3$?

■ A ■ B ■ C ■ D

1/4

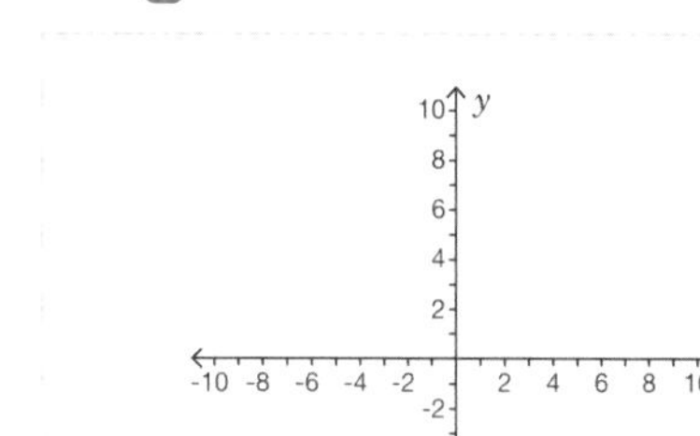

Level 3: Reasoning: With graphs of linear functions.

Required: 5/5 **Pupil Navigation:** on

Randomised: off

21. Nicki says that the line on the graph is $y = 3x + 1$. Is Nicki correct? Explain your answer.

a b c

- Open question, no set answer

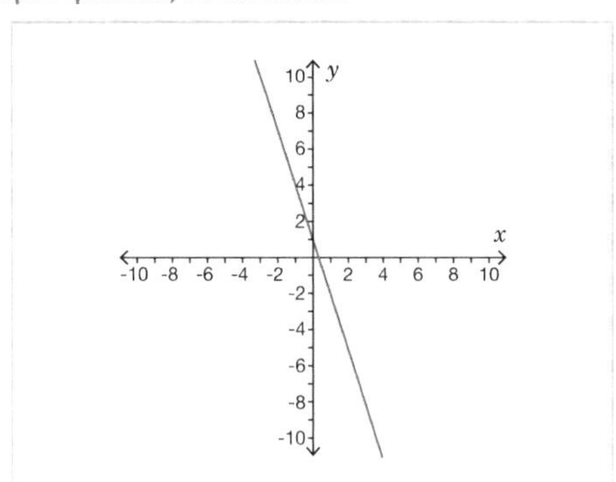

22. Spencer completes a table of values for the line of $y = -2x + 3$. However, he makes one mistake. What is the correct value for the mistake he has made?

1 2 3

- 5

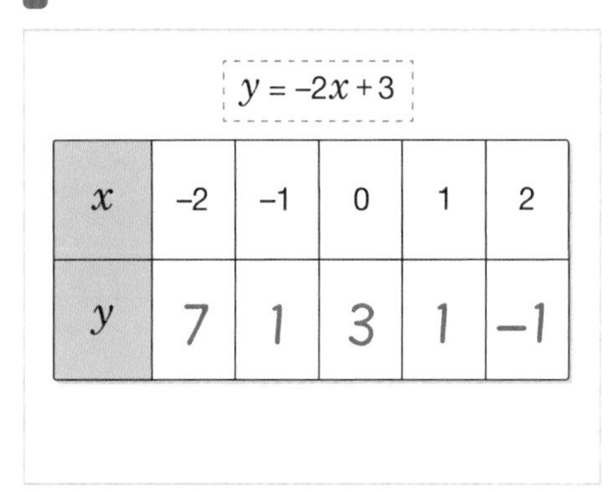

$y = -2x + 3$

x	–2	–1	0	1	2
y	7	1	3	1	–1

23. Brad had two attempts at drawing one of the lines for his homework, but can't remember which line he drew incorrectly. Which line does he need to erase?

1/3

- A
- B
- C

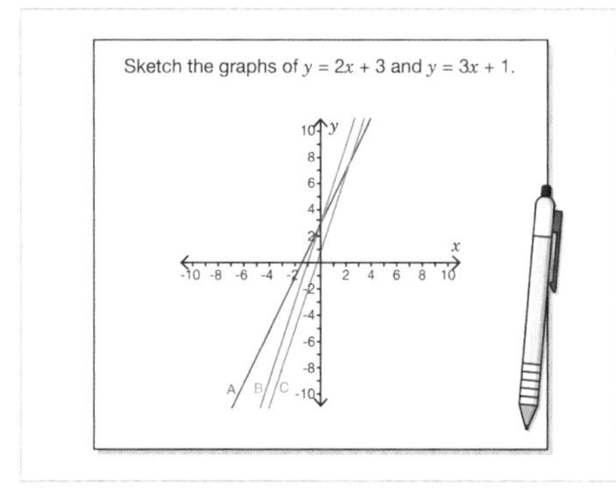

24. Ashraf completes a table of values for the line of $y = 4x + 8$. Is the table correct? Explain your answer.

a b c

- Open question, no set answer

$y = 4x + 8$

x	5	3	–7	0
y	28	20	–20	8

25. Ilana says that point P lies on the lines of $y = 2x + 3$, $y = 4x - 3$ and $y = x + 6$. Is she correct? Explain your answer.

a b c

- Open question, no set answer

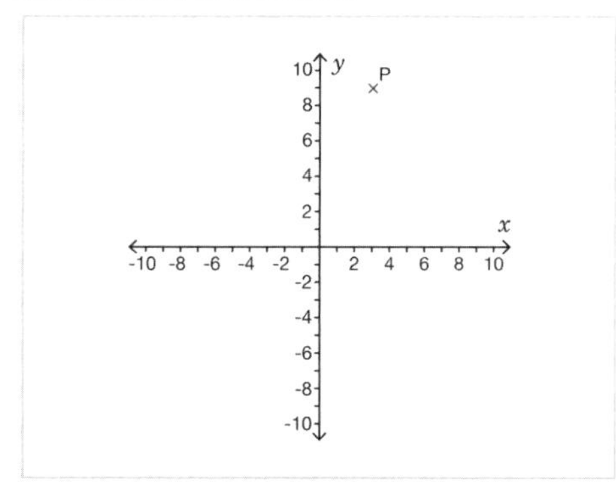

Level 4: Problem Solving: With graphs of linear functions.

Required: 5/5 **Pupil Navigation:** on

Randomised: off

26. What are the coordinates for the point which lies on the lines $y = 3x + 2$ and $y = 2x + 4$?

a b c

- 2, 8
- (2, 8)
- (2 8)
- (8, 2)
- (8 2)
- 8, 2

27. Select all of the lines that point P **does not** lie on.

2/5

- y = 2x + 16
- y = –2x
- y = 3x +18
- y = –3x – 4
- y = 4x + 26

Level 4 *continued*

28. What is the sum of the pair of coordinates that lie on the lines $y = 2x - 6$ and $y = -x$?

1 2 3

- 0
- 4

29. Andrew picks a point on the line $y = 2x - 7$.

a b c

He says that:
The x-coordinate is an integer greater than 11 and less than 17.
The y-coordinate is a square number.
What are the coordinates of Andrew's point?

- 16, 25
- (25, 16)
- (16, 25)
- 25, 16
- (16 25)
- (25 16)

30. Humira picks a point on the line $y = 3x - 2$.

She says that:
The x-coordinate is an integer greater than 0 and less than 11.
The y-coordinate is a prime number.
The difference between the x and y-coordinates is a multiple of 3.
What are the coordinates of Humira's point?

- (7, 19)
- (7 19)
- (19, 7)
- 3, 7
- 5, 13
- 19, 7
- (3 7)
- (5, 13)
- 7, 19
- (19 7)
- (5 13)
- (3, 7)

Use Graphs to Find Approximate Solutions to Contextual Problems

Objective: I can use graphs to answer problems in context.

Quick Search Ref: 10561

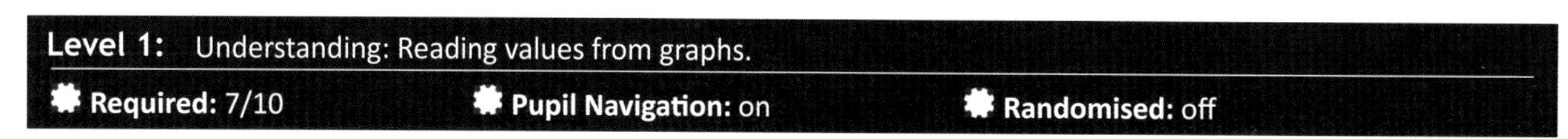

1. Use the graph to convert 40 kilometres to miles.
Don't include the units in your answer.

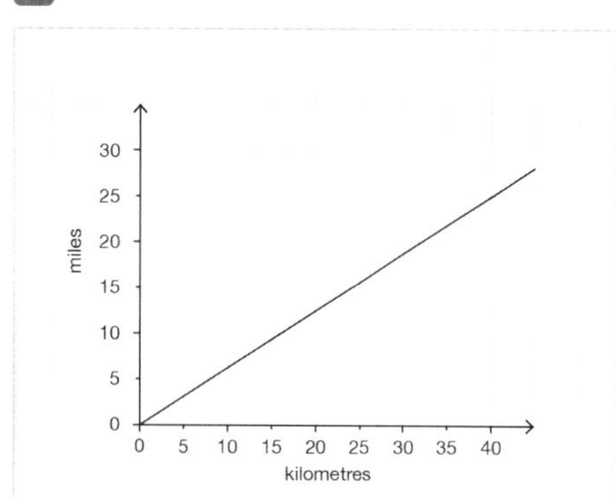

2. The distance-time graph shows Brian's journey to collect his daughter from university. What time did Brian arrive back home?
Give your answer using the 24-hour clock in the format hh:mm.

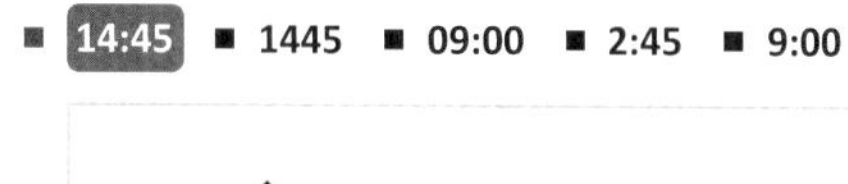

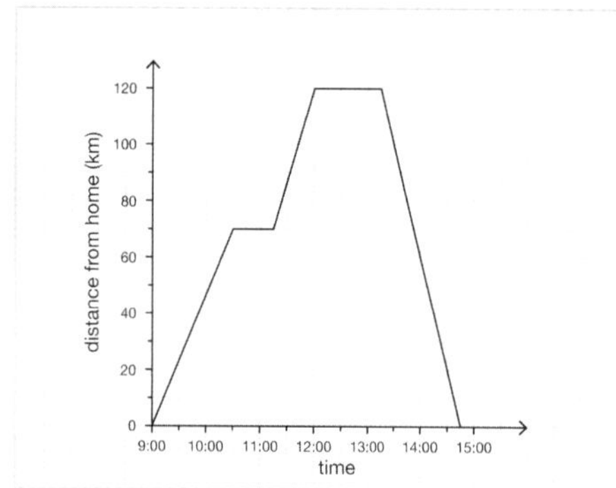

3. Use the graph to convert 50 degrees Fahrenheit to Celsius.
Don't include the units in your answer.

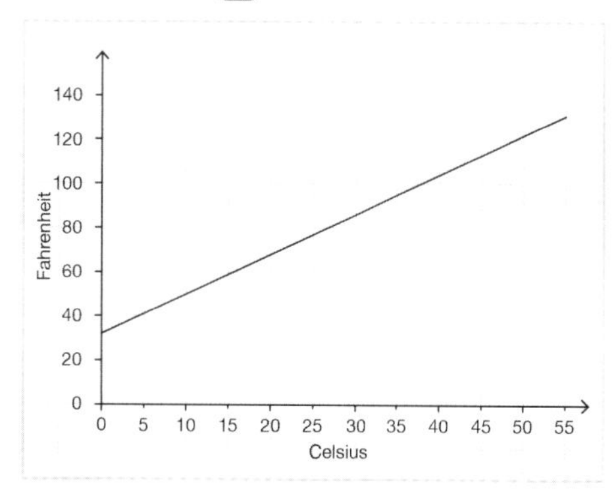

4. Use the graph to convert 20 miles to kilometres.
Include the units km (kilometres) in your answer.

■ 12.5 kilometres

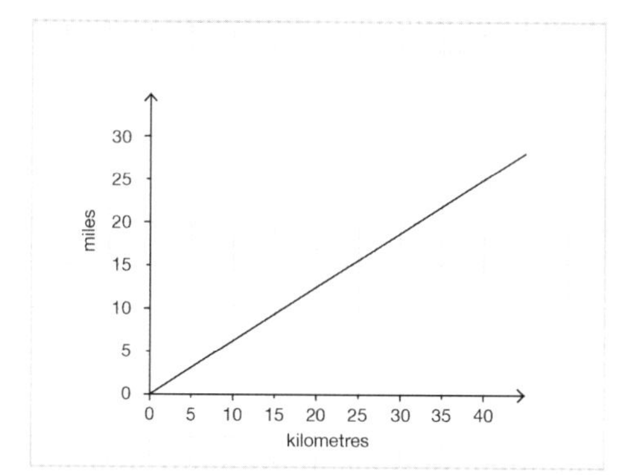

5. The graph shows how the typical stopping distance for a car changes with speed. What is the typical stopping distance in feet for a car travelling at 50 miles per hour?
Give your answer to the nearest foot and don't include the units.

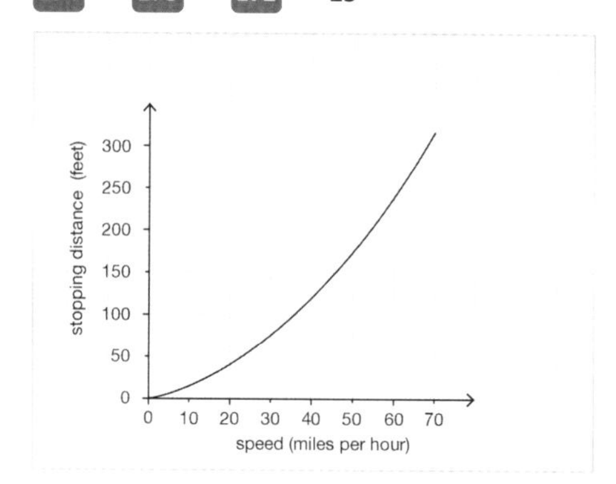

Level 1 *continued*

6. a b c The distance-time graph shows Brian's journey to collect his daughter from university. How many kilometres away is the university?
Include the units km (kilometres) in your answer.

■ 70 kilometres ■ 120 km ■ 120 kilometres
■ 70 km ■ 120

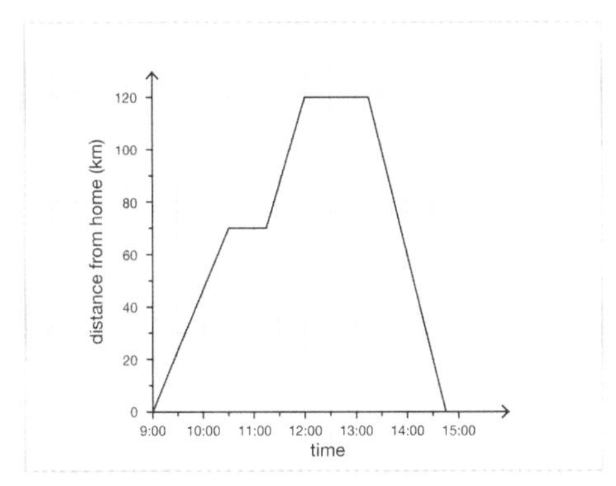

7. 1 2 3 The graph shows how the gravitational force acting between two meteorites changes depending on the distance between them. What is the strength of the gravitational force acting between two meteorites that are 4 km from each other?
Give your answer to the nearest newton and don't include the units.

■ 26 ■ 25 ■ 24 ■ 22 ■ 10

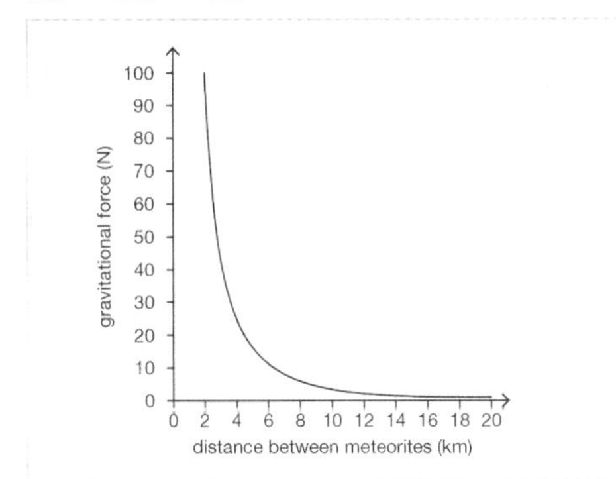

8. 1 2 3 Use the graph to convert 20 degrees Celsius to Fahrenheit.
Don't include the units in your answer.

■ 68 ■ 62

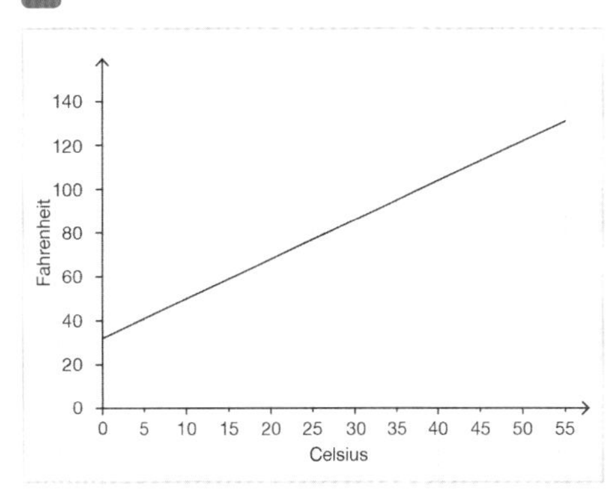

9. a b c The distance-time graph shows Brian's journey to collect his daughter from university. What time did Brian arrive at his daughter's university?
Give your answer using the 24-hour clock in the format hh:mm.

■ 12:00 ■ 1200

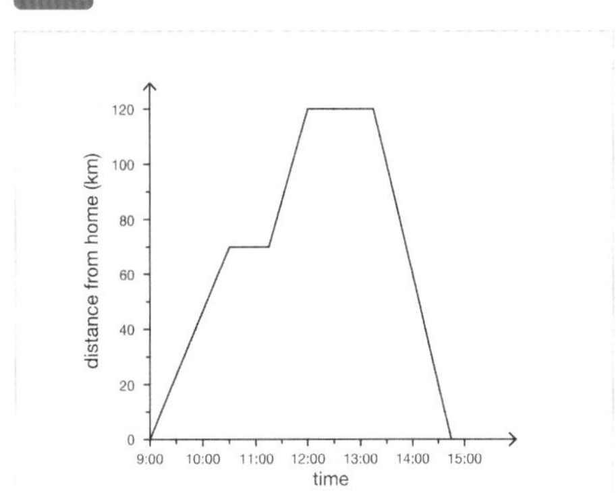

10. 1 2 3 The graph shows how the typical stopping distance for a car changes with speed. How many miles per hour is a car travelling at if it has a stopping distance of 75 feet?
Give your answer to the nearest mile per hour and don't include the units.

■ 30 ■ 29 ■ 31

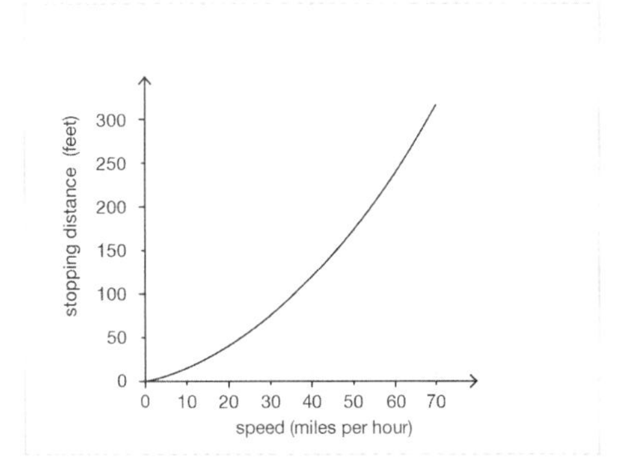

Level 2: Fluency: Make inferences by reading graphs.

Required: 7/10 **Pupil Navigation:** on

Randomised: off

11. Use the graph to convert 120 miles to kilometres.
Include the units km (kilometres) in your answer.

- 192 kilometres
- 192 km
- 192

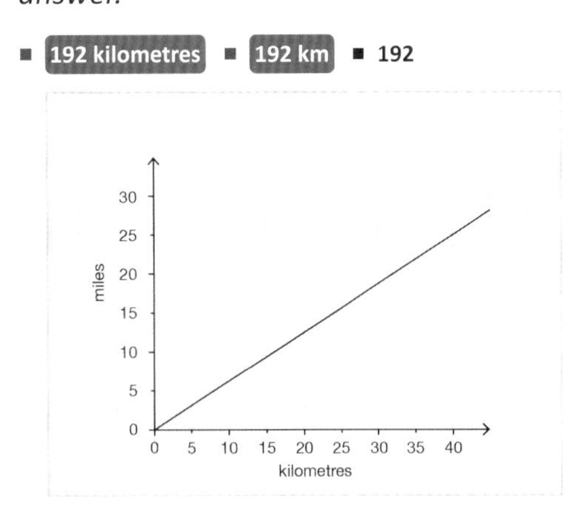

12. The distance-time graph shows Brian's journey to collect his daughter from university.
He stopped on the way to the university.
How many minutes did Brian stop for?
Don't include the units in your answer.

- 45
- 75

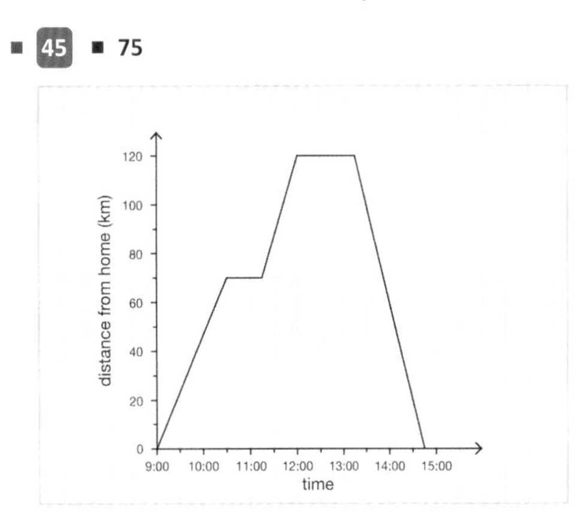

13. The container shown is filled using a tap with a constant flow. Select the graph that shows how the depth of water changes with time.

1/4

- (i)
- (ii)
- (iii)
- (iv)

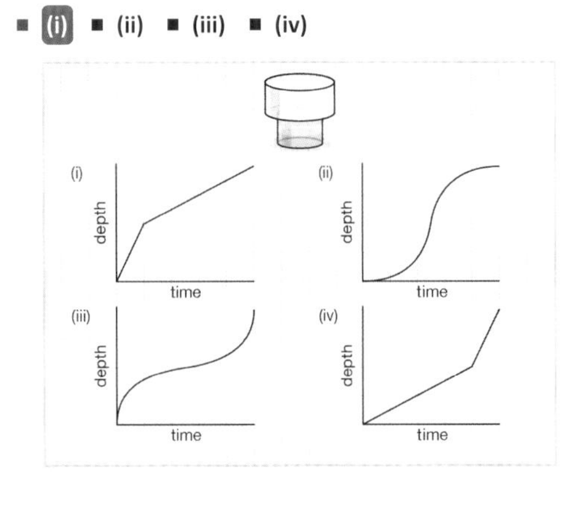

14. Eddy and Fazel are training for a race. Eddy runs 15 miles in a week. Fazel runs 28 kilometres in a week. How much further does Fazel run than Eddy?
Include the units km (kilometres) in your answer.

- 4 kilometres
- 4 km
- 4

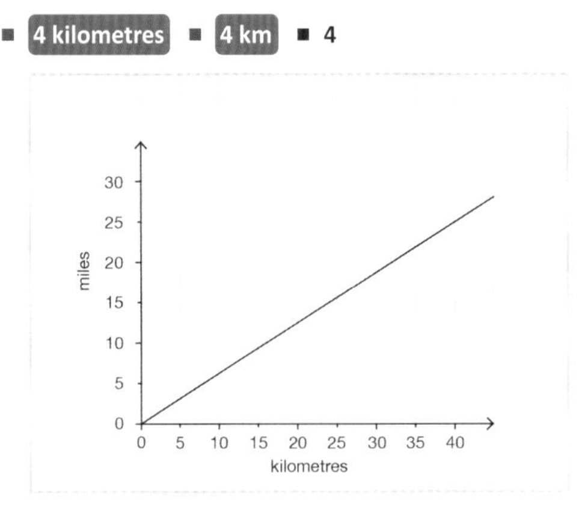

15. Whilst on holiday in America, Verity sees a phone on sale for \$120. The same phone is £87 in the UK. How much could Verity save if she buys the phone in America?
Give your answer in pounds and include the £ sign.

- 7
- £7.00
- £7
- 7.00

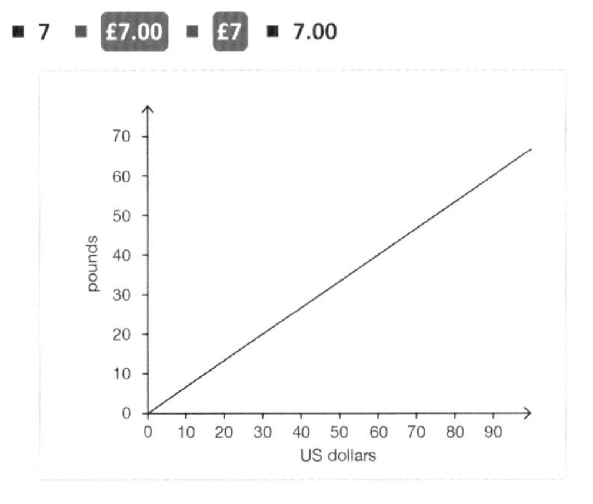

16. The distance-time graph shows Brian's journey to collect his daughter from university. How many minutes did Brian stop at the university for?
Don't include the units in your answer.

- 75
- 45

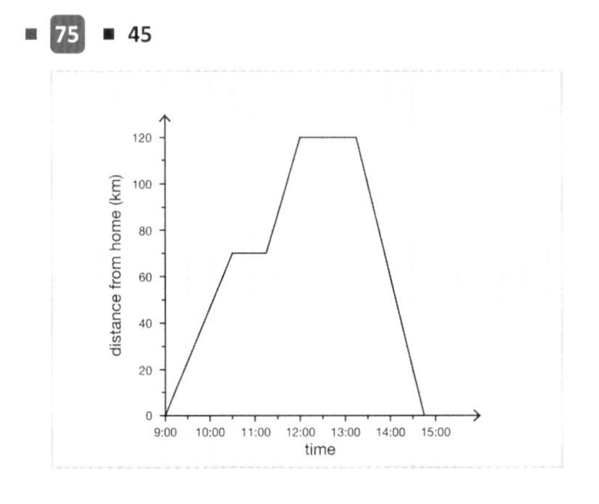

Level 2 *continued*

17. Three containers are filled using a tap with a constant flow. A graph is drawn for each container to show how the depth of water changes with time.
Sort the letters of the containers to match the order of the graphs.
Put the letter of the container that matches graph (i) first.

- (c)
- (a)
- (b)

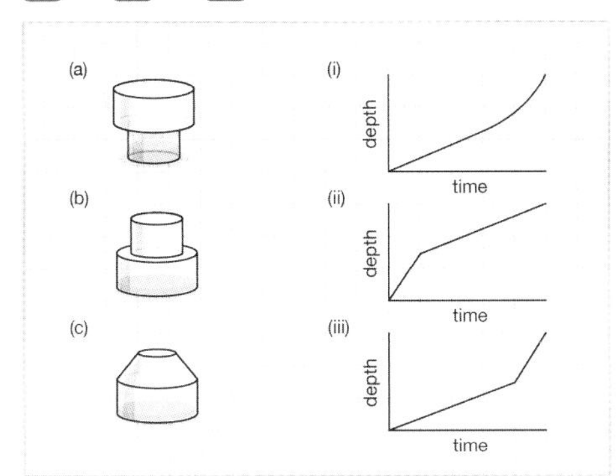

18. Use the graph to convert 150 litres to gallons.
Don't include the units in your answer.

- 33

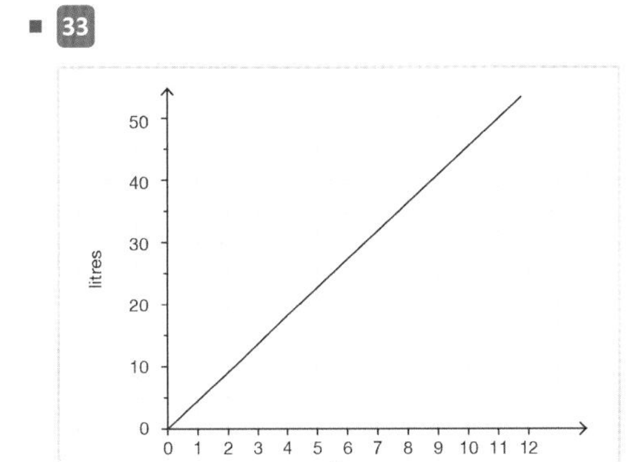

19. The distance-time graph shows Brian's journey to collect his daughter from university.
At what speed, in kilometres per hour, did Brian travel at on his journey home?
Don't include the units in your answer.

- 80

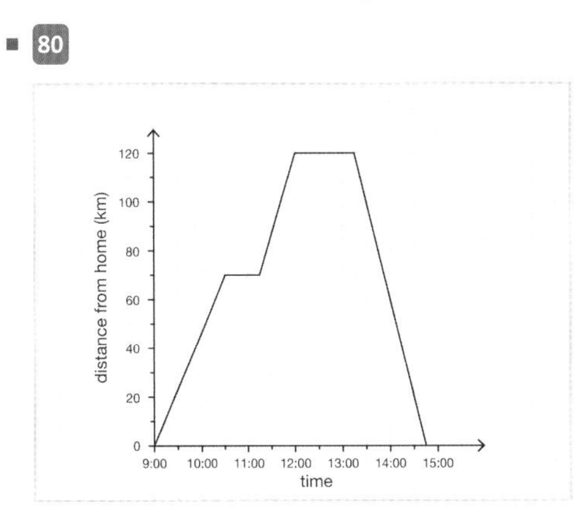

20. The containers shown are filled using a tap with a constant flow. The graph shows how the depth of water changes with time. Which container matches the graph?

1/4

- (i)
- (ii)
- (iii)
- (iv)

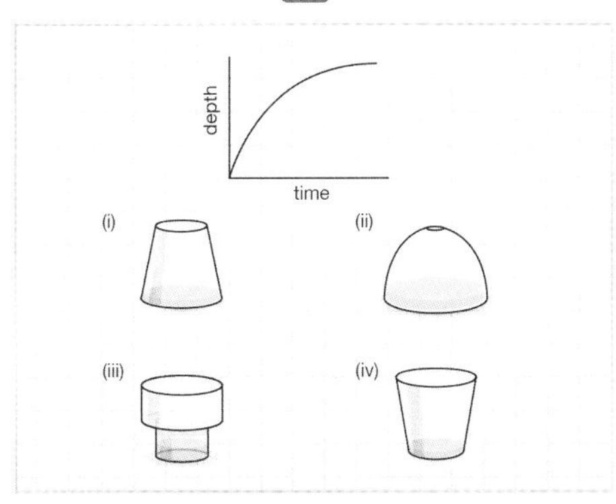

Level 3: Reasoning: Describe parts of distance-time graphs.

Required: 5/5 **Pupil Navigation:** on
Randomised: off

21. The diagram shows a distance-time graph for a race between Ahmed, Brontie and Carl.
Sort the names into the order they finished the race, from first to last.

- Brontie
- Carl
- Ahmed

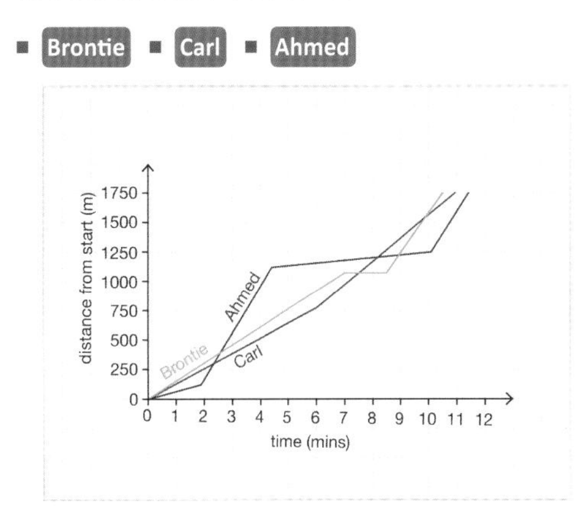

22. The diagram shows a distance-time graph for Sara cycling to school.
What happened in section C of the graph?

1/4

- Sara accelerated.
- Sara cycled over a jump.
- Sara stopped.
- Sara decelerated.

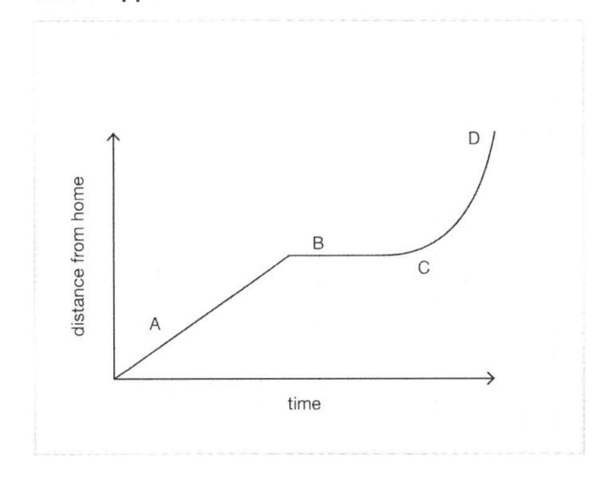

Level 3 *continued*

23. Harry is converting 60 °C into degrees Fahrenheit. Harry says that 30 °C is 86 °F, so 60 °C must be 172 °F. Is Harry correct? Explain your answer.

- Open question, no set answer

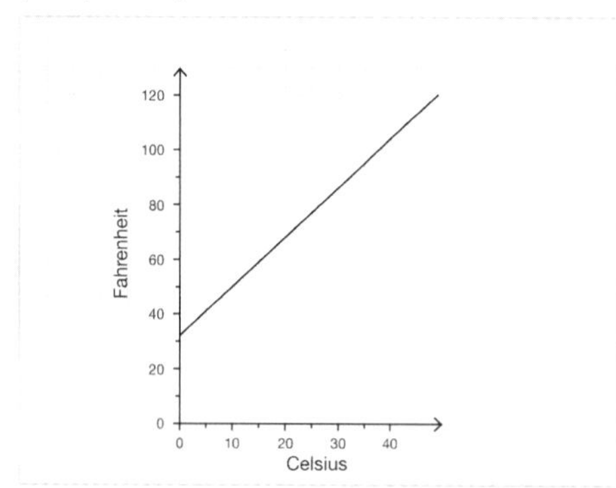

24. The diagram shows a distance-time graph for Rafael's journey to school. Describe Rafael's journey in words.

- Open question, no set answer

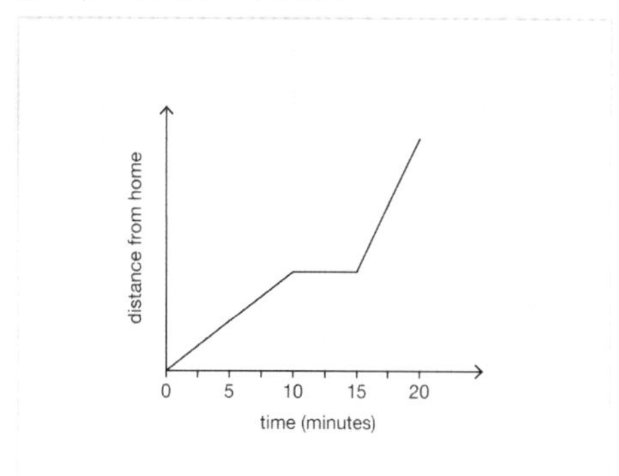

25. The diagram shows a distance-time graph for a race between Ahmed, Brontie and Carl. Select the **two** statements that describe what happened in the race between 6 and 8 minutes.

2/6

- **Carl was winning the race.**
- **Brontie and Carl crashed into each other.**
- **Ahmed was winning the race.**
- **Ahmed stopped for a rest.**
- **Brontie was winning the race.**
- **Carl overtook Brontie.**

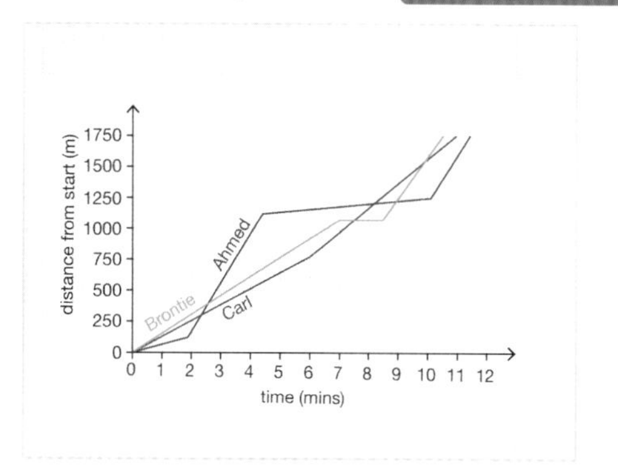

Level 4: Problem Solving: Solve multi-step problems involving interpreting graphs.

Required: 5/5 **Pupil Navigation:** on
Randomised: off

26. The graph shows the cost of hiring a bicycle. Give a formula for the total cost in pounds, C, of hiring a bike for h hours.
Give your answer in the form C = . . .

- C = 6h + 10
- C = 10 + 6h

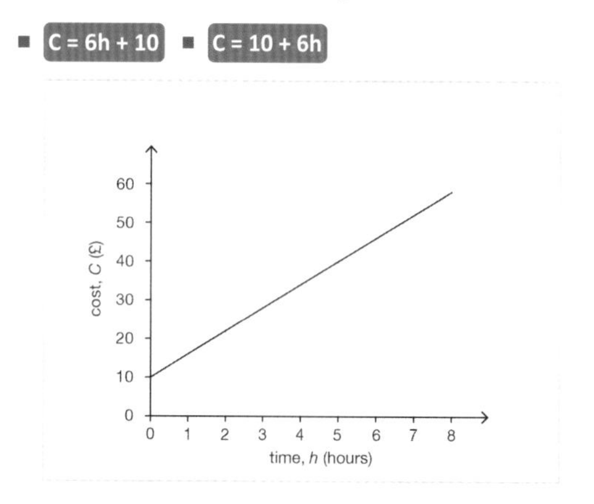

27. Four containers are filled using a tap with a constant flow. A graph is sketched for each container to show how the depth of water changes with time.
Arrange the letters of the graphs in the same order as the containers, starting with the graph of container (i).

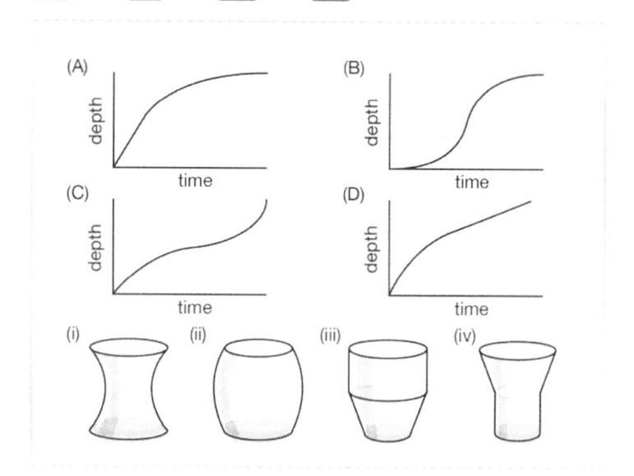

Level 4 *continued*

28. The population of bacteria increases by a multiplier each day.
The graph shows the spread of bacteria over three days.
Use the graph to estimate the population of bacteria after four days.

- 405

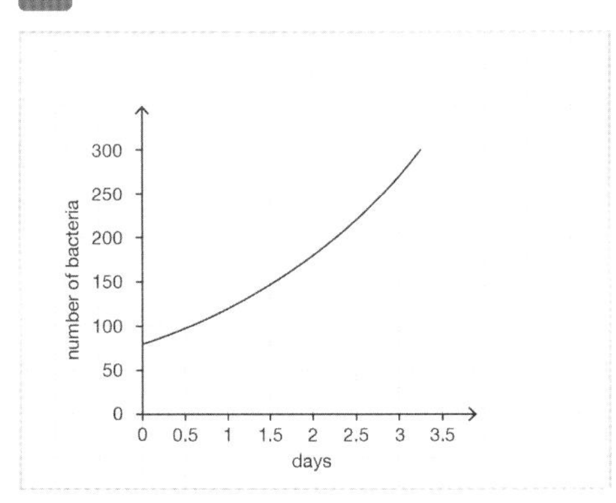

29. Use the graph to convert 60 °C to Fahrenheit.
Don't include the units in your answer.

- 140

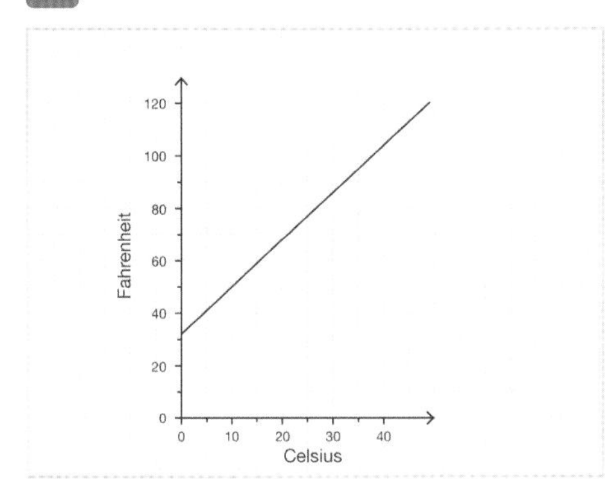

30. The graph shows the amount a plumber charges for work depending on time taken.
Give an equation for the total cost in pounds, *C*, in terms of the number of hours taken, *h*.
Give your answer in the form *C* – . . .

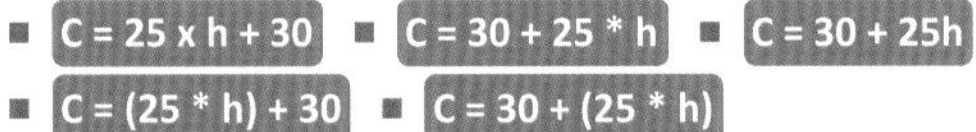

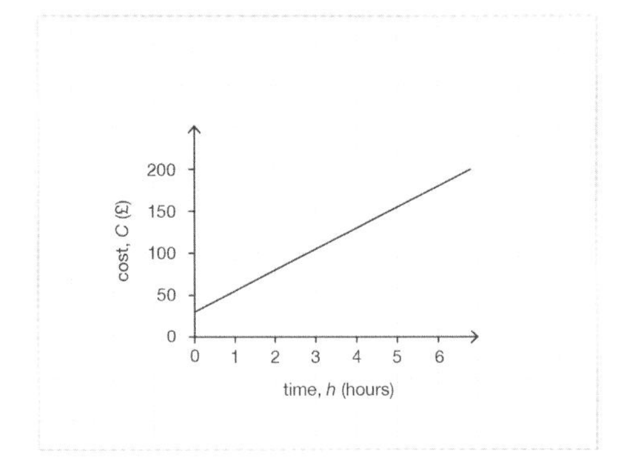

Use Graphs to Solve Simultaneous Equations

Objective: I can use graphs to solve simultaneous equations.

Quick Search Ref: 10556

Level 1: Understanding: Identify the graph of a line by calculating the coordinates of two of its points.

Required: 6/8 **Pupil Navigation:** on **Randomised:** off

1. Which of the following is a linear equation?

■ xy = 5 ■ y = x² ■ **3x – 5 = y/2** ■ 1/y = 3x + 2

1/4

2. For the equation $4y - 3x = 24$, find the value of y when $x = 0$.

■ **6**

$$4y - 3x = 24$$

3. For the equation $4y - 3x = 24$, find the value of x when $y = 0$.

■ **-8** ■ 8

$$4y - 3x = 24$$

4. Which graph represents the equation $2x - 3y = 18$?

■ (i) ■ (ii) ■ **(iii)** ■ (iv)

1/4

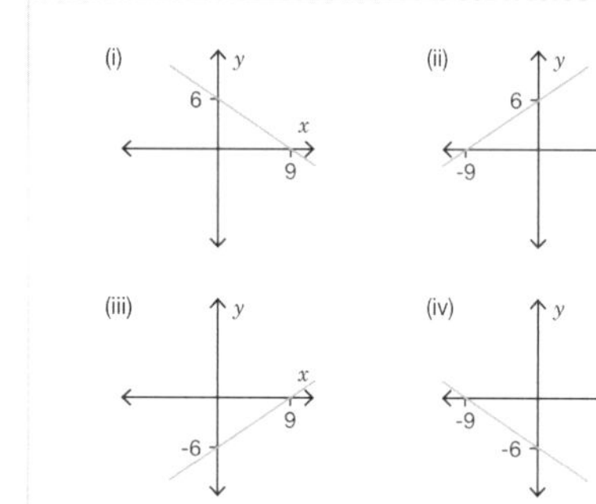

5. For the equation $2y + 7x = 23$, find the value of x when $y = 1$.

■ **3**

$$2y + 7x = 23$$

6. For the equation $3x - 5y = 30$, find the value of y when $x = 0$.

■ **-6** ■ 6

$$3x - 5y = 30$$

7. Which graph shows the equation $2y - 5x = 20$?

■ **(i)** ■ (ii) ■ (iii) ■ (iv)

1/4

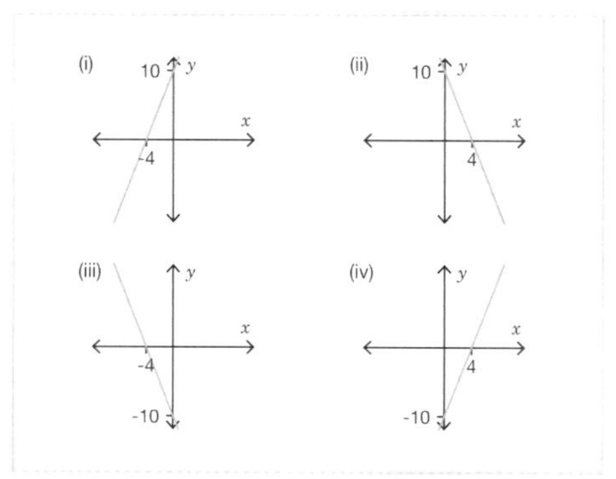

Level 1 *continued*

8. For the equation $9x + 4y = 40$, find the value of x when $y = 1$.

1 2 3

■ 4

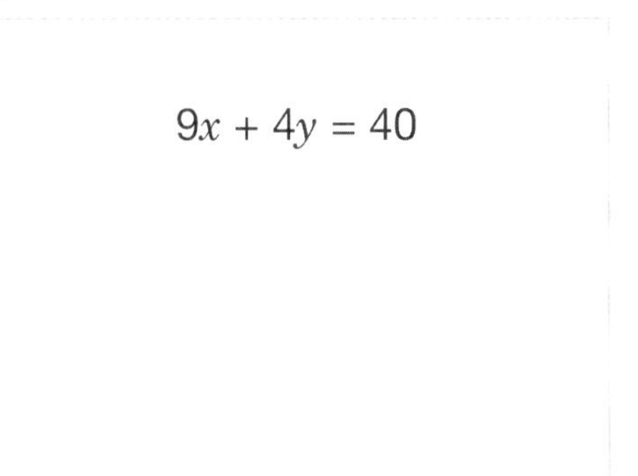

Level 2: Fluency: Use graphs to solve simultaneous equations.

Required: 6/8 **Pupil Navigation:** on

Randomised: off

9. Use the graph to solve these simultaneous equations:

$x + 2y = 4$

$2y - 3x = 12$

1/6

■ x = 2, y = 3 ■ x = 3, y = 2 ■ x = −2, y = 3

■ x = 3, y = −2 ■ x = 2, y = v3 ■ x = −3, y = 2

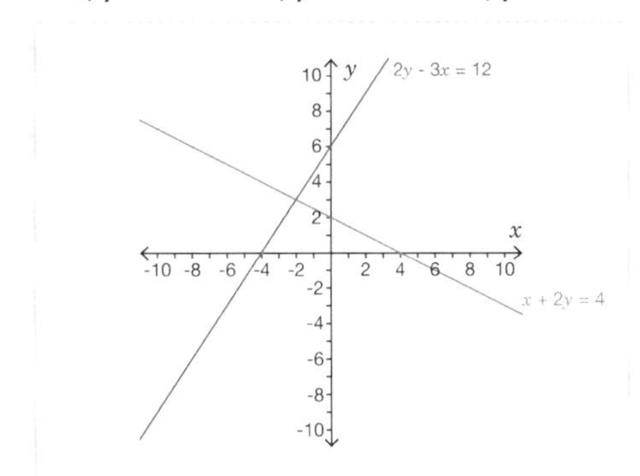

10. Use the graph to solve these simultaneous equations:

$x - y = 9$

$2y + 3x = 12$

1/5

■ x = 0, y = 6 ■ x = 4, y = 0 ■ x = 6, y = −3

■ x = 2, y = −7 ■ x = v3, y = 1

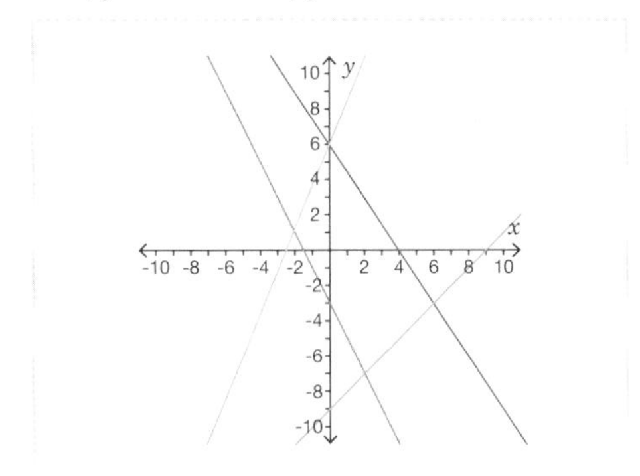

11. Use the graph to solve these simultaneous equations:

a b c

$2y - x = 8$

$3x + y + 3 = 0$

Give your answer as a pair of coordinates in the form (x, y).

■ (3, -2) ■ (-2, 3) ■ -2, 3 ■ (-2 3)

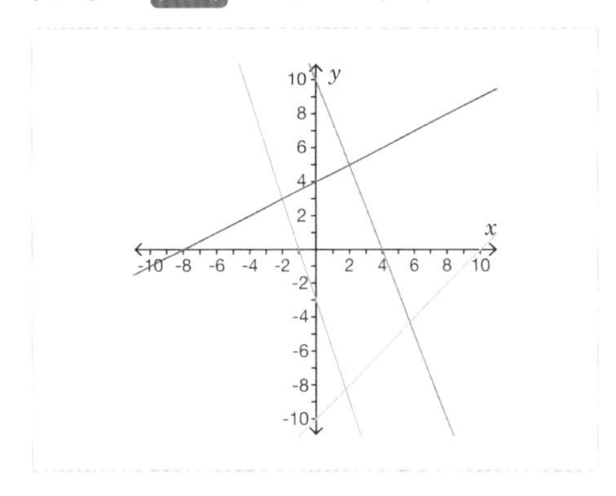

12. Use the graph to solve these simultaneous equations:

$2y - x = 4$

$y - 2x = 5$

1/6

■ x = 1, y = −2 ■ x = 2, y = −1 ■ x = −1, y = 2

■ x = −2, y = 1 ■ x = −1, y = −2 ■ x = −2, y = −1

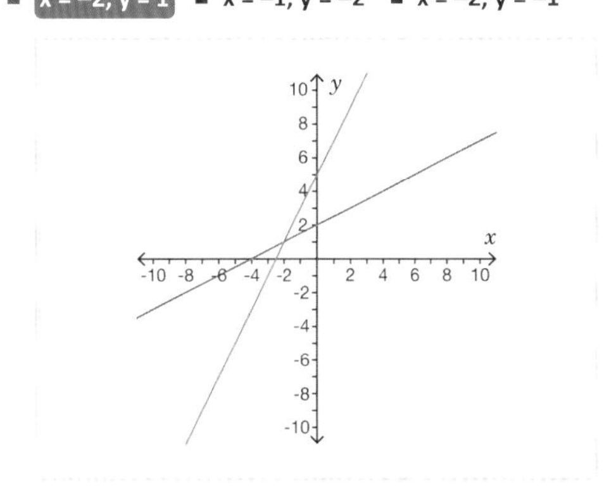

13. Use the graph to solve these simultaneous equations:

$x - y = 9$

$2x + y = 9$

1/6

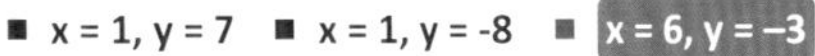

■ x = 1, y = 7 ■ x = 1, y = -8 ■ x = 6, y = −3

■ x = −5, y = 4 ■ x = −3, y = 5 ■ x = 1, y = 6

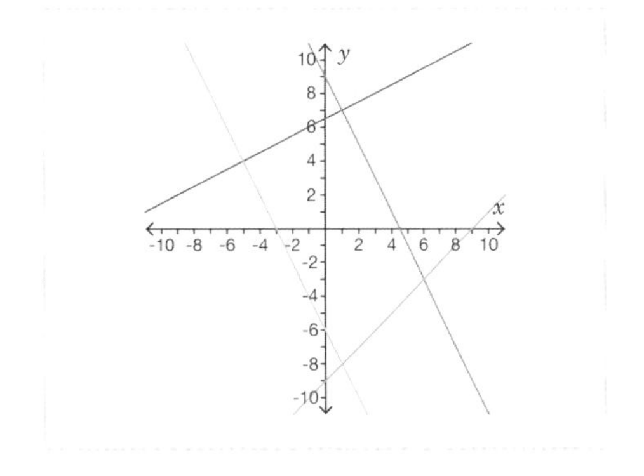

Level 2 *continued*

14. Use the graph to solve these simultaneous equations:
$5y - 12x = 30$
$3x + 2y = 12$
Give your answer as a pair of coordinates in the form (x, y).

- (6 0)
- (0, 6)
- (6, 0)
- 6, 0

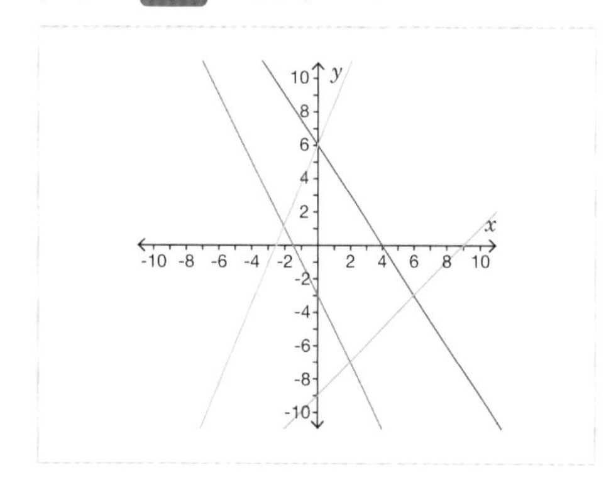

15. Use the graph to solve these simultaneous equations:
$x - y = 11$
1/5 $5x + 2y = 20$

- x = –5, y = 2
- x = 2, y = –9
- x = –2, y = 3
- x = 6, y = –5
- x = 2, y = 5

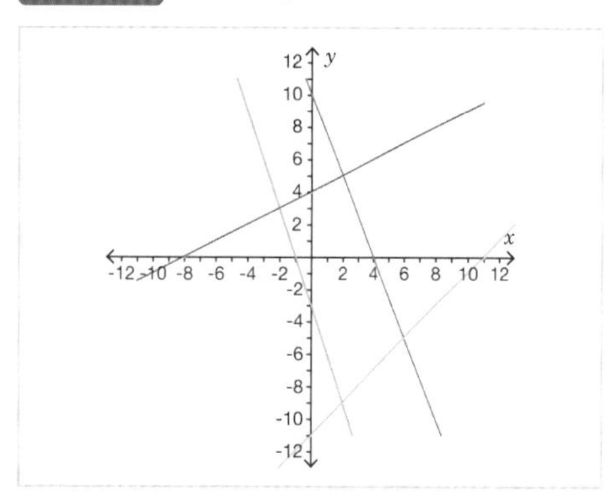

16. Use the graph to solve these simultaneous equations:
$2x + y + 6 = 0$
$2y - x = 13$
Give your answer as a pair of coordinates in the form (x, y).

- (-5 4)
- (-5, 4)
- (4, -5)
- -5, 4

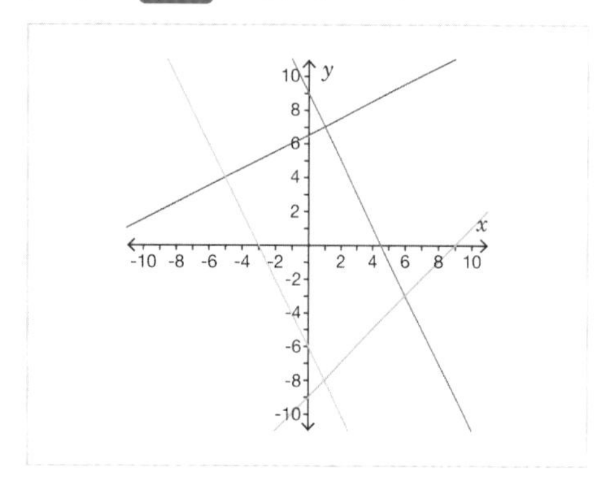

Level 3: Reasoning: Make inferences by graphing simultaneous equations.

Required: 3/3 **Pupil Navigation:** on
Randomised: off

17. Maya says that these simultaneous equations have a single solution:
$2y - x = 3$
$6 + 2x = 4y$
Is Maya correct? Explain your answer.

- Open question, no set answer

18. Frankie has correctly plotted a graph on the axes. What is the missing coefficient of *y*?

- 4

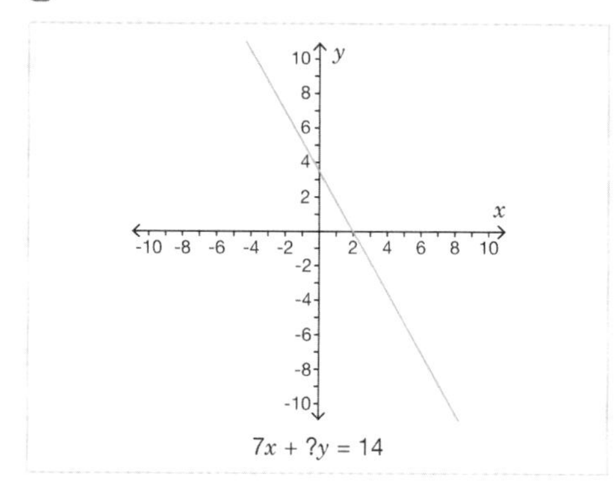
$7x + ?y = 14$

19. Mikey says these simultaneous equations don't have a solution:
$4x - 2y = 10$
$3y - 6x = 24$
Is Mikey correct? Explain your answer.

- Open question, no set answer

Level 4: Problem solving: Multi-step problems graphing simultaneous equations.

Required: 3/3 **Pupil Navigation:** on
Randomised: off

20. The diagram shows the graph of $y = x^2 + 2$.
Select the two solutions to these simultaneous equations:

2/7 $y = x^2 + 2$
$y = 3x$

- (0, 3)
- **(1, 3)**
- (3, 0)
- (6, 2)
- (3, 1)
- **(2, 6)**
- (–1, 3)

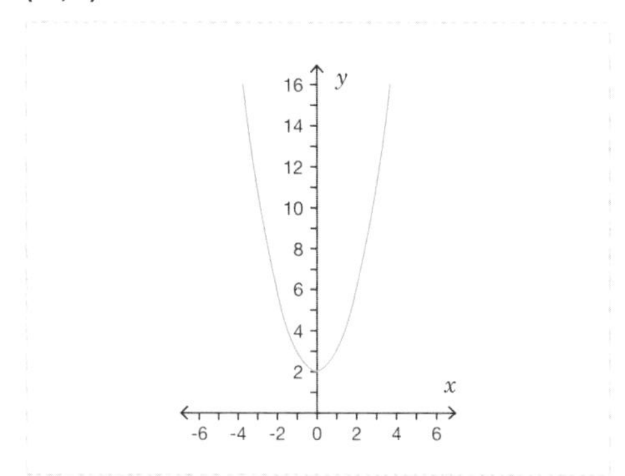

21. These simultaneous equations have a single solution:

$2y - 3x = 12$
2/5 $y + x = 1$
Which two other equations have the same solution?

- **y + 2x + 1 = 0**
- 3x – 2y = 6
- 4x – y + 8 = 0
- **2y – x = 8**
- 3y – 6x = 5

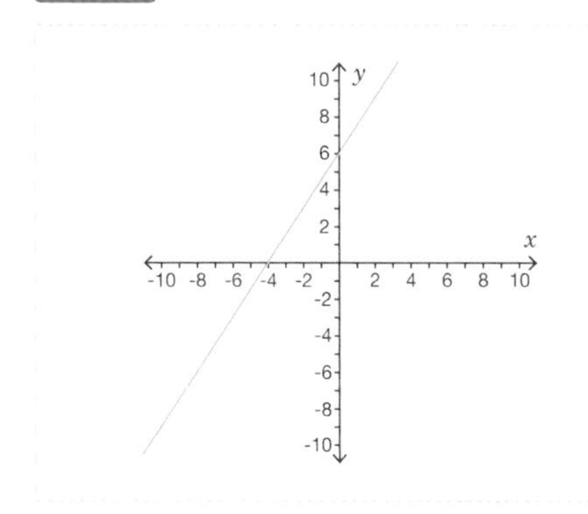

22. The graph shows how much four plumbers charge per job depending on the time taken, but the names are missing.
Ralph charges a £30 call-out fee plus £16 per hour for labour.
Sanya charges a £20 call-out fee plus £20 per hour for labour.
Identify the lines that represent Ralph's and Sanya's costs, and use these to identify how many minutes a job would have taken if Ralph and Sanya charged the same.
Don't include the units in your answer.

- **150**

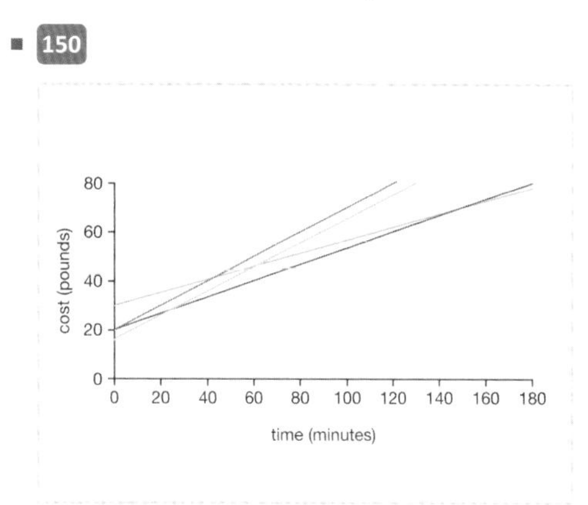

Use Linear and Quadratic Graphs to Estimate Values

Objective: I can find and estimate x- and y- values from linear and quadratic equations.

Quick Search Ref: 10558

Level 1: Understanding: Find x- and y-values from linear graphs.

Required: 7/10 **Pupil Navigation:** on **Randomised:** off

1. Using the graph, find the value of y when x = 6.

1 2 3

- 7
- 5

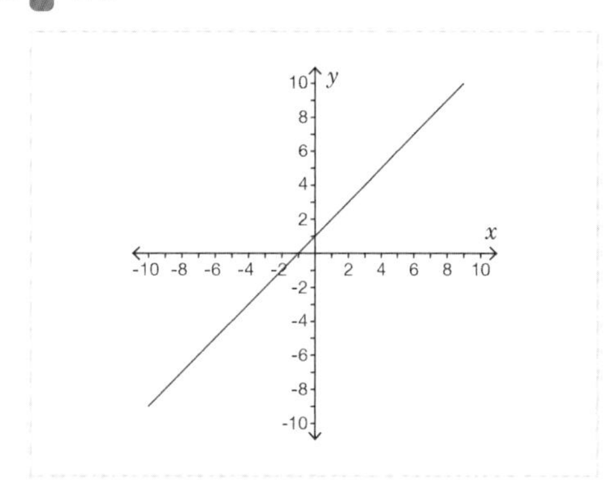

2. Using the graph, find the value of x when y = 8.

1 2 3

- 5
- 10

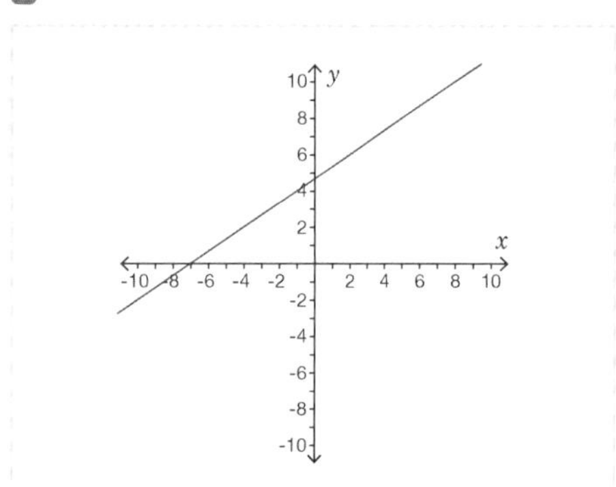

3. Using the graph, find the value of y when x = 4.

1 2 3

- -2
- -8

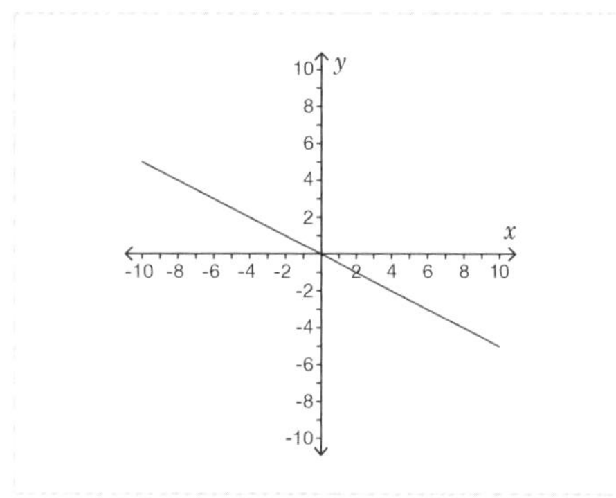

4. Using the graph, find the value of y when x = −8.

1 2 3

- -6
- -9

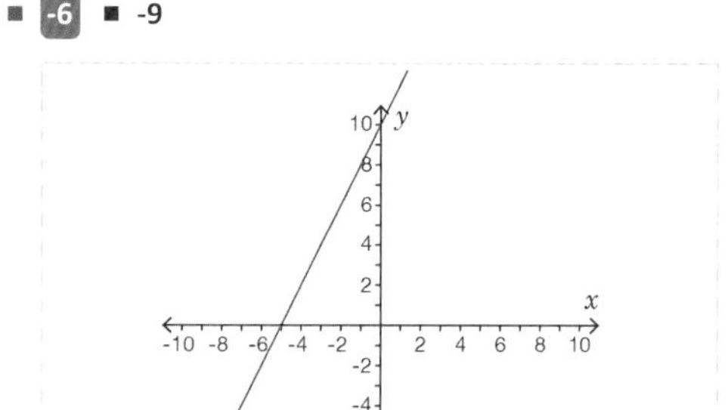

5. Using the graph, find the value of y when x = 6.

1 2 3

- 8

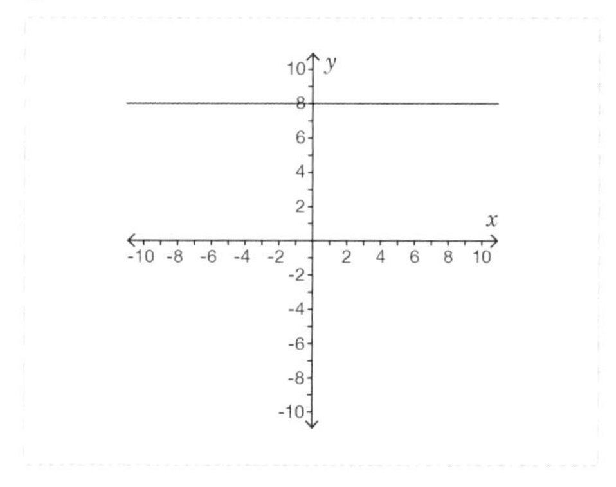

6. Using the graph, find the value of x when y = −5.

1 2 3

- -4

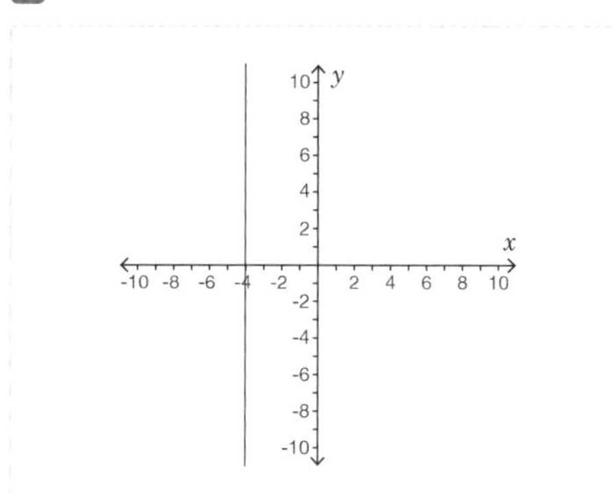

Level 1 ***continued***

7. Using the graph, find the value of x when y = 6.5.

1 2 3

■ -7 ■ 1

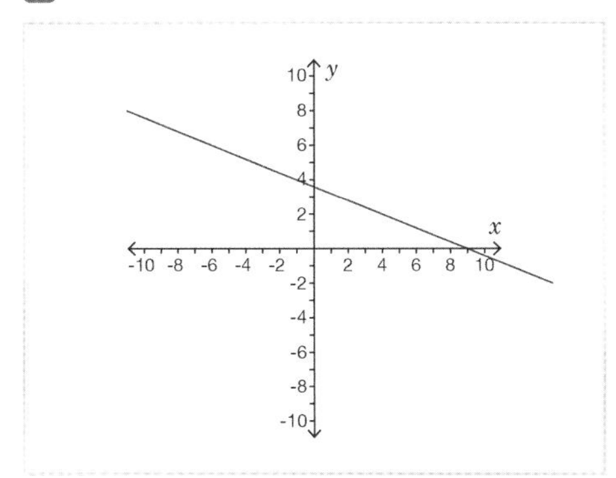

8. Using the graph, find the value of x when y = −8.5.

1 2 3

■ 8 ■ 2

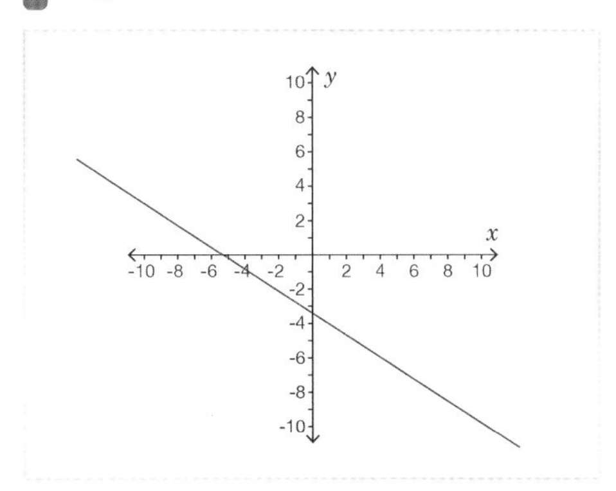

9. Using the graph, find the value of x when y = −7.

1 2 3

■ -9 ■ -5

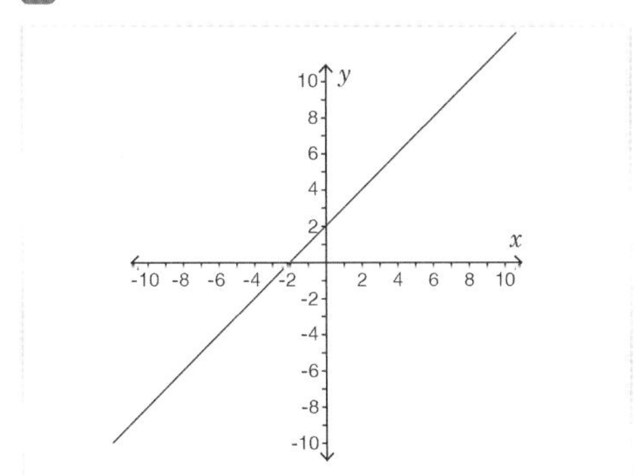

10. Using the graph, find the value of x when y = −6.

1 2 3

■ 7

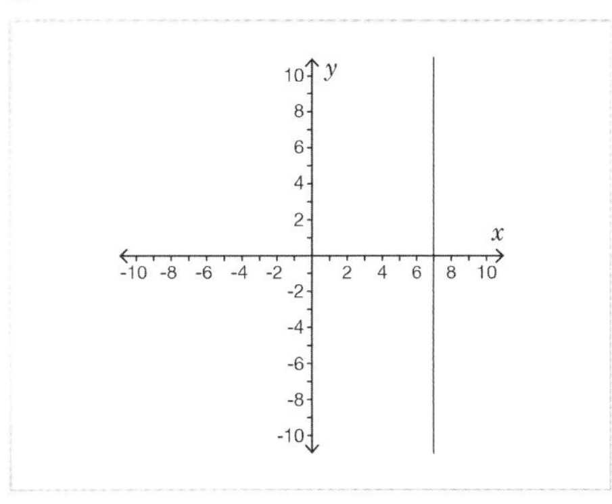

Level 2: Fluency: Estimate x- and y-values from quadratic graphs.

Required: 7/10 **Pupil Navigation:** on

Randomised: off

11. Using the graph, find the value of y when x = 2.

1 2 3

■ 5 ■ -4 ■ 4

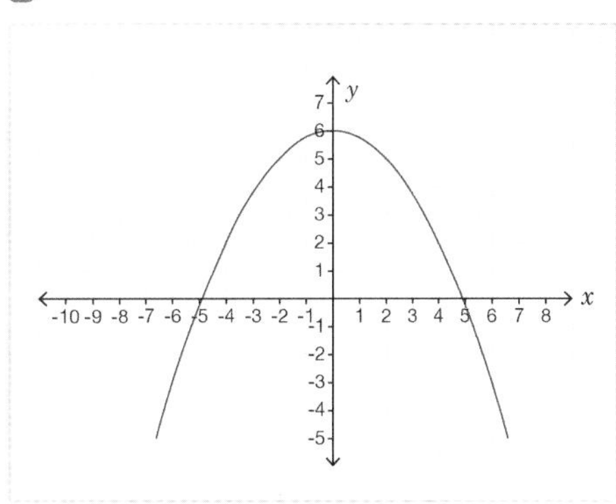

12. Using the graph, estimate the value of y when x = −3.
Give your answer to one decimal place.

a b c

■ 3.9 ■ -6.0 ■ 3.7 ■ 3.8 ■ -6 ■ 6 ■ 6.0

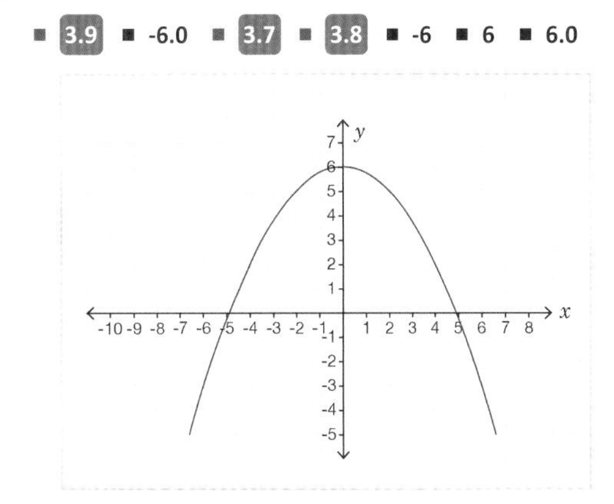

Level 2 *continued*

13. Using the graph, estimate the value of y when $x = 1$.
Give your answer to one decimal place.

1 2 3

- 3.7 ■ -2.1 ■ 3.6 ■ -6.1 ■ -2.2 ■ -2.3 ■ -6.3
- 3.8 ■ -6.2 ■ -6.4

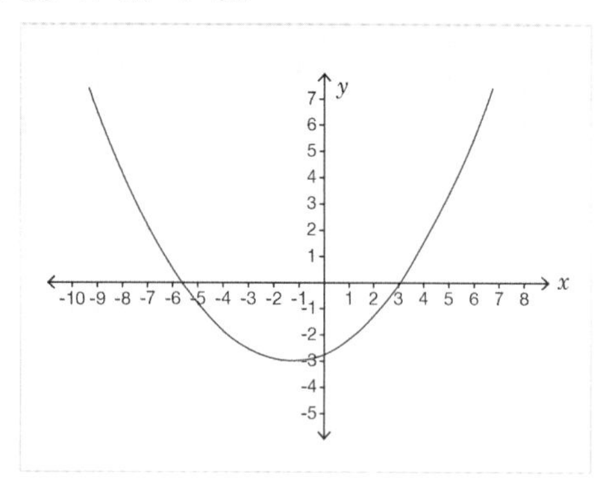

14. For the line shown on the graph, there are two possible values of x when $y = 6$. One of them is 6.2. Using the graph, estimate the other value.
Give your answer to one decimal place.

1 2 3

- -8.8 ■ 5.5 ■ -8.9 ■ -8.7 ■ 5.4 ■ 5.3

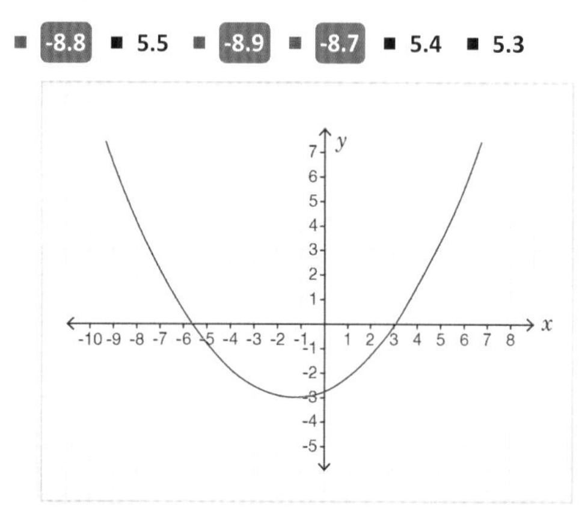

15. Using the graph, estimate the value of y when $x = -3$.
Give your answer to one decimal place.

1 2 3

- -1.4 ■ -1

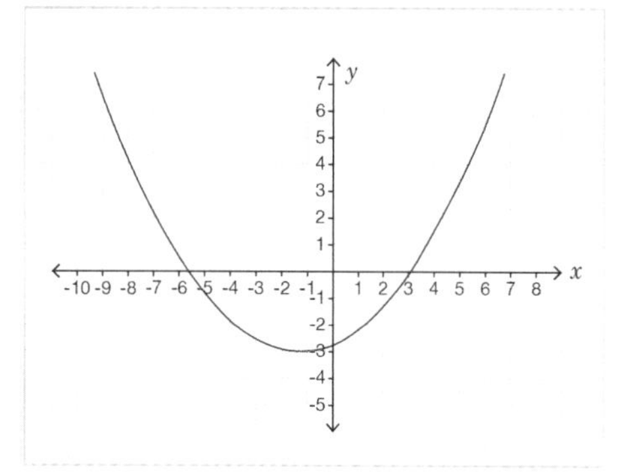

16. Using the graph, estimate the negative value of x when $y = -4.5$.
Give your answer to one decimal place.

a b c

- 0.9 ■ 6.5 ■ 1.0

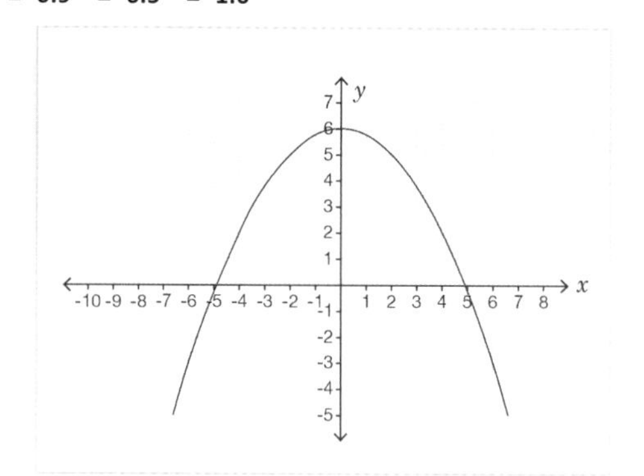

17. The graph shows the height of a drone during a flight. Select the best estimates for the two times at which the drone was at a height of 40 metres.

2/6

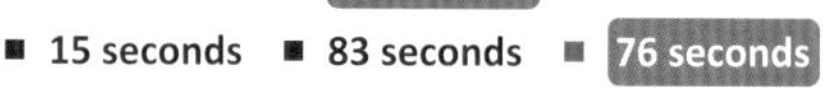

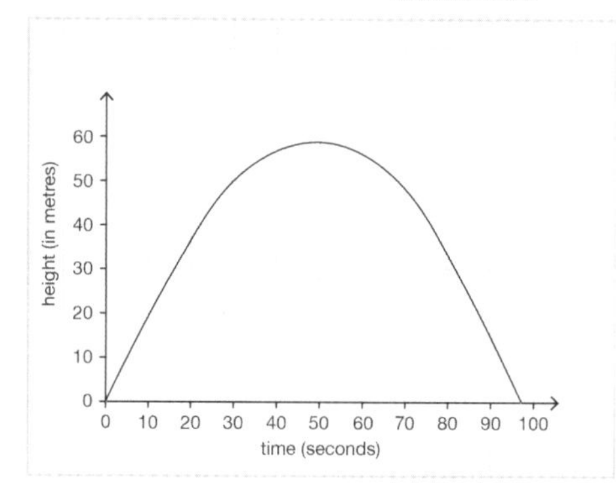

18. Select the best estimates for the two values of x when $y = -2$.

2/6

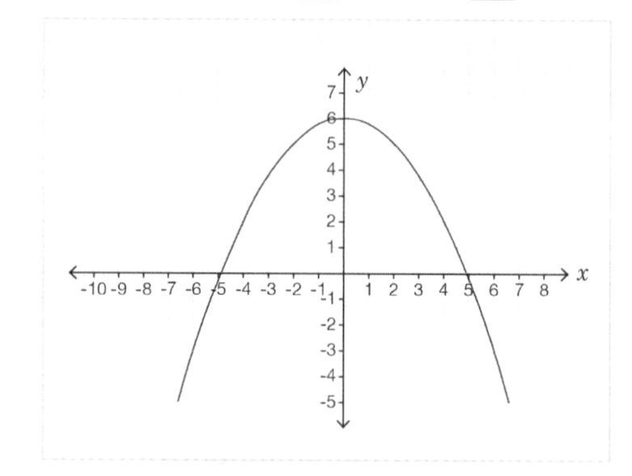

Level 2 *continued*

19. Using the graph, estimate the value of y when $x = 3$.
Give your answer to one decimal place.

- -4.3
- 5.2
- -4.5
- -3.8
- 5.3
- -4.4
- -3.7
- 5.1
- -3.9

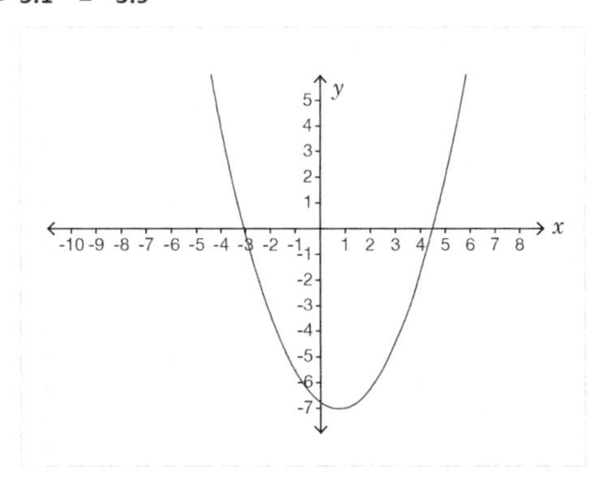

20. Using the graph, estimate the positive value of x when $y = -1$.
Give your answer to one decimal place.

- 4.2
- 4.1
- 4.3
- -2.8
- -5.7
- -2.9
- -5.6
- -2.7
- -5.5

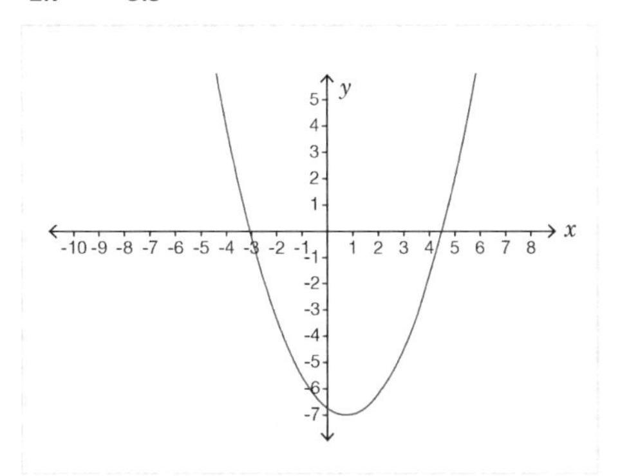

Level 3: Reasoning: Estimate x- and y-values from linear and quadratic graphs.

Required: 5/5 **Pupil Navigation:** on
Randomised: off

21. Steve says that for the equation shown on the graph, when $x = 2$, $y \approx 5.0$. Is he correct? Explain your answer.

- Open question, no set answer

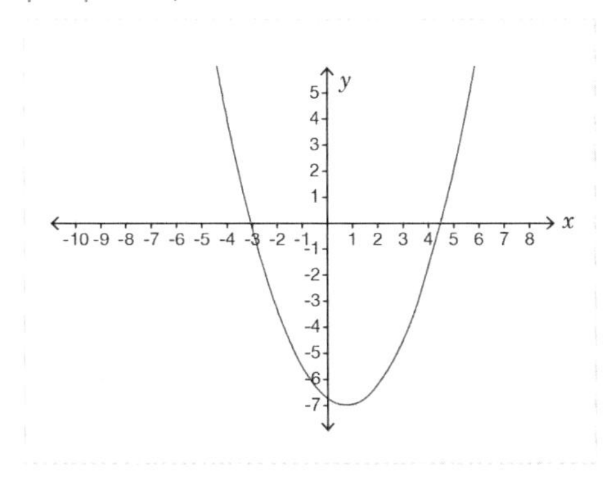

22. Estimate the minimum value of y for the graph.
Give your answer to one decimal place.

- -3
- -3.1
- -2.9

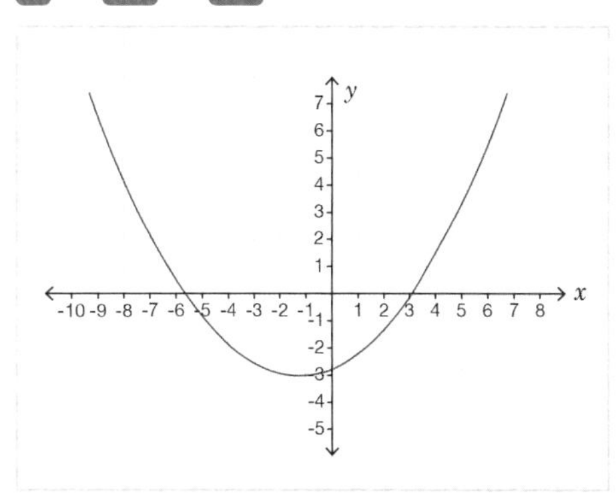

23. Estimate the maximum value of y for the graph.
Give your answer to one decimal place.

- 3.3
- 3.1
- 3.2

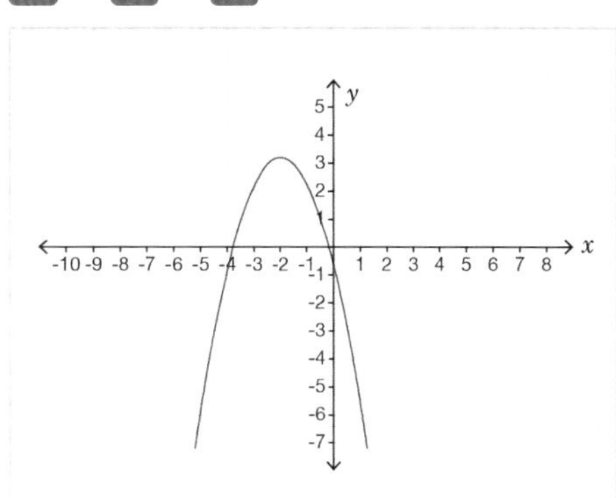

24. Rosie says that every y-value on the graph has two corresponding x-values. Rosie is incorrect. Which part of the graph shows that Rosie is incorrect? Explain your answer.

- Open question, no set answer

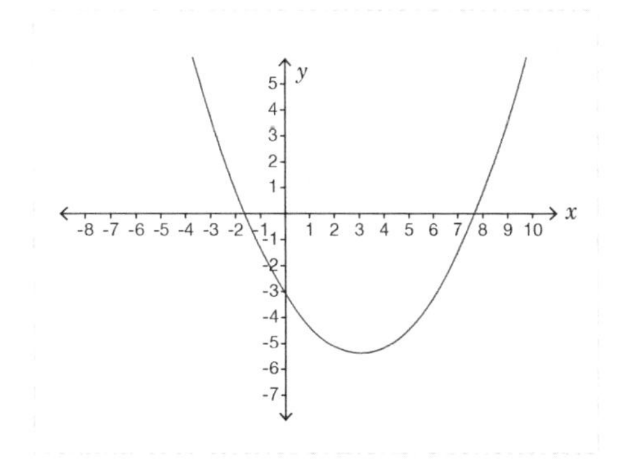

Level 3 *continued*

25. Using the graph, Daniel says that $x = -2$, $y = -1$ is a solution to these simultaneous equations:

$y = x^2 - 5$

$y = x + 1$

Is Daniel correct? Explain your answer.

- Open question, no set answer

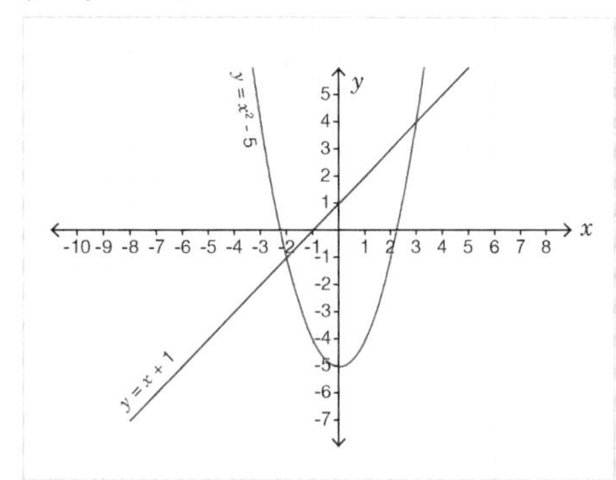

Level 4: Problem solving: Estimate x- and y-values from linear and quadratic graphs.

Required: 5/5 **Pupil Navigation:** on

Randomised: off

26. For the line of $x = y^2 - 3$, calculate the sum of the y-values when $x = 6$.

Give your answer to the nearest integer.

- 0

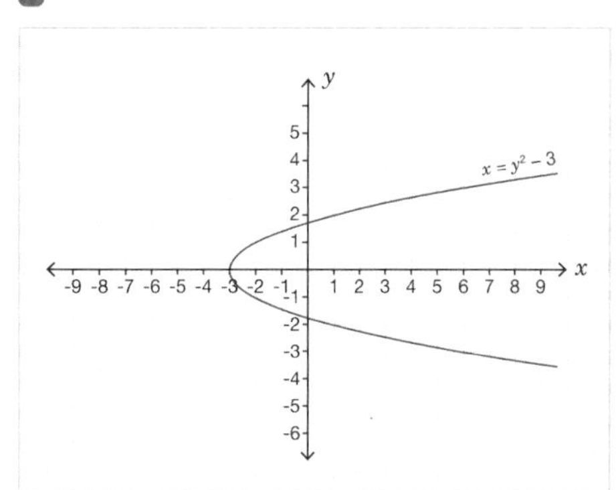

27. $x = -1$, $y = -5$ is one pair of solutions to these simultaneous equations:

$y = x^2 - 6$

$y = x - 4$

What is the product of the other pair of solutions?

- -4

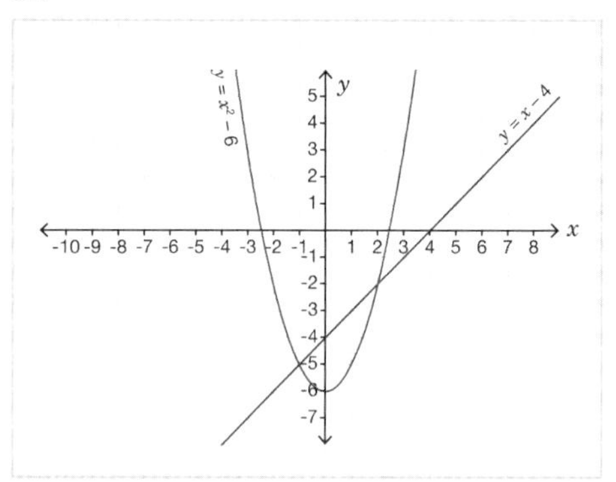

28. $x = 1$, $y = 0$ is one pair of solutions to these simultaneous equations:

1/6

$y = x^2 - 1$

$y = x + 1$

Select the other pair of solutions.

- x = 0, y = –1
- x = 0, y = 1
- **x = 2, y = 3**
- x = 3, y = 4
- x = 4, y = 5
- x = 0, y = 4

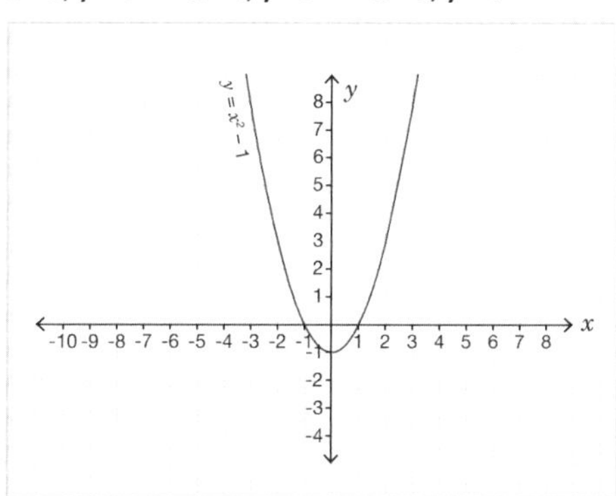

29. Using the graph, calculate the product of the two x-values when $y = 3$.

- -8

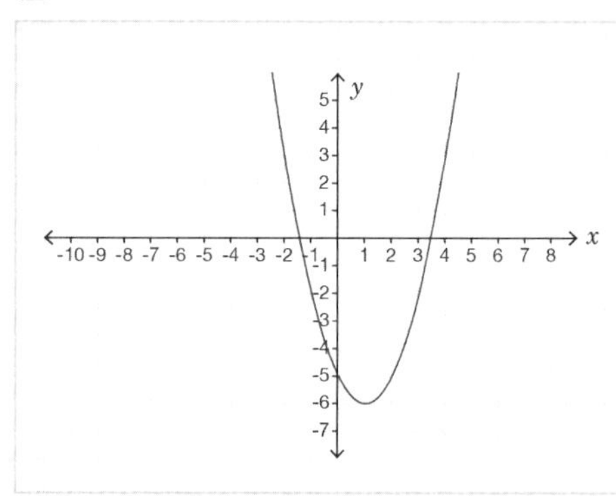

Level 4 *continued*

30. Use the graph to estimate the x-value of the positive solution to the equation $x^2 + 3x$ divide Numbers mentally 5 = 0.

Give your answer to one decimal place.

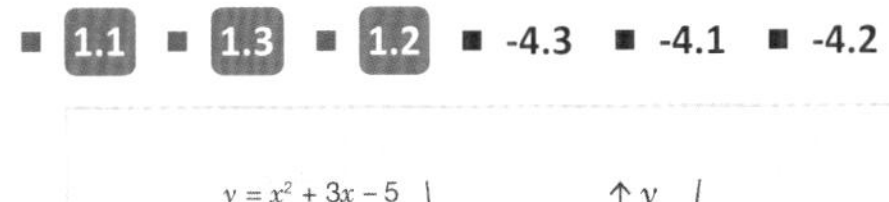

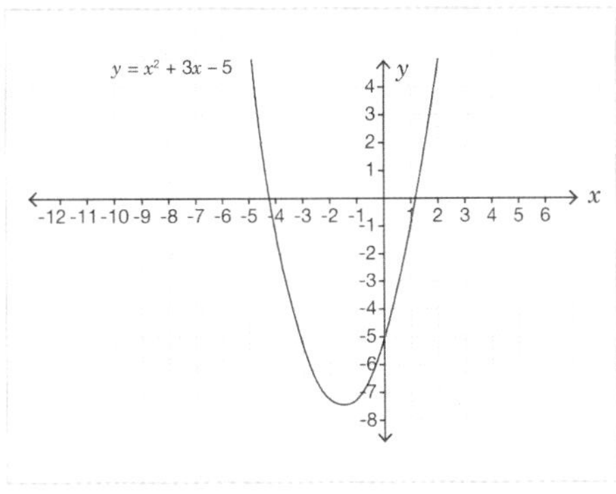

Recognise and Produce Graphs of Quadratic Functions

Objective: I can recognise and produce quadratic graphs.

Quick Search Ref: 10515

Level 1: Understanding: Complete tables of values with positive integer coefficients of x.

Required: 7/10 **Pupil Navigation:** on **Randomised:** off

1. This is a table of values for the equation $y = x^2 + 1$.
What are the coordinates of the point represented by the highlighted column?
Give your answer in the form (x, y).

- (2 5)
- **(2, 5)**
- 2, 5
- (5, 2)

x	0	1	2	3
x^2	0	1	4	9
+1	+1	+1	+1	+1
$y = x^2 + 1$	1	2	5	10

2. This is a table of values for the equation $y = x^2 - 1$.
What is the value of y when $x = 3$?

- 10
- **8**
- 5

x	0	1	2	3
x^2	0	1	4	9
-1	-1	-1	-1	-1
$y = x^2 - 1$	-1	0	3	?

3. This is a table of values for the equation $y = x^2 + 3$.
What is the value of y when $x = 1$?

- **4**
- 5

x	0	1	2	3
x^2	0			9
+3	+3			+3
$y = x^2 + 3$	3	?		12

4. This is a table of values for the equation $y = 2x^2 + 4$.
What is the value of y when $x = 3$?

- **22**
- 40
- 16

x	0	1	2	3	4
x^2	0				16
$2x^2$	0				32
+4	+4				+4
$y = 2x^2 + 4$	4			?	36

5. This is a table of values for the equation $y = x^2 + 3x$.
What is the value of y when $x = 4$?

- **28**
- 20

x	0	1	2	3	4
x^2					
$+3x$					
$y = x^2 + 3x$					?

6. This is a table of values for the equation $y = 2x^2 + 5x$.
What are the coordinates of the point on this line when $x = 4$?
Give your answer in the form (x, y).

- 4, 52
- **(4, 52)**
- (52, 4)
- 52
- (4, 84)
- (4 52)
- (4, 36)

x	0	1	2	3	4
x^2		1		9	
$2x^2$		2		18	
$+5x$		+5		+15	
$y = 2x^2 + 5x$		7		33	

Level 1 *continued*

7. a b c This is a table of values for the equation $y = 3x^2 - 2x$.
What are the coordinates of the point on this line when $x = 3$?
Give your answer in the form (x, y).

- (3 21)
- (3, 21)
- 3, 21
- (3, 75)
- (21, 3)
- 21
- (3, 12)

x	0	1	2	3
$y = 3x^2 - 2x$	0	1		

8. a b c This is a table of values for the equation $y = 5x^2 - 3x$.
What are the coordinates of the point on this line when $x = 4$?
Give your answer in the form (x, y).

- (4, 28)
- (4, 68)
- (4 68)
- 4, 68
- (68, 4)
- 68
- (4, 388)

x	0	1	2	3	4
$y = 5x^2 - 3x$	0			36	

9. a b c This is a table of values for the equation $y = 3x^2 + x$.
What are the coordinates of the point on this line when $x = 4$?
Give your answer in the form (x, y).

- (4, 52)
- 4, 52
- 52
- (4 52)
- (52, 4)
- (4, 148)
- (4, 28)

x	0	1	2	3	4
x^2		1		9	
$3x^2$		3		27	
$+x$		+1		+3	
$y = 3x^2 + x$		4		30	

10. 1 2 3 This is a table of values for the equation $y = x^2 - 5x$.
What is the value of y when $x = 3$?

- -16
- -9

x	0	1	2	3	4
x^2					
$-5x$					
$y = x^2 - 5x$				?	

Level 2: Fluency: Plot points with positive, negative and fractional coefficients of x.

Required: 7/10 **Pupil Navigation:** on
Randomised: off

11. This is a table of values for the equation $y = 2x^2 - 4$.
What are the coordinates of the point on this line when $x = -2$?
Give your answer in the form (x, y).

- (-2, 12) ■ (4, -2) ■ **(-2, 4)** ■ (-2, -12) ■ 4 ■ -2, 4 ■ (-2 4)

x	-3	-2	-1	0	1	2	3
x^2	9		1	0	1		9
$2x^2$	18		2	0	2		18
-4	-4		-4	-4	-4		-4
$y = 2x^2 - 4$	14		-2	-4	-2		14

12. This is a table of values for the equation $y = -x^2 + 4$.
What is the value of y when $x = 3$?

- **-5** ■ -2 ■ 13

x	-3	-2	-1	0	1	2	3
x^2		4	1	0	1		
$-x^2$		-4	-1	0	-1		
+4		+4	+4	+4	+4		
$y = -x^2 + 4$		0	3	4	3		?

13. This is a table of values for the equation $y = -2x^2 - 3$.
What are the coordinates of the point on this line when $x = $ v1?
Give your answer in the form (x, y).

- (-1, -1) ■ **(-1, -5)** ■ (-1, 1) ■ (-1 -5) ■ -5 ■ -1, -5 ■ (-5, -1)

x	-3	-2	-1	0	1	2	3
x^2	9	4		0		4	
$-2x^2$	-18	-8		0		-8	
-3	-3	-3		-3		-3	
$y = -2x^2 - 3$	-21	-11		-3		-11	

14. Bilal is going to plot the graph of $y = 3x^2 + 2x$.
Which letter shows the correctly plotted point for $x = -1$?

1/4 ■ A ■ **B** ■ C ■ D

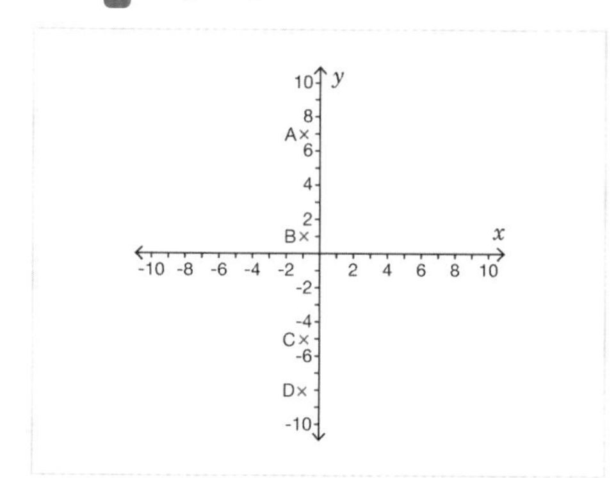

15. This is a table of values for the equation $y = 1/2\ x^2 + 3$.
What is the value of y when $x = 4$?

- **11** ■ 7

x	-2	-1	0	1	2	3	4
x^2	4			1	4	9	
$\frac{1}{2}x^2$	2			0.5	2	4.5	
+3	+3			+3	+3	+3	
$y = \frac{1}{2}x^2 + 3$	5			3.5	5	7.5	?

Level 2 *continued*

16. Mikkel is going to plot the graph of $y = -3x^2 - 4x$. Which letter shows the correctly plotted point for $x = -2$?

1/4

- A
- B
- **C** (selected)
- D

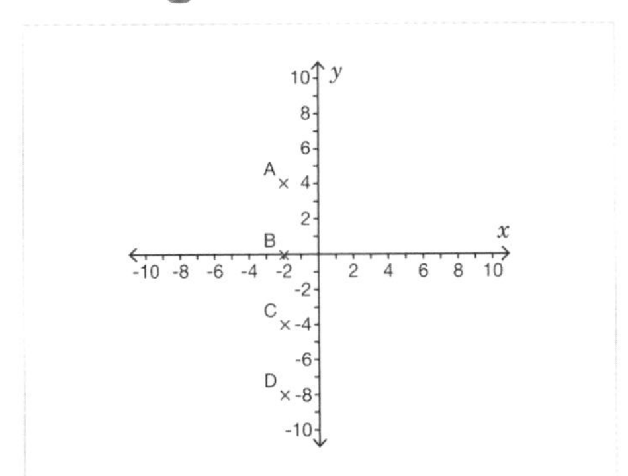

17. Manon is going to plot the graph of $y = 1/4x^2 - x$. Which letter shows the correctly plotted point for $x = -2$?

1/4

- **A** (selected)
- B
- C
- D

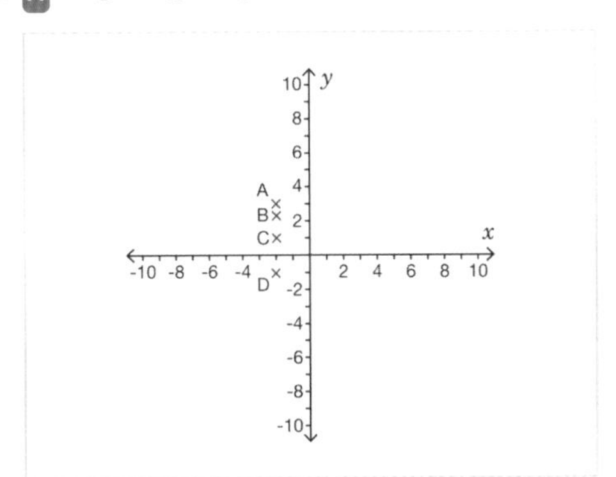

18. This is a table of values for the equation $y = -3x^2 + 2x$.
What are the coordinates of the point on this line when $x = -1$?
Give your answer in the form (x, y).

- (-1 -5)
- **(-1, -5)** (selected)
- -1, -5
- (-1, 1)
- -5
- (-5, -1)
- (-1, 7)

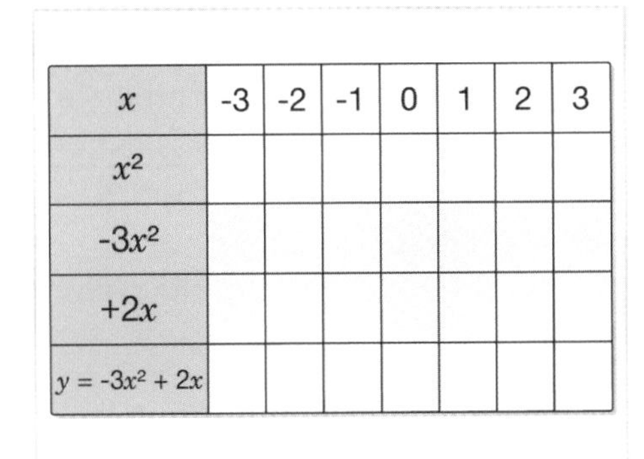

x	-3	-2	-1	0	1	2	3
x^2							
$-3x^2$							
$+2x$							
$y = -3x^2 + 2x$							

19. Russell is going to plot the graph of $y = 4x^2 - 3x$. Which letter shows the correctly plotted point for $x = -1$?

1/4

- **A** (selected)
- B
- C
- D

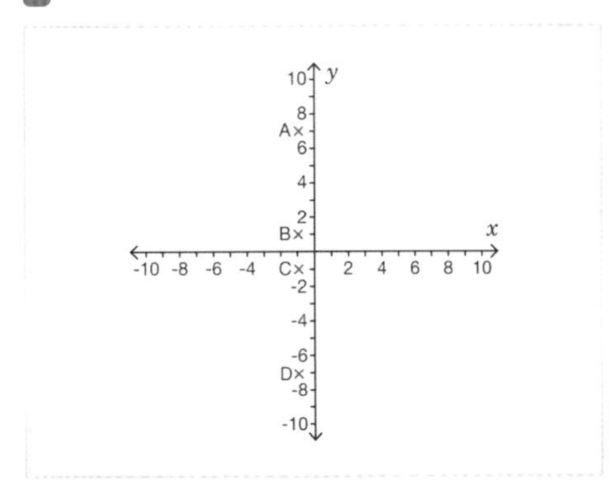

20. This is a table for values for the equation $y = 1/2x^2 + x$.
What are the coordinates of the point on this line when $x = -2$?
Give your answer in the form (x, y).

- 0
- **(-2, 0)** (selected)
- -2, 0
- (0, -2)
- (-2, -1)
- (-2, -4)
- (-2 0)

x	-3	-2	-1	0	1	2	3
x^2							
$\frac{1}{2}x^2$							
$+x$							
$y = \frac{1}{2}x^2 + x$							

Level 3: Reasoning: Use reason for graphs of quadratic functions.

Required: 5/5 **Pupil Navigation:** on
Randomised: off

21. Ben says that the equation of this line is $y = x^2 - 2$. Is Ben correct? Explain your answer.

- Open question, no set answer

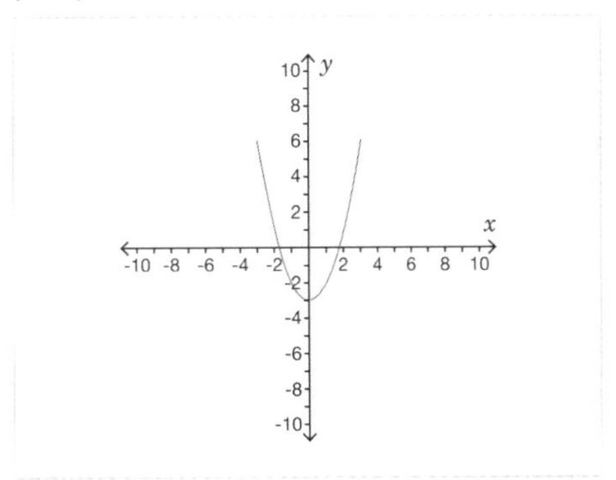

Level 3 *continued*

22. Emma made a mistake when completing this table of values for the equation $y = x^2 + 3$. Which column has she made a mistake in?

1/5

■ **x = -1** ■ x = 0 ■ x = 1 ■ x = 2 ■ x = 3

x	-1	0	1	2	3
x^2	-1	0	1	4	9
+3	+3	+3	+3	+3	+3
$y = x^2 + 3$	2	3	4	7	12

23. Sandeep produced this table of values for the equation $y = 3x^2 - 4x$. Has he worked out all of the values correctly? Explain your answer.

- Open question, no set answer

x	-8	-7	-6	-5
x^2	64	49	36	25
$3x^2$	192	147	108	75
$-4x$	-32	-28	-24	-20
$y = 3x^2 - 4x$	160	119	84	55

24. Verity says that the point with coordinates (1, 6) lies on all of the following lines:
$y = 2x^2 + 4$
$y = x^2 + 5$
$y = 4x^2 + 2$
$y = -x^2 + 7$
Is she correct? Explain your answer.

- Open question, no set answer

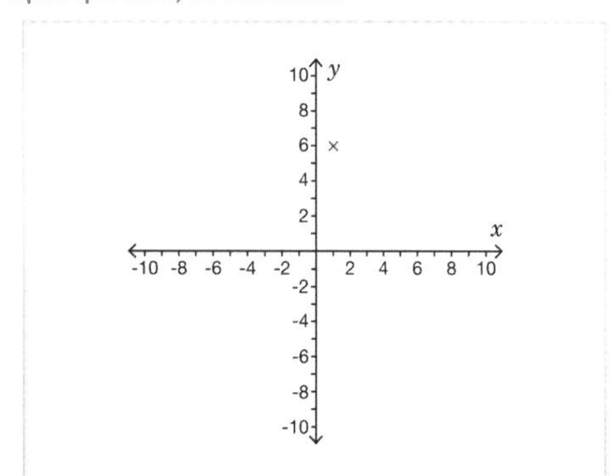

25. What are the coordinates of the point where the line $y = 3x^2 - 4$ intercepts the y-axis? *Give your answer in the form (x, y).*

■ 0, -4 ■ 0 ■ **(0, -4)** ■ (-4, 0) ■ -4 ■ (0 -4)

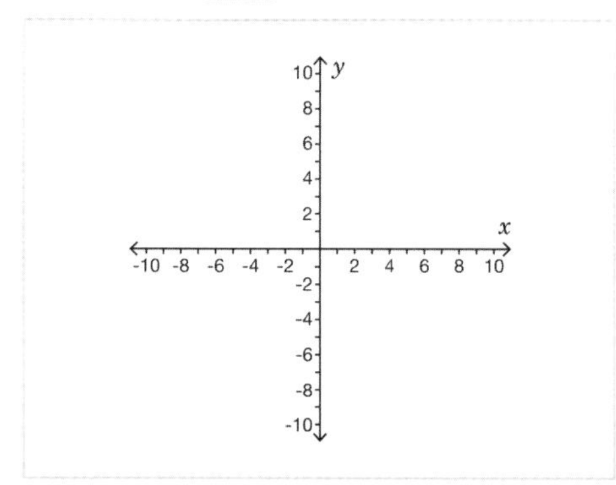

Level 4: Problem solving: Solve problems using graphs of quadratic functions.

Required: 5/5 **Pupil Navigation:** on
Randomised: off

26. Which of these lines does the point with coordinates (–2, 8) lie on?

1/4

■ **$y = x^2 + 4$** ■ $y = 2x^2 + 2$ ■ $y = 3x^2 - 6$ ■ **$y = 4x^2 - 8$**

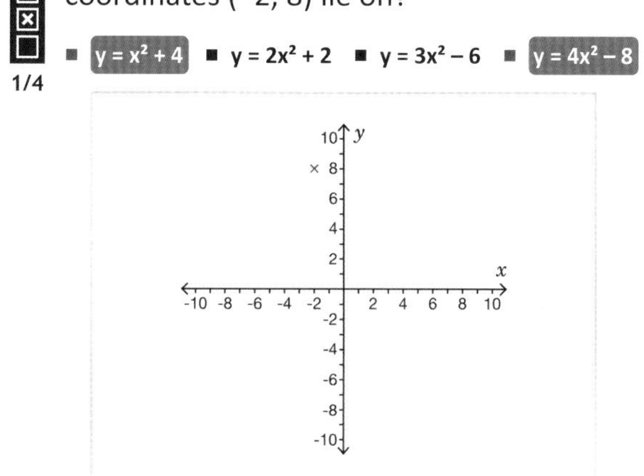

27. What is the sum of the two x-coordinates whose y-coordinate is 40 for the line $y = x^2 - 9$?

■ **0**

28. Henry is thinking of a point on the line $y = 3x^2 - 4x$. He says that the x-coordinate is an integer greater than –4 and less than 4, and the difference between the x- and y-coordinates is 8.
What are the coordinates of the point Henry is thinking of?

■ -1 ■ **(-1, 7)** ■ (-1 7) ■ -1, 7 ■ 7 ■ (7, -1)

29. On the line $y = 2x^2 + 4x$, what is the **negative** x-coordinate when $y = 70$?

■ **-7** ■ 5

Level 4 *continued*

30.  Freya is thinking of a point on the line $y = -2x^2 - 4x$. She says that the x-coordinate is an integer greater than -10 and less than v3, and the sum of the x- and y-coordinates is –54.

What are the coordinates of Freya's point?

- -6, -48
- (-48, -6)
- (-6, -48)
- -6
- -48
- (-6 -48)

Mathematics Y9

Geometry

Transformation
Similarity and Congruence
Pythagoras and Trigonometry

Describe Transformations

Objective: I can fully describe translations, rotations, reflections and enlargements.

Quick Search Ref: 10641

Level 1: Understanding: Categorise and describe translations, rotations, reflections and enlargements.

Required: 7/10 **Pupil Navigation:** on **Randomised:** off

1. What type of transformation maps shape V onto shape W?

1/5

- transformation
- translation
- reflection
- rotation
- **enlargement**

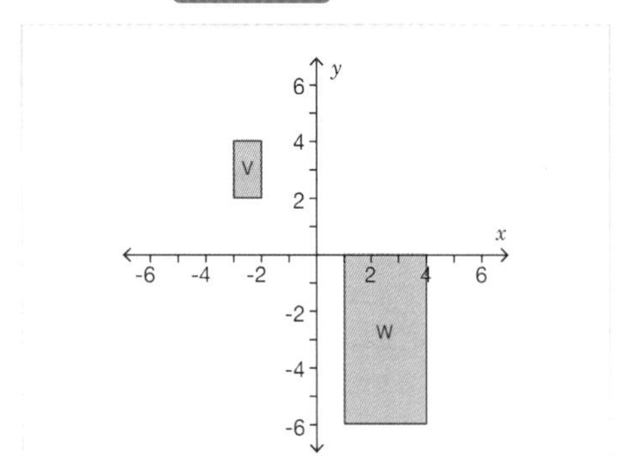

2. Select the **two statements** that together fully describe the rotation that maps shape G onto shape H.

2/6

- a rotation of 270° anti-clockwise
- **centre of rotation (0, 1)**
- a rotation of 90° clockwise
- centre of rotation (0, 0)
- **a rotation of 90° anti-clockwise**
- centre of rotation (1, 0)

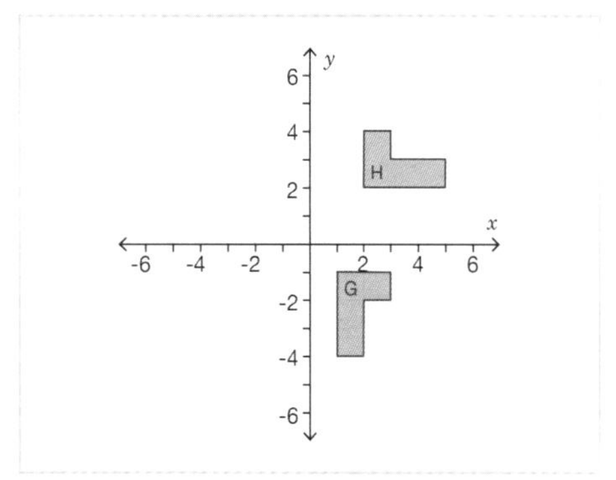

3. What type of transformation maps shape C onto shape D?

1/5

- transformation
- translation
- **reflection**
- rotation
- enlargement

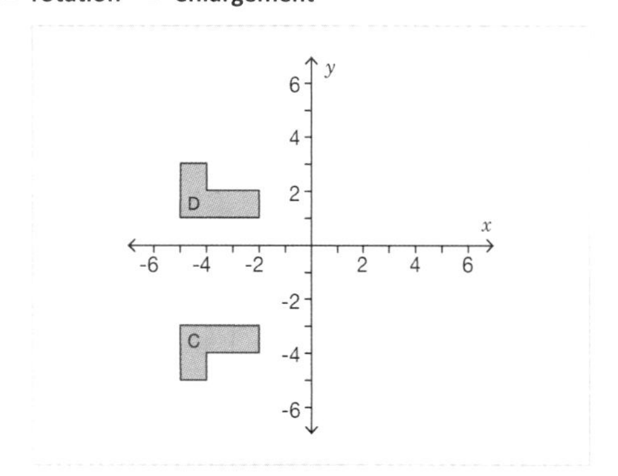

4. Enter the coordinates of the centre of enlargement that maps shape R onto shape S.
Give your answer in the form (x, y).

- (3, -3)
- **(4, -6)**
- 4, -6
- (4 -6)
- (-6, 4)

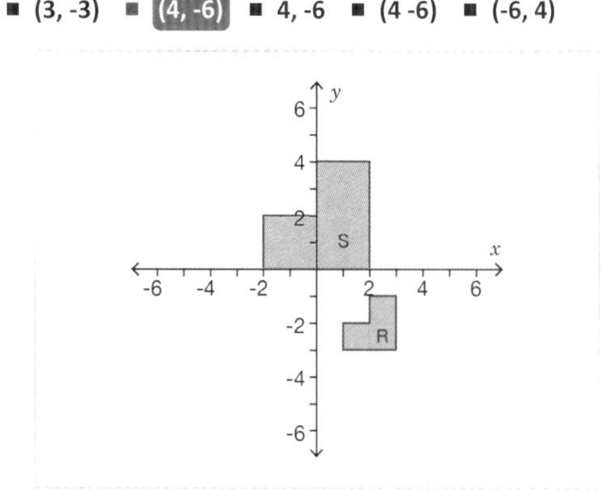

5. What type of transformation maps shape J onto shape K?

1/5

- transformation
- translation
- reflection
- **rotation**
- enalrgement

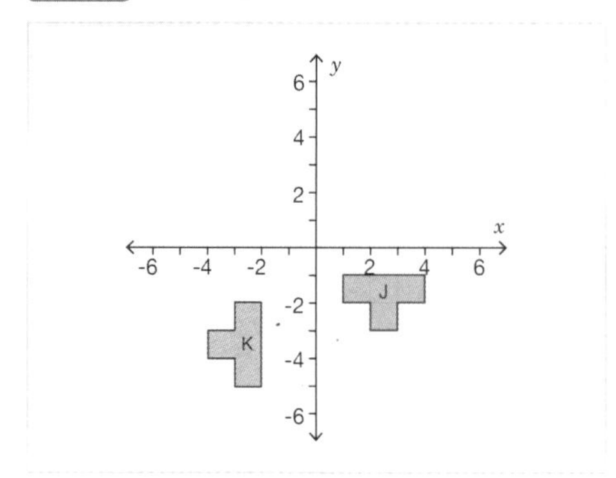

6. What number is missing from the bottom of the column vector that maps shape P onto shape Q?

- -7
- **7**
- 5
- -5

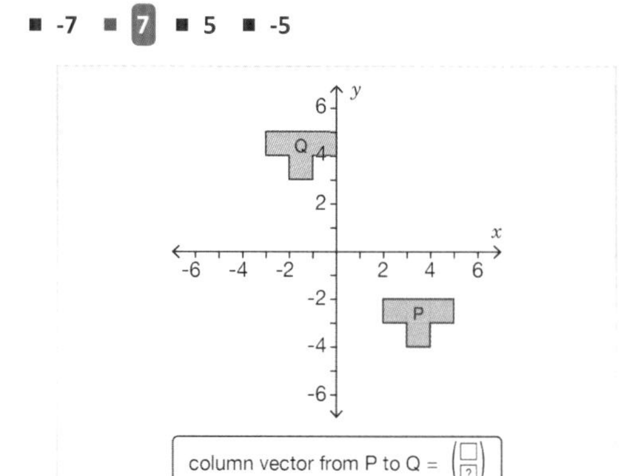

Level 1 *continued*

7. Shape A is reflected onto shape B. What is the equation of the line of reflection?
Give your answer in the form y = . . .

- y = -1x
- y = -x
- y = x
- x = y

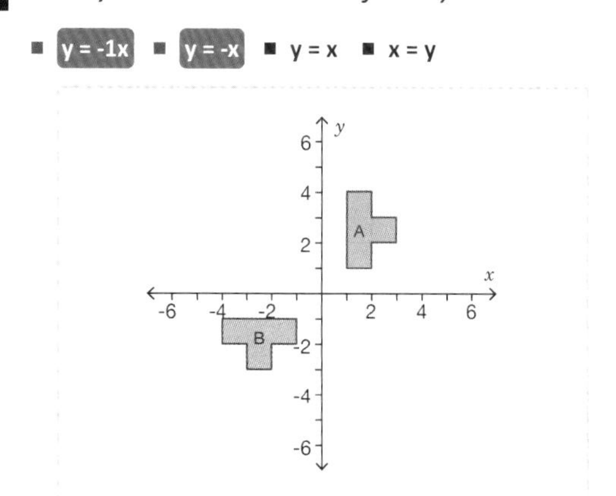

8. Select the **two statements** that together fully describe the enlargement that maps shape P onto shape Q.

2/6

- scale factor 2
- centre of enlargement (-6, 5)
- scale factor 4
- centre of enlargement (-6, 6)
- scale factor 3
- centre of enlargement (-5, 6)

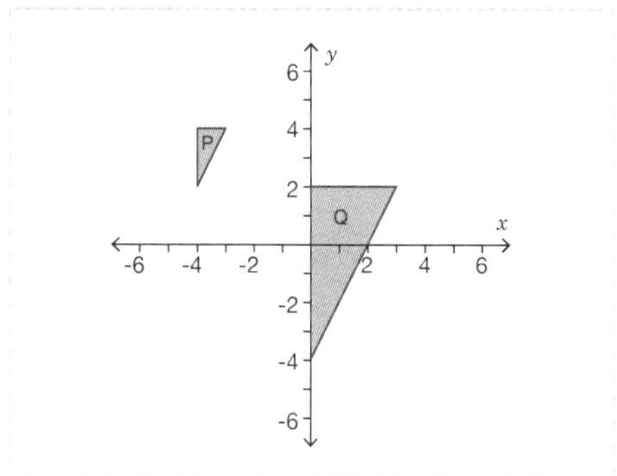

9. What type of transformation maps shape M onto shape N?

1/5

- transformation
- translation
- reflection
- rotation
- enlargement

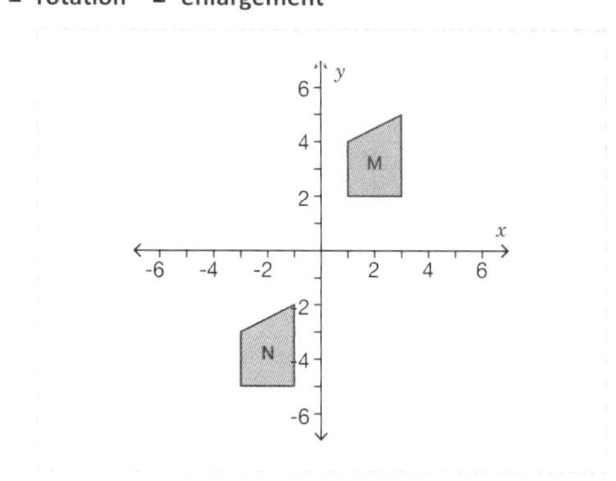

10. Shape U is reflected onto shape V. What is the equation of the line of reflection?

- x = -1
- y = -1

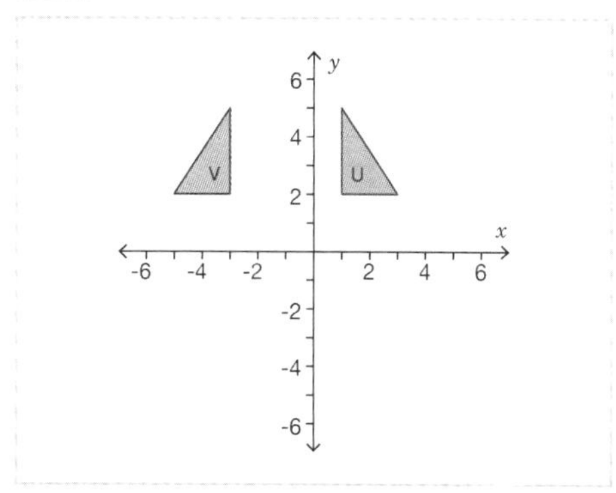

Level 2: Fluency: Fully describe transformations.

Required: 7/10 **Pupil Navigation:** on
Randomised: off

11. Describe fully the single transformation that maps shape E onto shape F.

- Open question, no set answer

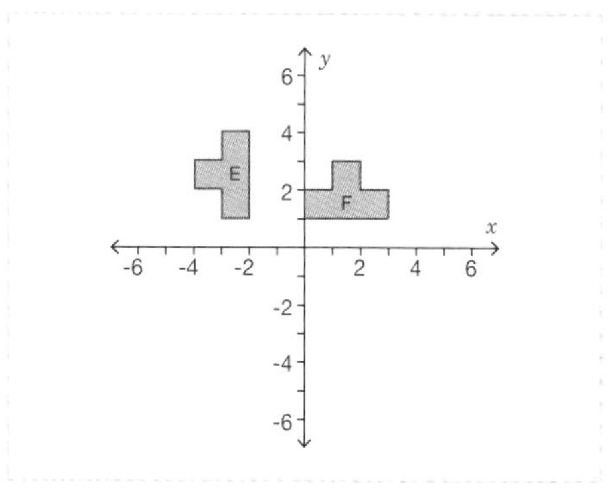

12. Each diagram shows a different transformation that maps shape P onto shape Q. Sort the descriptions into the order that the transformations are shown, from (i) to (iv).

- a reflection in the line y = -1
- a translation by a column vector with -6 as the top number and -4 as the bottom number
- a rotation of 90° anti-clockwise about the point (0, -1)
- a reflection in the line x = 1

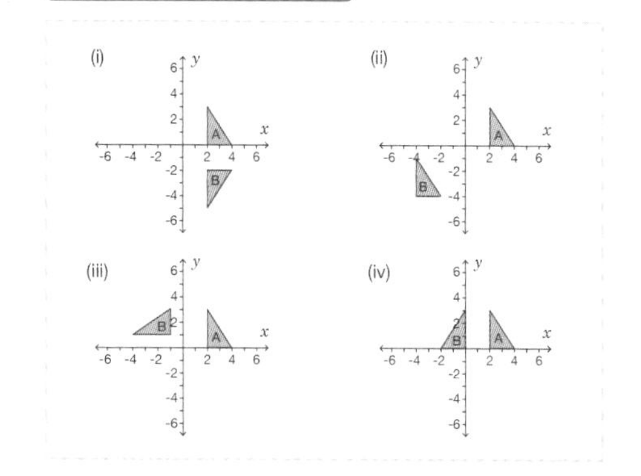

Level 2 *continued*

13. Describe fully the single transformation that maps shape U onto shape V.

a b c

- Open question, no set answer

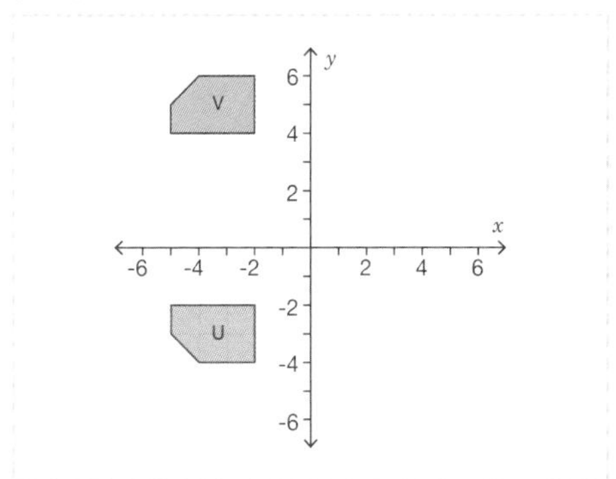

14. Each diagram shows a different transformation that maps shape C onto shape D. Sort the descriptions into the order that the transformations are shown, from (i) to (iv).

- a reflection in the line x = -1
- a reflection in the line y = x
- a rotation 180° about the point (0, 0)
- a reflection in the line y = -1

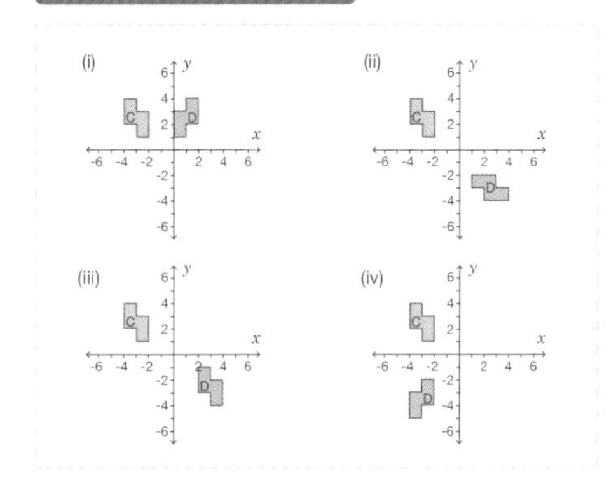

15. Select the **two transformations** which would map shape M onto shape N.

2/4

- a reflection in the line y = x
- a rotation of 90° anti-clockwise about the point (0, 0)
- a reflection in the line y = -x
- a rotation of 90° anti-clockwise about the point (-1, -1)

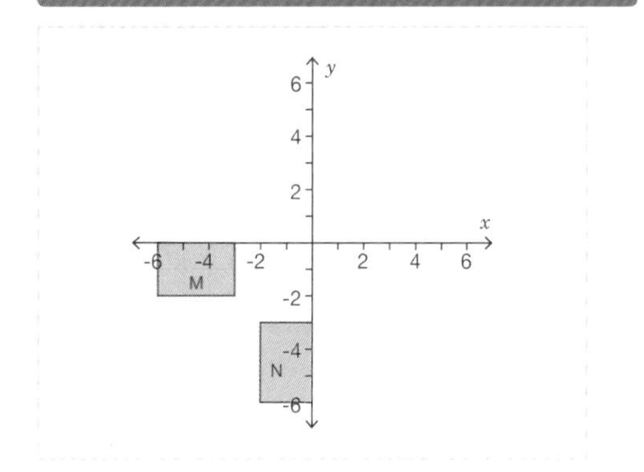

16. Describe fully the single transformation that maps shape J onto shape K.

a b c

- Open question, no set answer

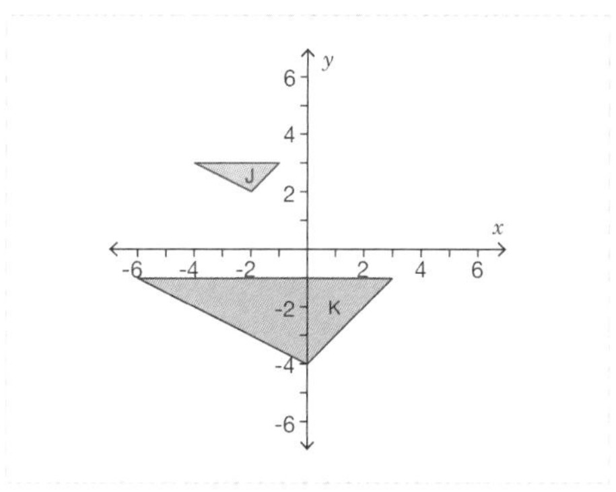

17. Each diagram shows a different transformation that maps shape P onto shape Q. Sort the descriptions into the order that the transformations are shown, from (i) to (iv).

- an enlargement with a centre of enlargement of (-6, -5) and a scale factor of 2
- a reflection in the line x = -1
- a translation by a column vector with 8 as the top number and 0 as the bottom number
- a rotation of 180° about the point (-1, -1)

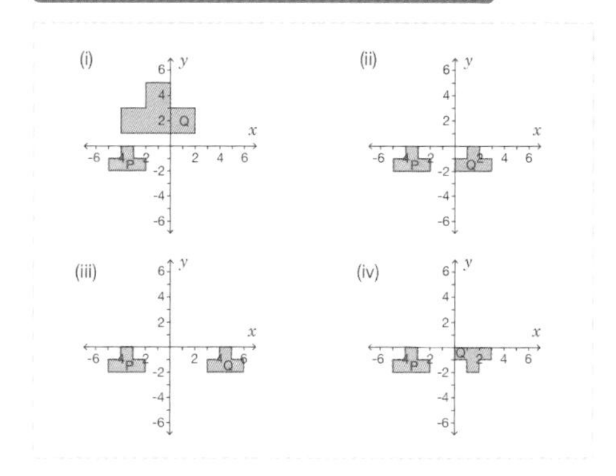

18. Select the transformation that would map shape S onto shape T.

1/4

- a translation by a column vector with -5 as the top number and 0 as the bottom number
- a reflection through the line x = -1
- a translation by a column vector with 0 as the top number and -2 as the bottom number
- a rotation of 180° about the point (-2, -1)

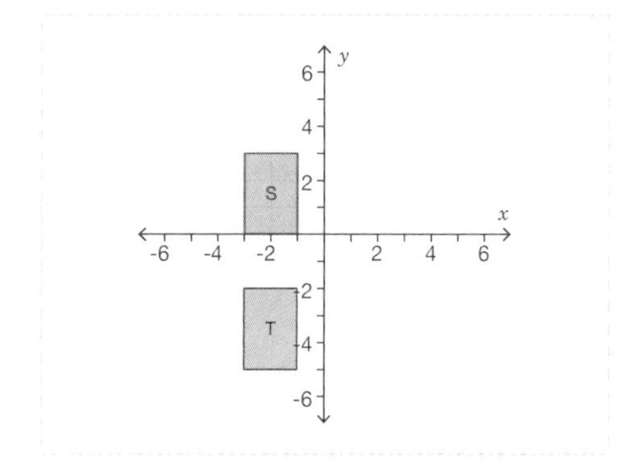

Level 2 *continued*

19. Describe fully the single transformation that maps shape G onto shape H.

a b c

- Open question, no set answer

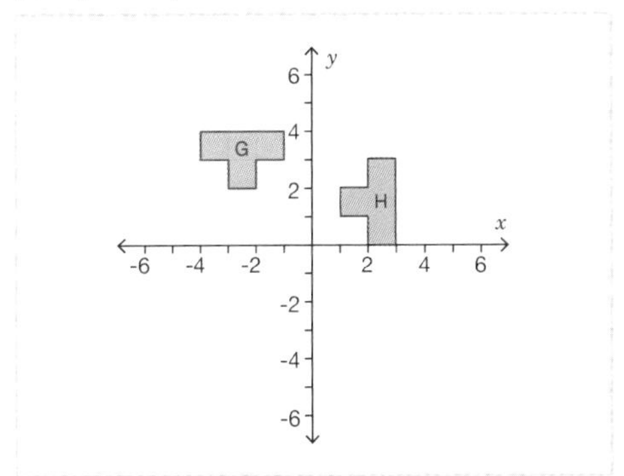

20. Each diagram shows a different transformation that maps shape T onto shape U. Sort the descriptions into the order that the transformations are shown, from (i) to (iv).

↑ ↓

- a translation by a column vector with -5 as the top number and 6 as the bottom number
- a reflection in the line y = x
- a rotation of 180° about the point (0, 0)
- a reflection in the line y = 1

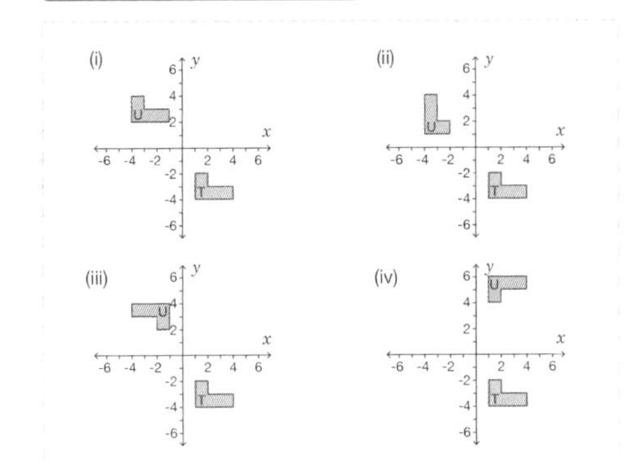

Level 3: Reasoning: Know what information is needed to fully describe each transformation.

Required: 5/5 **Pupil Navigation:** on
Randomised: off

21. What pieces of information do you need to give to fully describe the transformation that maps shape E onto shape F?

3/6

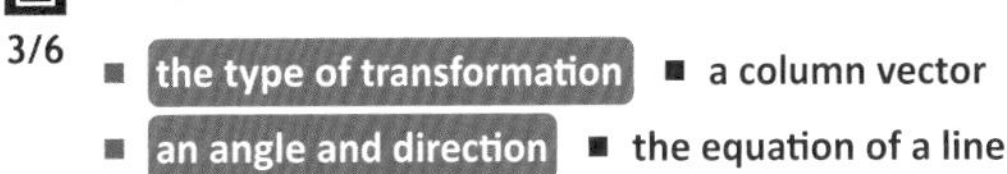

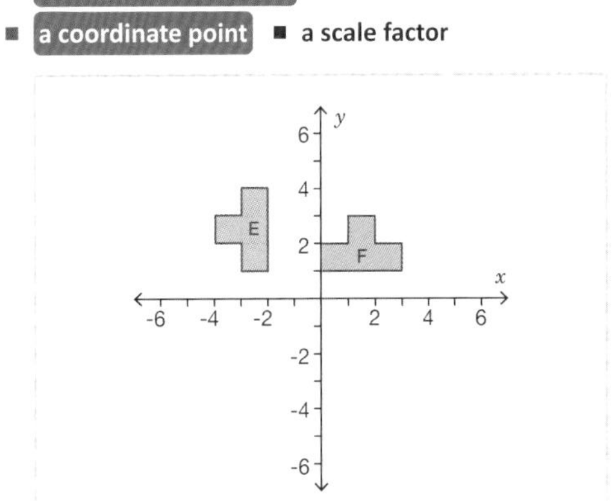

22. Harvey has made one mistake in his homework on transformations. What is the correct answer to the question he got wrong?

1 2 3

- 270

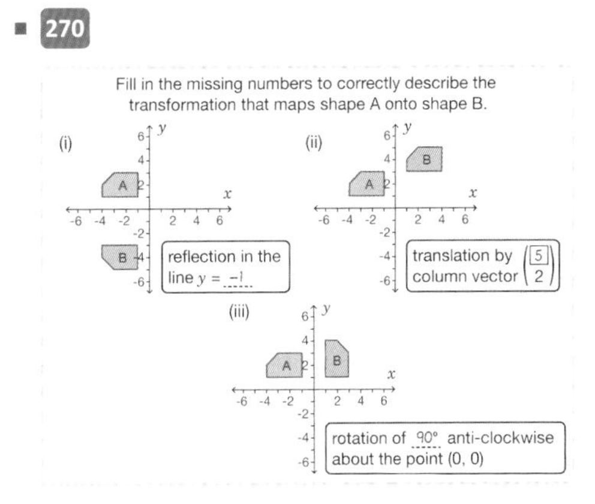

23. Select the type of transformation that couldn't map shape S onto shape T.

1/4

- translation
- rotation
- reflection
- enlargement

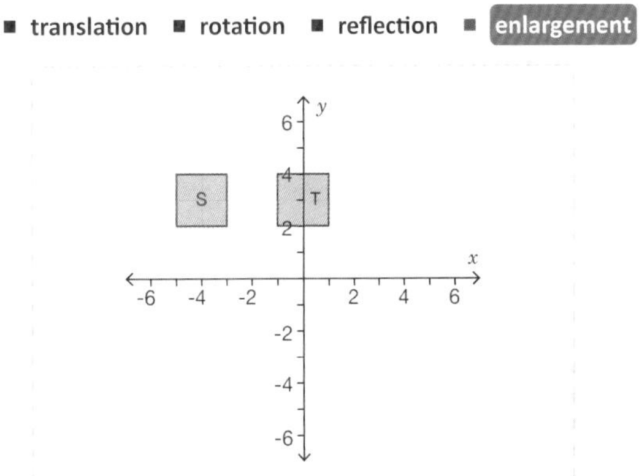

Level 3 *continued*

24. What pieces of information do you need to give to fully describe the transformation that maps shape C onto shape D?

2/6

- **the type of transformation**
- **a column vector**
- an angle and direction
- the equation of a line
- a coordinate point
- a scale factor

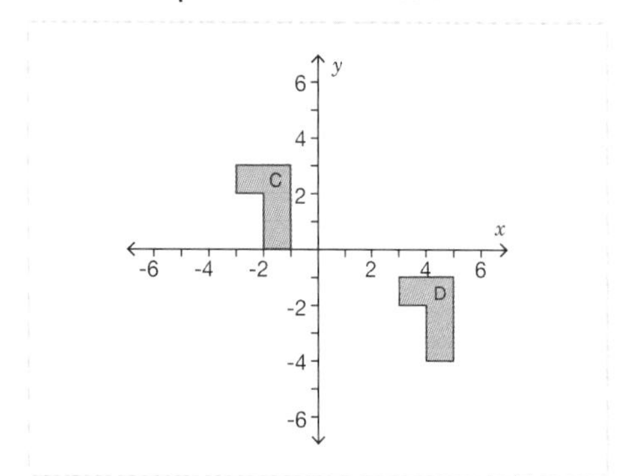

25. Bailey says the answer to the question in the image is 90° clockwise about the point (0, 1). Is Bailey correct? Explain your answer.

- Open question, no set answer

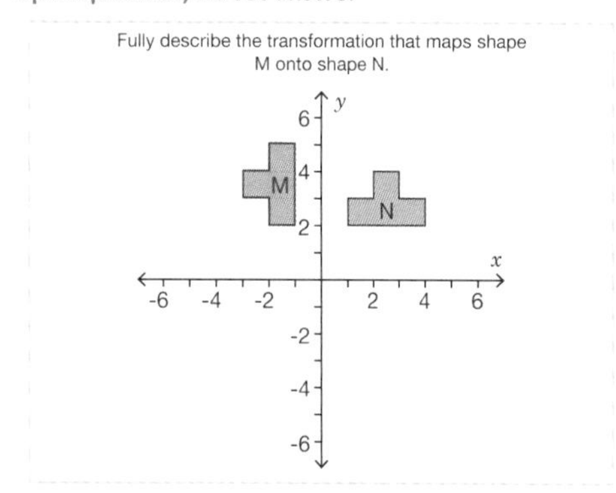

Level 4: Problem solving: Describe multi-step transformations with a single transformation.

Required: 5/5 **Pupil Navigation:** on
Randomised: off

26. Shape A is reflected through the line $x = 2$ onto shape B. Shape B is then reflected through the line $y = -1$ onto shape C.
Shape A can be mapped onto shape C by a rotation of 180°. What are the coordinates of the centre of rotation?
Give your answer in the form (x, y).

- **(2, -1)**

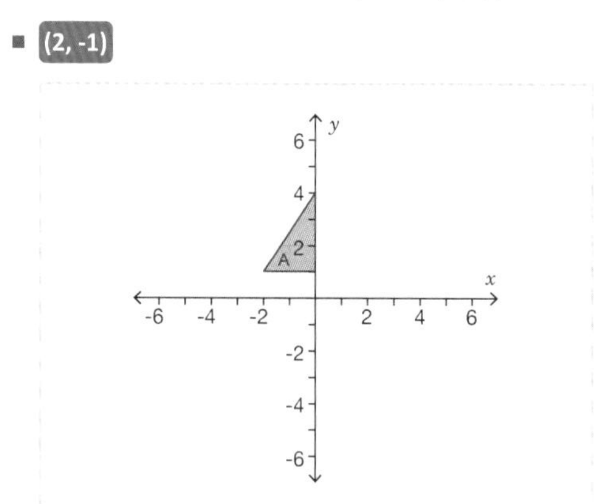

27. Point U is translated onto point V using a column vector with 0 as the top number and 6 as the bottom number. Point U can also be reflected in the line $y = -2$ onto point V. Point V has an x-coordinate of 4. What are the coordinates of point U?
Give your answer in the form (x, y).

- **(4, -5)**
- (4, 1)

A translation of $\begin{pmatrix} 0 \\ 6 \end{pmatrix}$ maps U onto V.

A reflection in the line $y = -2$ maps U onto V.

$V = (4, \square)$

$U = ?$

Level 4 *continued*

28. Rectangle P can be mapped onto rectangle Q by a rotation of 90° clockwise about the point (-1, 0). Rectangle P can also be mapped onto rectangle Q by a rotation of 90° anti-clockwise. What are the coordinates of the centre of this anti-clockwise rotation?
Give your answer in the form (x, y).

- (-5, -2)

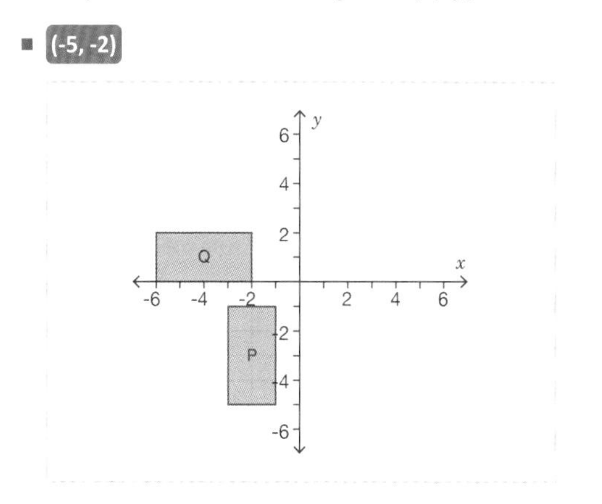

29. Triangle C is reflected onto triangle D in the line $y = 1$. Triangle C can also be mapped onto triangle D by a rotation. What are the coordinates of the centre of rotation?
Give your answer in the form (x, y).

- (-2, 1)

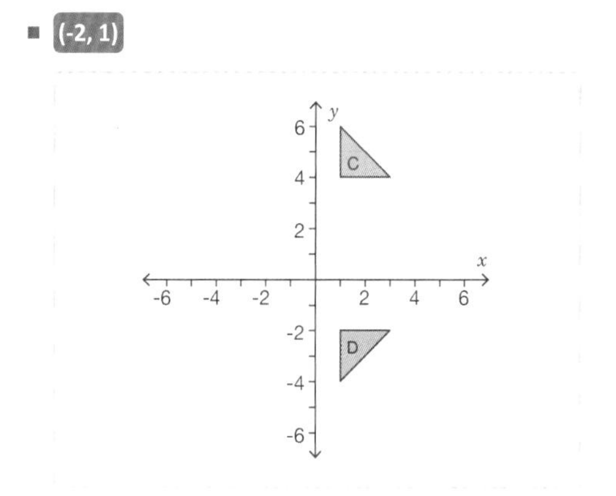

30. Shape R is translated onto shape S by a column vector with 0 as the top number and 8 as the bottom number. Shape S is then reflected in the line $y = -1$ onto shape T. Give the equation of the line of reflection that maps shape R onto shape T.

- y = -5

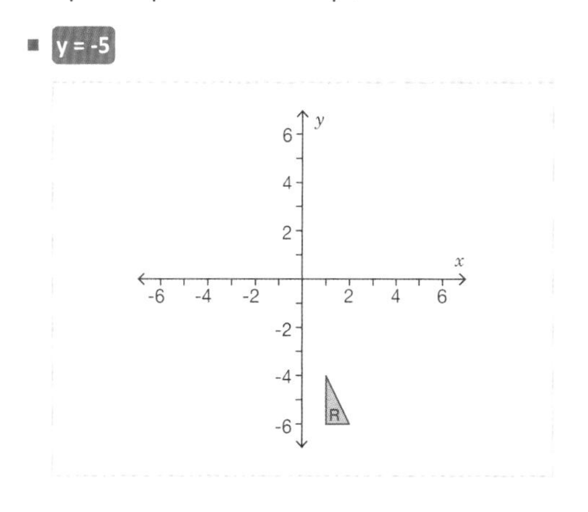

Describing Enlargements

Objective: I can fully describe enlargements.

Quick Search Ref: 10612

Level 1: Understanding: Describe enlargements supported by ray lines.

Required: 6/8 **Pupil Navigation:** on **Randomised:** off

1. What is the scale factor of enlargement from rectangle C to rectangle D?

- 8

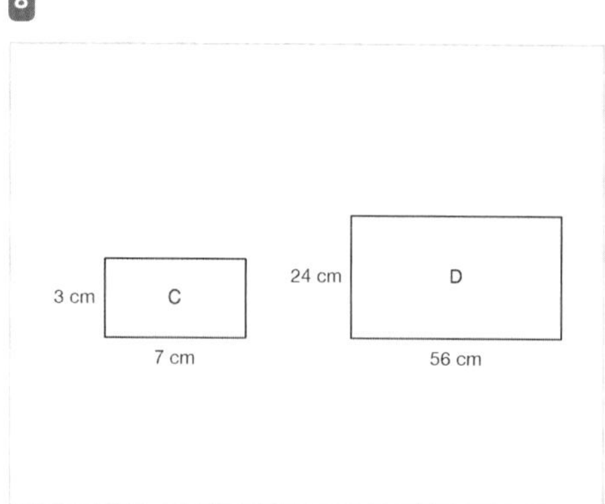

2. What is the scale factor of enlargement that maps triangle R onto triangle S?

- 4

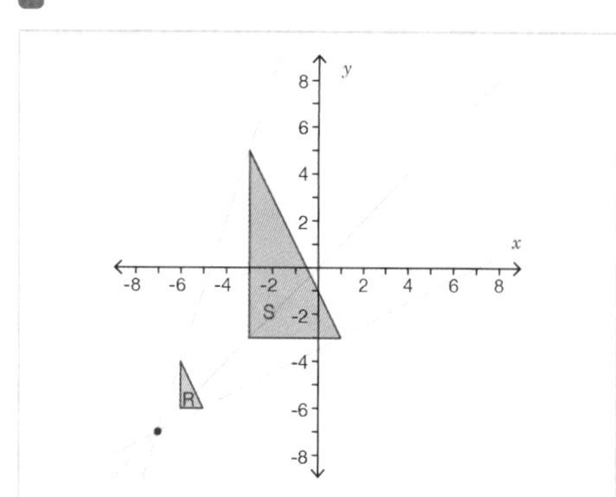

3. Select the coordinates of the centre of enlargement that maps triangle M onto triangle N.

1/5

- (-6, 1)
- (-3, 1)
- (-4, -7)
- (-7, 5)
- (2, -1)

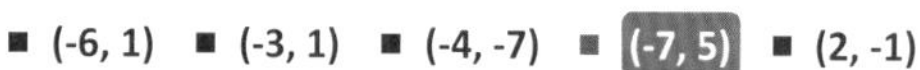

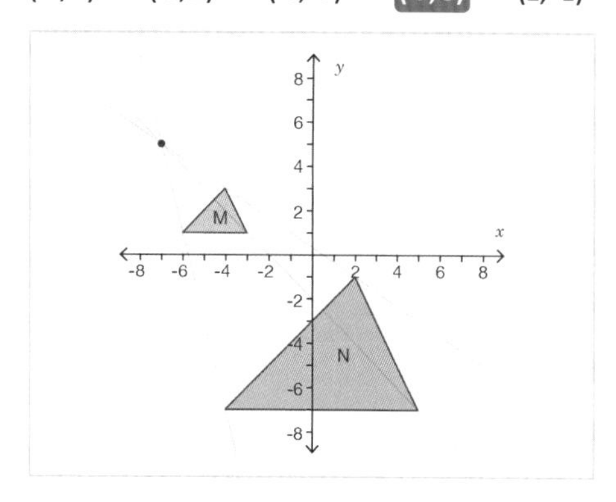

4. Select the **two statements** that together fully describe the enlargement that maps triangle V onto triangle W.

2/5

- scale factor 4
- centre of enlargement (6, -7)
- scale factor 2
- centre of enlargement (8, -8)
- scale factor 3

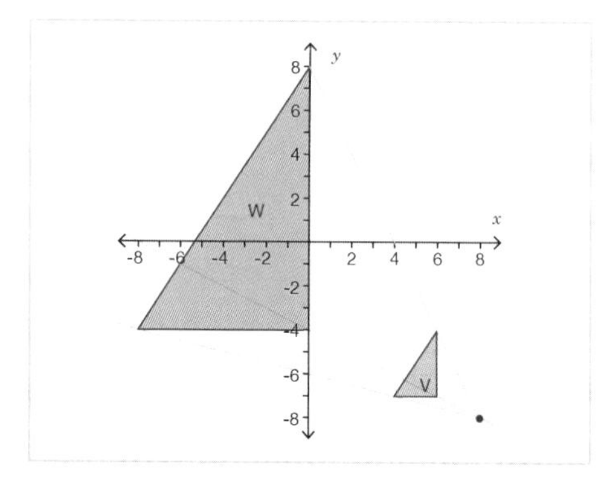

5. Give the coordinates of the centre of enlargement that maps triangle A onto triangle B.
Give your answer in the form (x, y).

- (5, 3)
- (6, 5)
- (2, 3)
- (5, 6)
- (3, 1)
- (5 6)
- 5, 6

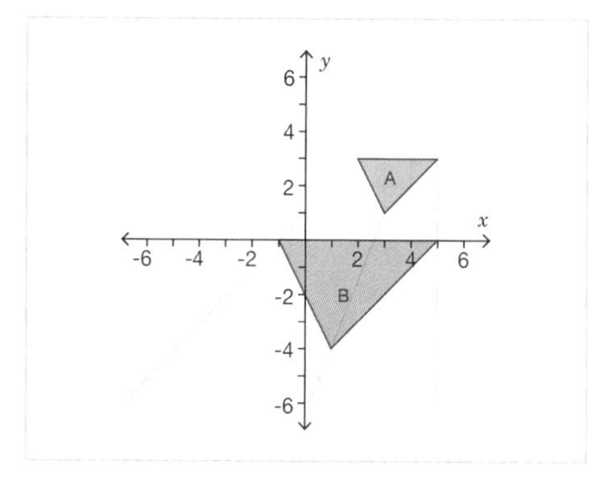

Level 1 *continued*

6. Give the coordinates of the centre of enlargement that maps rectangle J onto rectangle K.
Give your answer in the form (x, y).

- (-5, -4)
- (-2, -2)
- (-6, -6)
- (-2, -4)
- (-6 -6)
- (-5, -2)
- -6, -6

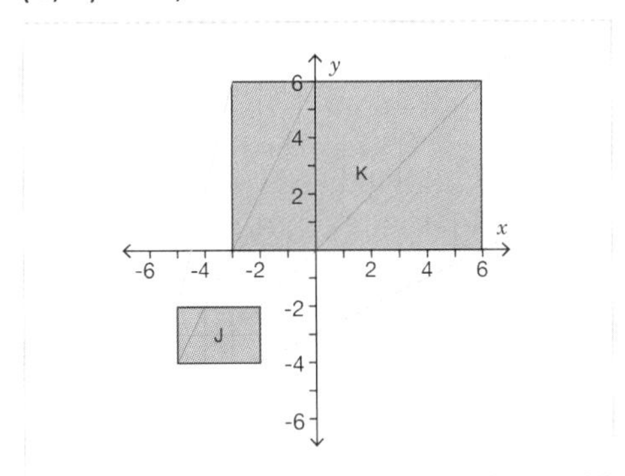

7. Select the **two statements** that together fully describe the enlargement that maps shape E onto shape F.

2/5

- scale factor 3
- centre of enlargement (3, 5)
- centre of enlargement (2, 2)
- scale factor 2
- centre of enlargement (4, 2)

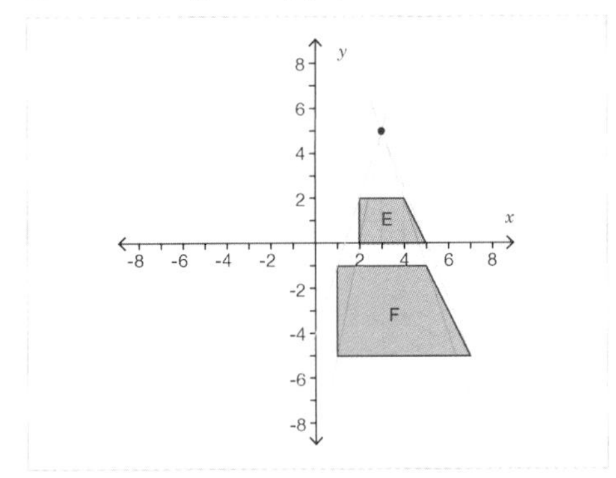

8. Give the coordinates of the centre of enlargement that maps triangle T onto triangle U.
Give your answer in the form (x, y).

- (-3, 1)
- (-2, 3)
- (-5, 4)
- -5, 4
- (-5 4)
- (4, -5)
- (-4, 3)

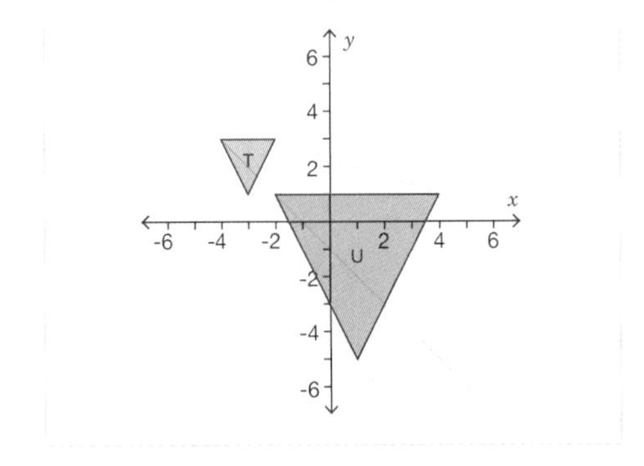

Level 2: Fluency: Describe enlargements without ray lines.

Required: 6/8 **Pupil Navigation:** on
Randomised: off

9. Select the **two statements** that together fully describe the enlargement that maps shape P onto shape Q.

2/6

- scale factor 3
- centre of enlargement (-4, -4)
- scale factor 2
- centre of enlargement (-4, -5)
- scale factor 4
- centre of enlargement (-6, -5)

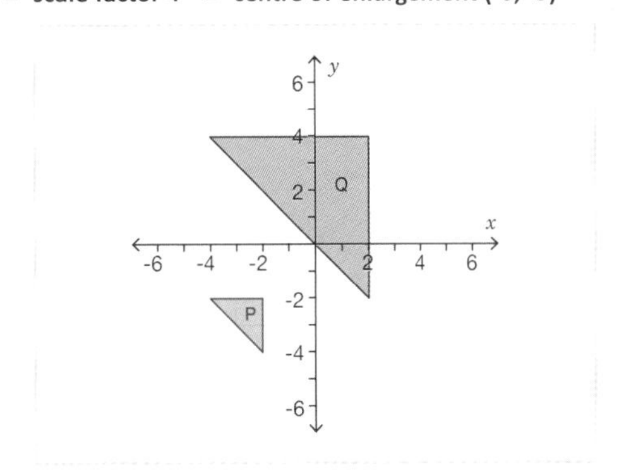

10. Give the coordinates of the centre of enlargement that maps shape R onto shape S.
Give your answer in the form (x, y).

- (6, -2)

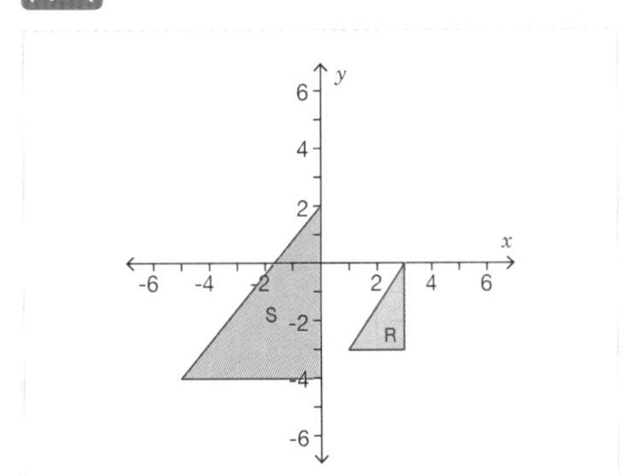

11. Give the coordinates of the centre of enlargement that maps shape G onto shape H.
Give your answer in the form (x, y).

- (-3, -6)

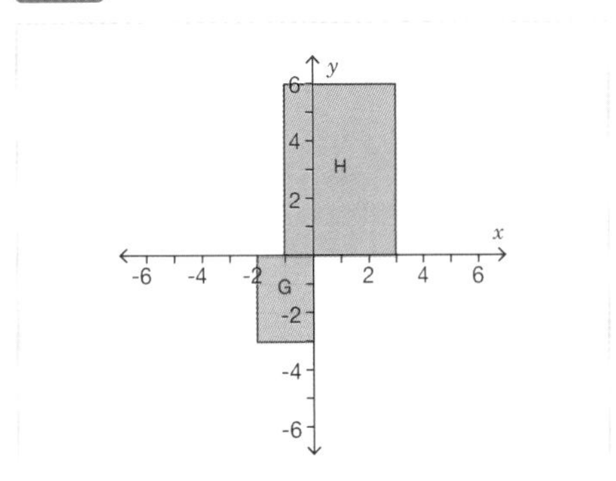

Level 2 *continued*

12. Select the **two statements** that together fully describe the enlargement that maps shape C onto shape D.

2/6

- centre of enlargement (2, -1)
- scale factor 5
- centre of enlargement (3, 0)
- scale factor 3
- scale factor 4
- centre of enlargement (3, -1)

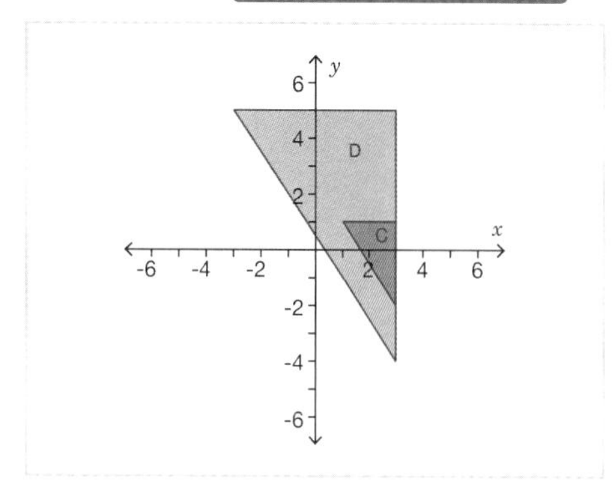

13. Give the coordinates of the centre of enlargement that maps shape M onto shape N.
Give your answer in the form (x, y).

- (5, -6)

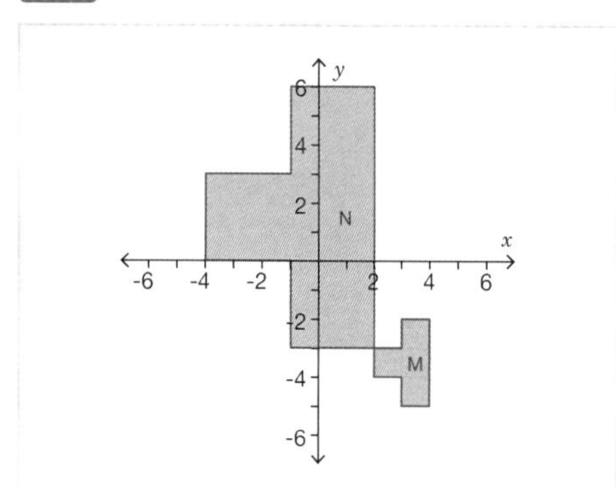

14. Select the **two statements** that together fully describe the enlargement that maps shape A onto shape B.

2/6

- scale factor 3
- scale factor 2
- scale factor 4
- centre of enlargement (-6, 3)
- centre of enlargement (-5, 3)
- centre of enlargement (-7, 4)

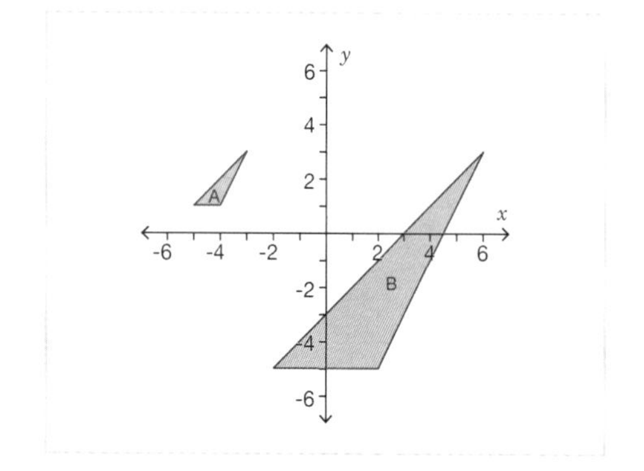

15. Give the coordinates of the centre of enlargement that maps shape V onto shape W.
Give your answer in the form (x, y).

- (-2, 3)

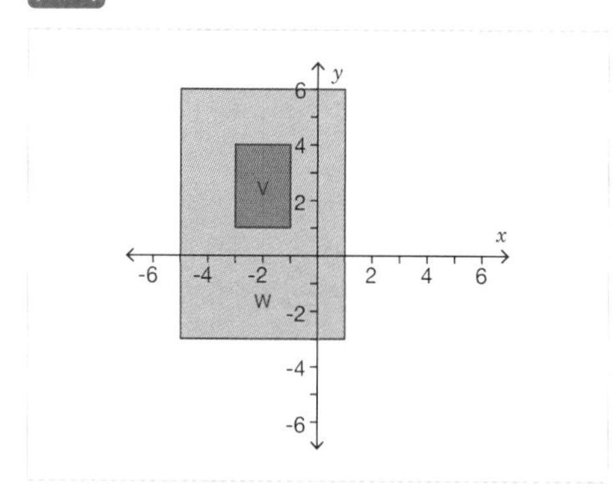

16. Give the coordinates of the centre of enlargement that maps shape J onto shape K.
Give your answer in the form (x, y).

- (-4, -5)

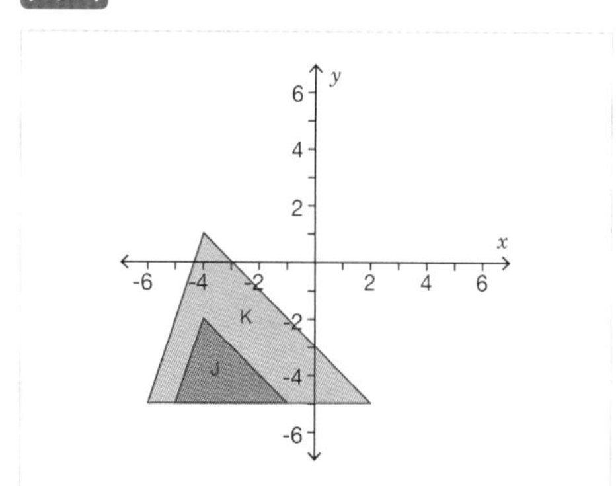

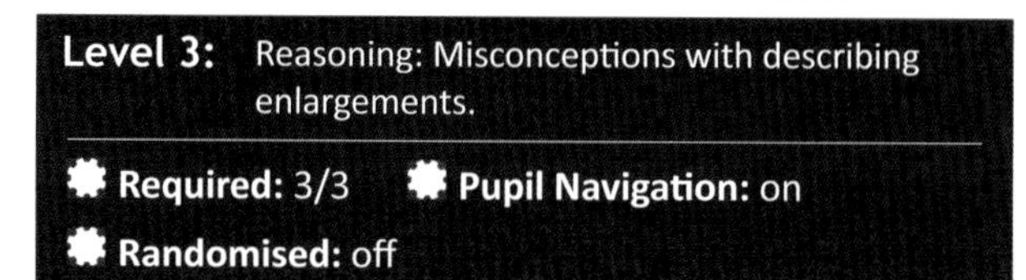

17. Tyler is calculating the coordinates of the centre of enlargement. Explain the mistake Tyler has made.

- Open question, no set answer

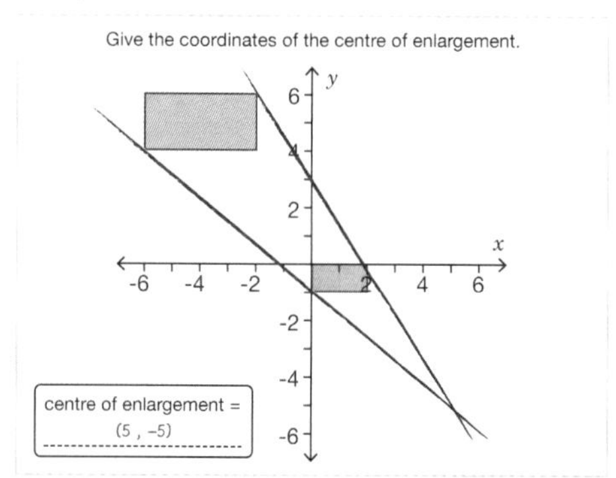

Level 3 *continued*

18. What is the scale factor of enlargement that maps shape D onto shape E?
Give your answer as a decimal.

- 1/2
- 0.5
- 2

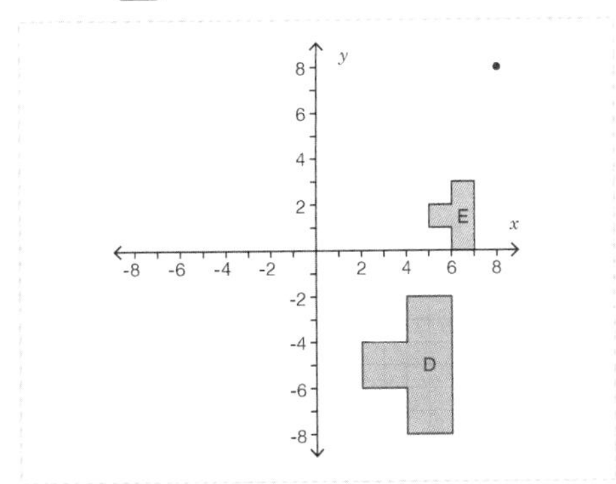

19. Scarlett has made a mistake on an enlargement question. What should the coordinates of the centre of enlargement be?
Give your answer in the form (x, y).

- (1, -6)

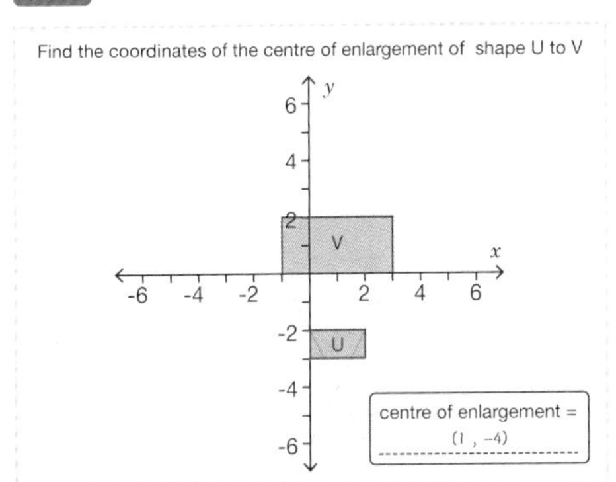

Level 4: Problem solving: Solve multi-step problems involving describing enlargements.

Required: 3/3 **Pupil Navigation:** on
Randomised: off

20. Give the coordinates of the centre of enlargement that maps shape M onto shape N.
Give your answer in the form (x, y).

- (1/2, 6)
- (0.5, 6)

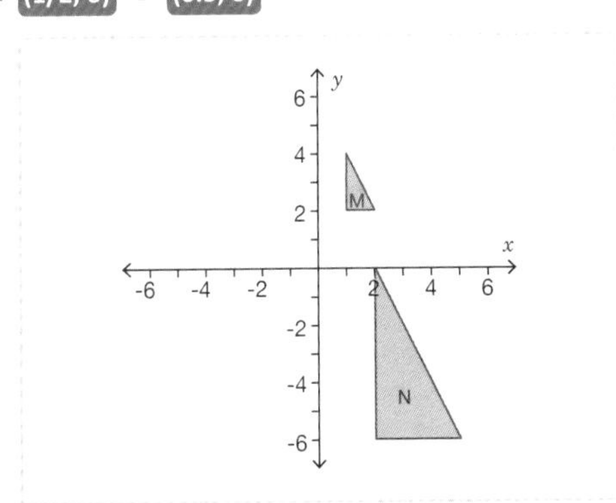

21. Katie is completing an enlargement task but has lost part of the question. She knows that triangle *ABC* is enlarged by a scale factor of 3. What are the coordinates of the centre of enlargement?
Give your answer in the form (x, y).

- (4, 9)

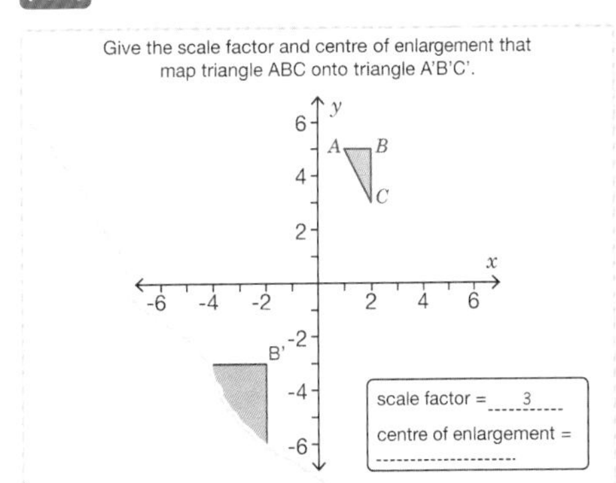

22. When triangle A is enlarged to map onto triangle B, the equations of two of the ray lines are $y = 2x + 1$ and $y = x - 1$. What are the coordinates of the centre of enlargement?
Give your answer in the form (x, y).

- (-2, -3)

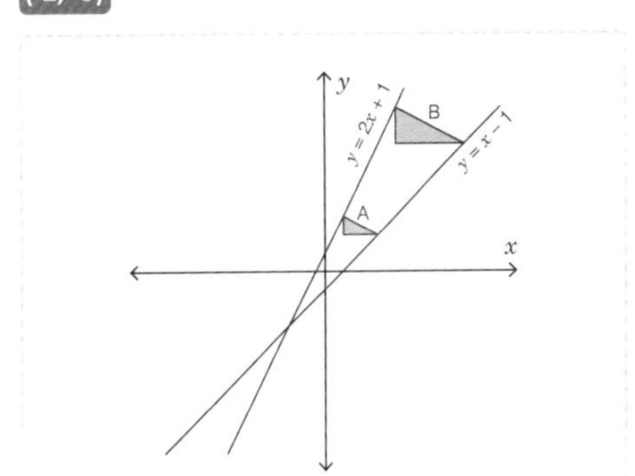

Describing Reflections

Objective: I can fully describe reflections.

Quick Search Ref: 10610

Level 1: Understanding: Give the equations of lines of reflections shown.

Required: 7/10 **Pupil Navigation:** on **Randomised:** off

1. What is the equation of the line shown in the diagram?

1/6

- y = -3
- **x = 3**
- y = 3x
- x = -3
- y = 3
- x = 3y

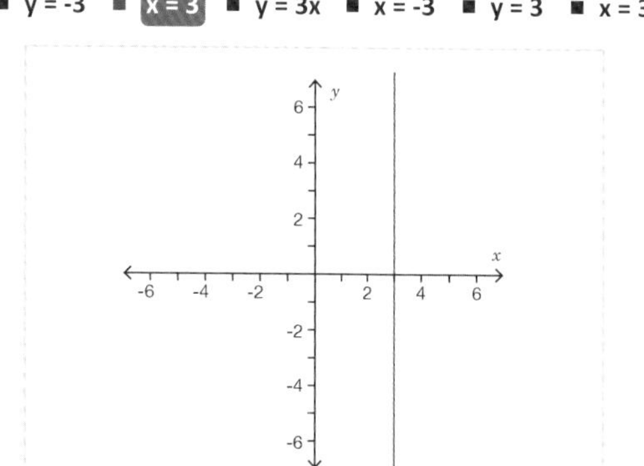

2. Which image shows the graph of the line $y = -2$?

1/4

- **(i)**
- (ii)
- (iii)
- (iv)

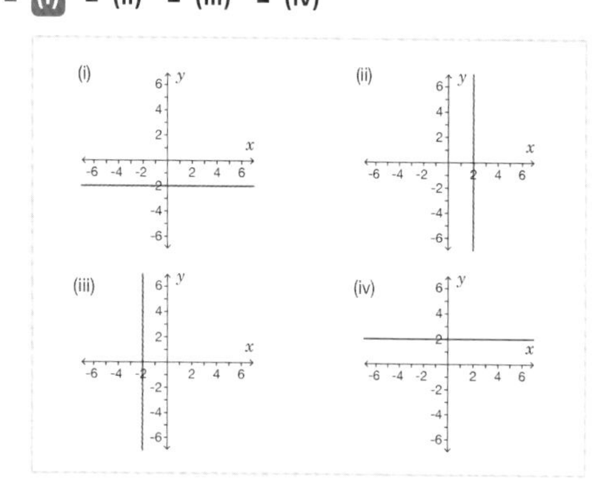

3. Shape R is reflected onto shape S. What is the equation of the line of reflection?

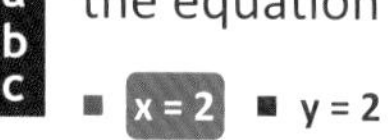

- **x = 2**
- y = 2

4. Shape U is reflected onto shape V. What is the equation of the line of reflection?

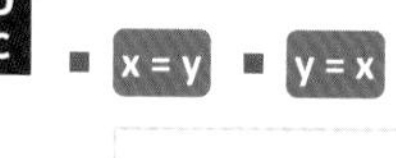

- **y = 5**
- x = 5

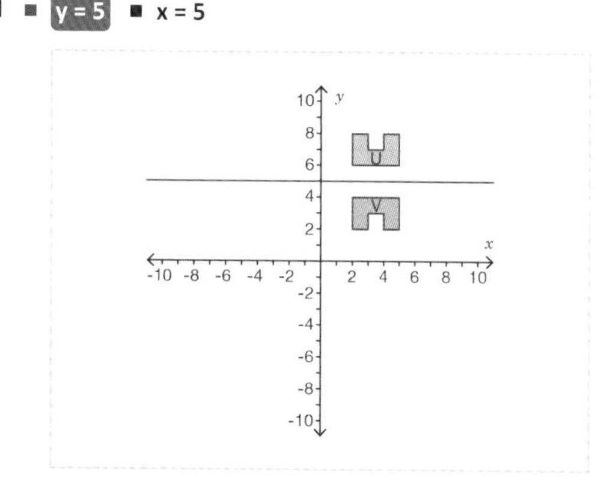

5. Shape E is reflected onto shape F. What is the equation of the line of reflection?

- **x = y**
- **y = x**

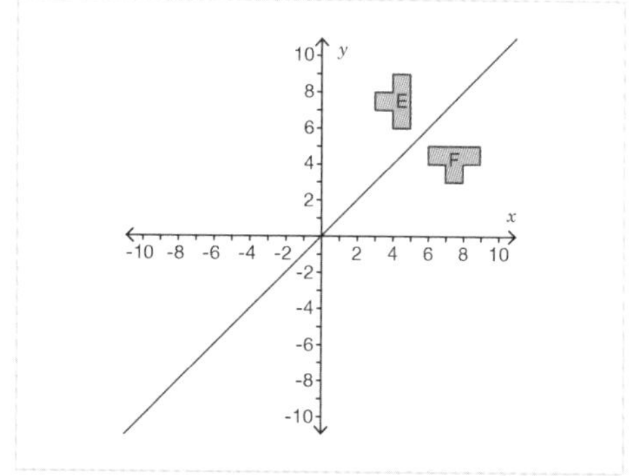

6. Select the diagram that shows shape M reflected in the line $x = -1$ onto shape N.

1/4

- (i)
- (ii)
- **(iii)**
- (iv)

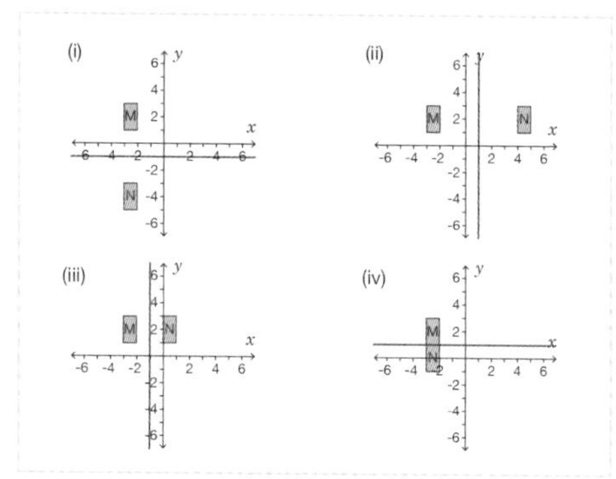

Level 1 *continued*

7. Shape C is reflected onto shape D. What is the equation of the line of reflection?
Give your answer in the form y = . . .

- y = -x

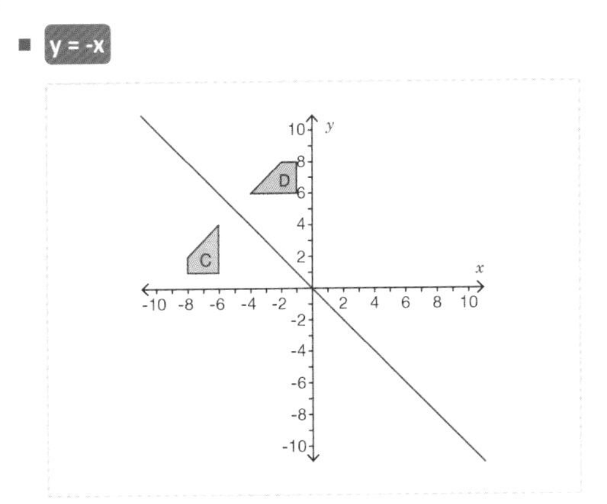

8. Select the diagram that shows shape J reflected onto shape K in the line $y = -1$.

1/4

- (i)
- (ii)
- (iii)
- (iv)

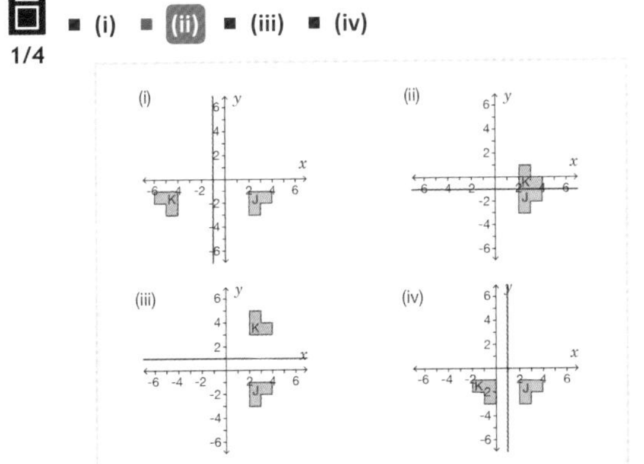

9. Shape A is reflected onto shape B. What is the equation of the line of reflection?

- y = -2
- x = -2
- x = 2

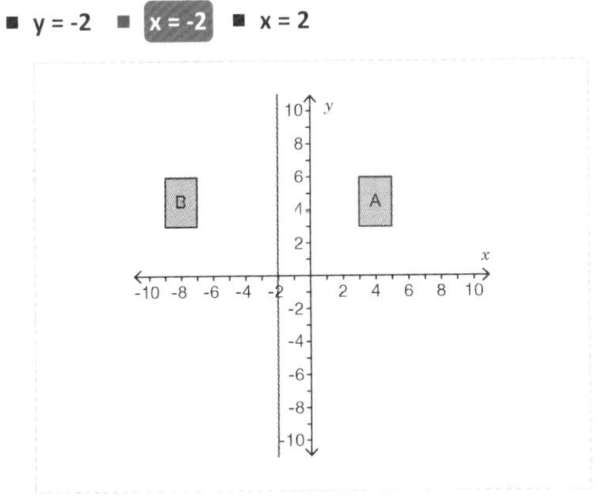

10. Shape P is reflected onto shape Q. What is the equation of the line of reflection?

- y = x
- x = y

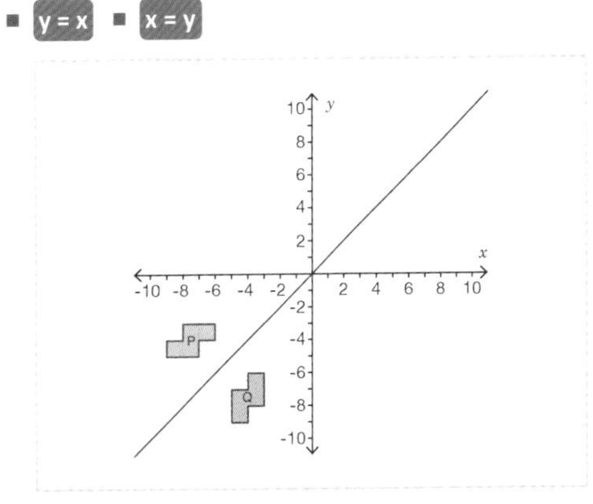

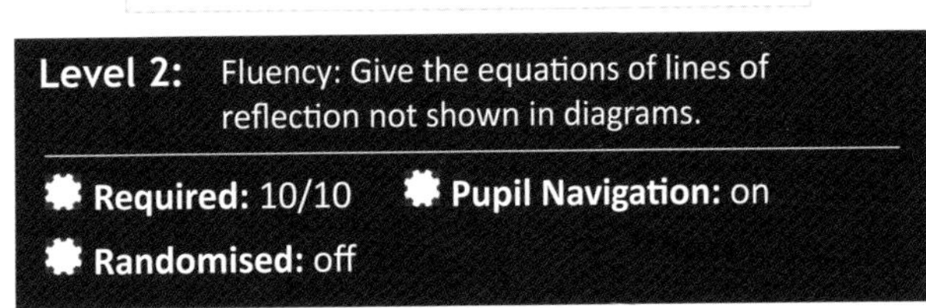

Level 2: Fluency: Give the equations of lines of reflection not shown in diagrams.

Required: 10/10 **Pupil Navigation:** on
Randomised: off

11. Select the **two descriptions** of the line of reflection used to reflect shape T onto shape U.

2/6

- x-axis
- y = 0
- x = y
- y-axis
- x = 0
- y = x

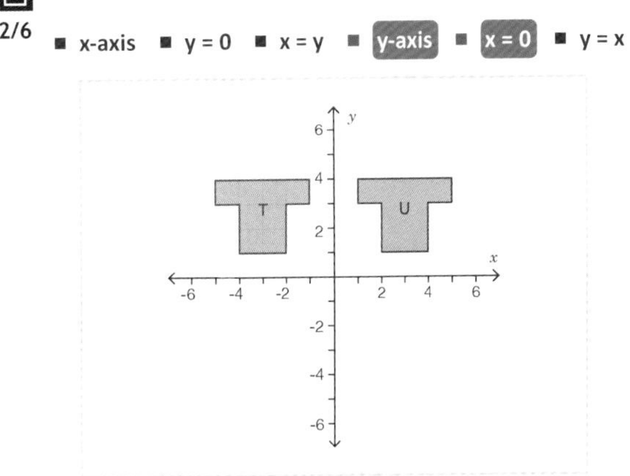

12. Shape G is reflected onto shape H. What is the equation of the line of reflection?

- x = 2
- y = 2

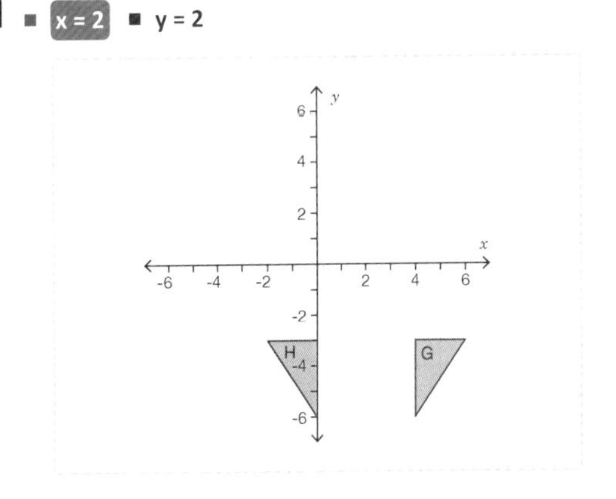

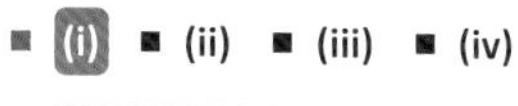

Level 2 *continued*

13. Which diagram shows the reflection of shape E onto shape F in the line $y = 1$?

1/4

- (i)
- (ii)
- (iii)
- (iv)

14. Shape R is reflected onto shape S. What is the equation of the line of reflection?

- x = y
- y = -x
- y = x
- x = -y
- -y = x
- -x = y

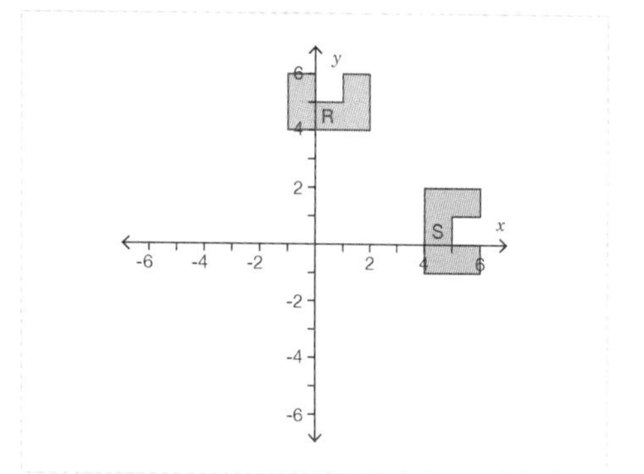

15. Which diagram shows the reflection of shape C onto shape D in the line $x = -1$?

1/4

- (i)
- (ii)
- (iii)
- (iv)

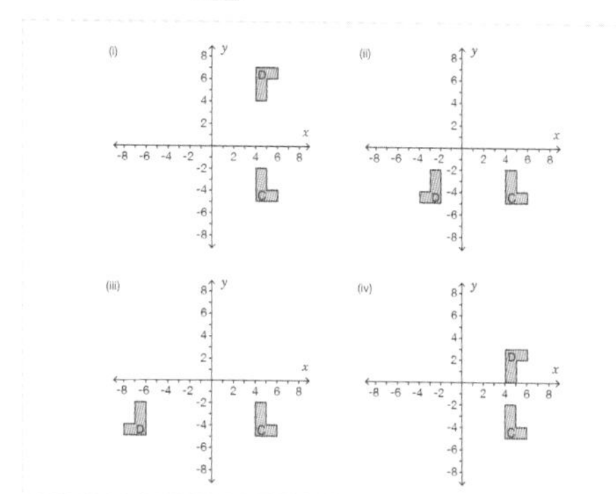

16. Shape U is reflected onto shape V. What is the equation of the line of reflection?

- y = 1
- x = 1

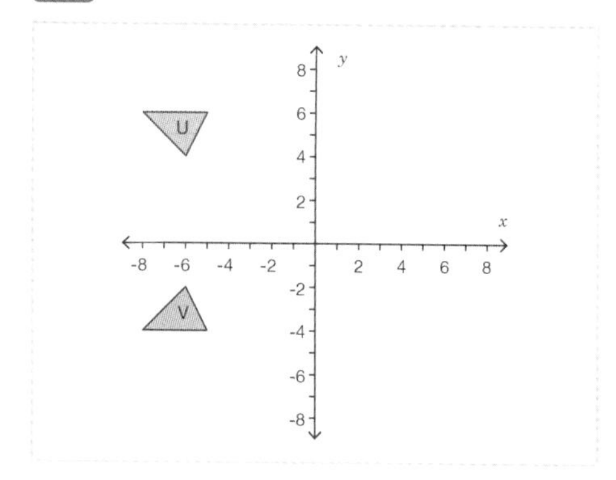

17. Shape A is reflected onto shape B. What is the equation of the line of reflection? *Give your answer in the form y = . . .*

- y = -x
- y = x
- x = y

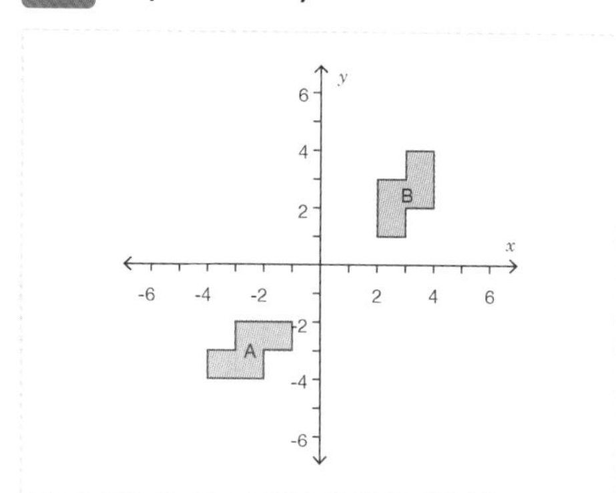

18. Shape M is reflected onto shape N. What is the equation of the line of reflection?

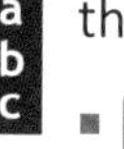

- x = -1
- y = -1

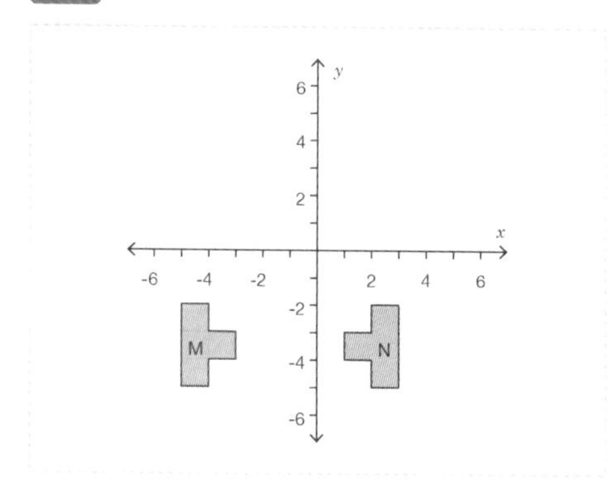

Level 2 *continued*

19. Shape P is reflected onto shape Q. What is the equation of the line of reflection?

- $y = -2$
- $x = -2$

20. Which diagram shows the reflection of shape J onto shape K in the line $y = x$?

1/4

- (i)
- (ii)
- (iii)
- (iv)

Level 3: Reasoning: Make inferences and avoid misconceptions when describing reflections.

Required: 5/5 **Pupil Navigation:** on

Randomised: off

21. Shape C is reflected onto shape D. What is the equation of the line of reflection?

- $y = 0.5$
- $y = 1/2$

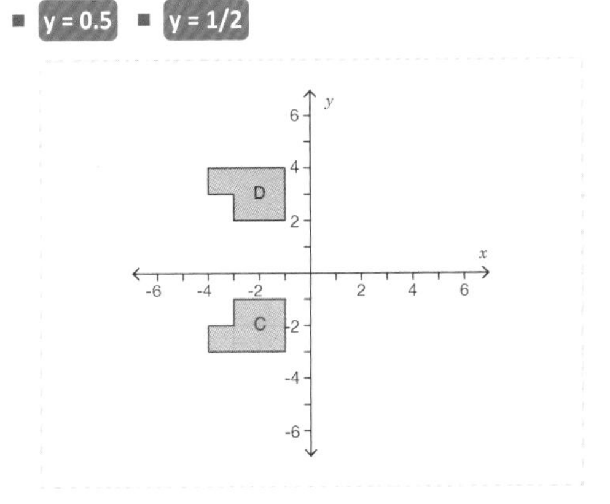

22. The rectangle is reflected in the lines $x = 1$ and then $x = 3$. Layla says that you can carry out the reflections in the opposite order and still get the same result. Is Layla correct? Explain your answer.

- Open question, no set answer

23. David has made a mistake completing a homework question on describing reflections. What is the correct answer to the question?

- y-axis
- $x = 0$
- y axis

Give the equation of the line which reflects shape U onto shape V

answer : $y = 0$

24. The shape $ABCD$ is reflected in the lines $x = -2$ and then $y = 1$. Katie says that you can carry out the reflections in the opposite order and still get the same result. Is Katie correct? Explain your answer.

- Open question, no set answer

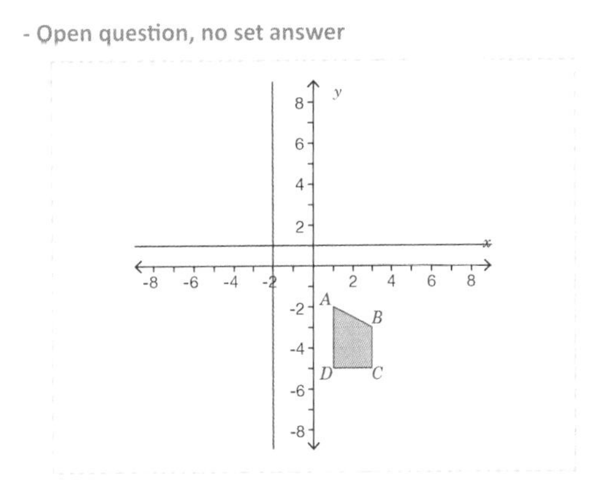

Level 3 *continued*

25. Point G is reflected onto point H in the line AB. Select two properties of the line HG.

2/4

- Line HG intersects line AB at right angles.
- Line HG bisects line AB into two equal parts.
- Line HG has a gradient of 1.
- Line HG intersects line AB at the point (1, 1).

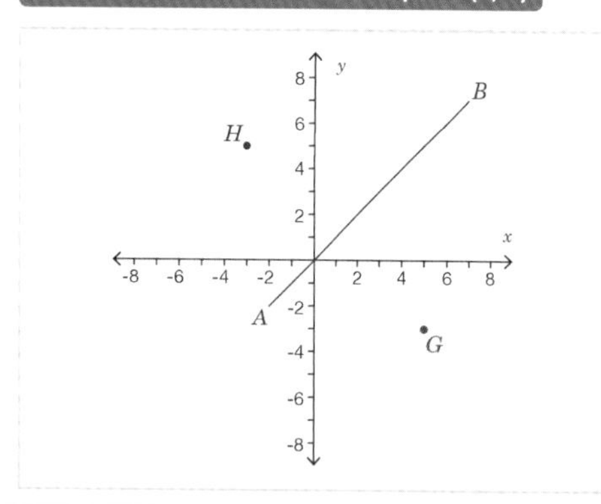

Level 4: Problem solving: Solve multi-step problems on describing reflections.

Required: 5/5 **Pupil Navigation:** on
Randomised: off

26. Shape S is reflected onto shape T. What is the equation of the line of reflection?
Give your answer in the form y = mx + c.

- y = x - 3
- y = 1x - 3

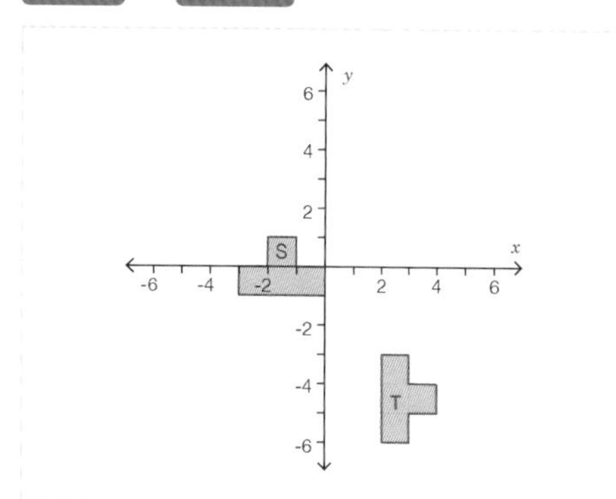

27. Point J = (-2, 3) is reflected onto point K = (6, 3). Give the equation of the line of reflection.

- x = 2
- y = 2

28. Shape E is reflected onto shape F. What is the equation of the line of reflection?
Give your answer in the form y = mx + c.

- y = -x + 1
- y = -1x + 1

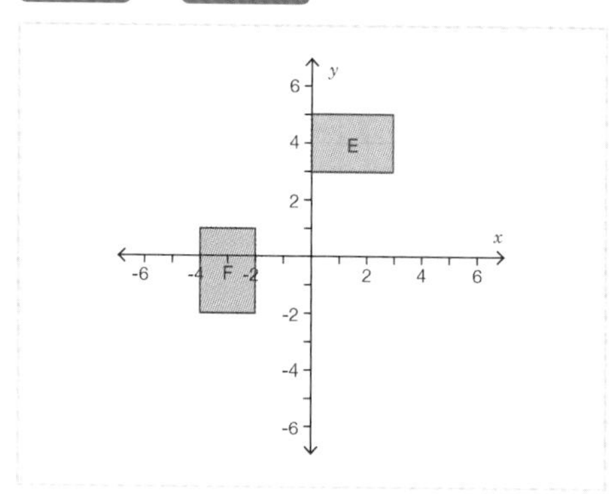

29. Shape A is reflected onto shape B, which is then reflected onto shape C. Give the coordinates of the point where the two lines of reflection intersect.
Give your answer in the form (x, y).

- (1, -1)

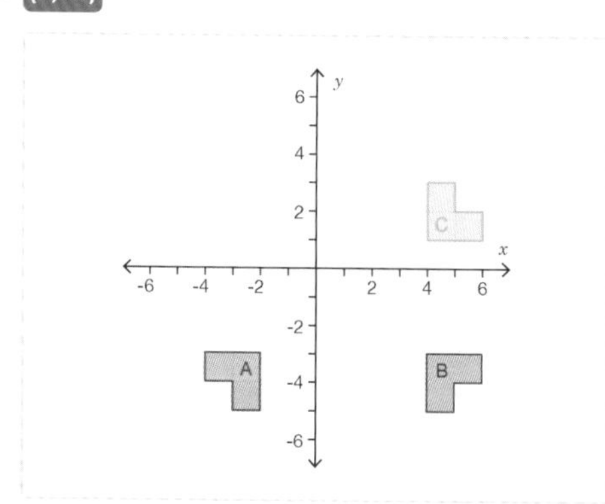

30. Shape V is reflected onto shape W. What is the equation of the line of reflection?
Give your answer in the form y = mx + c.

- y = 2x + 0
- y = 2x

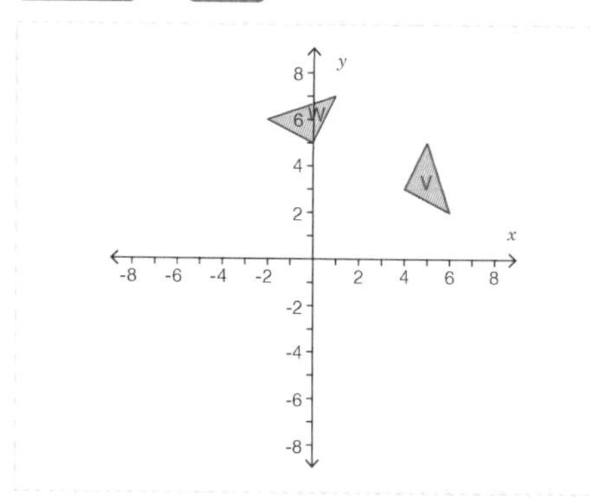

Enlarging Shapes

Objective: I can find scale factors and enlarge shapes.

Quick Search Ref: 10588

Level 1: Understanding: Find scale factors and enlarge shapes using drawn ray lines.

Required: 6/8 **Pupil Navigation:** on **Randomised:** off

1. Shape S is an enlargement of shape R. What is the scale factor of this enlargement?

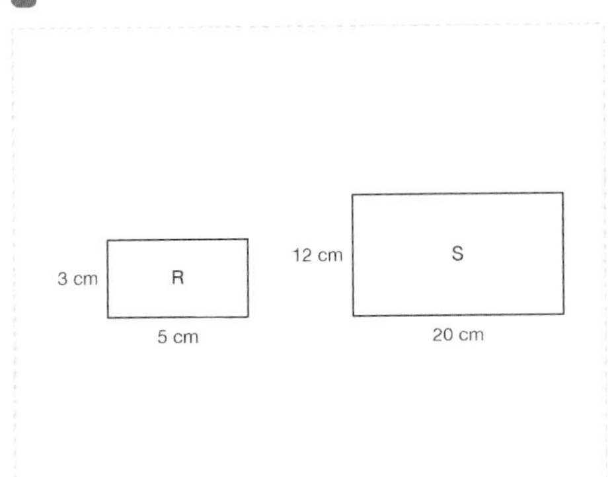

2. Rectangle F is an enlargement of rectangle E. What is the height, x, of rectangle F? *Include the units cm (centimetres) in your answer.*

■ 9 cm ■ 9 centimetres ■ 9

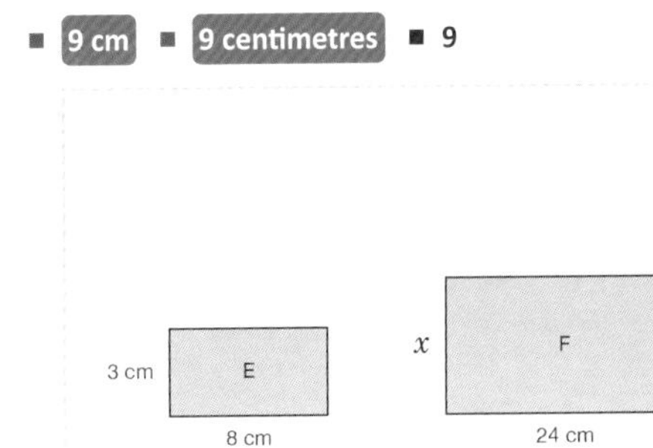

3. The shape is enlarged by a scale factor of 2 using centre of enlargement (0, 0). What are the coordinates of the image of point A?

1/4

■ (6, 2) ■ (3, 6) ■ (2, 4) ■ (6, 4)

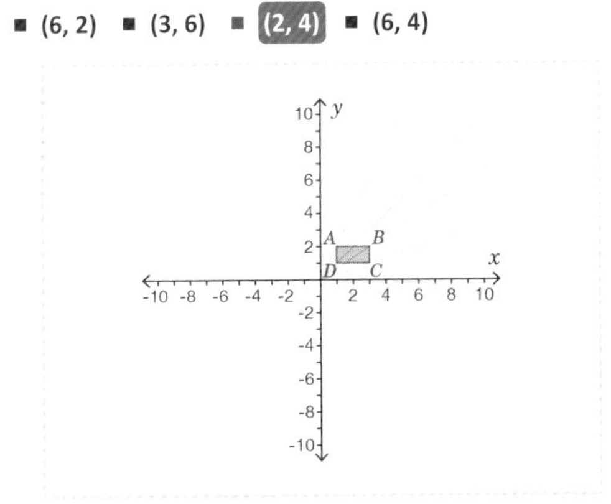

4. The shape is enlarged by a scale factor of 3 using centre of enlargement (0, 0). What are the coordinates of the image of point P? *Give your answer in the form (x, y).*

■ (-9 3) ■ (-9, 3) ■ (-12, 4) ■ -9, 3 ■ (3, -9)

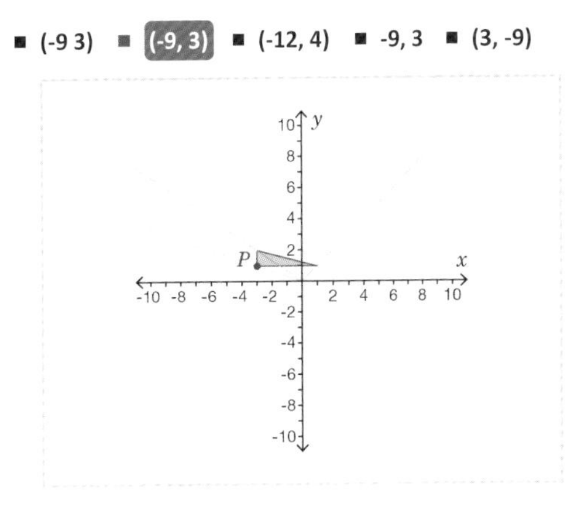

5. The shape is enlarged by a scale factor of 4 using centre of enlargement (8, 6). What are the coordinates of the image of point D? *Give your answer in the form (x, y).*

■ -4, -2 ■ (-4, -2) ■ (-4 -2) ■ (-2, -4) ■ (-7, -4)

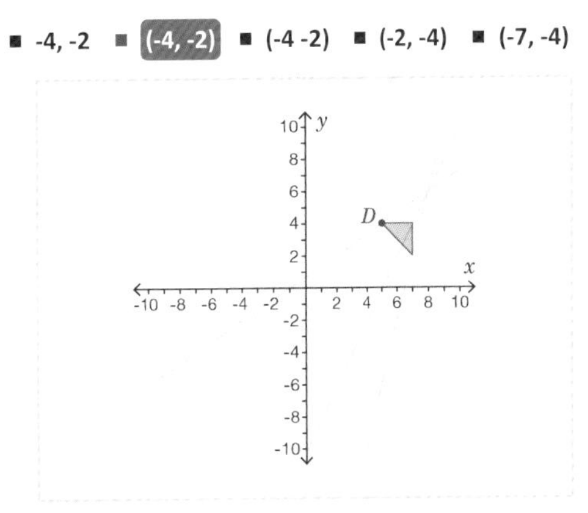

6. The shape is enlarged by a scale factor of 2 using centre of enlargement (-1, 1). What are the coordinates of the image of point T? *Give your answer in the form (x, y).*

■ (3, 5) ■ (5 3) ■ (5, 3) ■ (8, 4) ■ 5, 3

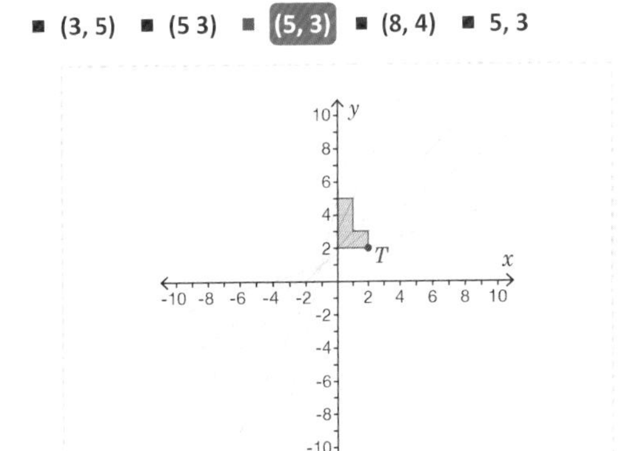

Level 1 continued

7. The shape is enlarged by a scale factor of 3 using centre of enlargement (5, -5). What are the coordinates of the image of point E?
Give your answer in the form (x, y).

- -1, 7
- (-1, 7)
- (-1 7)
- (7, -1)
- (-3, 11)

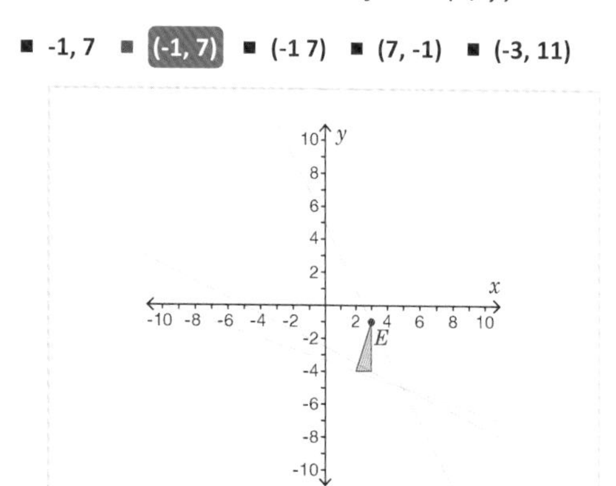

8. The shape is enlarged by a scale factor of 2 using centre of enlargement (-2, 1). What are the coordinates of the image of point Z?
Give your answer in the form (x, y).

- (-6, -3)
- (-8, -5)
- (-6 -3)
- -6, -3
- (-3, -6)

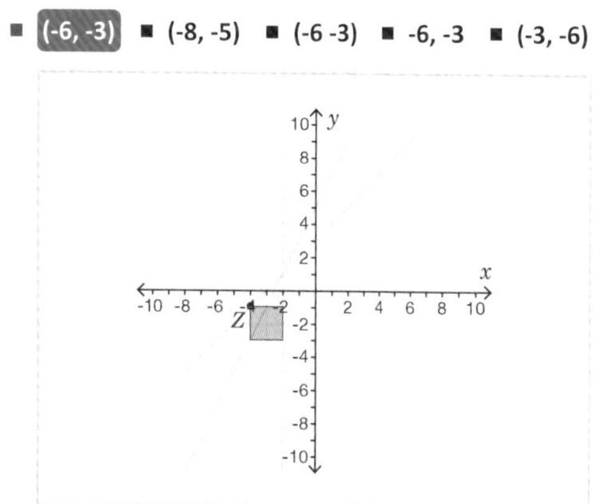

Level 2: Fluency: Enlarge shapes without using ray lines.

Required: 8/8 **Pupil Navigation:** on
Randomised: off

9. Rectangle R is enlarged by a scale factor of 2 using centre of enlargement (0, 0). Which grid shows the correct enlargement of R?

1/4

- (ii)
- (iii)
- (iv)
- (i)

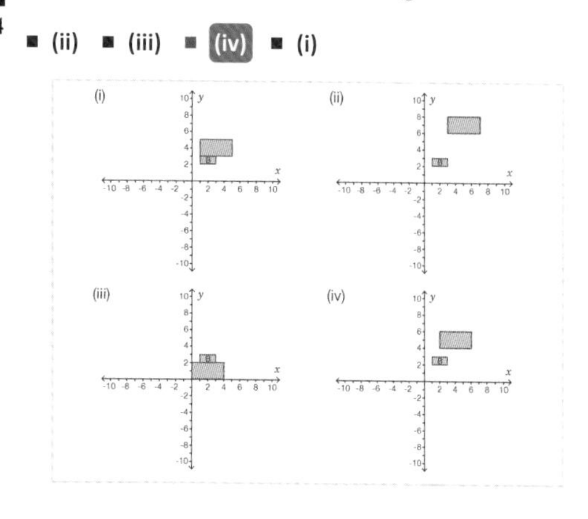

10. The shape is enlarged by a scale factor of 5 using centre of enlargement (0, 0). What are the coordinates of the image of point H?
Give your answer in the form (x, y).

- (-5, 10)
- (-6, 12)

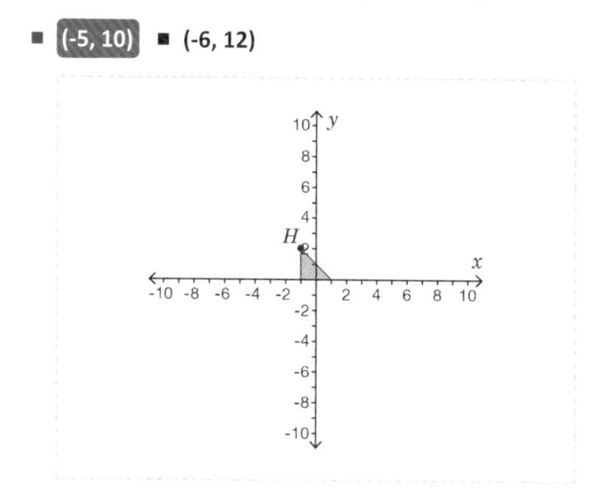

11. The shape is enlarged by a scale factor of 3 using centre of enlargement (0, 0). What are the coordinates of the image of point C?
Give your answer in the form (x, y).

- (-3, -9)
- (-4, -12)

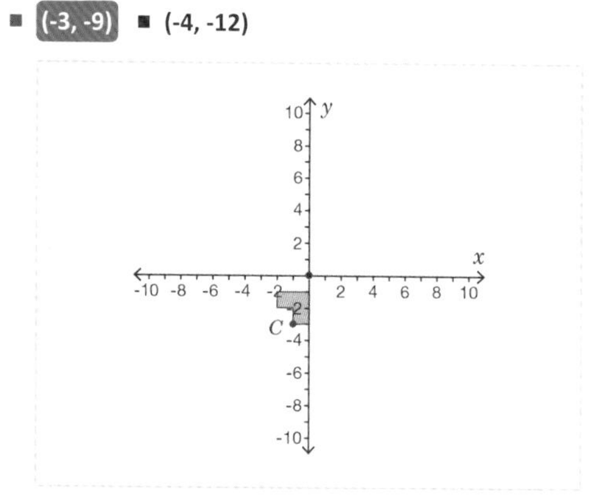

12. The shape is enlarged by a scale factor of 4 using centre of enlargement (2, -1). What are the coordinates of the image of point S?
Give your answer in the form (x, y).

- (-10, 7)
- (-13, 9)

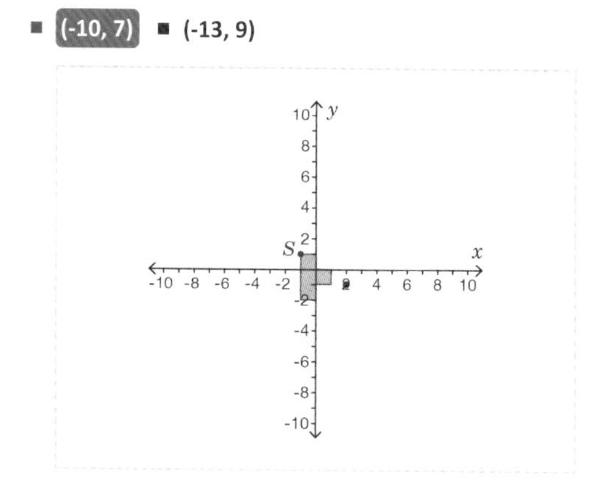

Level 2 *continued*

13. Shape Q is enlarged by a scale factor of 2 using centre of enlargement (-8, 6). Which image shows the correct enlargement of Q?

1/4

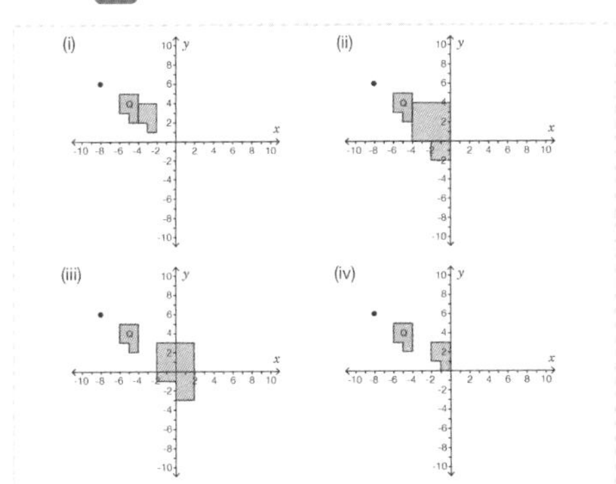

14. The shape is enlarged by a scale factor of 3 using centre of enlargement (9, 6). What are the coordinates of the image of point *D*? *Give your answer in the form (x, y).*

■ (-3, -3) ■ (-7, -6)

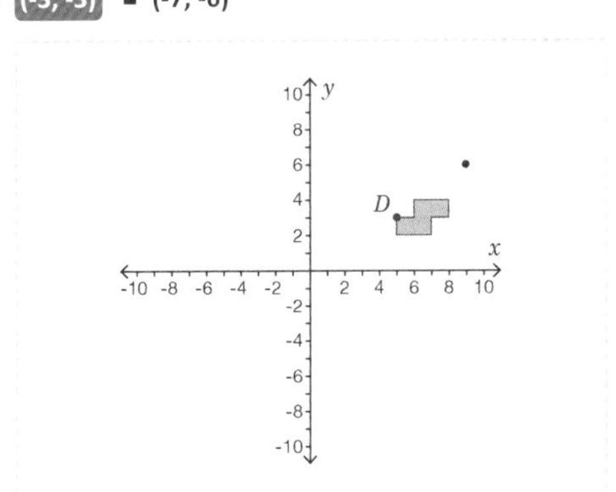

15. The shape is enlarged by a scale factor of 5 using centre of enlargement (-10, 8). What are the coordinates of the image of point *S*? *Give your answer in the form (x, y).*

■ (-5, -7) ■ (-4, -10)

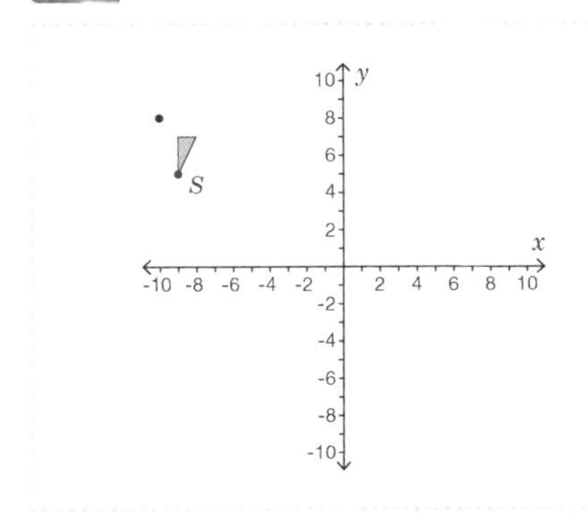

16. The shape is enlarged by a scale factor of 2 using centre of enlargement (4, -6). What are the coordinates of the image of point *W*? *Give your answer in the form (x, y).*

■ (0, 0) ■ (-2, 3)

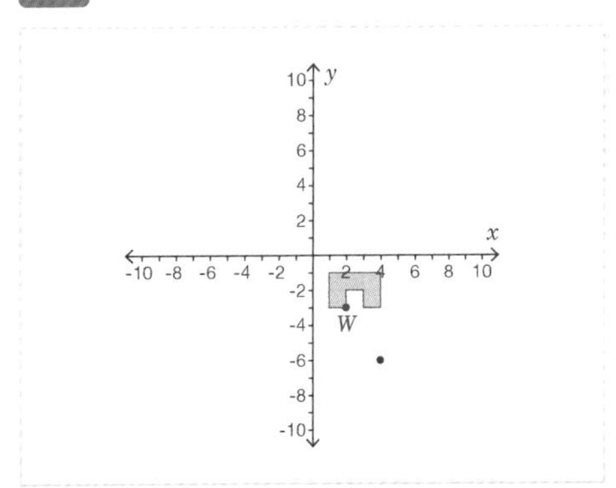

Level 3: Reasoning: Reason about enlargements.

Required: 3/3 **Pupil Navigation:** on
Randomised: off

17. Ray has enlarged the rectangle by a scale factor of 4 using centre of enlargement (-10, -10). Explain the mistake that Ray has made.

- Open question, no set answer

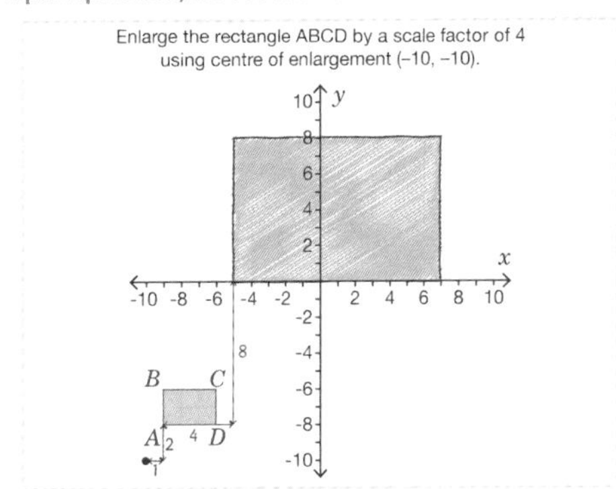

18. Select the shape that is an enlargement of shape A.

1/5

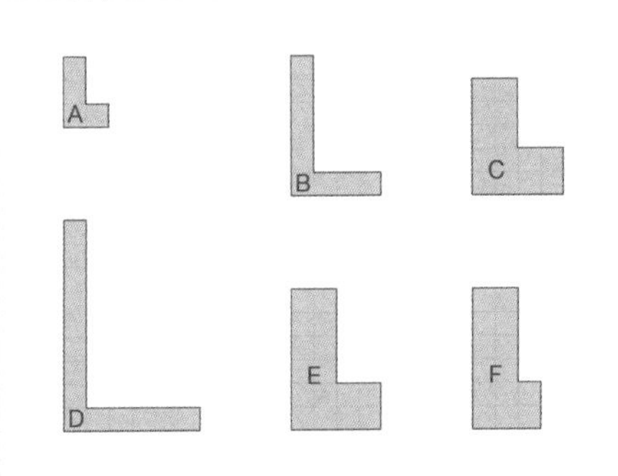

Level 3 *continued*

19. Shape T is enlarged by a scale factor of 3 using centre of enlargement (-8, 6). How many times larger is the image than shape T?

- **9**
- 3

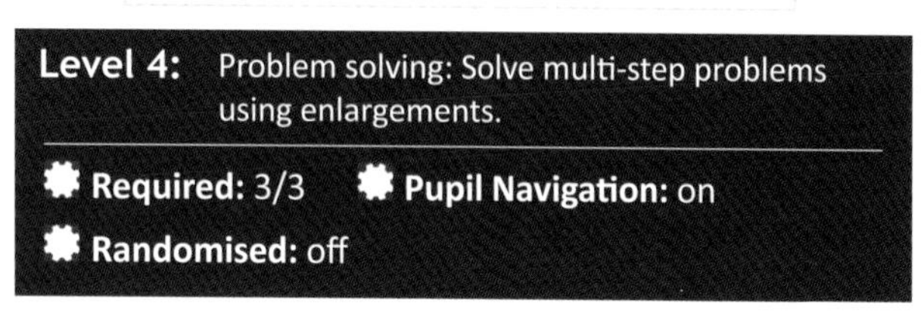

Level 4: Problem solving: Solve multi-step problems using enlargements.

Required: 3/3 **Pupil Navigation:** on

Randomised: off

20. The shape is enlarged by a scale factor of 2.5 using centre of enlargement (10, -8). What are the coordinates of the image of point *V*?
Give your answer in the form (x, y).

- **(-5, 7)**

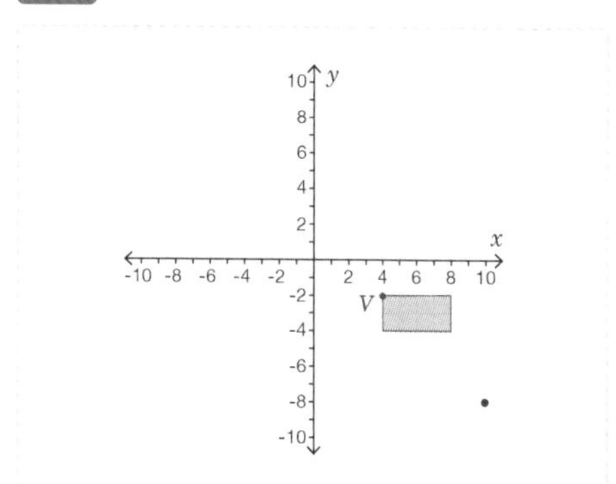

21. Shape N is an enlargement of shape M. What are the coordinates of the centre of enlargement?
Give your answer in the form (x, y).

- **(-8, 7)**

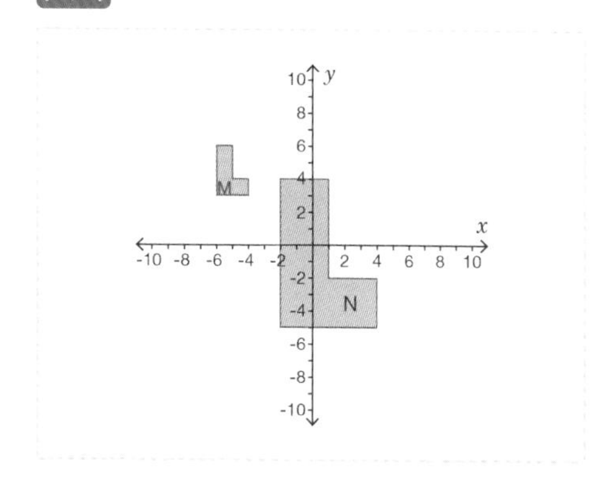

22. The shape is enlarged by a scale factor of 3 using centre of enlargement (8, 6). The image is then enlarged by a scale factor of 2 using centre of enlargement (6, -8). What are the coordinates of the image of point *P* after the two enlargements?
Give your answer in the form (x, y).

- **(-8, 8)**

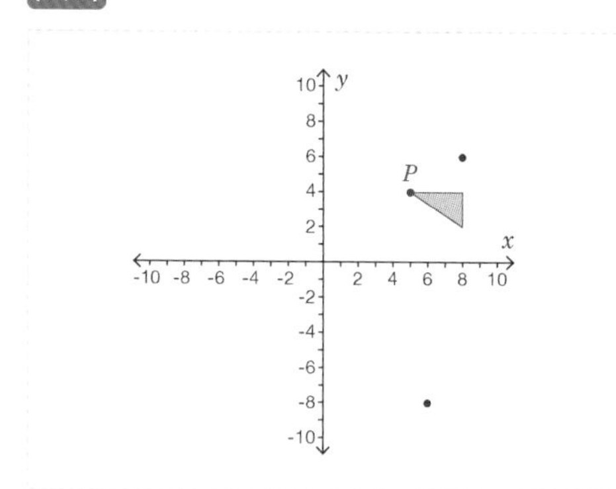

Reflecting Shapes

Objective: I can reflect shapes in a given line.

Quick Search Ref: 10582

Level 1: Understanding: Reflect shapes in a mirror line.

Required: 8/10 **Pupil Navigation:** on **Randomised:** off

1. Which image shows the correct reflection of shape A in the x-axis?

■ (i) ■ (ii) ■ (iii) ■ (iv)

1/4

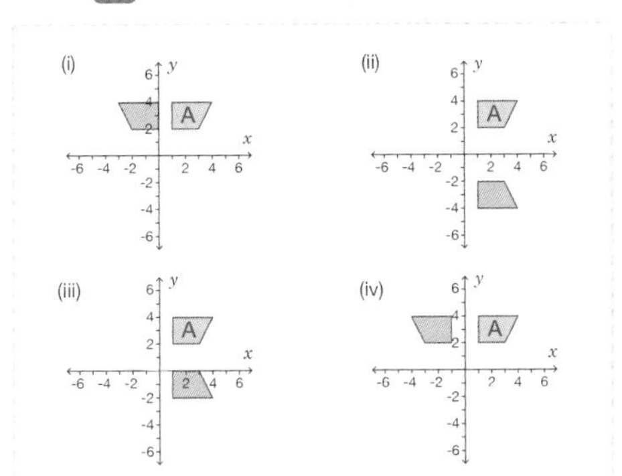

2. Point V is reflected in the y-axis. What are the coordinates of the point that V is mapped to?

1/6

■ (-3, 4) ■ (3, -4) ■ (4, -3) ■ (-3, -4) ■ -3, 4 ■ (-3 4)

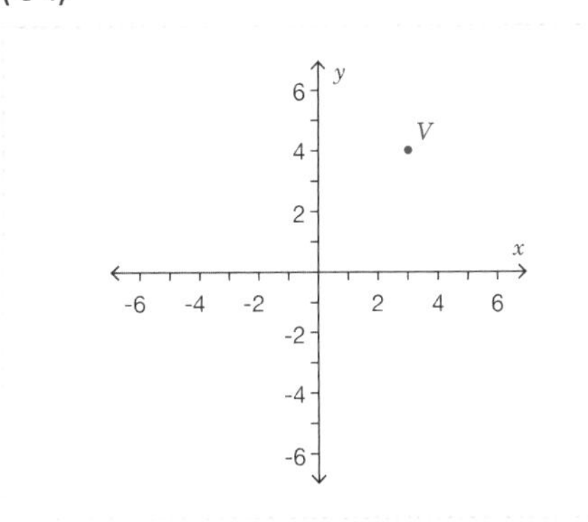

3. Point R is reflected in the line $x = 5$. What are the coordinates of the point that R is mapped to?

Give your answer in the form (x, y).

■ 7, 2 ■ (7, 2) ■ (2, 7) ■ (7 2)

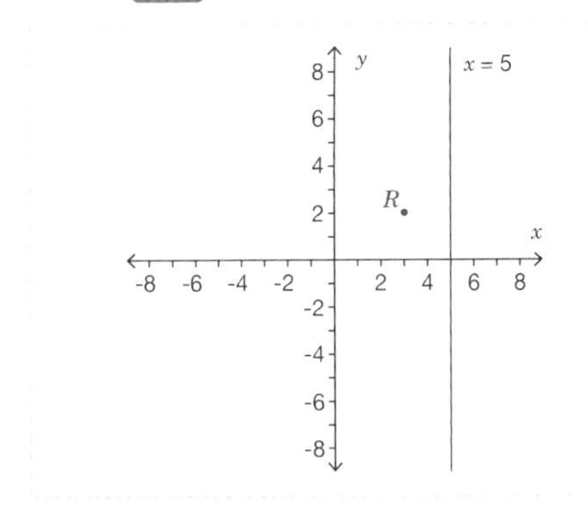

4. The shape is reflected in the line $y = -2$. What are the coordinates of the point that T is mapped to?

Give your answer in the form (x, y).

■ (-3, -5) ■ -3, -5 ■ (-3 -5) ■ (-5, -3)

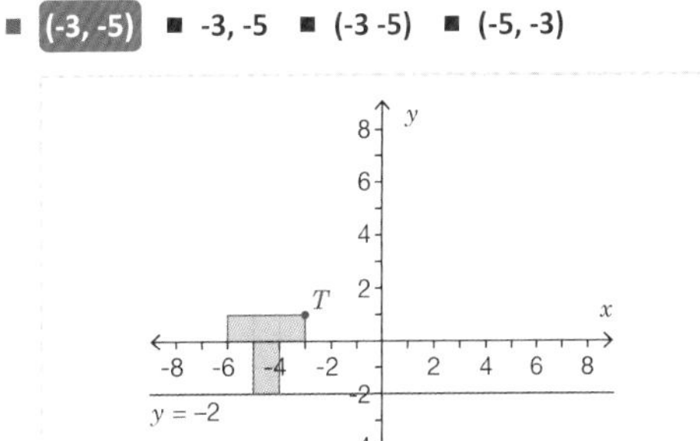

5. Which image shows the correct reflection of shape M in the line $y = x$?

■ (i) ■ (ii) ■ (iii) ■ (iv)

1/4

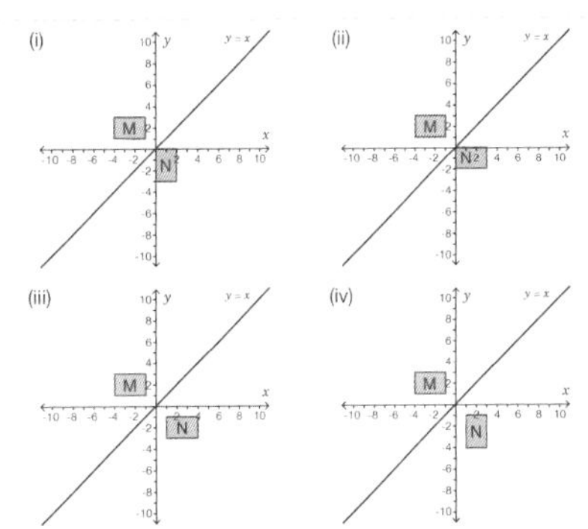

6. The shape is reflected in the line $x = -1$. What are the coordinates of the point that B is mapped to?

Give your answer in the form (x, y).

■ (-6, -3) ■ (-3, -6) ■ (-6 -3) ■ -6, -3

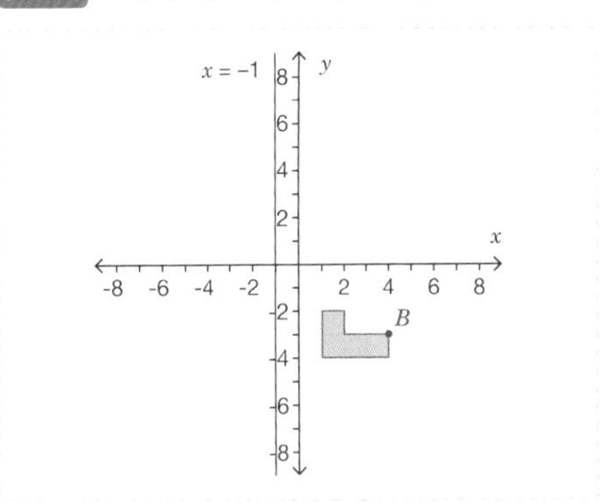

Level 1 *continued*

7. The shape is reflected in the line $y = -x$. What are the coordinates of the point that R is mapped to?
Give your answer in the form (x, y).

- (-2 6) ■ **(-2, 6)** ■ -2, 6 ■ (6, -2)

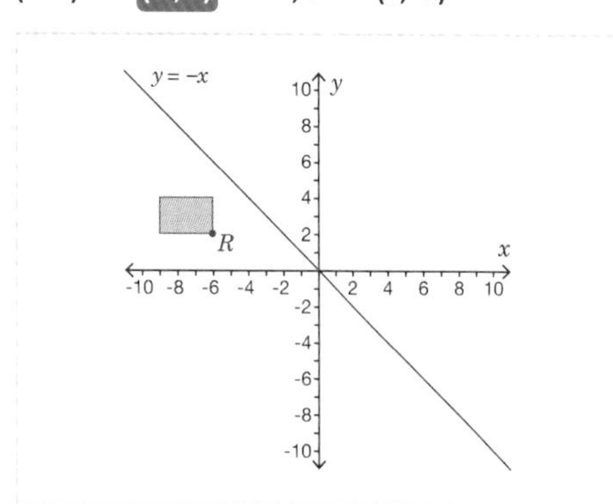

8. Which image shows the correct reflection of shape S in the y-axis?

1/4

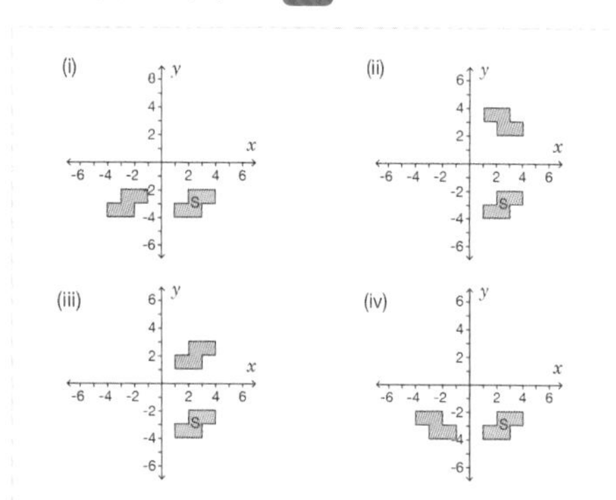

9. Point Q is reflected in the line $y = -4$. What are the coordinates of the point that Q is mapped to?
Give your answer in the form (x, y).

- 1, -2 ■ **(1, -2)** ■ (1 -2) ■ (-2, 1)

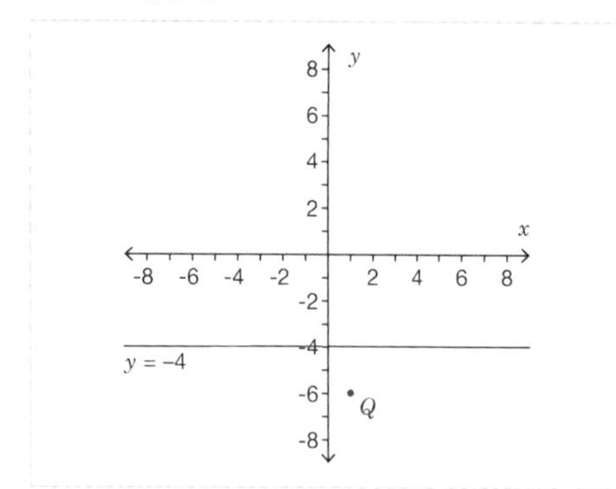

10. The shape is reflected in the line $y = x$. What are the coordinates of the point that G is mapped to?
Give your answer in the form (x, y).

- **(7, 1)** ■ 7, 1 ■ (7 1) ■ (1, 7)

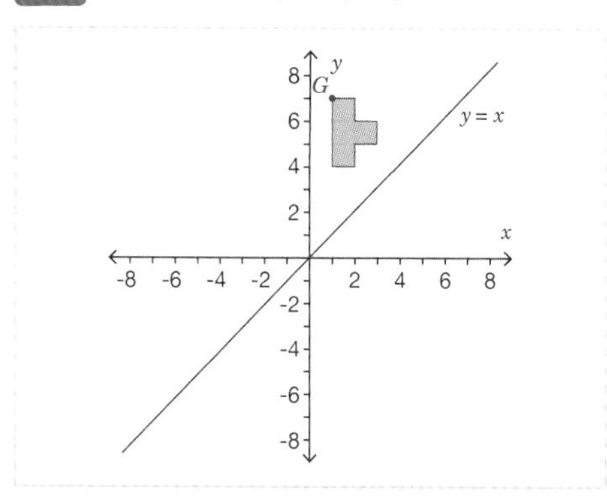

Level 2: Fluency: Reflect shapes given the equation of a mirror line.

Required: 7/10 **Pupil Navigation:** on
Randomised: off

11. Which graph shows the line $x = 3$?

1/4

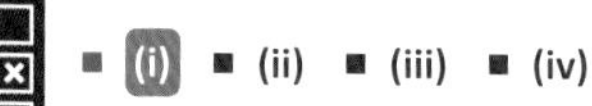

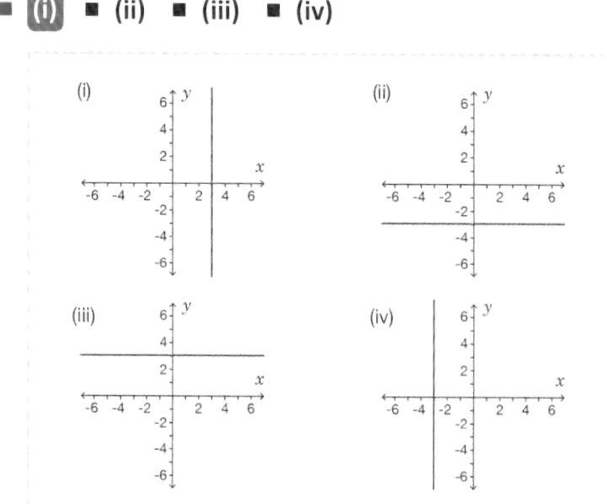

12. Which image shows the correct reflection of shape L in the line $y = -1$?

1/4

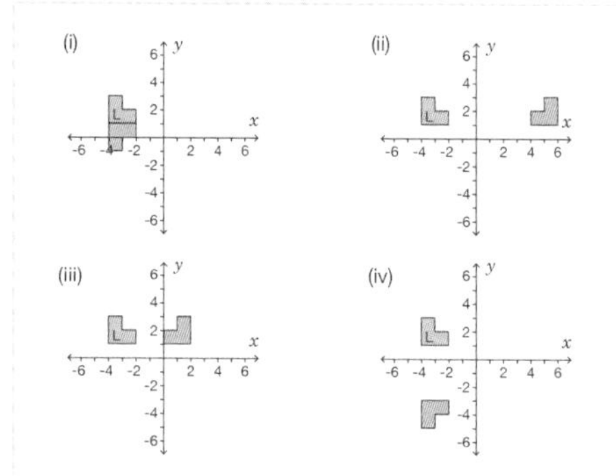

Level 2 *continued*

13. Point D is reflected in the line $x = 1$. What are the coordinates of the point that D is mapped to?
Give your answer in the form (x, y).

- (0, 3)
- (2, -1)

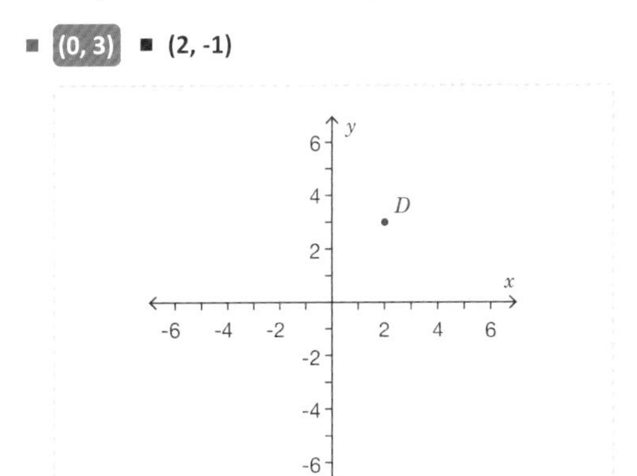

14. The shape is reflected in the line $y = -3$. What are the coordinates of the point that E is mapped to?
Give your answer in the form (x, y).

- (-2, -8)
- (-4, 2)

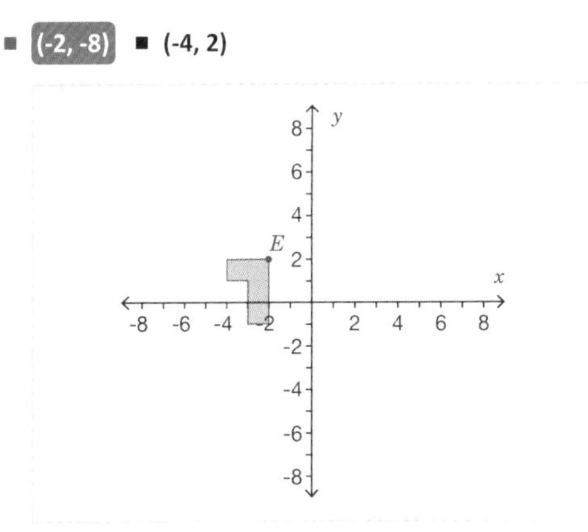

15. Which image shows a reflection of shape C in the line $y = x$?

1/4

- (i)
- (ii)
- (iii)
- (iv)

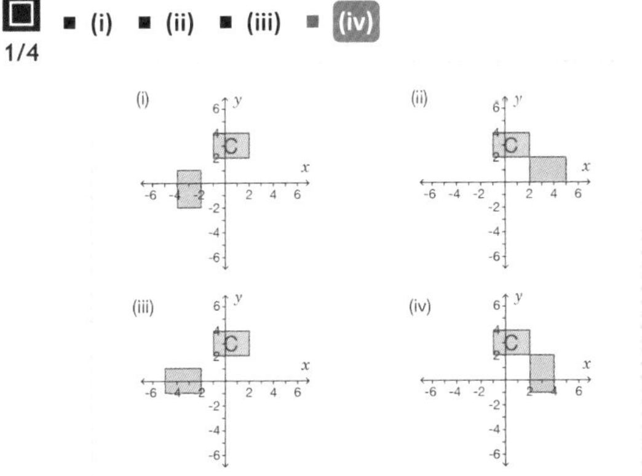

16. The shape is reflected in the line x = -2. What are the coordinates of the point that W is reflected to?
Give your answer in the form (x, y).

- (-8, -4)
- (4, 0)

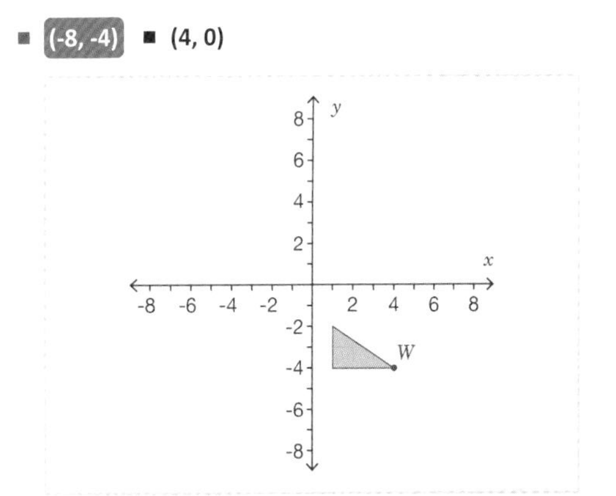

17. The shape is reflected in the line $y = -x$. What are the coordinates of the point that P is mapped to?
Give your answer in the form (x, y).

- (-4, 1)
- (4, -1)

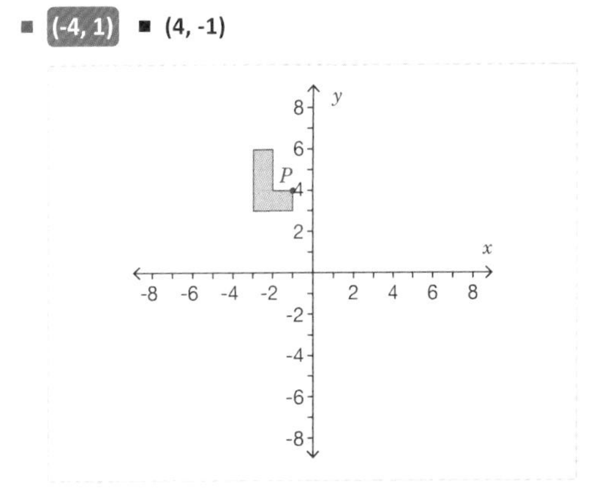

18. Which image shows a reflection of shape L in the line $x = -2$?

1/4

- (i)
- (ii)
- (iii)
- (iv)

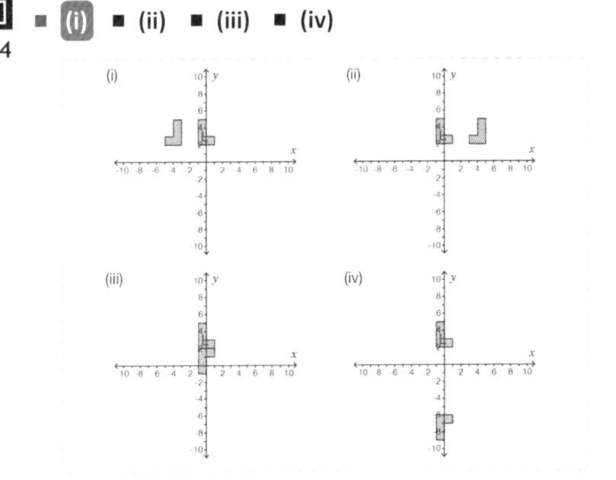

Level 2 continued

19. The shape is reflected in the line $y = 3$. What are the coordinates of the point that E is reflected to?
Give your answer in the form (x, y).

- (2, 6)
- (4, 0)

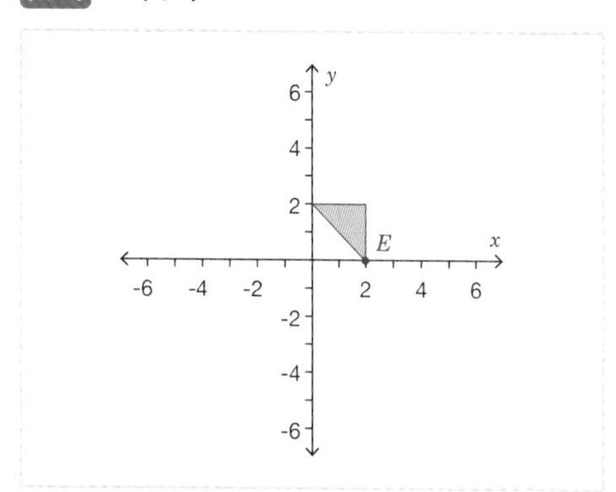

20. The shape is reflected in the line $y = -x$. What are the coordinates of the point that M is reflected to?
Give your answer in the form (x, y).

- (-1, -4)
- (1, 4)

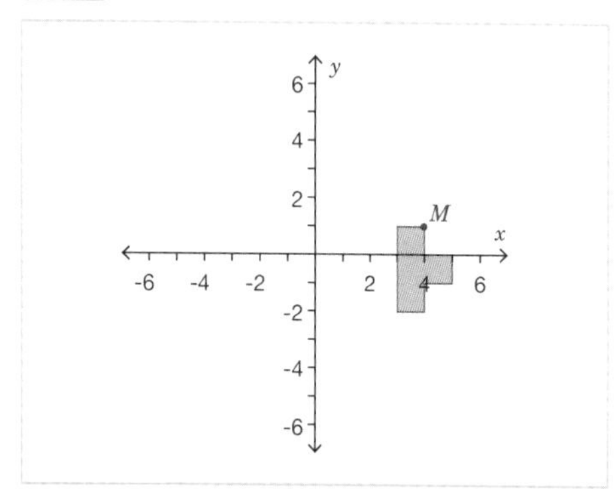

Level 3: Reasoning: Make inferences using lines of reflection.

Required: 5/5 **Pupil Navigation:** on
Randomised: off

21. Which of the four images does **not** show a reflection of the shape on the left?

1/4

- (i)
- (ii)
- (iii)
- (iv)

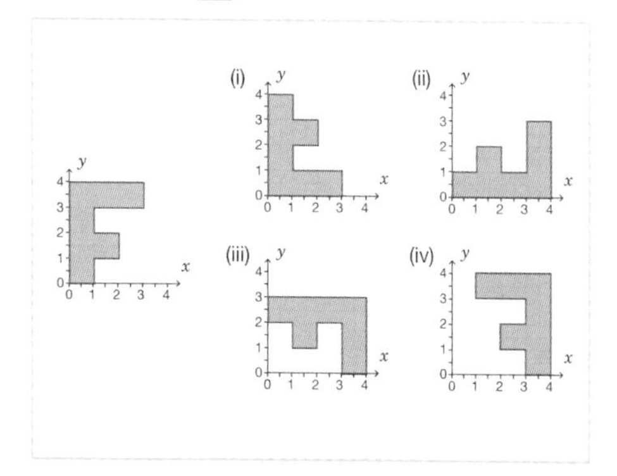

22. Point S is reflected in a line and maps onto point T. Point S is 3 cm from the line of reflection.
What is the distance between points S and T in centimetres?
Include the units in your answer.

- 6 cm
- 6 centimetres
- 3 centimetres
- 3 cm
- 6

23. Omar has reflected rectangle $ABCD$ in the line $y = x$ and labelled the reflection $A'B'C'D'$.
Explain the mistake that Omar has made.

- Open question, no set answer

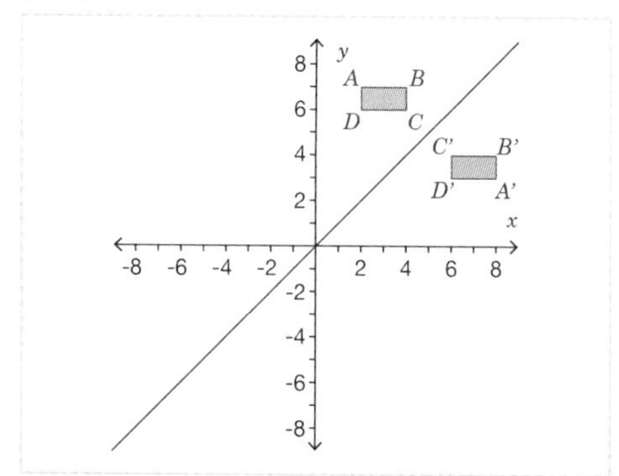

24. Point $D = (2, -3)$ is reflected in the line $y = -1$.
This point is then reflected in the line $y = -1$.
What are the new coordinates of point D?
Give your answer in the form (x, y).

- (2, -3)

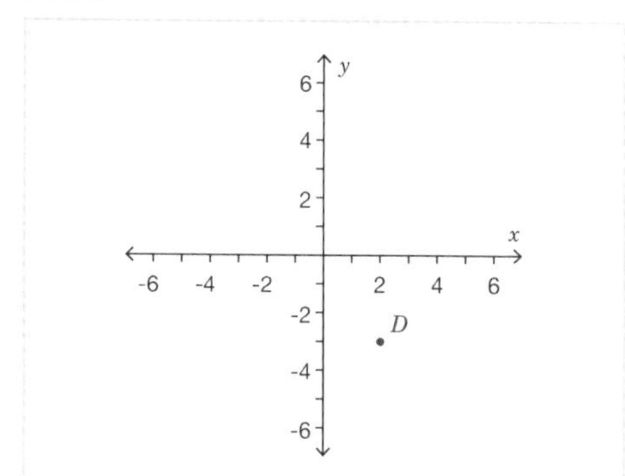

Level 3 *continued*

25. Point *A* is reflected onto point *B* using line *CD* as the line of reflection. Select **two properties** of line *CD*.

2/4

- Line CD has a length of 8 squares.
- Line CD intersects line AB at right angles.
- Line CD has a gradient of 1.
- Line CD bisects line AB into two equal parts.

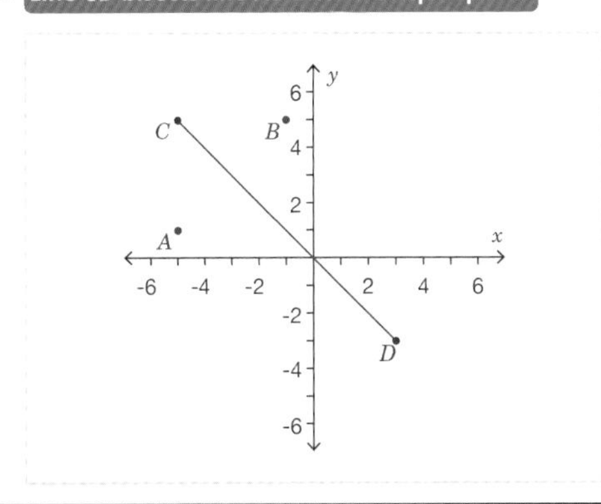

Level 4: Problem solving: Find the equations of lines of reflection and solve multi-step problems.

Required: 5/5 **Pupil Navigation:** on

Randomised: off

26. Shape U is reflected and maps onto shape V. What is the equation of the line of reflection?

- x = 2

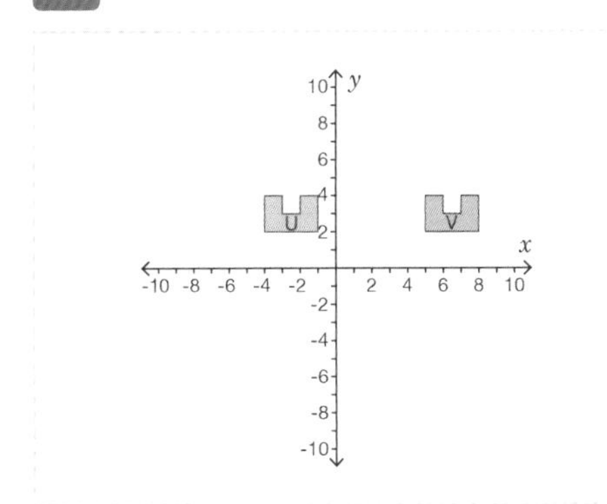

27. The shape is reflected in the line $x = 3$, and then its image is reflected in the line $y = -1$. What are the coordinates of point G after the two reflections?

Give your answer in the form (x, y).

- (-1, -5)

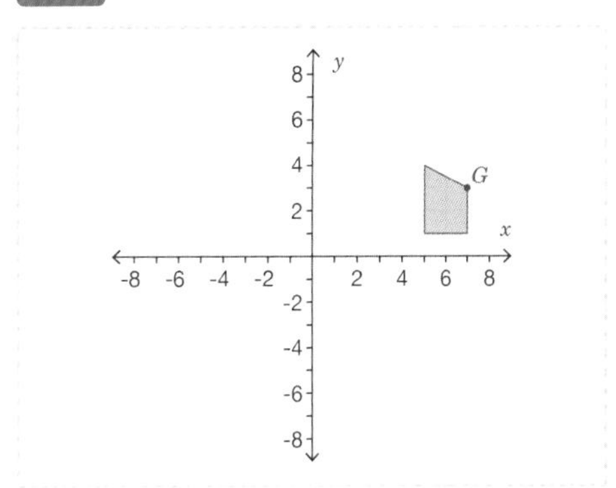

28. Shape J is reflected onto shape K. What is the equation of the line of reflection?

- y = -2

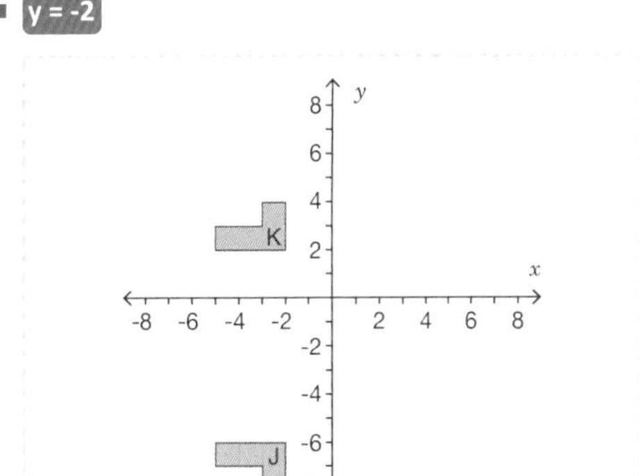

29. Select the **two reflections** that together map shape S onto shape T.

1/6

- y = -1
- y = x
- x = 1
- y = 1
- y = -x
- x = -1

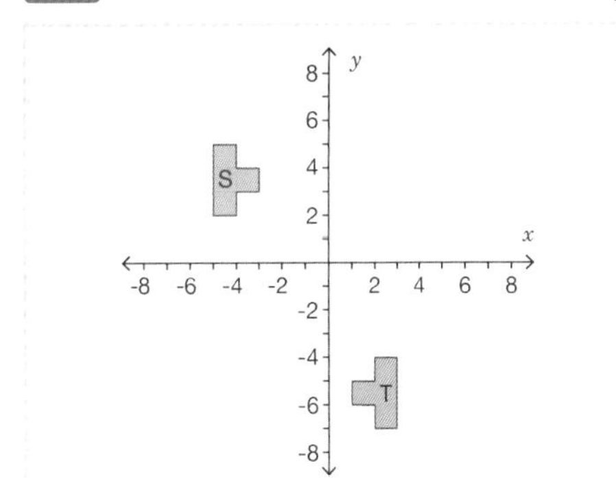

Level 4 *continued*

30. The shape is reflected in the line $y = 2x$.

a b c

What are the coordinates of the point that A is mapped to?

Give your answer in the form (x, y).

- (-1, 8)

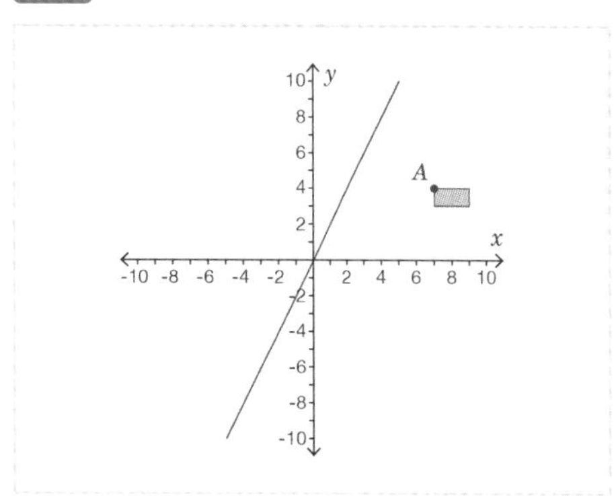

Rotating Shapes and Describing Rotations

Objective: I can rotate shapes and describe rotations.

Quick Search Ref: 10616

Level 1: Understanding: Rotating points and shapes about the origin.

Required: 7/10 **Pupil Navigation:** on **Randomised:** off

1. Peter is facing north. He rotates 90 degrees clockwise, then 180 degrees, then 90 degrees anti-clockwise. What direction is he now facing?

1/4

- north
- east
- south
- west

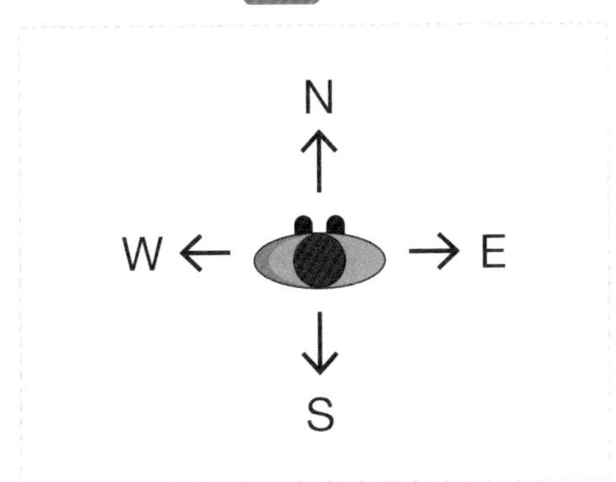

2. Shape T is rotated 90° clockwise about the origin (0, 0) to map onto shape U. Which diagram shows the correct rotation?

1/4

- (i)
- (ii)
- (iii)
- (iv)

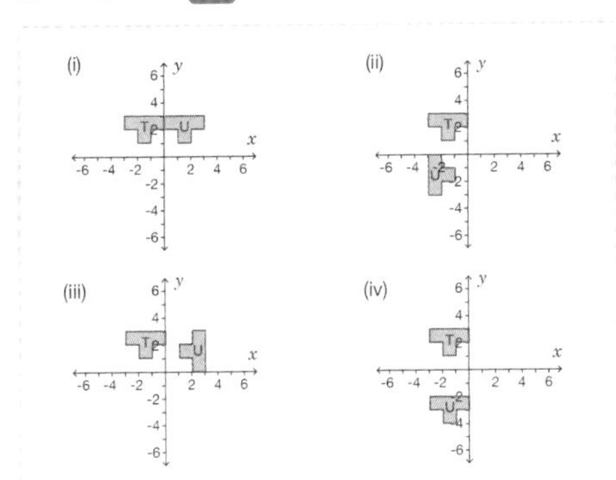

3. Point P is rotated 180° about the origin (0, 0). What are the coordinates that point P is rotated to?
Give your answer in the form (x, y).

- (-1, -3)
- (1, -3)
- (-3, -1)
- (-1 -3)
- -1, -3

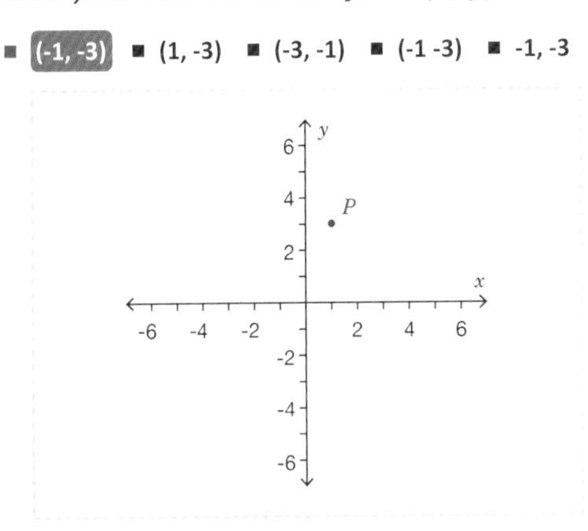

4. The shape is rotated 90° clockwise about the origin (0, 0). What are the coordinates that point T is rotated to?
Give your answer in the form (x, y).

- (4, -2)
- (-2, 4)
- (-4, 2)
- 4, -2
- (4 -2)

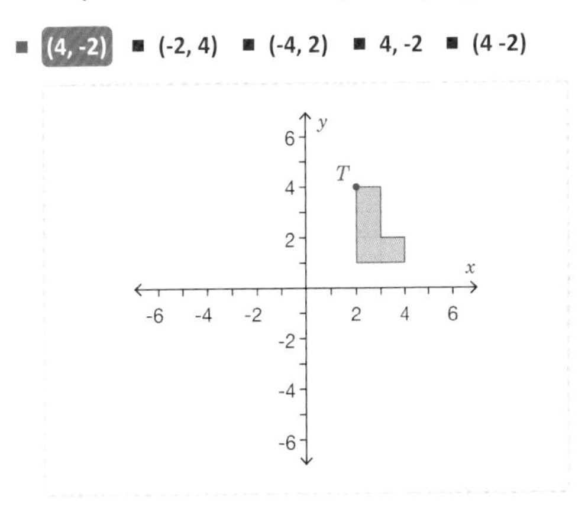

5. The shape is rotated 180° about the origin (0, 0). What are the coordinates that point B is rotated to?
Give your answer in the form (x, y).

- -6, 3
- (-6, 3)
- (-6 3)
- (3, -6)
- (-6, -3)

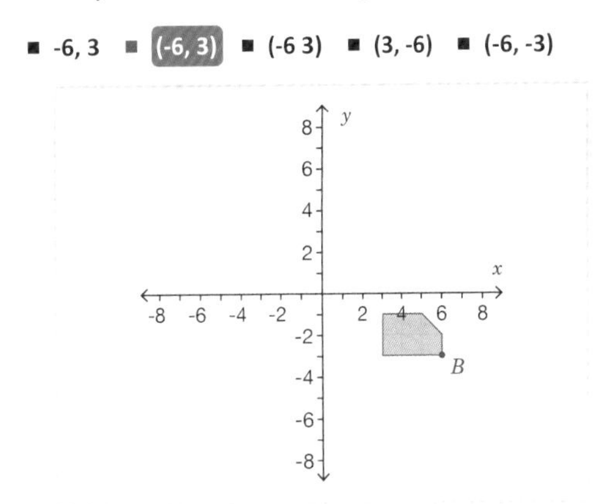

6. The shape is rotated 90° anti-clockwise about the origin (0, 0). What are the coordinates that point R is rotated to?
Give your answer in the form (x, y).

- (-5, 0)
- -5, 0
- (0, -5)
- (5, 0)
- (-5 0)

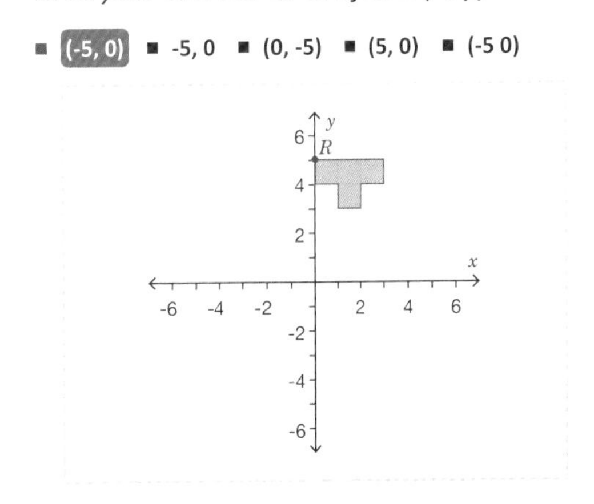

Level 1 *continued*

7. The shape is rotated 270° anti-clockwise about the origin (0, 0). What are the coordinates that point *U* is rotated to?
Give your answer in the form (x, y).

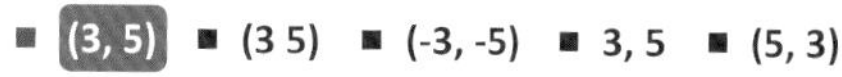

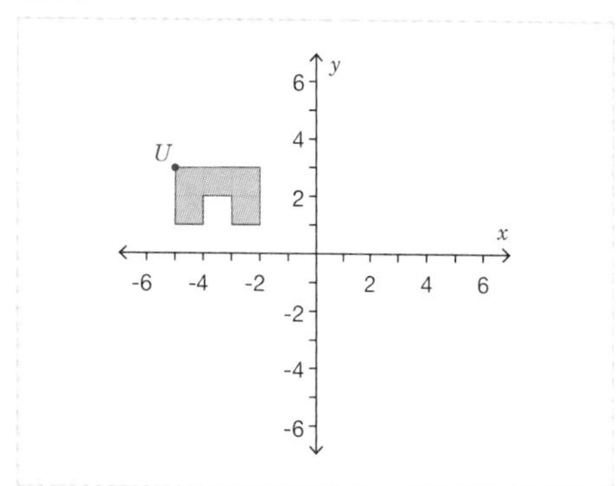

8. The shape is rotated 180° about the origin (0, 0). What are the coordinates that point *L* is rotated to?
Give your answer in the form (x, y).

- **(4, -3)**
- **(4 -3)**
- **(-4, -3)**
- **(-3, 4)**
- **4, -3**

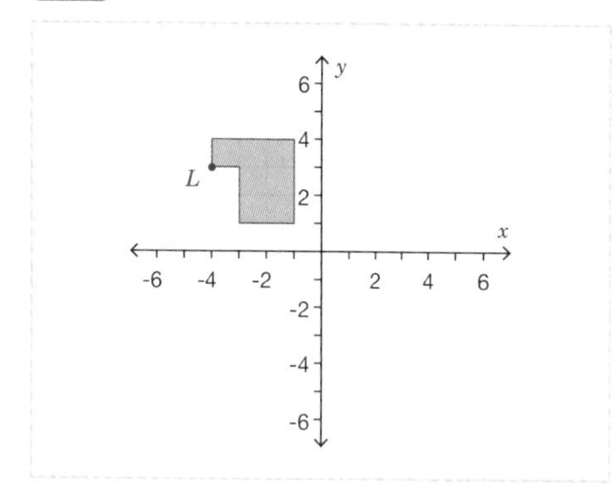

9. The shape is rotated 90° anti-clockwise about the origin (0, 0). What are the coordinates that point *R* is rotated to?
Give your answer in the form (x, y).

- **(-3, 1)**
- **(1, -3)**
- **-3, 1**
- **(-3 1)**

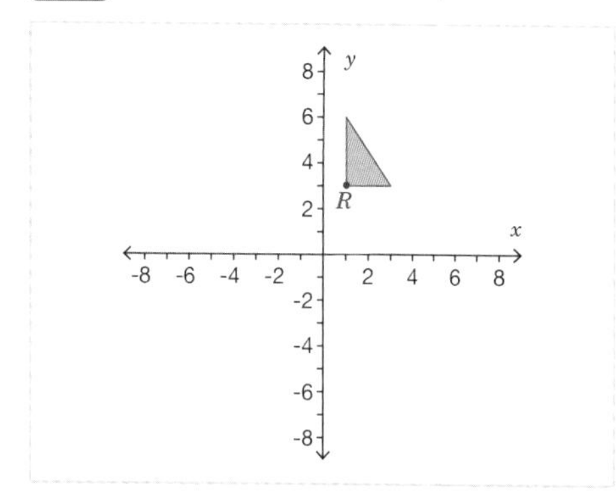

10. The shape is rotated 90° clockwise about the origin (0, 0). What are the coordinates that point *C* is rotated to?
Give your answer in the form (x, y).

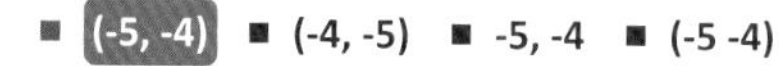

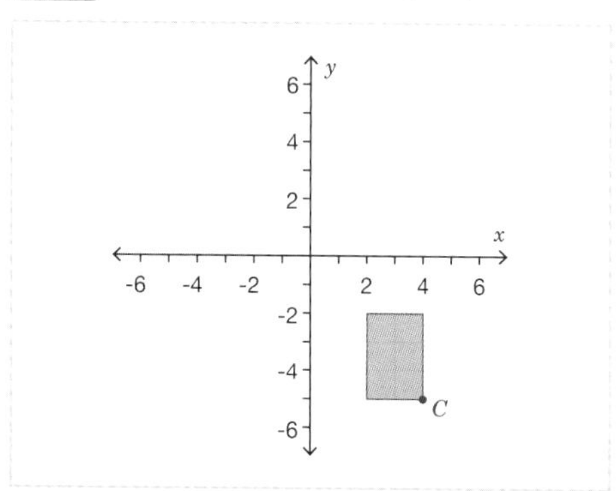

11. Shape A has been rotated about the point (0, 0) to map onto shape B. Which statement describes the angle and direction of rotation?

1/4

- **90° clockwise**
- **270° anticlockwise**
- **180°**
- 

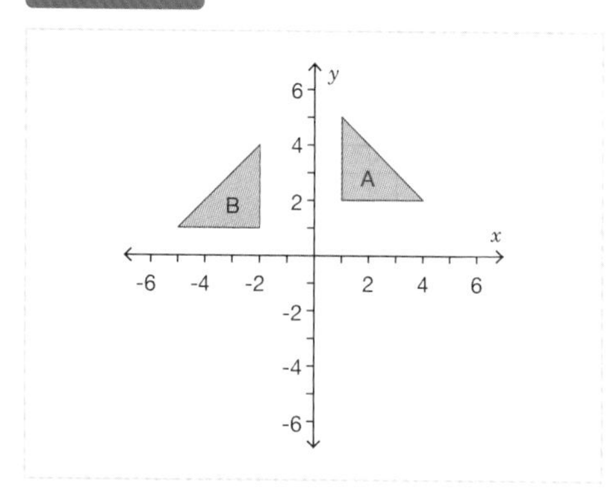

12. Shape U is rotated 180° to map onto shape V. What are the coordinates of the centre of rotation?
Give your answer in the form (x, y).

- **(3, 1)**

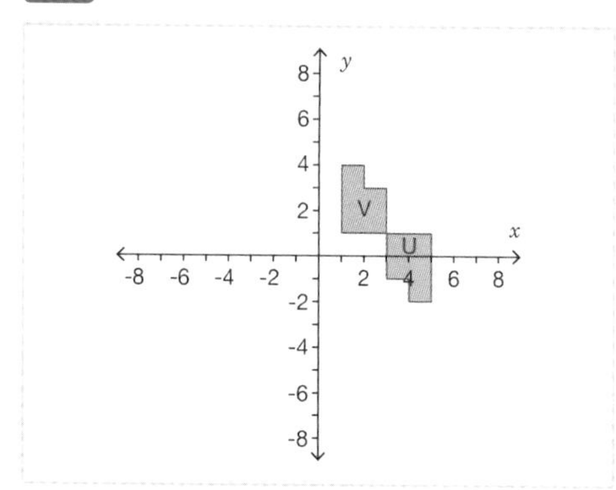

Level 2 *continued*

13. Select the **two statements** that describe the rotation that maps shape E onto shape F.

2/6

- centre of rotation (-3, 2)
- rotation of 90° anti-clockwise
- rotation of 180°
- centre of rotation (0, 0)
- rotation of 90° clockwise
- centre of rotation (-4, -3)

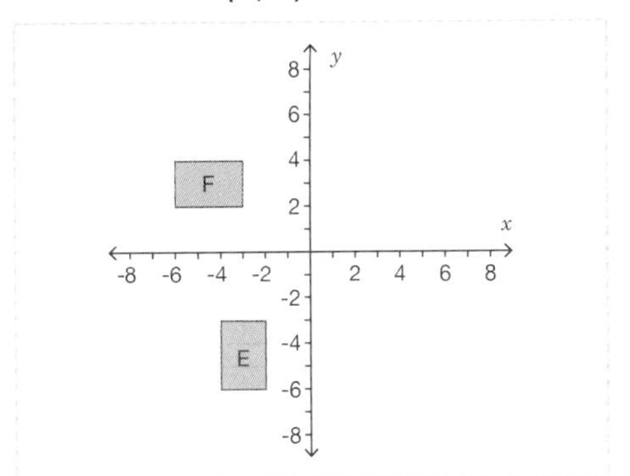

14. Shape R is rotated 180° to map onto shape S. What are the coordinates of the centre of rotation?
Give your answer in the form (x, y).

- (0, 1)

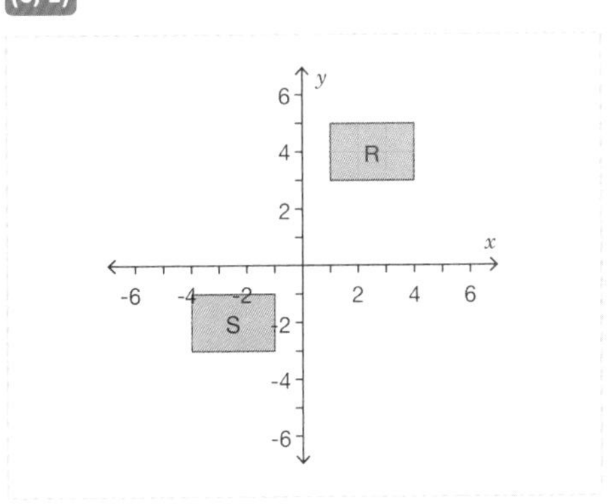

15. Shape M is rotated to map onto shape N. What are the coordinates of the centre of rotation?
Give your answer in the form (x, y).

- (-1, 0)

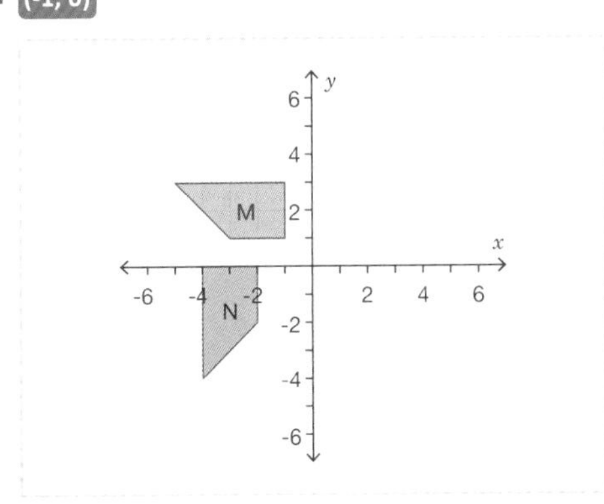

16. Select the **two statements** that describe the rotation that maps shape G onto shape H.

2/6

- centre of rotation (2, 0)
- rotation of 90° clockwise
- centre of rotation (1, 0)
- rotation of 180°
- rotation of 90° anti-clockwise
- centre of rotation (0, 1)

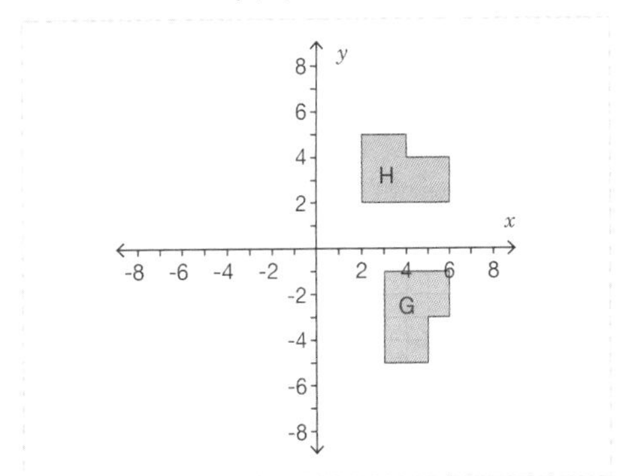

17. Shape A has been rotated to map onto shape B. What are the coordinates of the centre of rotation?
Give your answer in the form (x, y).

- (0, 2)

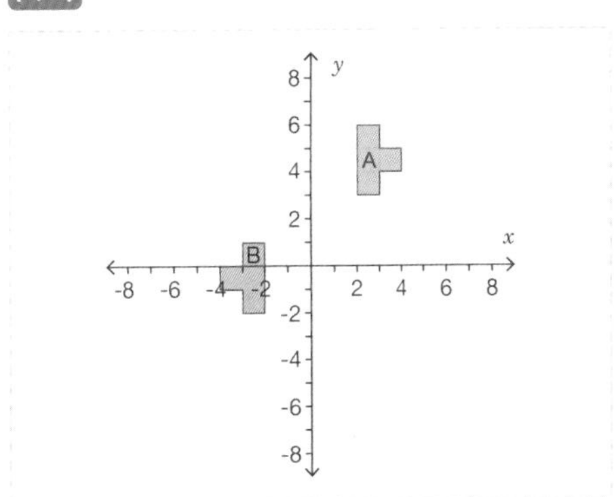

18. Select the **two statements** that describe the rotation that maps shape S onto shape T.

2/6

- centre of rotation (0, 1)
- rotation of 90° clockwise
- centre of rotation (1, 1)
- rotation of 180°
- rotation of 90° anti-clockwise
- centre of rotation (0, 0)

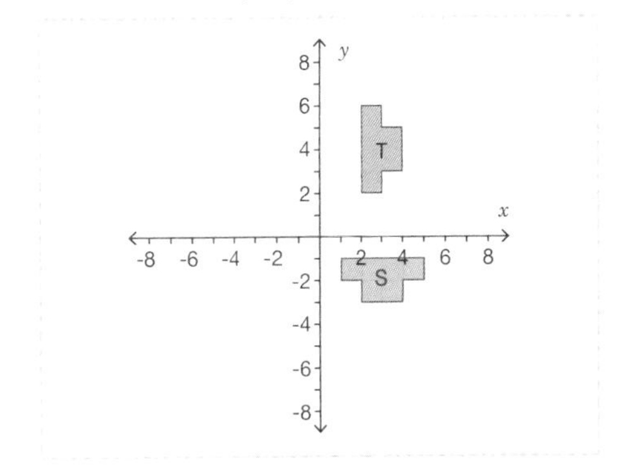

Level 2 *continued*

19. Shape P is rotated to map onto shape Q. What are the coordinates of the centre of rotation?
Give your answer in the form (x, y).

- (-1, 0)

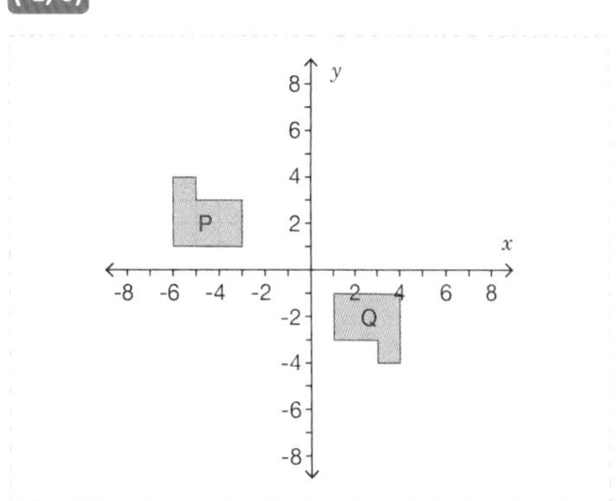

20. Shape C is rotated to map onto shape D. What are the coordinates of the centre of rotation?
Give your answer in the form (x, y).

- (0, -1)

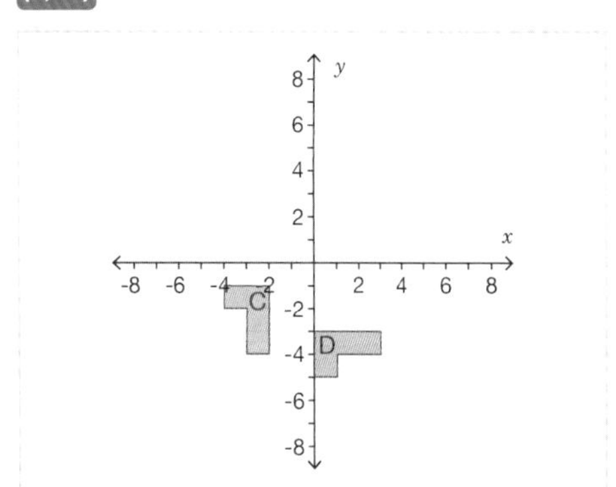

Level 3: Reasoning: Visualise rotations.

Required: 5/5 **Pupil Navigation:** on
Randomised: off

21. Which shape can shape A **not** be mapped onto by rotation?

1/4

- (i)
- (ii)
- (iii)
- (iv)

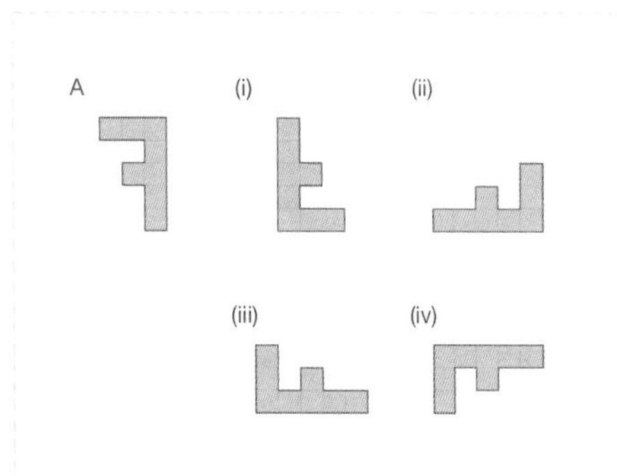

22. The point (0, 4) is rotated by 90° anti-clockwise about the origin. What are the coordinates of the new point?
Give your answer in the form (x, y).

- (-4, 0)
- (4, 0)

23. Petra is rotating the rectangle 90° clockwise about the origin using tracing paper. Explain the mistake Petra has made.

- Open question, no set answer

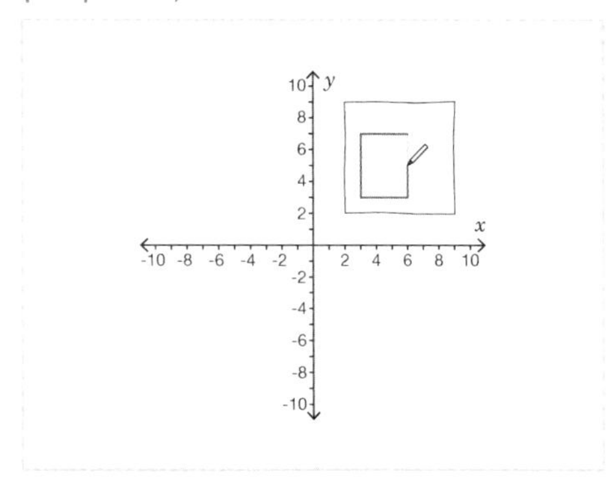

24. The point (2, 3) is rotated 180° about the origin. What are the coordinates of the new point?
Give your answer in the form (x, y).

- (-2, -3)

25. Sort the descriptions of the four rotations that map shape A onto shape B into the same order as the diagrams, from rotation (i) to rotation (iv).

- a rotation of 90° clockwise about (0, 0)
- a rotation of 180° about (-1, 0)
- a rotation of 90° anti-clockwise about (0, 0)
- a rotation of 90° clockwise about (1, 0)

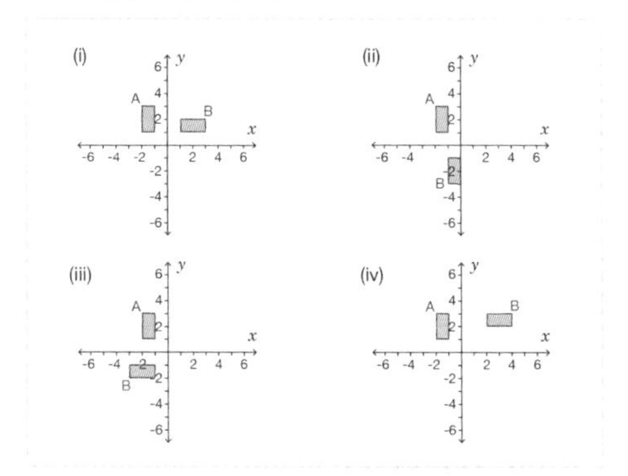

Level 4: Problem solving: Understand rotations about different centres and equivalent transformations.

Required: 5/5 **Pupil Navigation:** on
Randomised: off

26. The shape is rotated 90° clockwise about the point (1, 2). What are the coordinates that point *T* is rotated to?
Give your answer in the form (x, y).

- (4, -2)

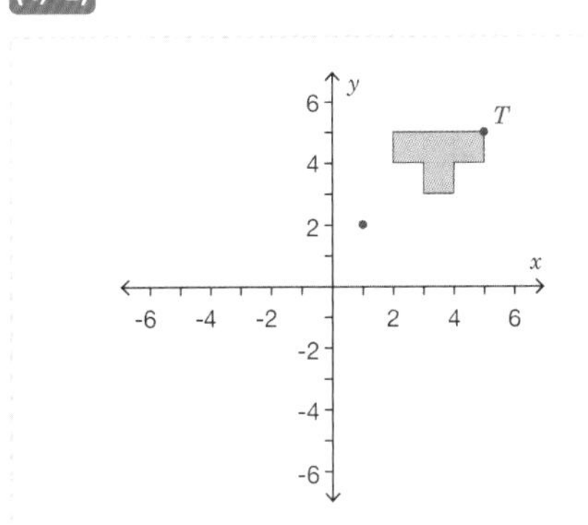

27. Point *P* = (-1, 3) is rotated 90° anti-clockwise about the origin (0, 0). What is the missing bottom number of the column vector that would translate point *P* to the same position?

- -4 ▪ -2 ▪ 4

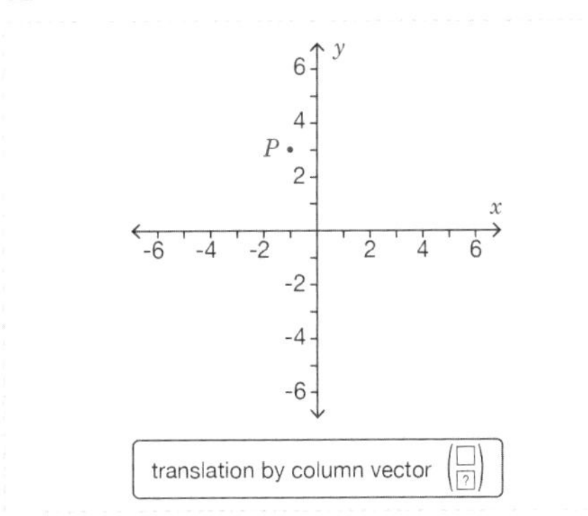

28. The shape is rotated 90° anti-clockwise about the point (1, -1). What are the coordinates that point *F* is rotated to?
Give your answer in the form (x, y).

- (5, -4)

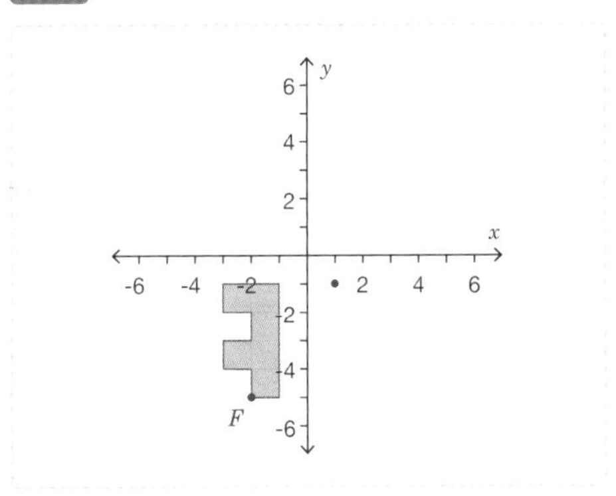

29. Shape B is rotated 90° clockwise about (-1, 0) and then 90° anti-clockwise about the point (1, 0). What is the missing top number of the column vector that would translate B to the same position?

- 2 ▪ -2

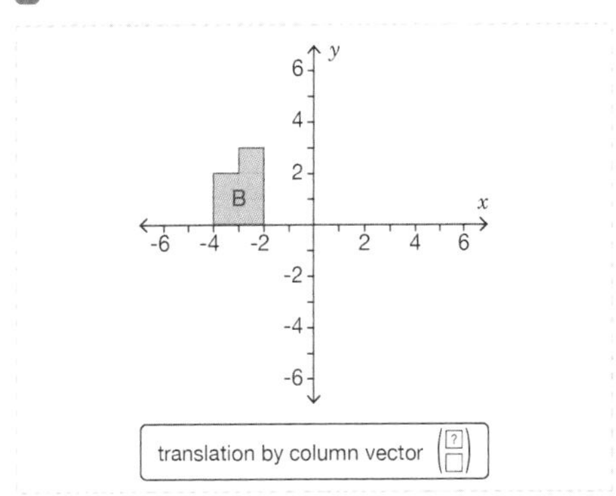

30. Rectangle A can be rotated clockwise 90° about the origin (0, 0) to map onto rectangle B. Rectangle A can also be rotated onto rectangle B using an anti-clockwise rotation of 90°. What are the coordinates of the centre of rotation for the anti-clockwise rotation?
Give your answer in the form (x, y).

- (7, 1)

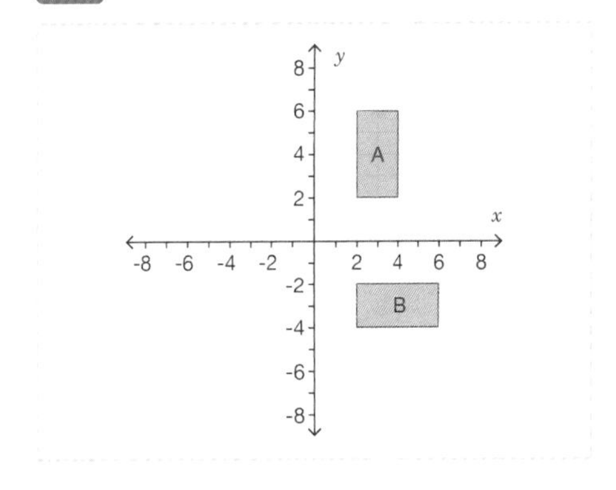

Translating Shapes and Describing Translations

Objective: I can translate shapes and describe translations through column vectors.

Quick Search Ref: 10576

Level 1: Understanding: Translate a point or shape through a column vector.

Required: 7/10 **Pupil Navigation:** on **Randomised:** off

1. A shape is translated by the column vector shown. Select the **two statements** that together describe this translation.

2/7

- 5 squares down
- 2 squares right
- 5 squares up
- **2 squares up**
- 5 squares left
- 2 squares down
- **5 squares right**

$$\begin{pmatrix} 5 \\ 2 \end{pmatrix}$$

2. A shape is translated by the column vector shown. Select the **two statements** that together describe this translation.

2/7

- 1 square up
- 3 squares left
- 1 square left
- **3 squares right**
- **1 square down**
- 3 squares up
- 1 square right

$$\begin{pmatrix} 3 \\ -1 \end{pmatrix}$$

3. Point P is translated by the column vector shown in the diagram. Select the coordinates of the point that P is translated to.

1/6

- A
- **B**
- C
- D
- E
- F

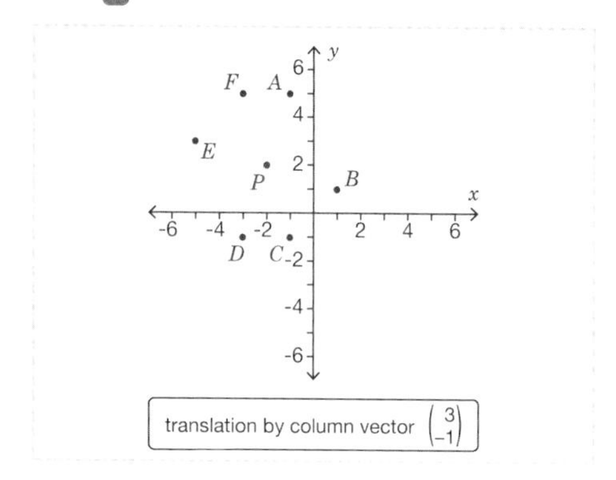

4. Point A is translated by the column vector shown in the diagram. What are the coordinates of the point that A is translated to?
Give your answer in the form (x, y).

- 5, -1
- **(5, -1)**
- (0, 4)
- (-1, 5)
- (5 -1)

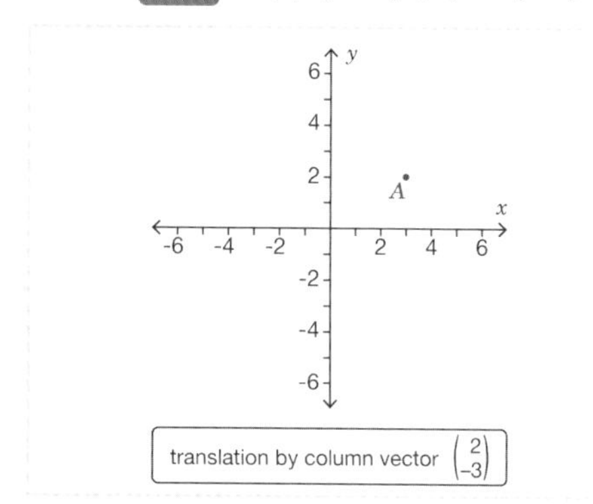

5. Shape B is mapped onto shape C using the column vector shown. Which image shows the correct position of shape C?

1/4

- (i)
- **(ii)**
- (iii)
- (iv)

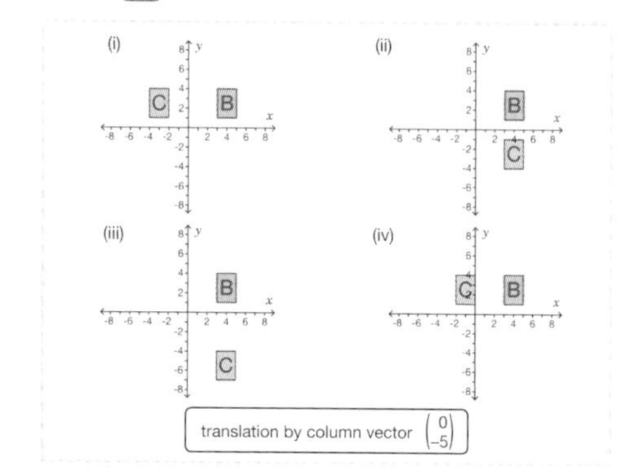

Level 1 *continued*

6. Triangle ABC is translated by the column vector shown in the diagram. What are the coordinates of the point that A is translated to?
Give your answer in the form (x, y).

- (-5, -1)
- (-1, -5)
- (-11, 5)
- -1, -5
- (-1 -5)

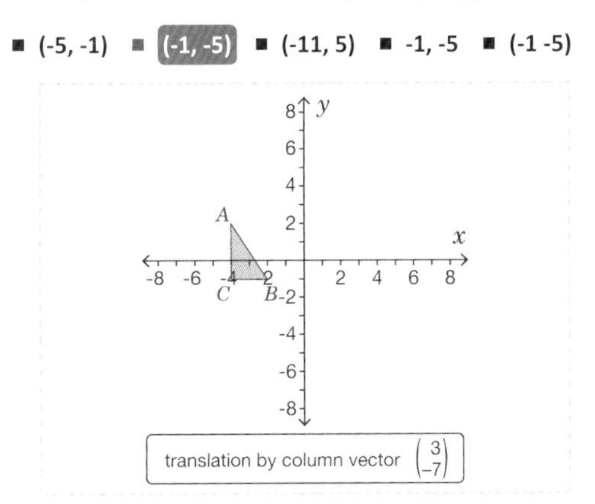

7. Shape T is translated by the column vector shown in the diagram. What are the coordinates of the point that P is translated to?
Give your answer in the form (x, y).

- (4, -1)
- (-5, 8)
- 4, -1
- (-1, 4)
- (4 -1)

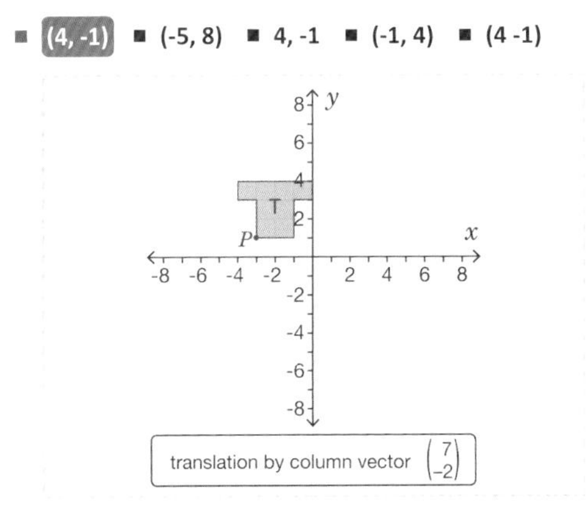

8. Shape S is mapped onto shape T using the column vector shown in the diagram. Which image shows the correct position of T?

1/4

- (i)
- (ii)
- (iii)
- (iv)

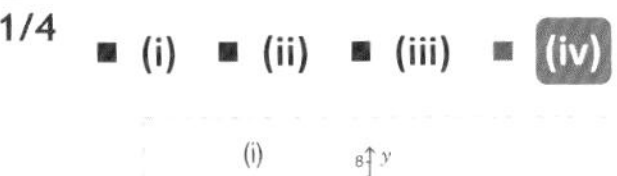

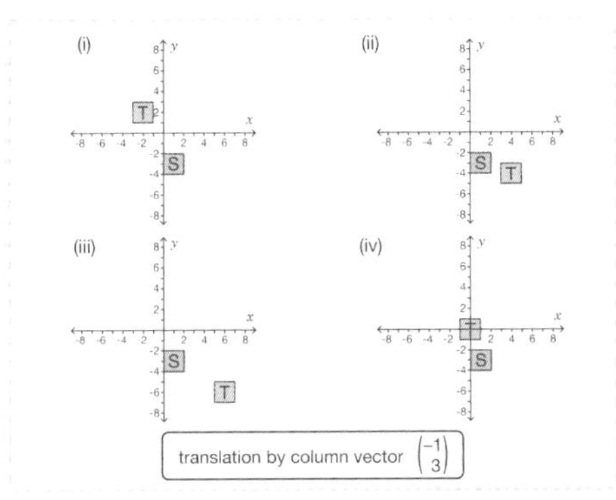

9. Rectangle $ABCD$ is translated by the column vector shown in the diagram. What are the coordinates of the point that C is translated to?
Give your answer in the form (x, y).

- (-1, 2)
- (2, -1)
- (2 -1)
- 2, -1
- (9, -8)

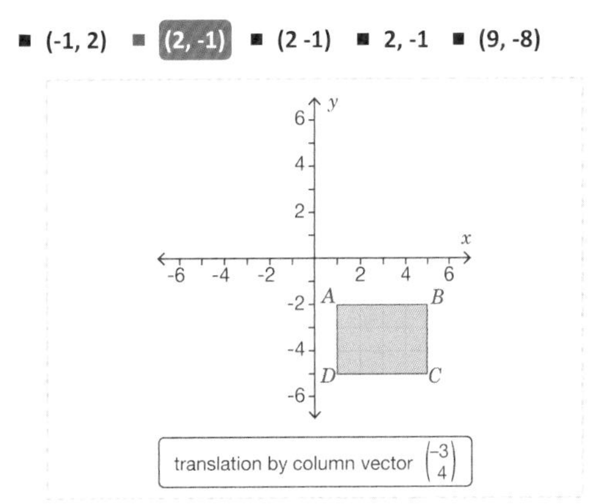

10. Shape A is translated by the column vector shown in the diagram. What are the coordinates of the point that P is translated to?
Give your answer in the form (x, y).

- 2, 2
- (2, 2)
- (2 2)
- (-3, 7)

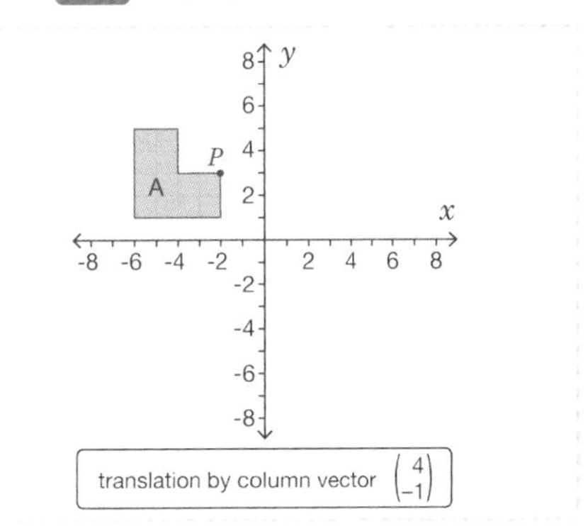

Level 2: Fluency: Describe a translation using a column vector.

Required: 7/10 **Pupil Navigation:** on

Randomised: off

11. Select the **two statements** that together describe the translation that maps shape A onto shape B.

2/7

- 1 square right
- 5 squares right
- 1 square up
- 5 squares up
- 1 square down
- 5 squares down
- 1 square left

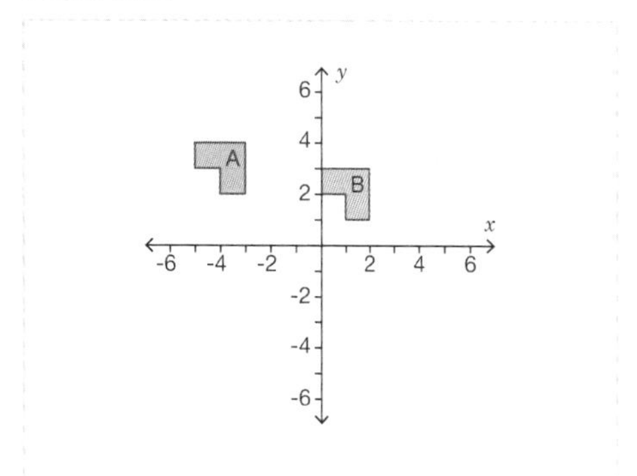

12. Select the column vector that maps point P onto point Q.

1/4

- (i)
- (ii)
- (iii)
- (iv)

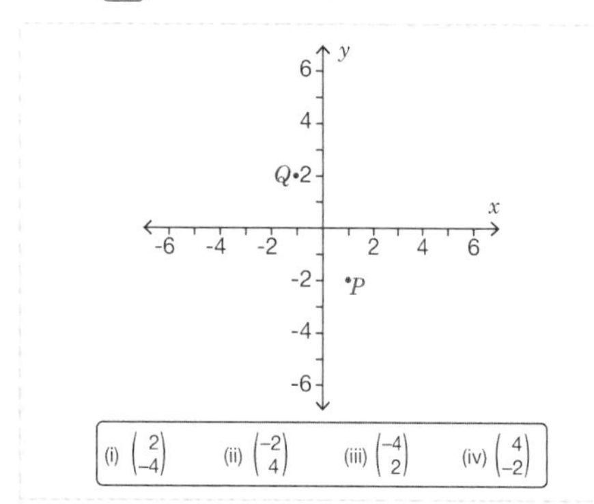

13. What number is missing from the top of the column vector that maps point C onto point D?

- 0
- 3

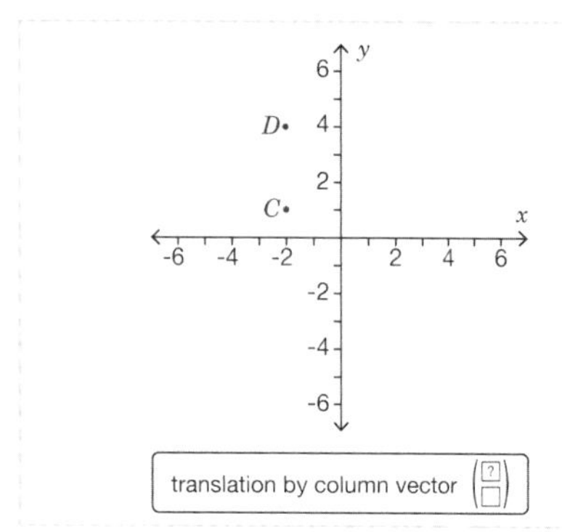

14. What number is missing from the bottom of the column vector that maps point A onto point B?

- 5
- -5
- 2

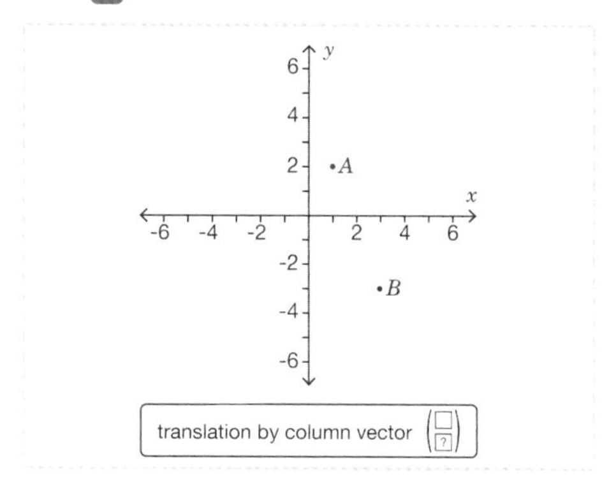

15. Select the column vector that maps shape R onto shape S.

1/4

- (i)
- (ii)
- (iii)
- (iv)

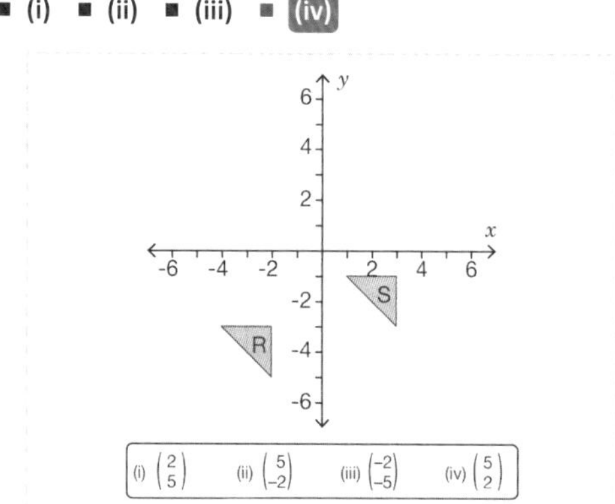

16. What number is missing from the top of the column vector that maps shape E onto shape F?

- -3
- -1
- 4

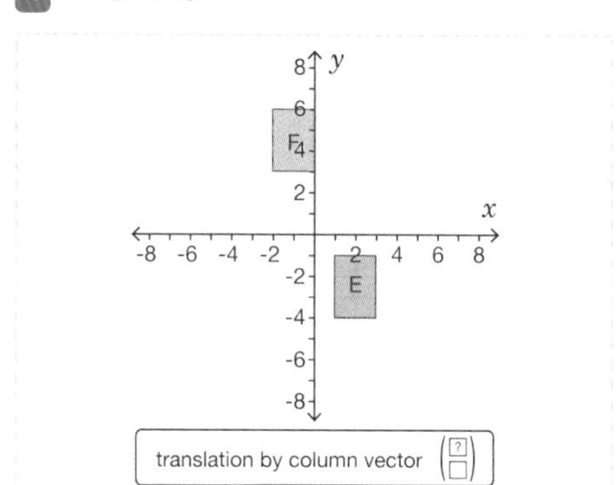

Level 2 *continued*

17. What number is missing from the bottom of the column vector that maps shape T to shape U?

▪ -4 ▪ -2 ▪ -3

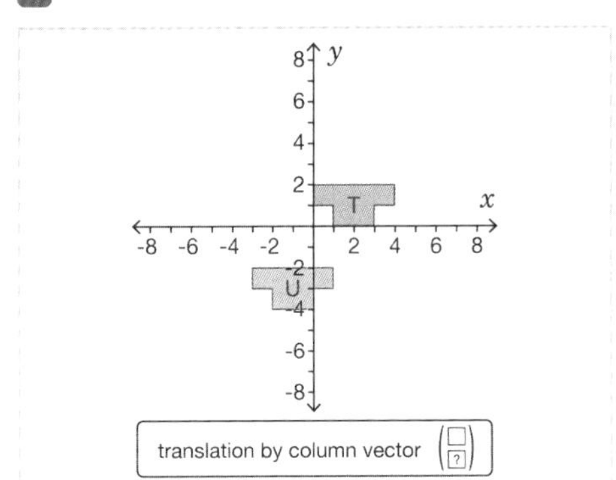

18. Select the column vector that maps shape A onto shape B.

1/4

▪ (i) ▪ (ii) ▪ (iii) ▪ (iv)

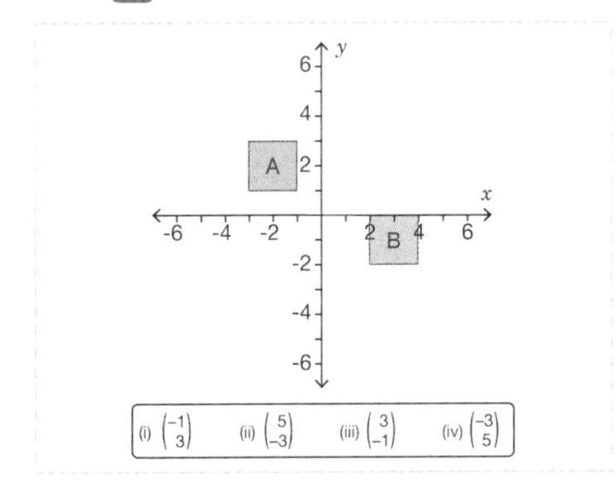

19. What number is missing from the top of the column vector that maps point R onto point S?

▪ -3 ▪ 8 ▪ 3

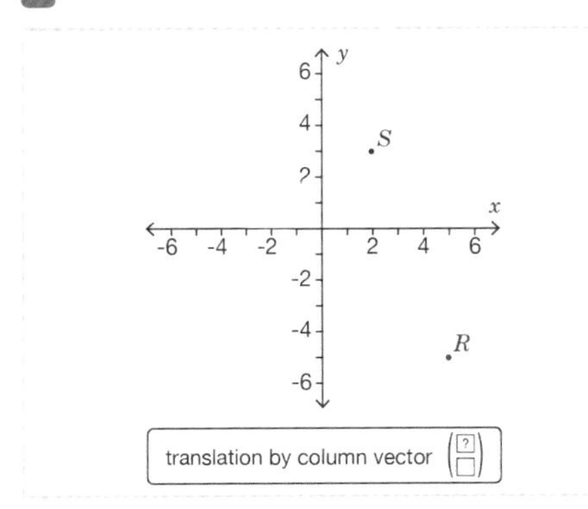

20. What number is missing from the bottom of the column vector that maps shape U onto shape V?

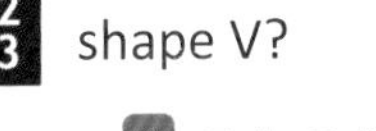

▪ -4 ▪ 4 ▪ 5

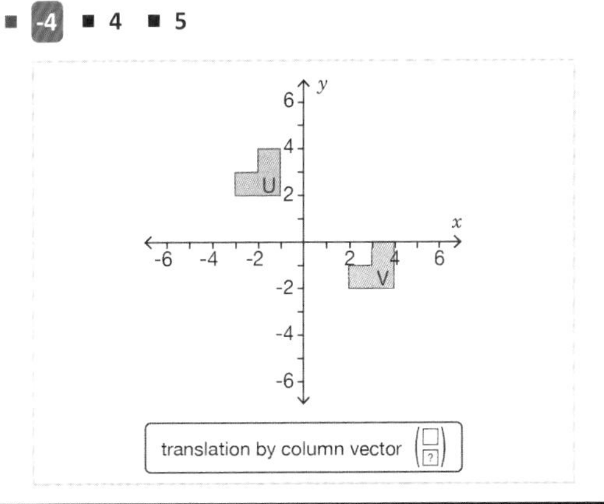

Level 3: Reasoning: Find the inverse of translations.

Required: 5/5 **Pupil Navigation:** on

Randomised: off

21. The column vector that maps shape A onto shape B is shown in the diagram. Select the column vector that maps shape B onto shape A.

1/4

▪ (i) ▪ (ii) ▪ (iii) ▪ (iv)

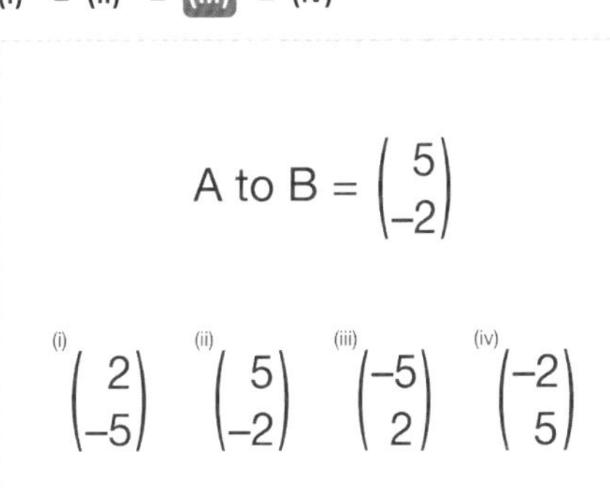

22. Tahir is answering a homework question on translations. Explain the mistake that Tahir has made.

- Open question, no set answer

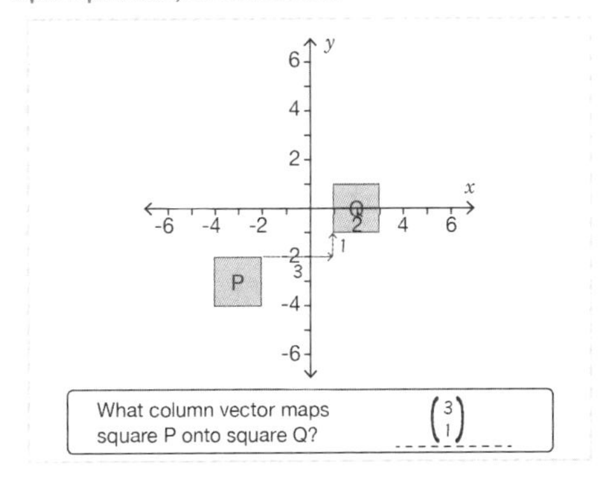

Level 3 *continued*

23. The column vector that maps shape M onto shape N is shown in the diagram. What number is missing from the bottom of the column vector that maps shape N onto shape M?

- **-3**
- -5

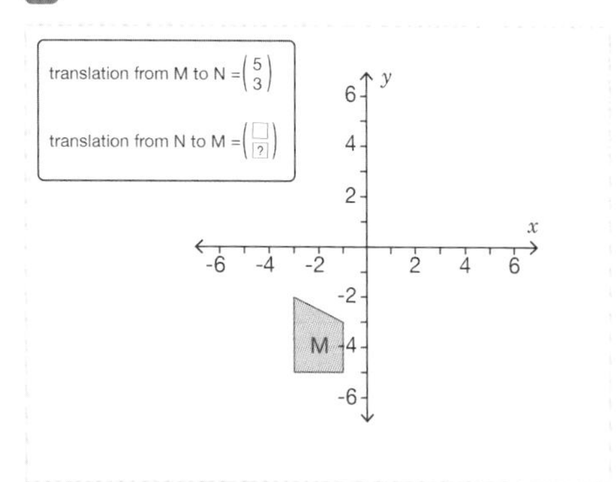

24. Renee says if you translate the point *P* by column vector **a** then column vector **b**, you will arrive at the same point as if you translate the point *P* by column vector **b** then column vector **a**. Is Renee correct? Explain your answer.

- Open question, no set answer

$$a = \begin{pmatrix}3\\-1\end{pmatrix} \quad b = \begin{pmatrix}-2\\5\end{pmatrix}$$

25. Point *C* is mapped onto point *D* by the column vector shown in the diagram. What is the mid-point of the line *CD*?
Give your answer in the form (x, y).

- **(1, -1)**
- (-1, 1)
- 1, -1
- (1 -1)

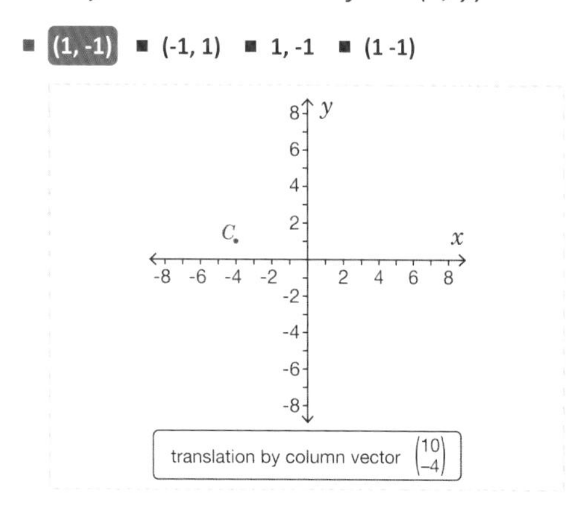

Level 4: Problem solving: Combine translations.

Required: 5/5 **Pupil Navigation:** on
Randomised: off

26. The diagram shows the column vectors that map shape A onto shape B, shape B onto shape C and shape C onto shape D.
What number is missing from the top of the column vector that maps shape A onto shape D?

- **3**
- 2

A to B = $\begin{pmatrix}4\\1\end{pmatrix}$ B to C = $\begin{pmatrix}-2\\2\end{pmatrix}$ C to D = $\begin{pmatrix}1\\-1\end{pmatrix}$ A to D = $\begin{pmatrix}?\\\square\end{pmatrix}$

27. The column vectors that map shape R onto shapes S and T are shown in the diagram. What number is missing from the bottom of the column vector that maps shape S onto shape T?

- **-5**
- 5

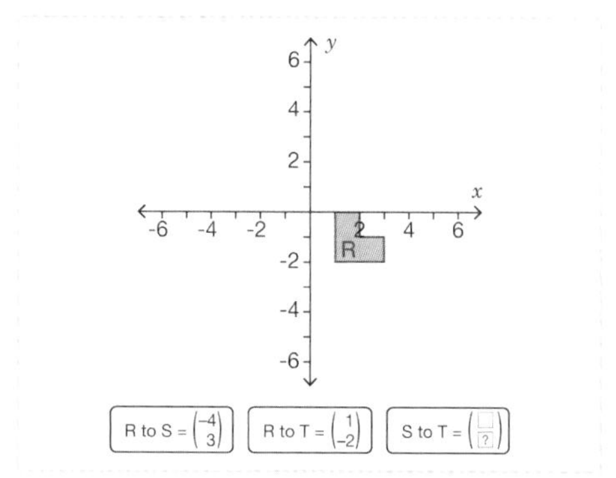

Level 4 *continued*

28. Point *P* is repeatedly translated by the column vector shown in the diagram. If all the images of point *P* are joined in a straight line, what is the equation of the line?
Give your answer in the form y = mx + c.

- **y = 2x + 5**

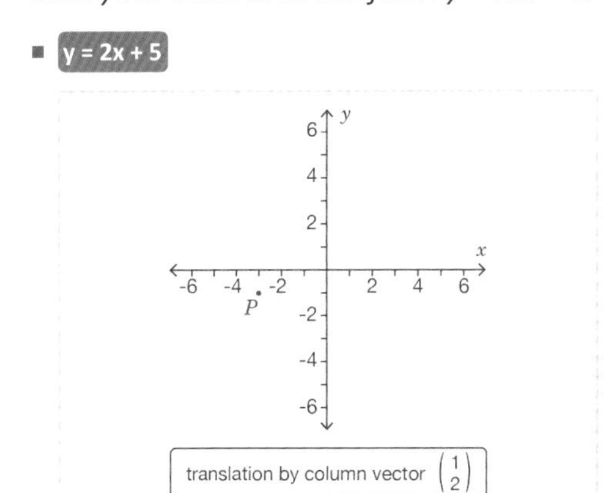

29. Shape R is reflected in the *x*-axis and then the *y*-axis. In its new position, it is labelled shape S. What is the missing top number in the column vector that maps shape R onto shape S?

- **9** ■ **4**

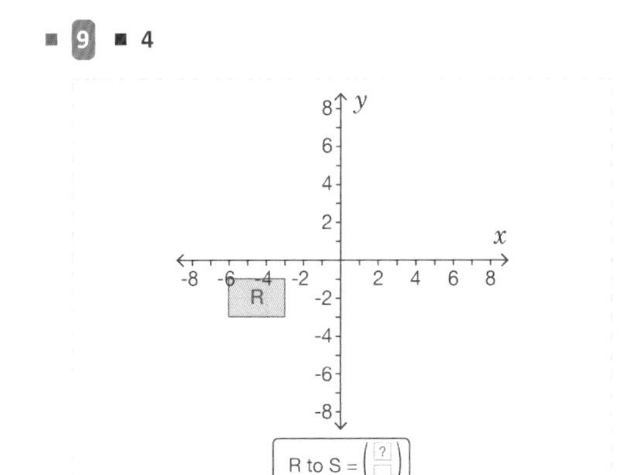

30. Which combination of column vectors **a**, **b**, and/or **c** would map the pirate onto the treasure?

1/3

- **a, a, a, b, c** ■ **a, b, b, c, c** ■ **a, b, b, b, c**

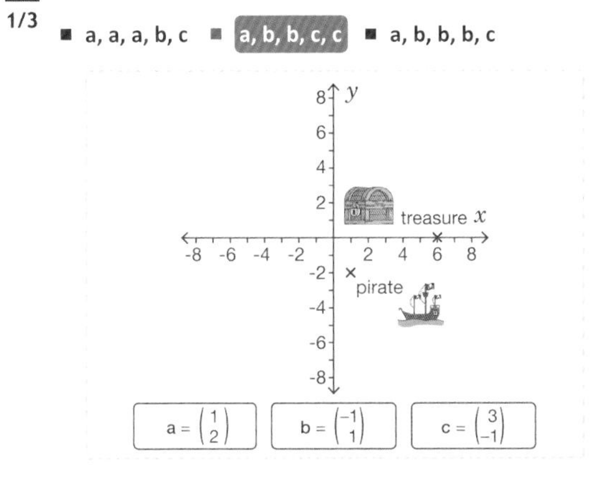

Find Missing Lengths in Similar Shapes

Objective: I can find missing lengths and angles in similar shapes.

Quick Search Ref: 10592

Level 1: Understanding: Finding missing sides and angles from two similar shapes of the same orientation.

Required: 7/10 **Pupil Navigation:** on **Randomised:** off

1. Select the **two statements** that define similar shapes.

2/4

- **Corresponding angles are equal.**
- Corresponding angles are different sizes but are all proportional.
- All pairs of corresponding sides are equal.
- **All pairs of corresponding sides are in the same ratio.**

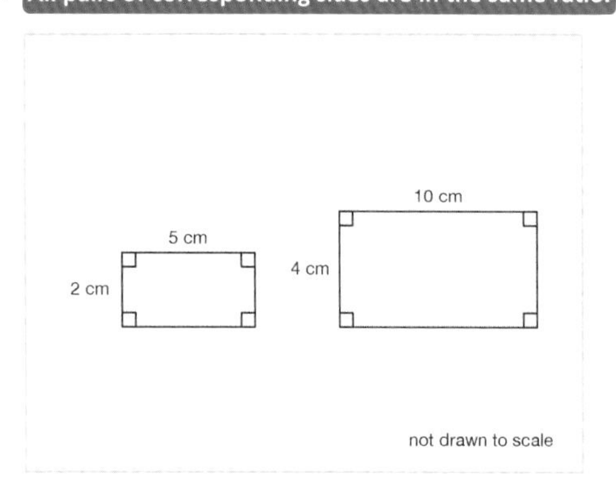

2. Select the scale factor for the enlargement of rectangle A to rectangle B.

1/4

- 14
- 12
- 2
- **3**

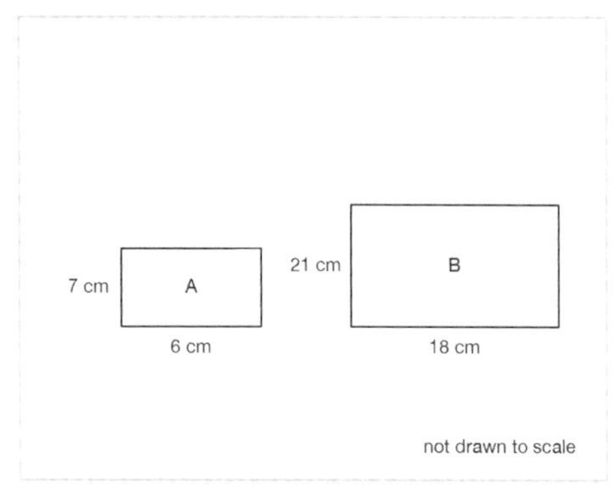

3. Select the shape that is similar to shape A.

1/5

- A
- B
- **C**
- D
- E

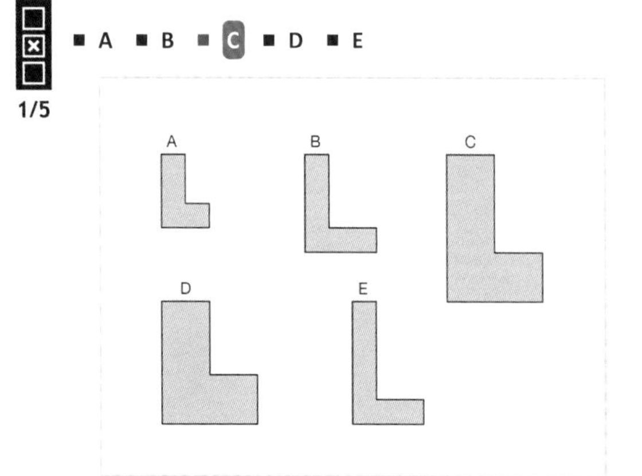

4. The diagram shows two similar rectangles. What is the value of x?
Include the units cm (centimetres) in your answer.

- **10 cm**
- **10 centimetres**
- 2.5 cm
- 9
- 10
- 9 cm
- 9 centimetres
- 2.5 centimetres
- 2.5

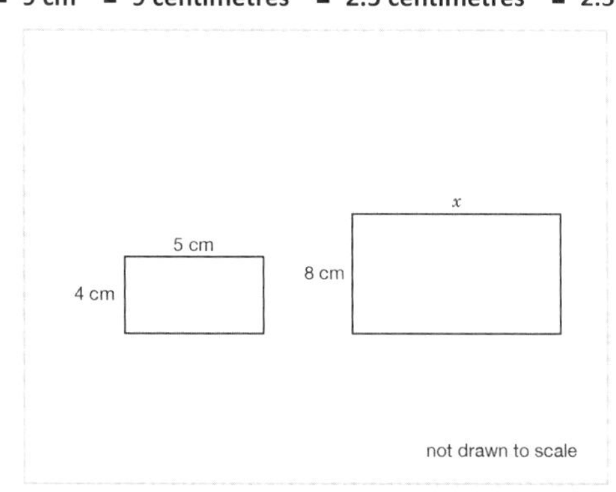

5. The diagram shows two similar shapes. What is the value of y in degrees?
Don't include the units in your answer.

- **33**
- 66

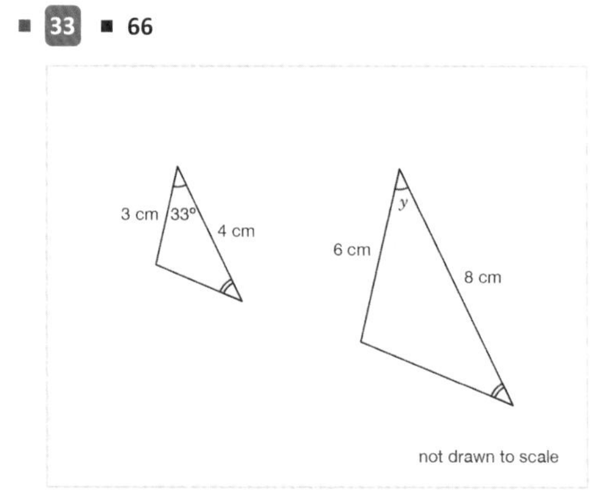

Level 1 *continued*

6. The diagram shows two similar shapes. What is the value of x?
Include the units cm (centimetres) in your answer.

- 35 centimetres ▪ 35 cm ▪ 1.4 cm
- 23 centimetres ▪ 1.4 ▪ 23 ▪ 1.4 centimetres
- 23 cm ▪ 35

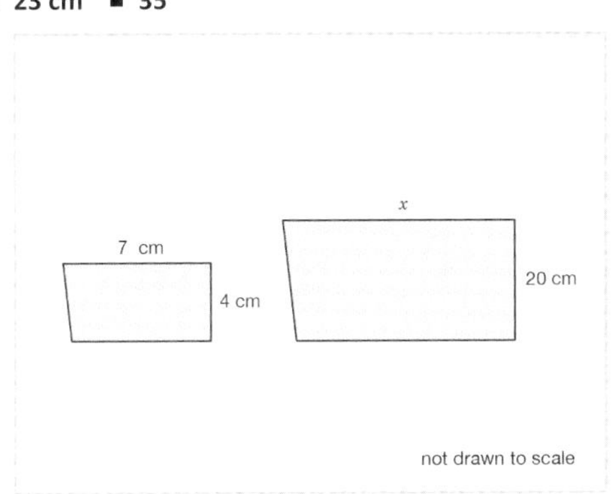

7. The diagram shows two similar parallelograms. What is the value of y?
Include the units cm (centimetres) in your answer.

- 18 centimetres ▪ 2 centimetres ▪ 2 cm ▪ 18 cm
- 2 ▪ 18

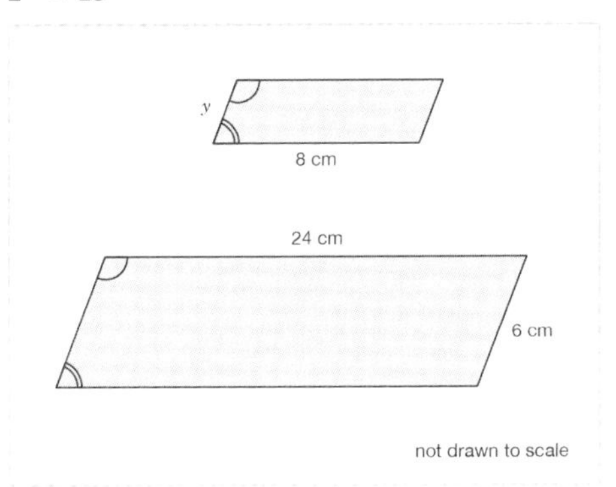

8. The diagram shows two similar shapes. What is the value of y in degrees?
Don't include the units in your answer.

- 20 ▪ 40

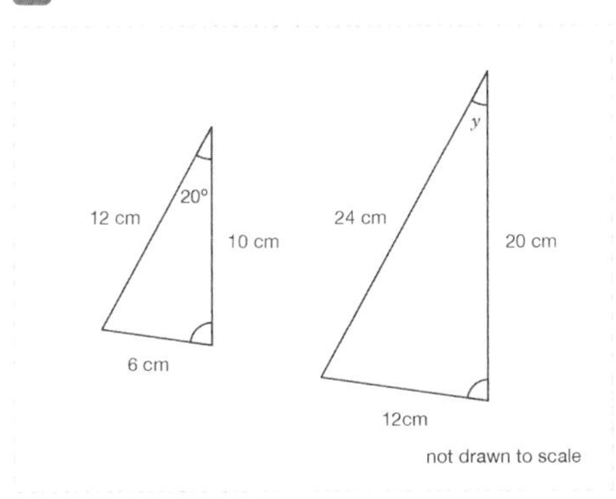

9. The diagram shows two similar rectangles. What is the value of y?
Include the units cm (centimetres) in your answer.

- 7 cm ▪ 7 centimetres ▪ 252 cm ▪ 252 ▪ 7
- 252 centimetres ▪ 27 ▪ 27 cm ▪ 27 centimetres

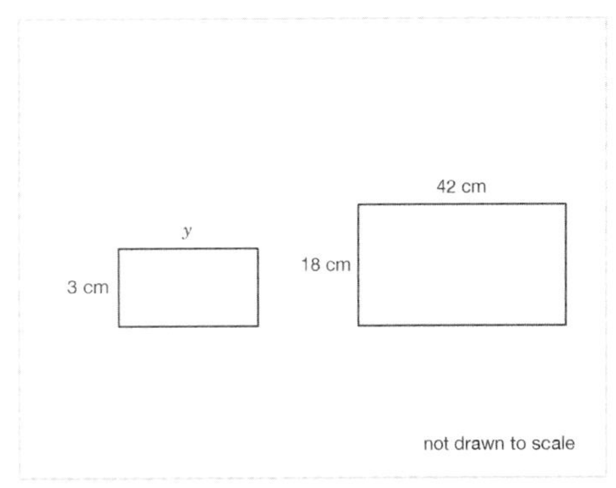

10. The diagram shows two parallelograms. What is the value of y?
Include the units cm (centimetres) in your answer.

- 168 ▪ 168 cm ▪ 10.5 ▪ 60 cm
- 168 centimetres ▪ 10.5 cm ▪ 60 ▪ 60 centimetres
- 10.5 centimetres

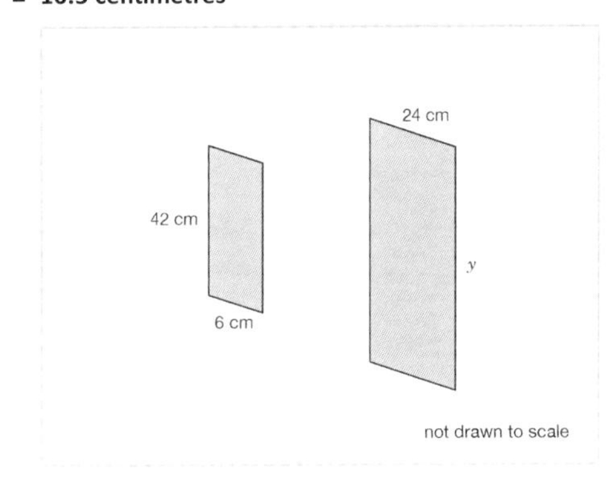

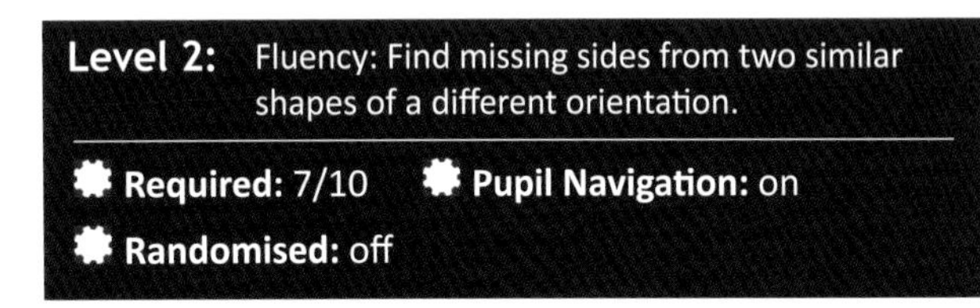

Level 2: Fluency: Find missing sides from two similar shapes of a different orientation.

Required: 7/10 **Pupil Navigation:** on

Randomised: off

11. The diagram shows two similar shapes. What is the value of x?
Include the units cm (centimetres) in your answer.

- **27 centimetres** ■ **3** ■ **27 cm** ■ **27** ■ **3 cm**
- **18 cm** ■ **18 centimetres** ■ **18** ■ **3 centimetres**

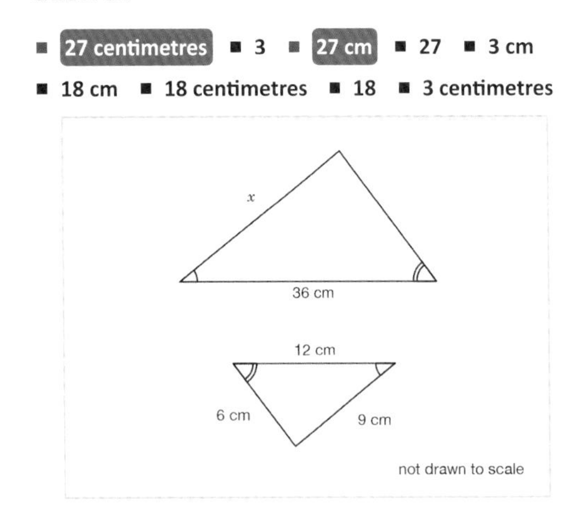

12. The diagram shows two similar shapes. What is the value of y?
Include the units cm (centimetres) in your answer.

- **3 centimetres** ■ **3 cm** ■ **36** ■ **4 centimetres**
- **4 cm** ■ **36 centimetres** ■ **3** ■ **36 cm** ■ **4**

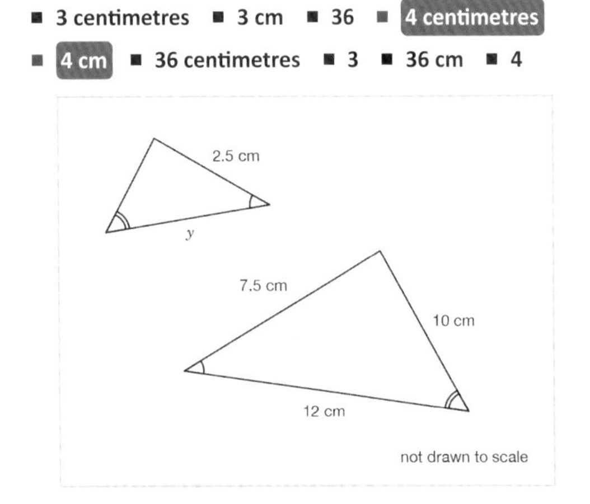

13. The diagram shows two similar shapes. What is the value of y?
Include the units cm (centimetres) in your answer. Give your answer as a decimal.

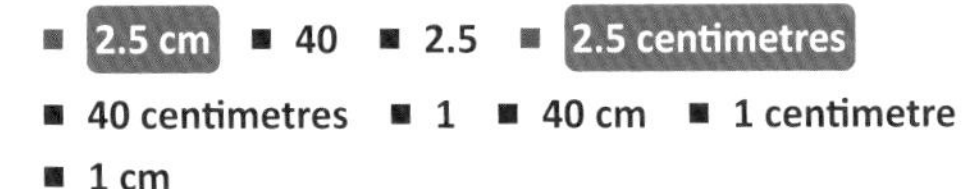

- **2.5 cm** ■ **40** ■ **2.5** ■ **2.5 centimetres**
- **40 centimetres** ■ **1** ■ **40 cm** ■ **1 centimetre**
- **1 cm**

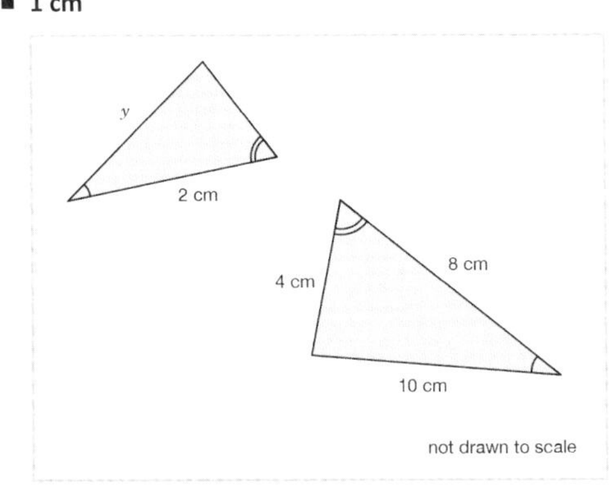

14. Side BE is parallel to side CD. Sides ABC and AED are straight lines. What is the length of side CD?
Include the units cm (centimetres) in your answer.

- **4 centimetres** ■ **9 cm** ■ **9 centimetres** ■ **4 cm**
- **3 cm** ■ **4** ■ **9** ■ **3** ■ **3 centimetres.**

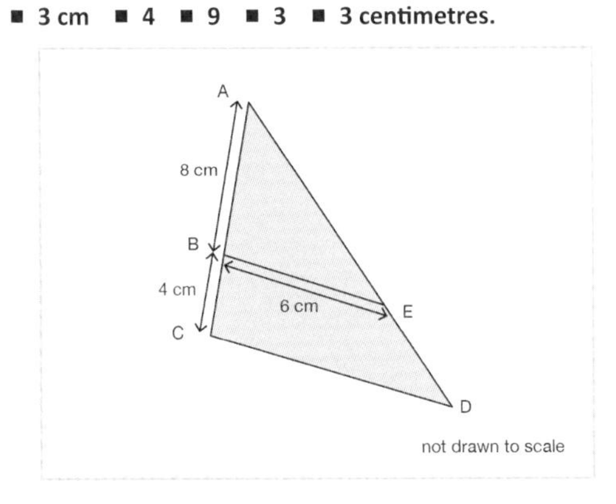

15. Side AB is parallel to side DE. ACE and BCD are straight lines. What is the length of side BC?
Include the units cm (centimetres) in your answer.

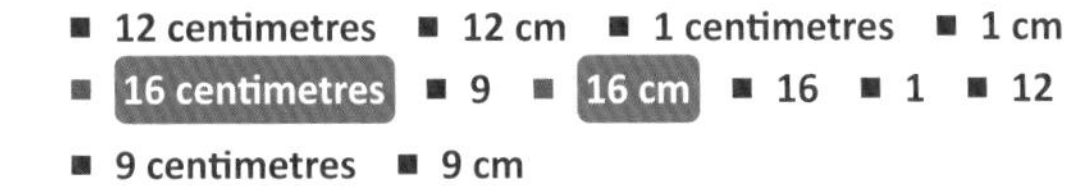

- **12 centimetres** ■ **12 cm** ■ **1 centimetres** ■ **1 cm**
- **16 centimetres** ■ **9** ■ **16 cm** ■ **16** ■ **1** ■ **12**
- **9 centimetres** ■ **9 cm**

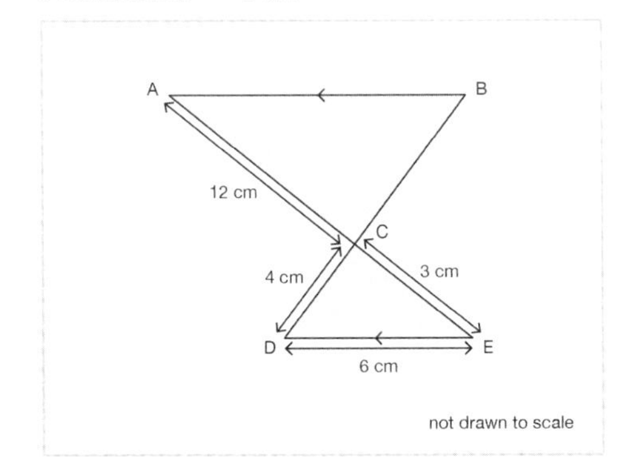

Level 2 *continued*

16. A builder constructs a timber frame as shown in the diagram. Line BE is parallel to side CD. Sides ABC and AED are straight lines. What is the length of side AC?
Give your answer to 1 decimal place and include the units m (metres).

- 4.4 m ■ 2.2 ■ 4.4 ■ 4.4 metres ■ 1.1 ■ 1.1 m
- 2.2 metres ■ 1.1 metres ■ 2.2 m

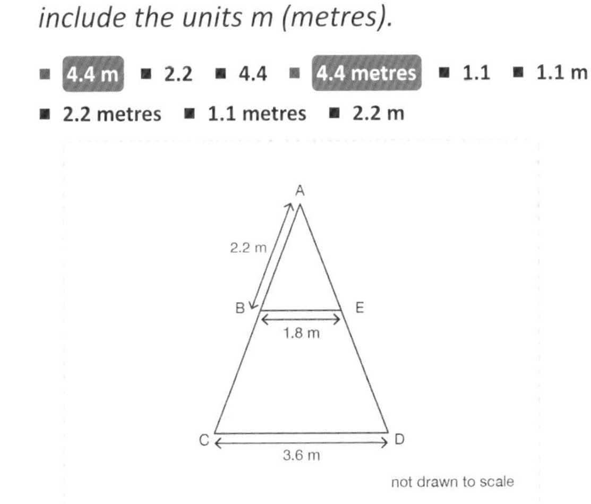

17. Side AB is parallel to side DE. ACE and BCD are straight lines. What is the length of line AE?
Give your answer to 1 decimal place and include the units cm (centimetres).

- 10 centimetres ■ 45 cm ■ 45 centimetres ■ 45
- 10 ■ 10 cm

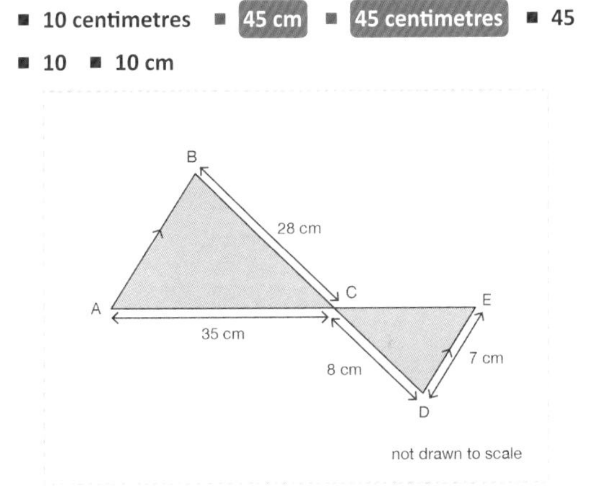

18. Line BE is parallel to side CD. ABC and AED are straight lines. What is the length of BE?
Include the units cm (centimetres) in your answer.

- 10 centimetres ■ 8 cm ■ 8 centimetres ■ 200
- 200 centimetres ■ 10 ■ 10 cm ■ 8 ■ 200 cm

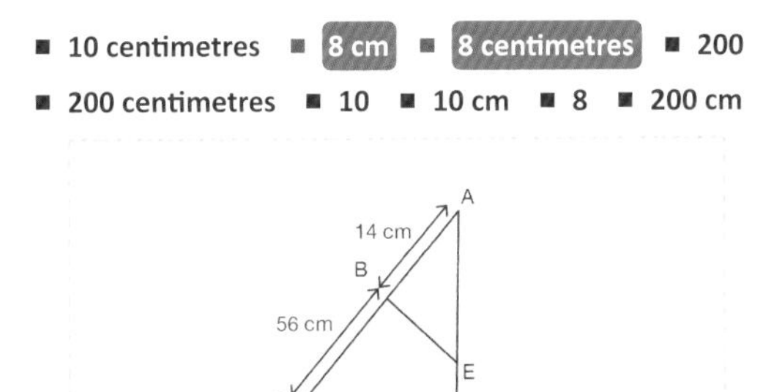

19. The diagram shows two similar shapes. What is the length of side y?
Include the units cm (centimetres) in your answer.

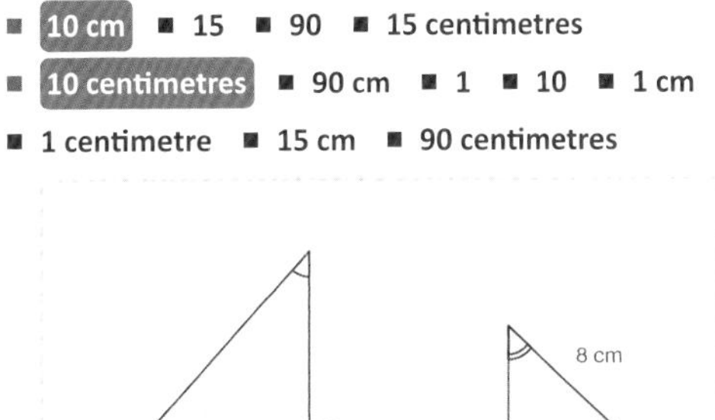

- 10 cm ■ 15 ■ 90 ■ 15 centimetres
- 10 centimetres ■ 90 cm ■ 1 ■ 10 ■ 1 cm
- 1 centimetre ■ 15 cm ■ 90 centimetres

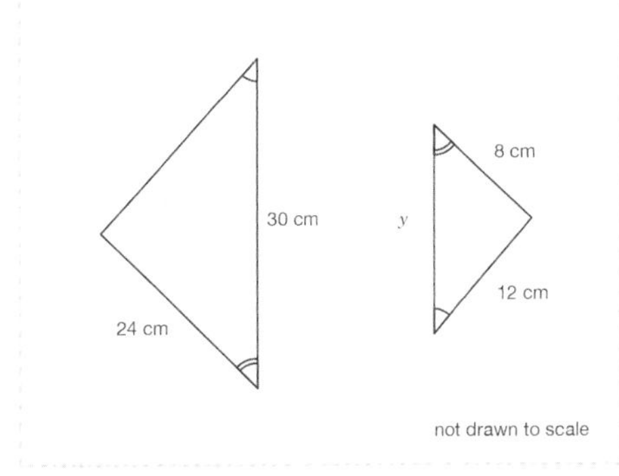

20. Side AB is parallel to side DE. ACE and BCD are straight lines. What is the length of side DE?
Include the units cm (centimetres) in your answer.

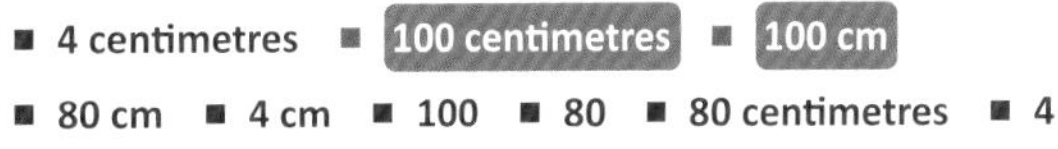

- 4 centimetres ■ 100 centimetres ■ 100 cm
- 80 cm ■ 4 cm ■ 100 ■ 80 ■ 80 centimetres ■ 4

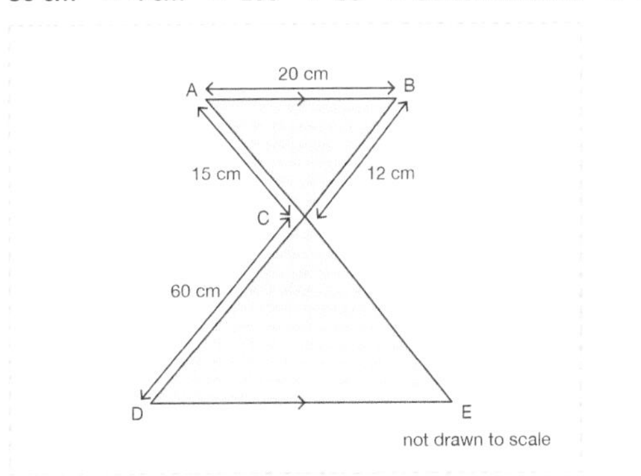

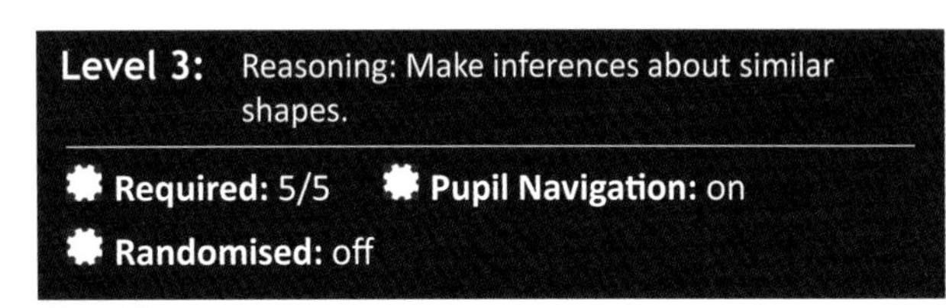

21. AB is parallel to DE. ACE and BCD are straight lines. Select the two sentences that explain how you know that triangle ABC is similar to triangle EDC.

2/5

- Corresponding angles are equal.
- Alternate angles are equal.
- Angles at a point add to 360°.
- Angles on a straight line add to 180°.
- Vertically opposite angles are equal.

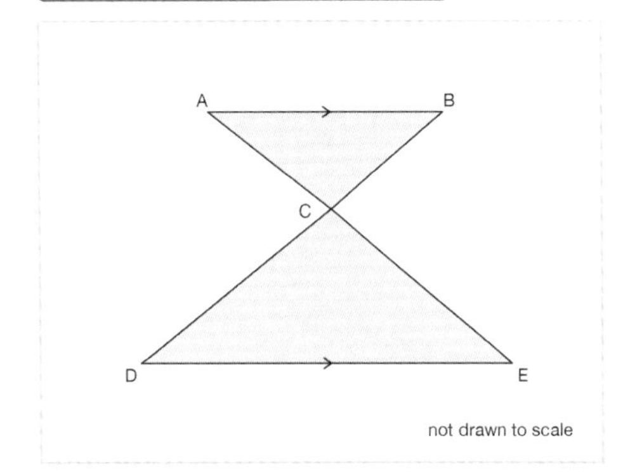

22. The diagram shows the dimensions of two rectangular bathroom tiles. Prove that they are similar.

- Open question, no set answer

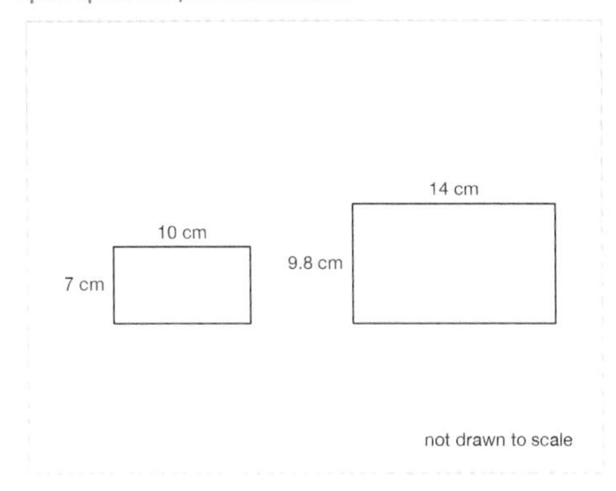

23. Are these two triangles similar? Explain your answer.

- Open question, no set answer

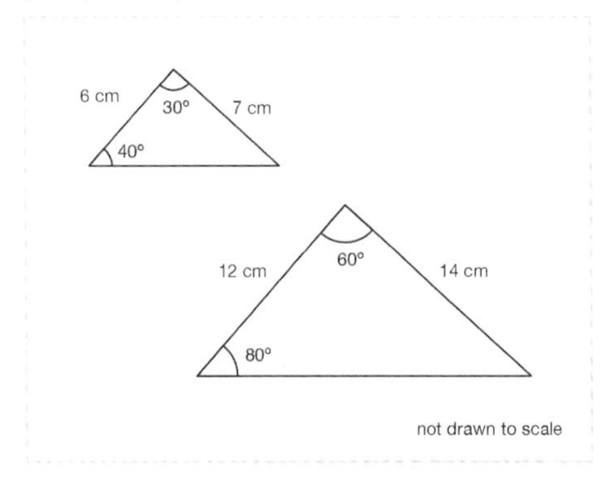

24. The diagram shows the dimensions for two plots of land. Are the plots of land similar? Explain your answer.

- Open question, no set answer

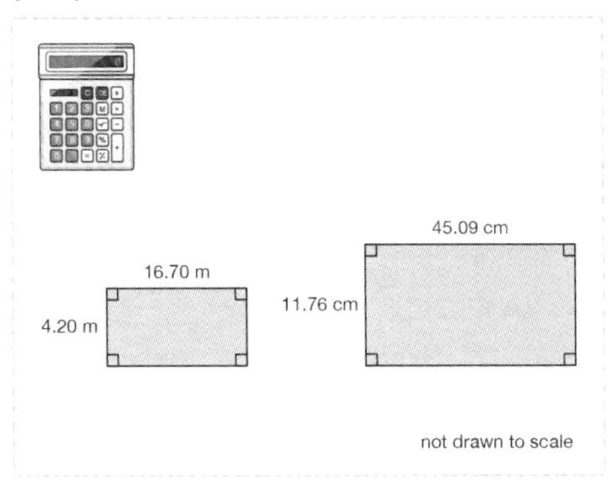

25. The diagram shows the dimensions of two rectangles. Zara says that the corresponding sides have a scale factor of 2, so the area of rectangle B must be twice as large as the area of rectangle A. Is she correct? Explain your answer.

- Open question, no set answer

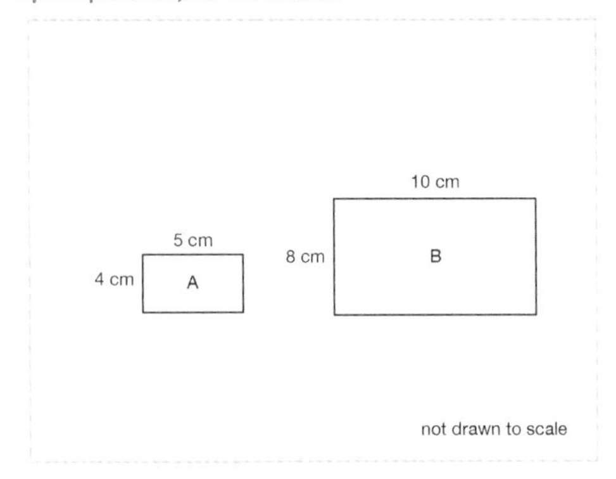

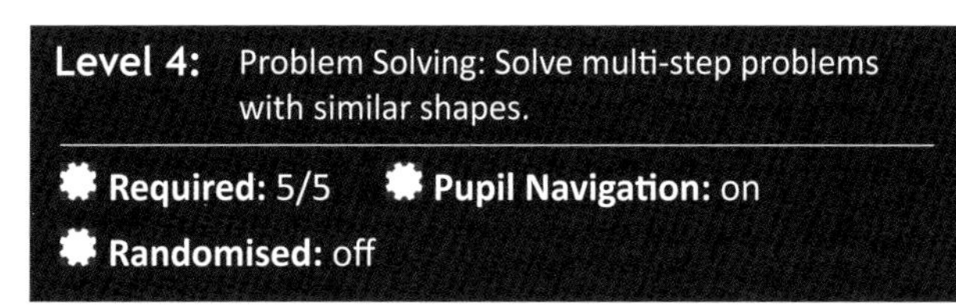

26. AB is parallel to DE. ACE and BCD are straight lines. What is the total perimeter of both triangles?
Give your answer to 2 decimal places and include the units cm (centimetres).
Hint: use the scale factor as an improper fraction and not a decimal.

- 4.71
- 5.73
- 4.71 cm
- 40.44
- **40.44 cm**
- **40.44 centimetres**
- 40.4
- 40.4 cm
- 5.73 centimetres.
- 4.71 centimetres
- 40.4 centimetres
- 5.73 cm

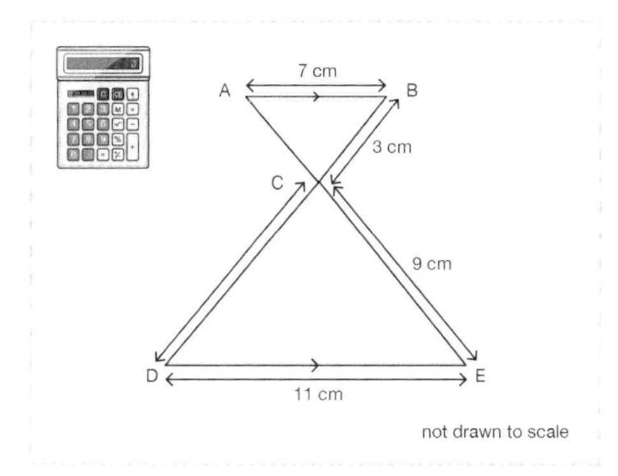

27. A builder constructs a timber frame as shown in the diagram. Line BE is parallel to side CD. ABC and AED are straight lines. What is the length of DE?
Give your answer as an improper fraction and include the units m (metres).

- 90/7 m
- **48/7 m**
- **48/7 metres**
- 90/7
- 48/7
- 90/7 metres

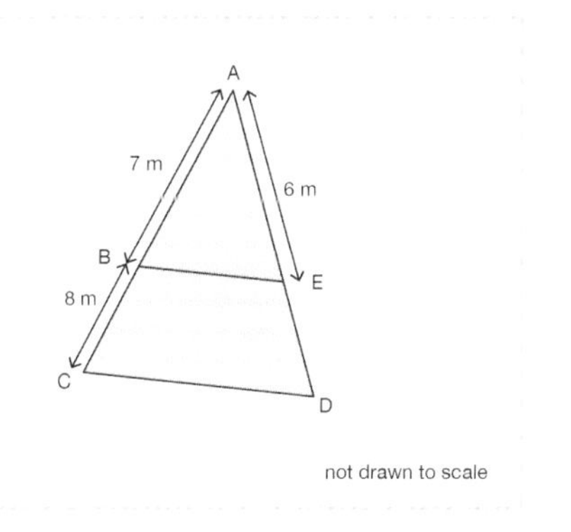

28. The dimensions of an image are 13 cm wide by 22 cm tall. A teacher wants to photocopy the image so that its dimensions are 18.2 cm wide by 30.8 cm tall. By what percentage should the teacher increase the image?
Give your answer as a percentage and include the % sign in your answer.

- 140
- **140%**
- 1.4
- 1.4%
- 40
- 40%

13 cm, 22 cm, 18.2 cm, 30.8 cm

not drawn to scale

29. Line BE is parallel to side CD. ABC and AED are straight lines. What is the length of line AB?
Include the units cm (centimetres) in your answer.

- **3 centimetres**
- **3 cm**
- 3

C, B, A, E, D; 7.5 cm, 31.5 cm, 9 cm, 6 cm

not drawn to scale

30. The diagram shows the dimensions of two similar rectangles. Calculate the area of rectangle B in cm^2.
Don't include the units in your answer.

- **48**
- 24

A: 4 cm, 3 cm; B: 8 cm

not drawn to scale

Identify Congruent Triangles

Objective: I can identify congruent triangles using proof.

Quick Search Ref: 10598

Level 1: Understanding: Identify features of congruency.

Required: 7/10 **Pupil Navigation:** on **Randomised:** off

1. Select the **two properties** of congruent shapes.

2/4

- **Corresponding sides are equal in length.**
- Corresponding sides of different lengths are in the same proportion.
- The shapes have the same orientation.
- **Corresponding angles are equal.**

2. Which two triangles are congruent?

2/4

- **A** ▪ **B** ▪ C ▪ D

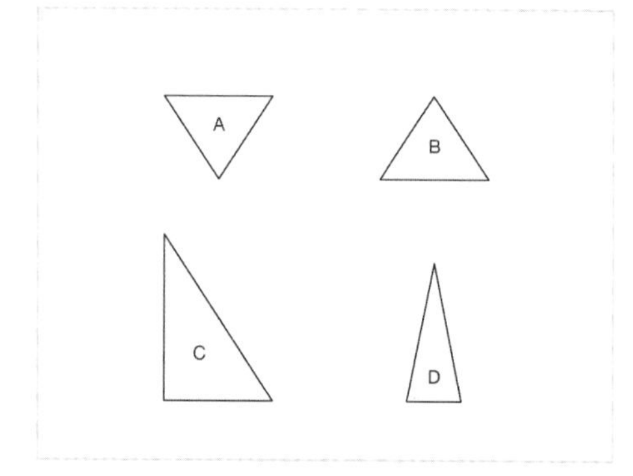

3. Which two triangles are congruent?

2/4

- A ▪ B ▪ **C** ▪ **D**

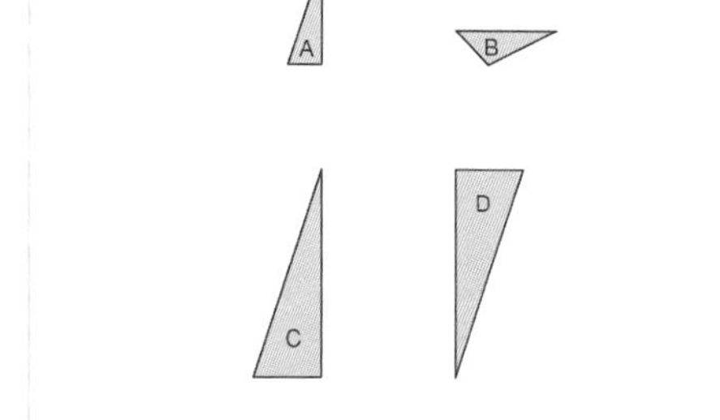

4. The diagram shows two congruent triangles. Select the point on triangle *DEF* that corresponds to point *A* on triangle *ABC*.

1/3

- D ▪ E ▪ **F**

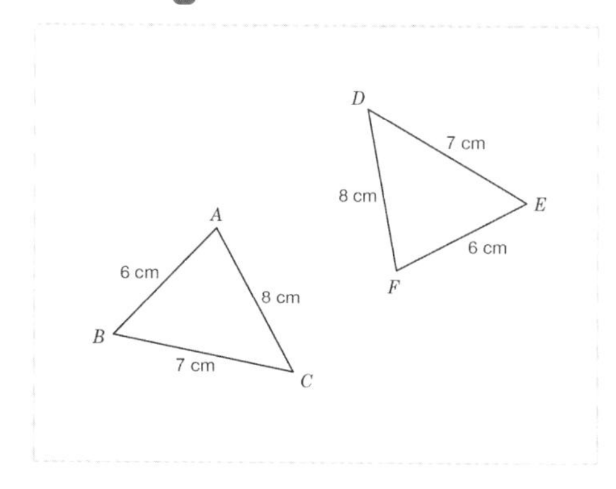

5. Are these two triangles congruent?

1/2

- **yes** ▪ no

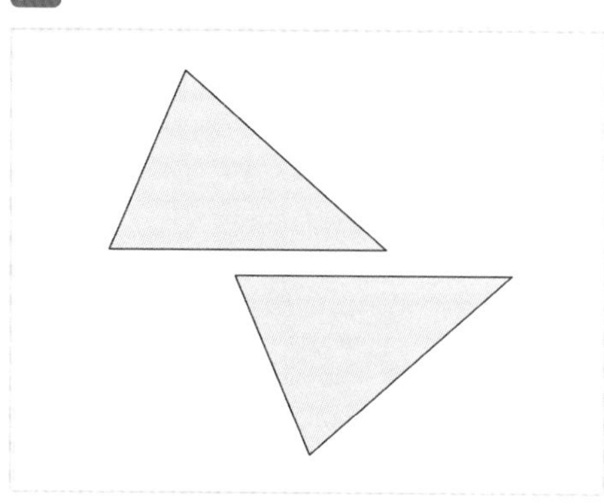

6. The diagram shows two congruent triangles with an equal angle marked. What is the length of side *DE*?
Include the units cm (centimetres) in your answer.

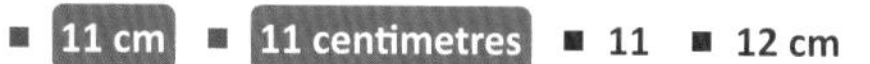

- **11 cm** ▪ **11 centimetres** ▪ 11 ▪ 12 cm ▪ 12 centimetres ▪ 12

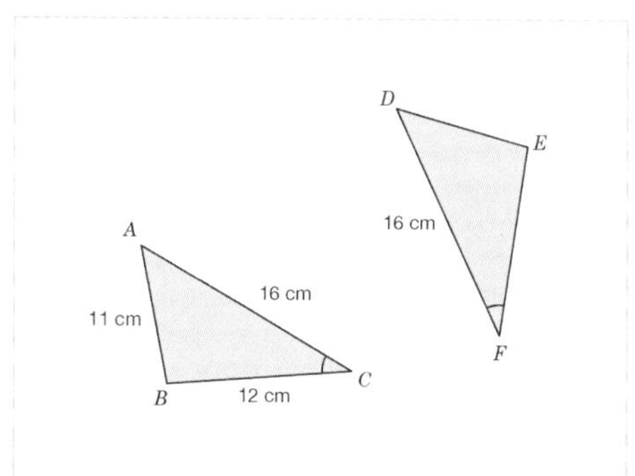

Level 1 *continued*

7. The diagram shows two congruent triangles. Which angle corresponds with angle *b*?

1/3

■ d ■ e ■ f

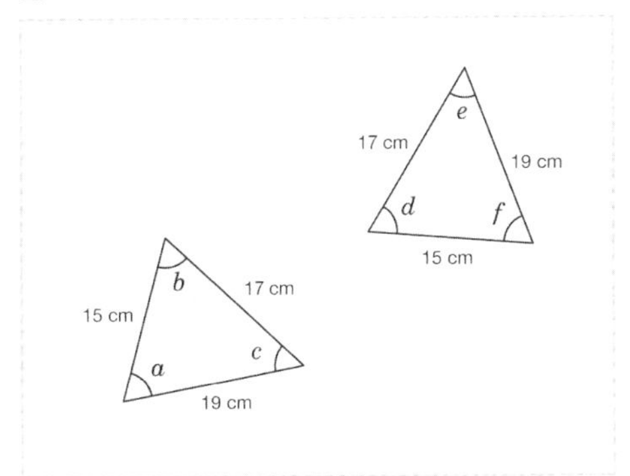

8. Which two triangles are congruent?

2/4

■ A ■ B ■ C ■ D

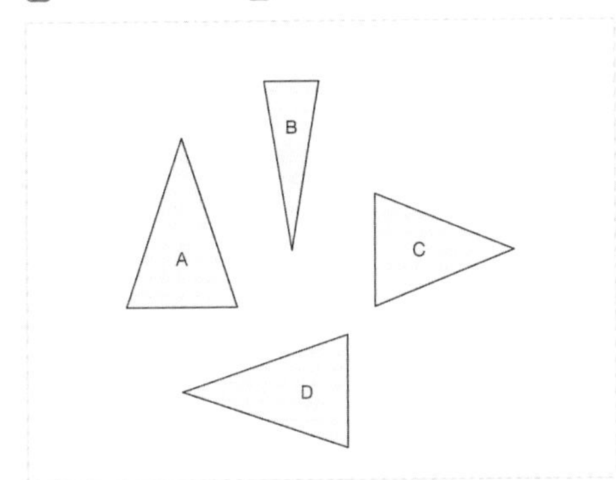

9. The diagram shows two congruent triangles. Select the point on triangle *DEF* that corresponds to point *A* on triangle *ABC*.

1/3

■ D ■ E ■ F

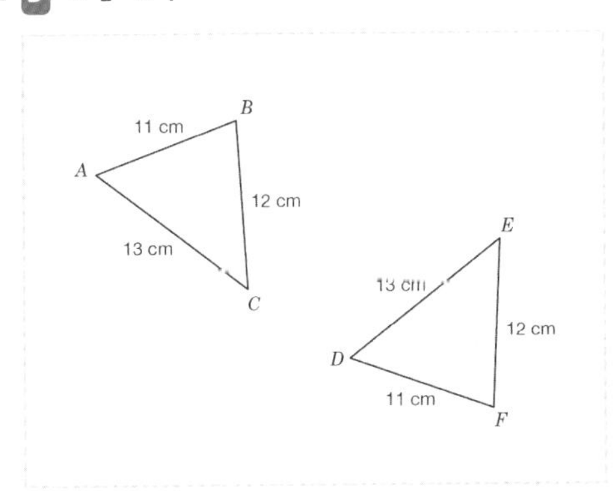

10. The diagram shows two congruent triangles. What is the length of side *DE*? *Include the units cm (centimetres) in your answer.*

■ 17 cm ■ 18 centimetres ■ 17 centimetres ■ 18 ■ 17 ■ 18 cm

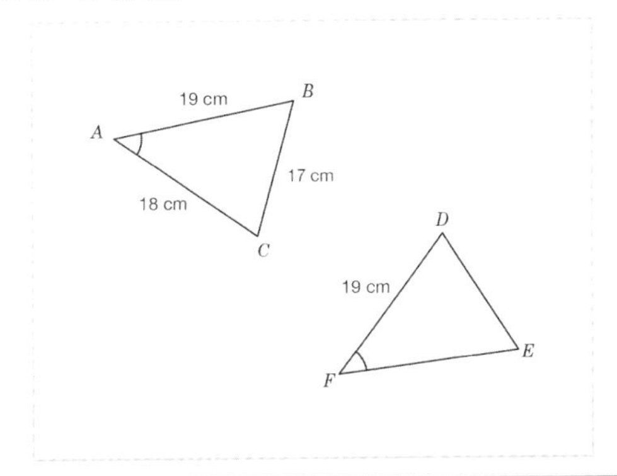

Level 2: Fluency: Identify the condition that proves two triangles are congruent.

Required: 7/10 **Pupil Navigation:** on

Randomised: off

11. Which condition proves that these triangles are congruent?

1/4

- The triangles are congruent by SSS.
- The triangles are congruent by SAS.
- The triangles are congruent by AAS.
- The triangles are congruent by RHS.

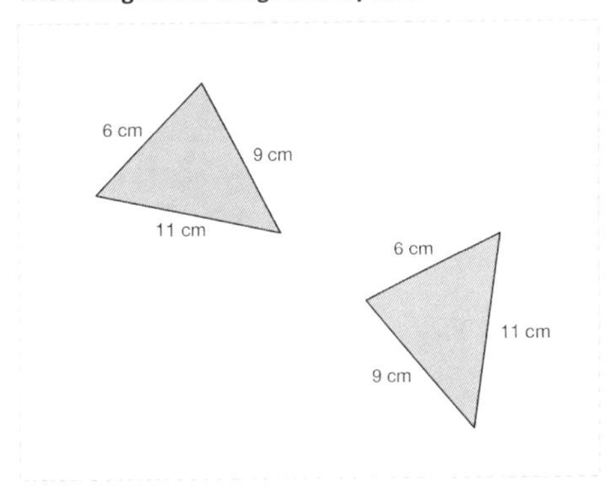

12. Which condition proves that these triangles are congruent?

1/4

- The triangles are congruent by SSS.
- The triangles are congruent by SAS.
- The triangles are congruent by AAS.
- The triangles are congruent by RHS.

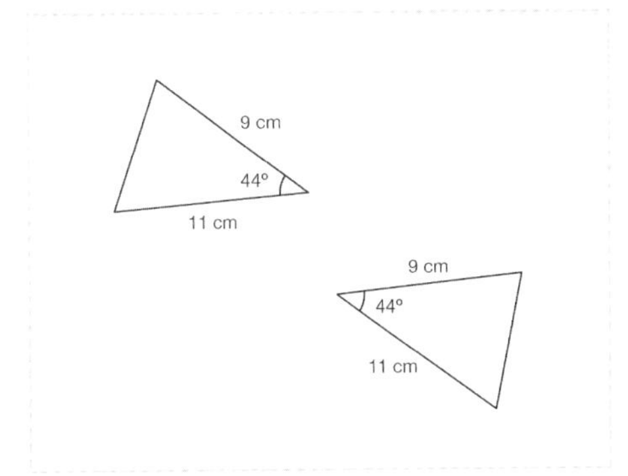

Level 2 *continued*

13. Which condition proves that these triangles are congruent?

1/4

- The triangles are congruent by SSS.
- The triangles are congruent by SAS.
- The triangles are congruent by AAS.
- **The triangles are congruent by RHS.**

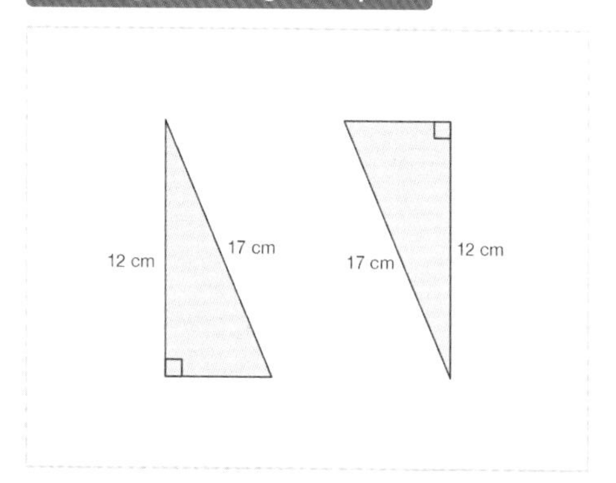

14. Are these triangles congruent?

1/5

- Yes, the triangles are congruent by SSS.
- Yes, the triangles are congruent by SAS.
- **Yes, the triangles are congruent by AAS.**
- Yes, the triangles are congruent by RHS.
- No, the triangles are not congruent.

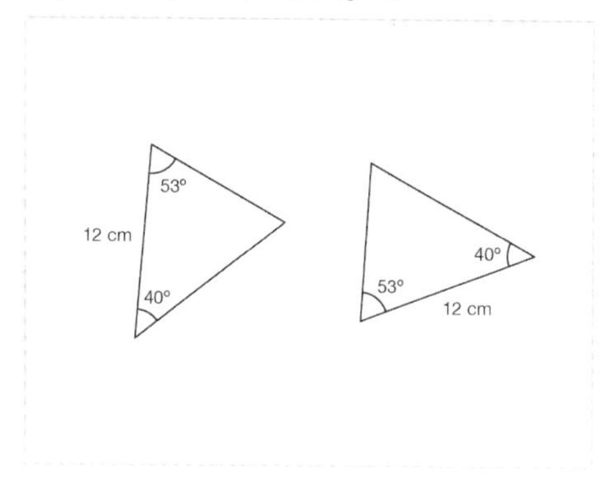

15. Are these triangles congruent?

1/5

- Yes, the triangles are congruent by SSS.
- Yes, the triangles are congruent by SAS.
- Yes, the triangles are congruent by AAS.
- Yes, the triangles are congruent by RHS.
- **No, the triangles are not congruent.**

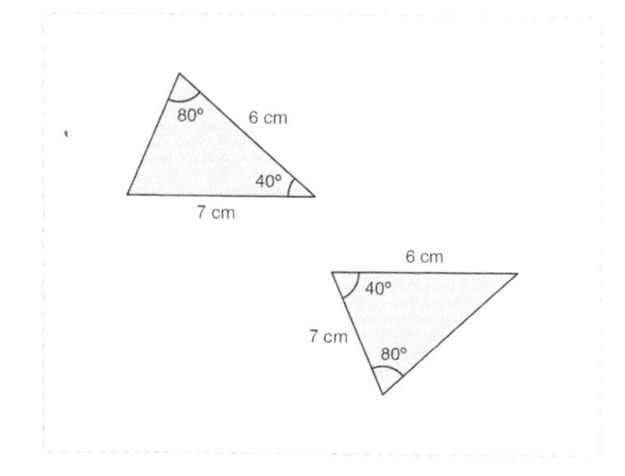

16. Which condition proves that these triangles are congruent?

1/4

- **The triangles are congruent by SSS.**
- The triangles are congruent by SAS.
- The triangles are congruent by AAS.
- The triangles are congruent by RHS.

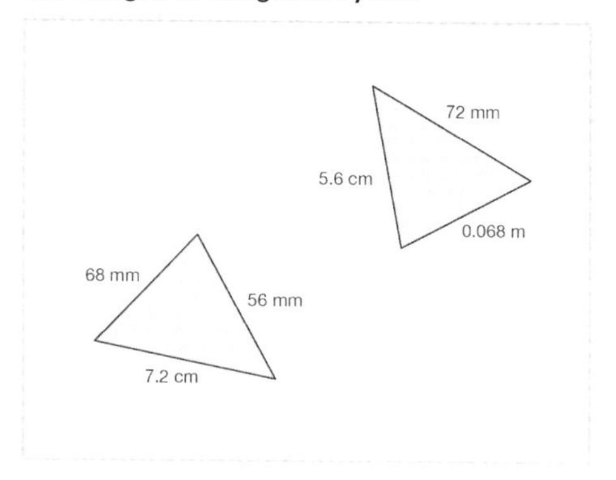

17. Which condition proves that these triangles are congruent?

1/4

- The triangles are congruent by SSS.
- The triangles are congruent by SAS.
- **The triangles are congruent by AAS.**
- The triangles are congruent by RHS.

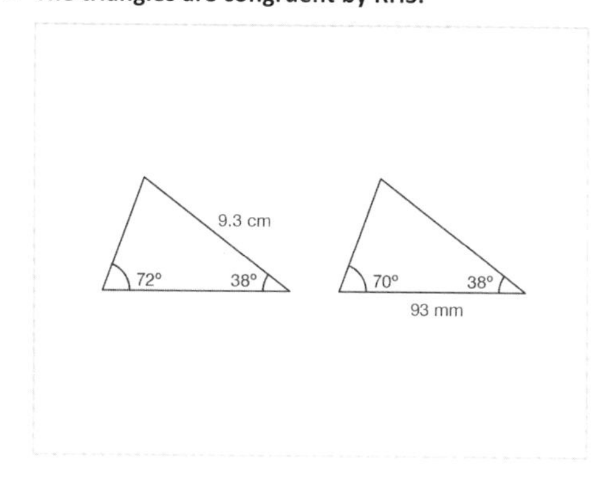

18. Which condition proves that these triangles are congruent?

1/4

- The triangles are congruent by SSS.
- The triangles are congruent by SAS.
- The triangles are congruent by AAS.
- **The triangles are congruent by RHS.**

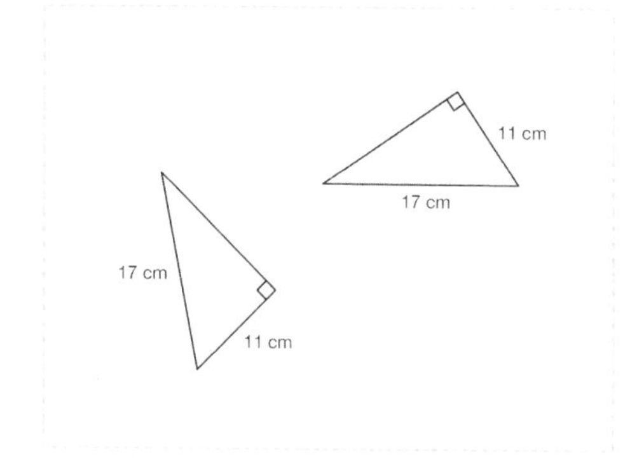

Level 2 *continued*

19. Are these triangles congruent?

1/5

- Yes, the triangles are congruent by SSS.
- Yes, the triangles are congruent by SAS.
- Yes, the triangles are congruent by AAS.
- Yes, the triangles are congruent by RHS.
- No, the triangles are not congruent.

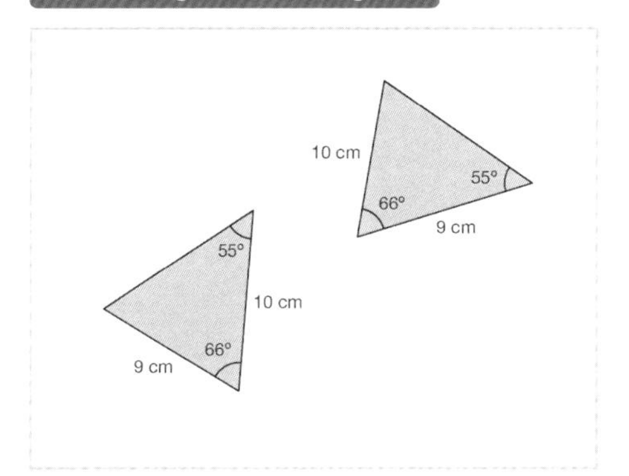

20. Which condition proves that these triangles are congruent?

1/4

- The triangles are congruent by SSS.
- The triangles are congruent by SAS.
- The triangles are congruent by AAS.
- The triangles are congruent by RHS.

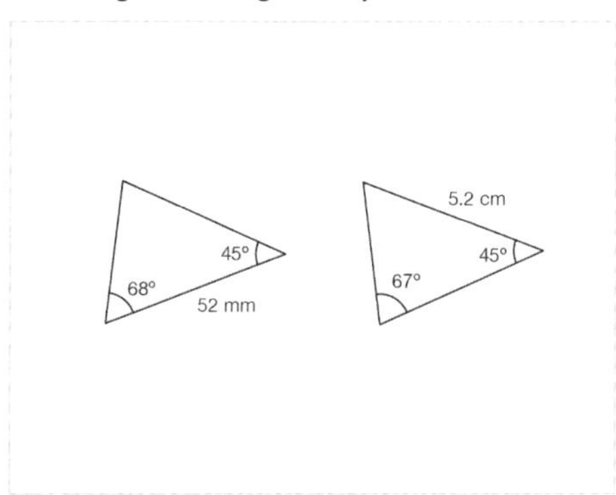

Level 3: Reasoning: Make inferences within congruent triangles.

Required: 5/5 **Pupil Navigation:** on
Randomised: off

21. Andrew and Neve each draw a triangle with a side of 8 cm, an angle of 40° and an angle of 50°. Andrew says that the two triangles are congruent. Explain why Andrew may be incorrect.

- Open question, no set answer

22. Select the triangle that is not congruent to the other three.

1/4

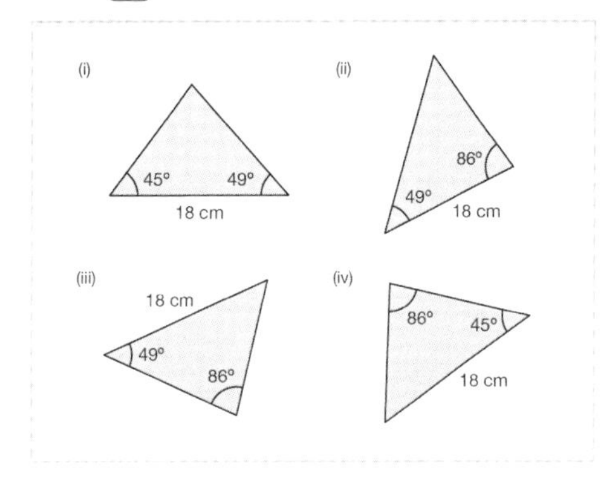

23. Select the condition that would not prove that two triangles, A and B, are congruent.

1/4

- A and B have three identical sides and B is a rotation of A.
- A and B have three identical sides and B is a reflection of A.
- A and B have three identical sides and B is a translation of A.
- A and B have three identical angles.

24. Kate and Ryan each draw a triangle with a side of 12 cm, a side of 9 cm and an angle of 40°. Kate says that the two triangles are congruent. Explain why Kate may be incorrect.

- Open question, no set answer

25. In the diagram, $PQ = QR = RS = PS$. Explain how you know that triangle PQR is congruent to triangle PRS.

- Open question, no set answer

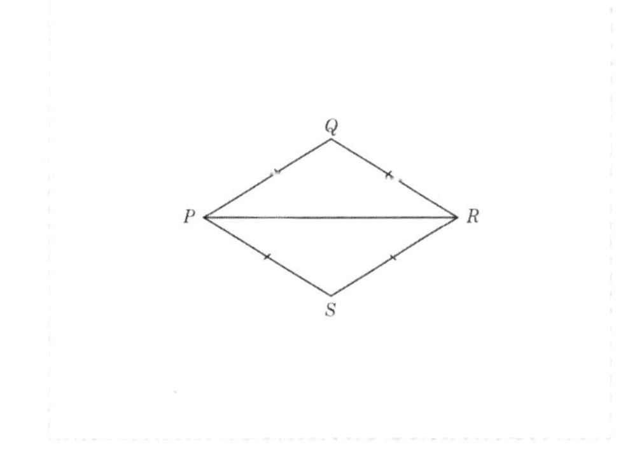

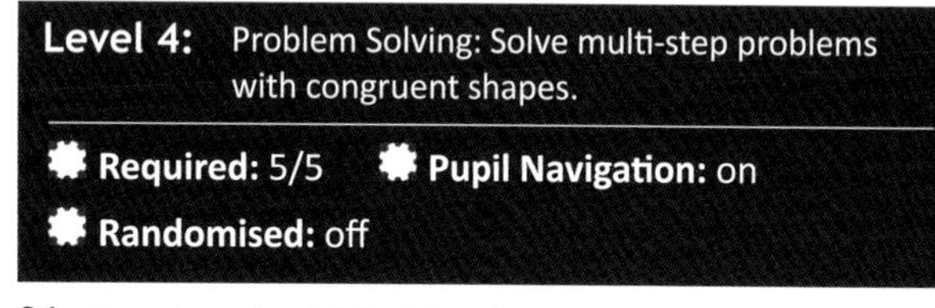
Level 4: Problem Solving: Solve multi-step problems with congruent shapes.

Required: 5/5 **Pupil Navigation:** on

Randomised: off

26. In triangle *ABC*, *AB* = 9 cm, ∠*BAC* = 60° and ∠*ABC* = 40°. In triangle *DEF*, *EF* = 9 cm, ∠*DEF* = 40° and ∠*DFE* = 60°. Are the triangles congruent?

1/5

- Yes, the triangles are congruent by SSS.
- Yes, the triangles are congruent by SAS.
- Yes, the triangles are congruent by AAS.
- Yes, the triangles are congruent by RHS.
- No, the triangles are not congruent.

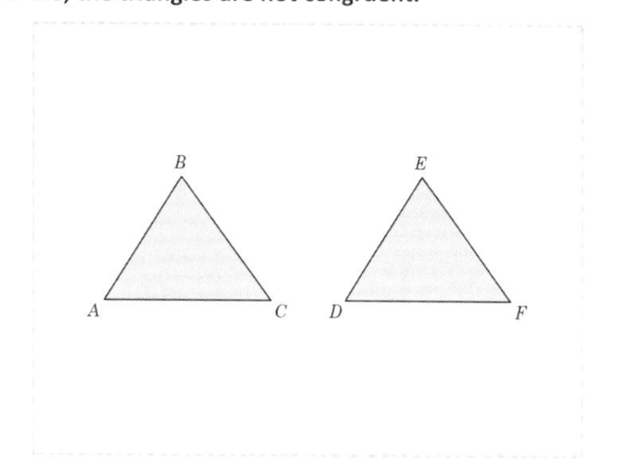

27. In the diagram, *AB* = *DE*. Which angle in triangle *ABC* corresponds to ∠*CDE* in triangle *CDE*?

1/3

- ∠CAB
- ∠BCA
- ∠CBA

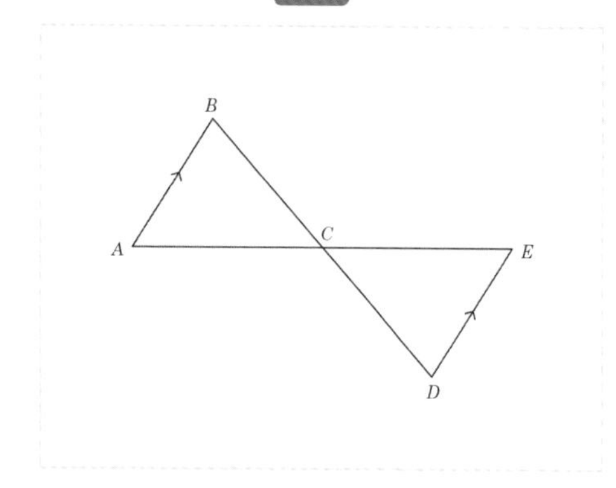

28. In triangle *ABC*, *AB* = 17 cm, ∠*ABC* = 40° and ∠ACB = 55°. In triangle *DEF*, *DF* = 17 cm, ∠*DFE* = 40° and ∠*EDF* = 55°. Are these triangles congruent?

1/5

- Yes, the triangles are congruent by SSS.
- Yes, the triangles are congruent by SAS.
- Yes, the triangles are congruent by AAS.
- Yes, the triangles are congruent by RHS.
- No, the triangles are not congruent.

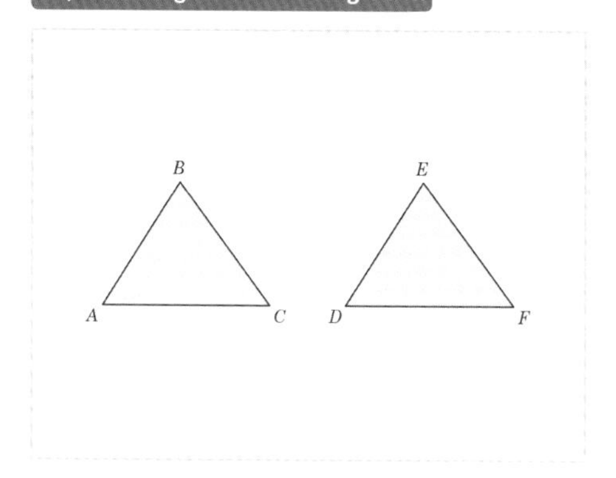

29. In triangle *ABC*, *AC* = 14 cm, ∠*ACB* = 38° and *BC* = 16 cm. In triangle DEF, DE = 16 cm, ∠*EDF* = 38° and *EF* = 14 cm. Are these triangles congruent?

1/5

- Yes, the triangles are congruent by SSS.
- Yes, the triangles are congruent by SAS.
- Yes, the triangles are congruent by AAS.
- Yes, the triangles are congruent by RHS.
- No, the triangles are not congruent.

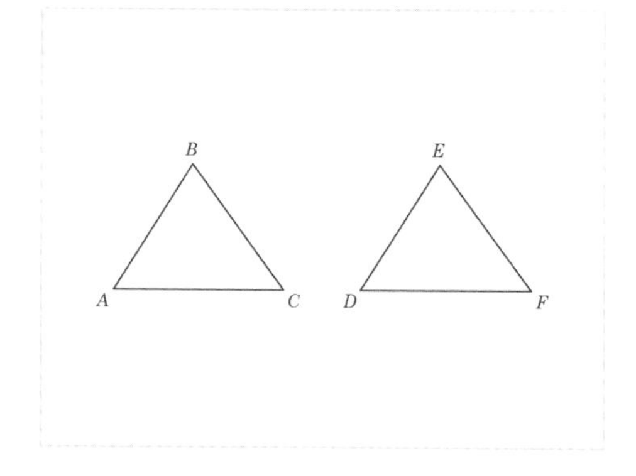

Level 4 *continued*

30.  The two triangles are congruent. What is the perimeter of triangle *DEF*?
Include the units cm (centimetres) in your answer.

- 23 cm
- 22
- 23 centimetres
- 23
- 22 centimetres
- 22 cm

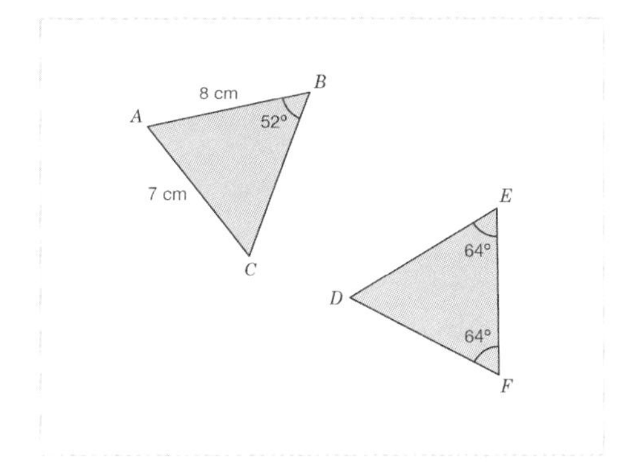

Identify the Hypotenuse, Adjacent and Opposite Sides in Right-Angled Triangles

Objective: I can identify the hypotenuse, adjacent and opposite sides in right-angled triangles.

Quick Search Ref: 10615

Level 1: I can identify the Hypotenuse, Adjacent and Opposite Sides in Right-Angled Triangles.

Required: 20/20 **Pupil Navigation:** on **Randomised:** off

1. Select the hypotenuse of the triangle.

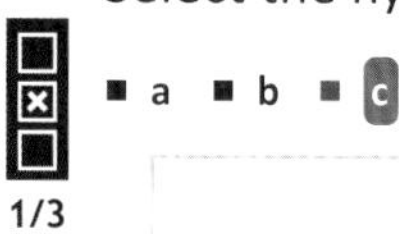

■ a ■ b ■ c

1/3

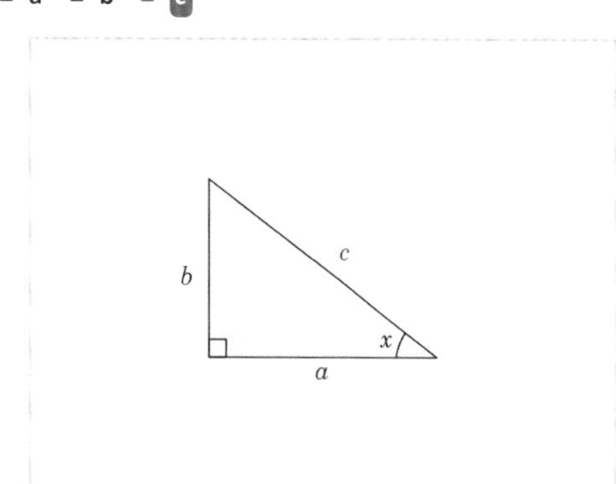

2. Select the side adjacent to angle x in the triangle.

■ d ■ e ■ f

1/3

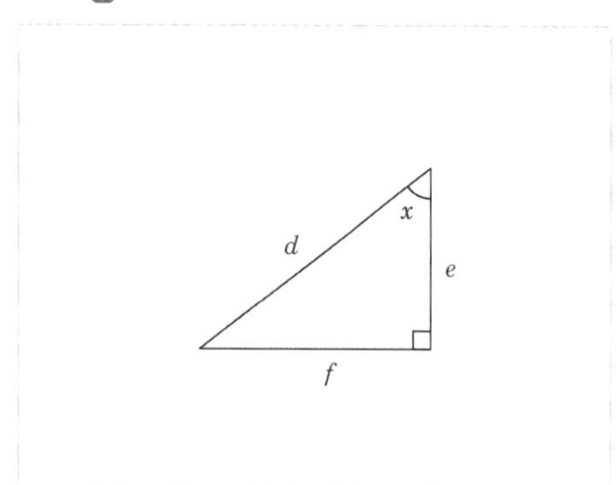

3. Select the side opposite angle x in the triangle.

■ r ■ s ■ t

1/3

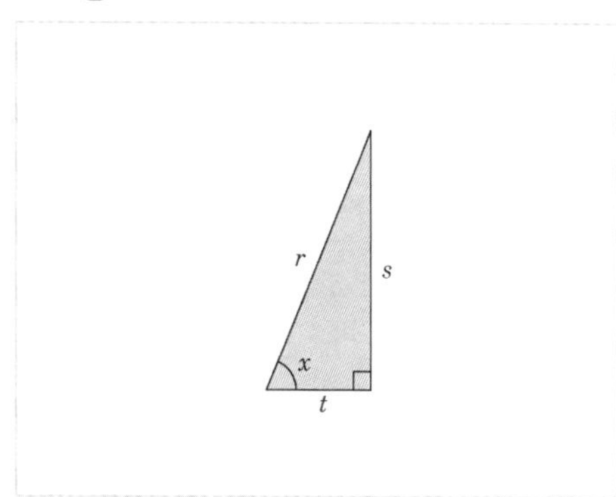

4. Select the hypotenuse of the triangle.

■ j ■ k ■ l

1/3

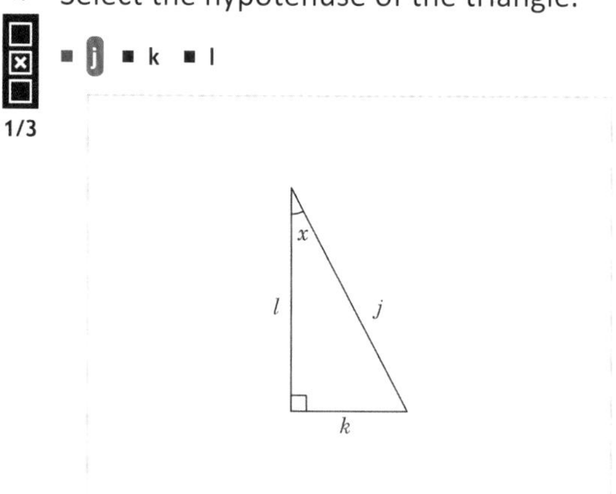

5. Select the side adjacent to angle x in the triangle.

■ PQ ■ QR ■ PR

1/3

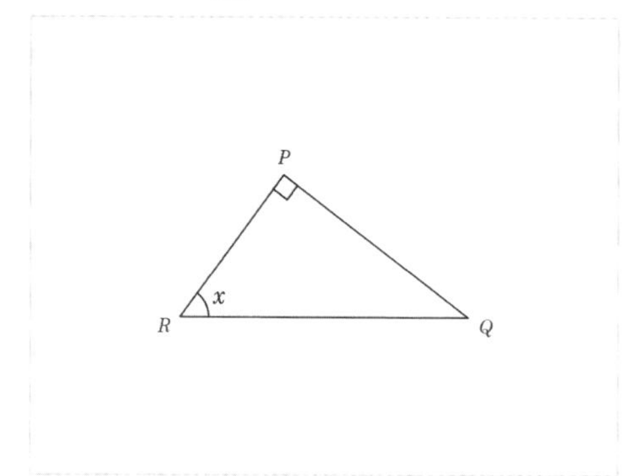

6. Select the side opposite angle x in the triangle.

■ l ■ m ■ n

1/3

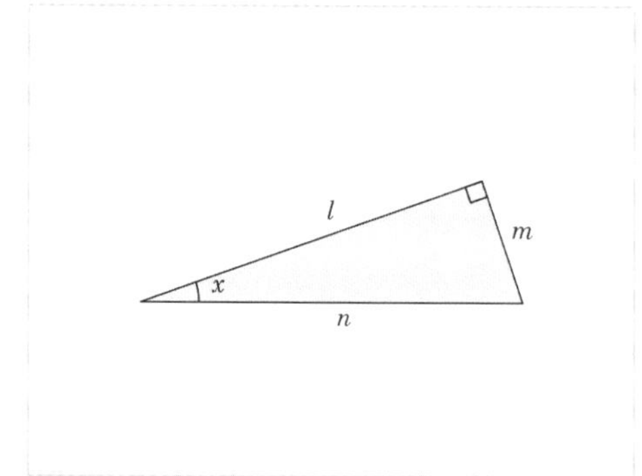

Level 1 *continued*

7. Select the hypotenuse of the triangle.

1/3

- FG
- GH
- FH

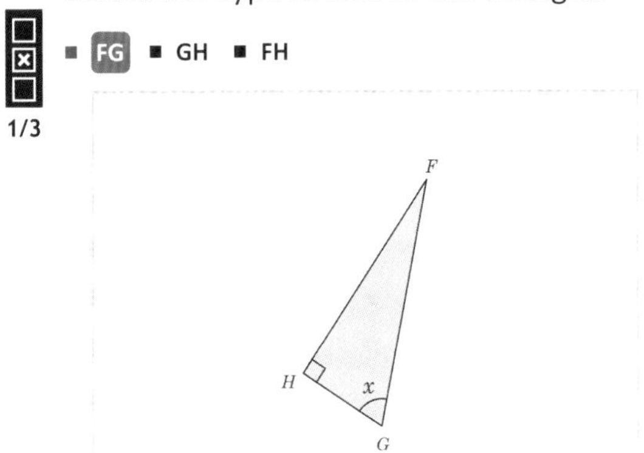

8. Select the side adjacent to angle x in the triangle.

1/3

- r
- s
- t

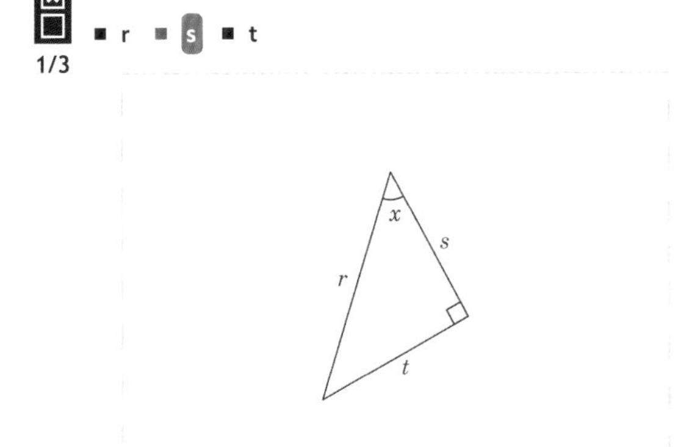

9. Select the side opposite angle x in the triangle.

1/3

- AB
- BC
- AC

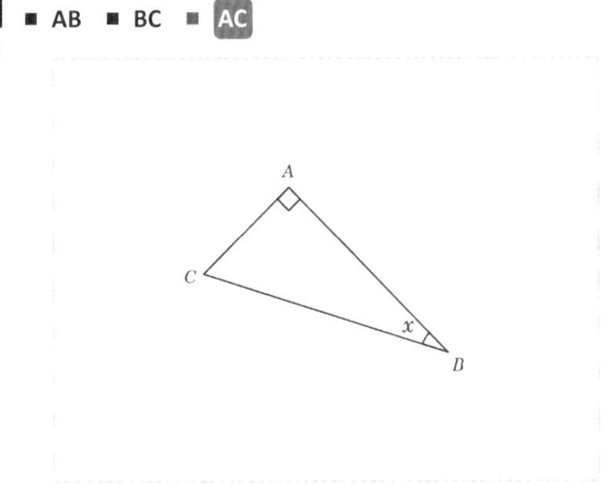

10. Select the hypotenuse of the triangle.

1/3

- x
- y
- z

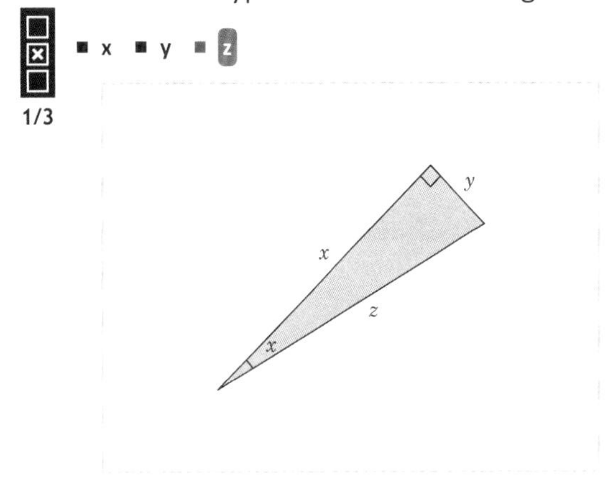

11. Enter the length of the hypotenuse of the triangle.
Include the units cm (centimetres) in your answer.

- 5 centimetres
- 3 centimetres
- 4 cm
- 5 cm
- 4 centimetres
- 3 cm
- 5
- 3
- 4

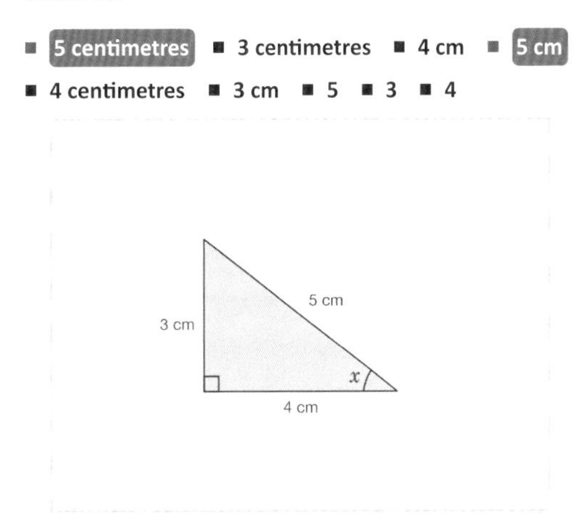

12. Enter the length of the side adjacent to angle y in the triangle.
Include the units m (metres) in your answer.

- 6 m
- 6
- 10
- 6 metres
- 8 metres
- 8
- 10 m
- 10 metres
- 8 m

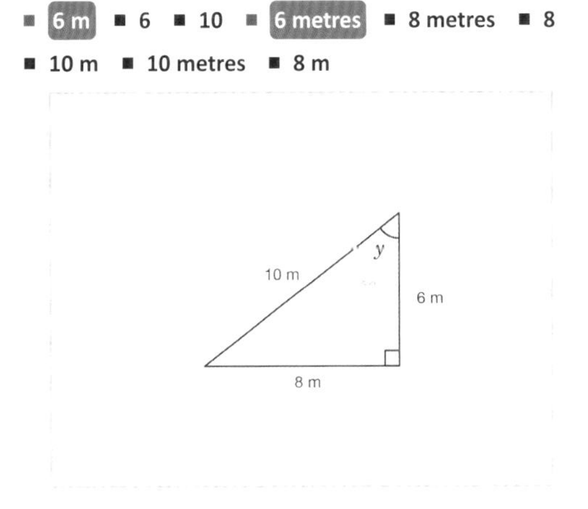

Level 1 *continued*

13. Enter the length of the side opposite angle *a* in the triangle.
Include the units mm (millimetres) in your answer.

- 12 ■ 13 ■ **12 millimetres** ■ **12 mm** ■ 5
- 13 millimetres ■ 5 mm ■ 13 mm ■ 5 millimetres

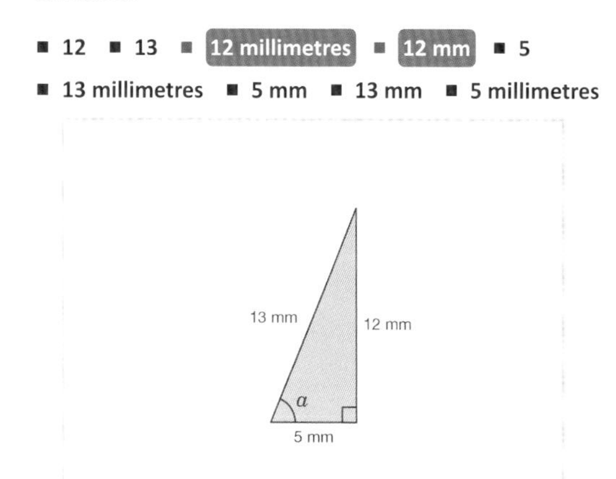

14. Enter the length of the hypotenuse of the triangle.
Include the units cm (centimetres) in your answer.

- **17 cm** ■ 15 cm ■ **17 centimetres** ■ 17 ■ 8 cm
- 15 ■ 15 centimetres ■ 8 ■ 8 centimetres

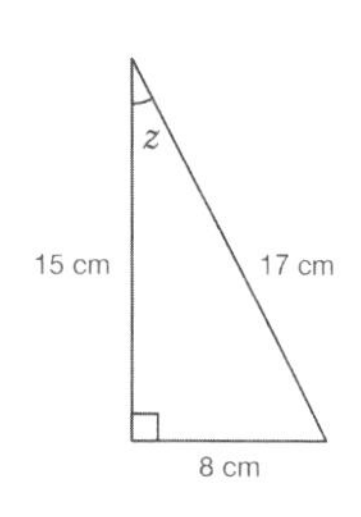

15. Enter the length of the side adjacent to angle *c* in the triangle.
Include the units m (metres) in your answer.

- 16 ■ **12 metres** ■ 16 metres ■ **12 m**
- 20 metres ■ 20 ■ 12 ■ 16 m ■ 20 m

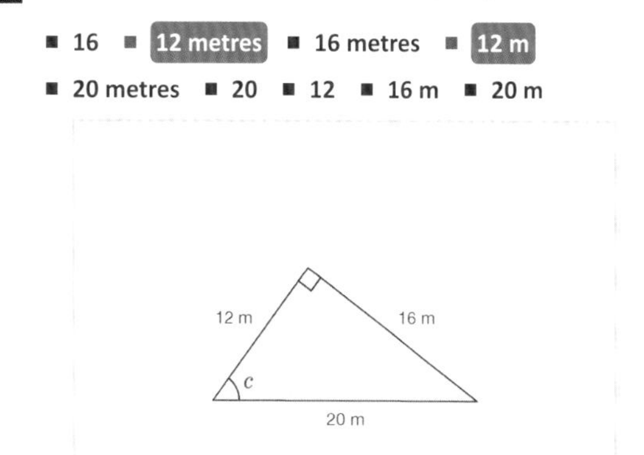

16. Enter the length of the side opposite angle *x* in the triangle.
Include the units mm (millimetres) in your answer.

- **9 millimetres** ■ 25 mm ■ 25 ■ 24 ■ **9 mm**
- 25 millimetres ■ 24 mm ■ 24 millimetres ■ 9

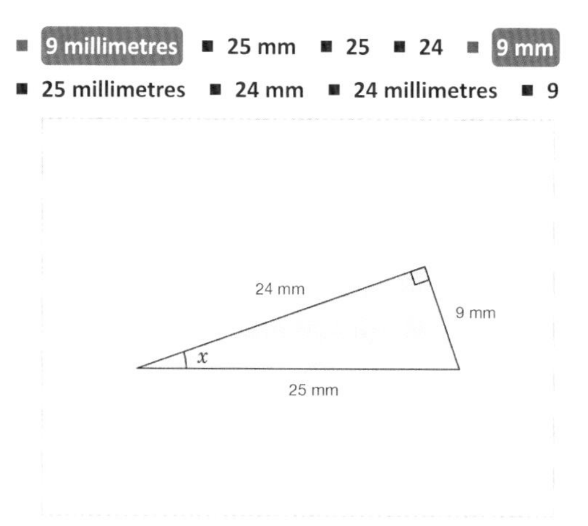

17. Enter the length of the hypotenuse of the triangle.
Include the units cm (centimetres) in your answer.

- 24 cm ■ 10 ■ **26 cm** ■ 24 centimetres
- **26 centimetres** ■ 10 centimetres ■ 24 ■ 26
- 10 cm

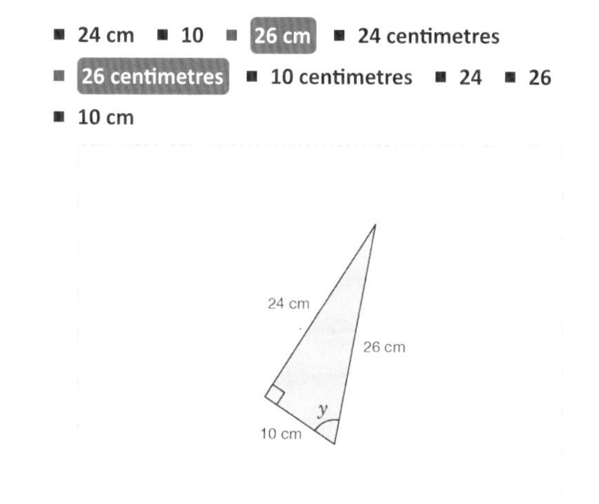

18. Enter the length of the side adjacent to angle *m* in the triangle.
Include the units m (metres) in your answer.

- 29 ■ 21 metres ■ **20 m** ■ **20 metres** ■ 20
- 21 m ■ 21 ■ 29 metres ■ 29 m

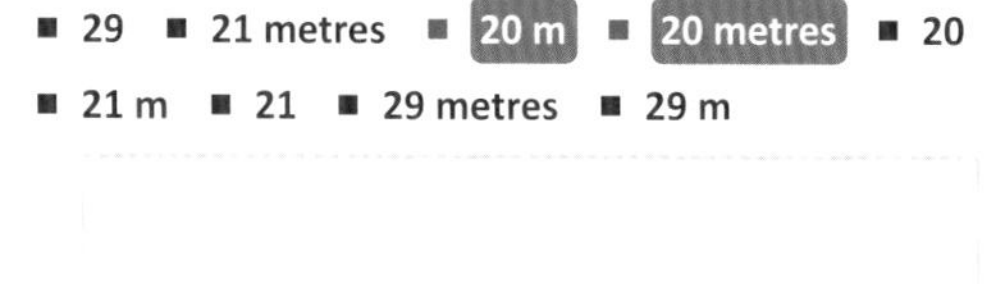

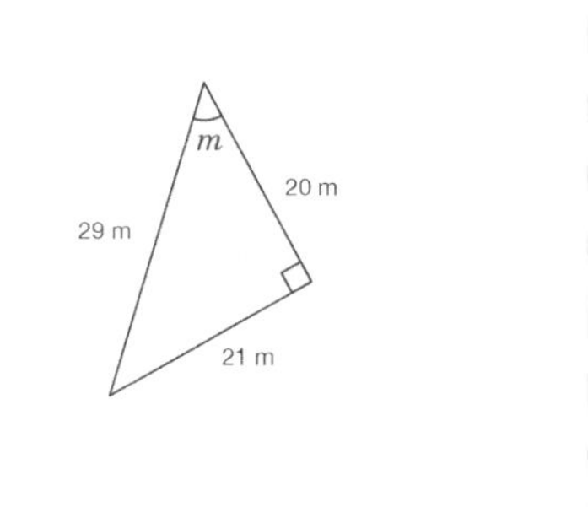

Level 1 *continued*

19. Enter the length of the side opposite angle *y* in the triangle.
Include the units cm (centimetres) in your answer.

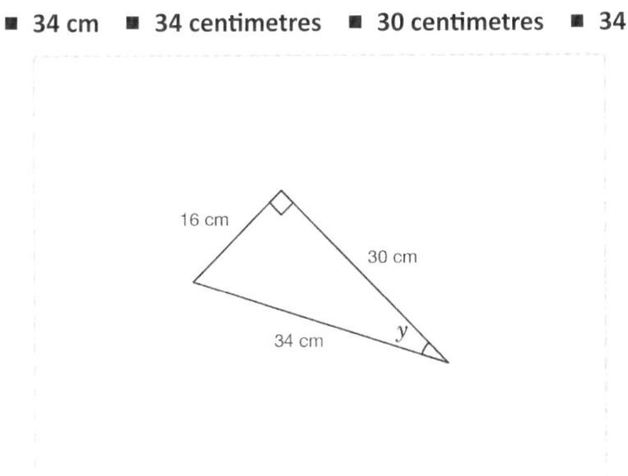

20. Enter the length of the hypotenuse of the triangle.
Include the units cm (centimetres) in your answer.

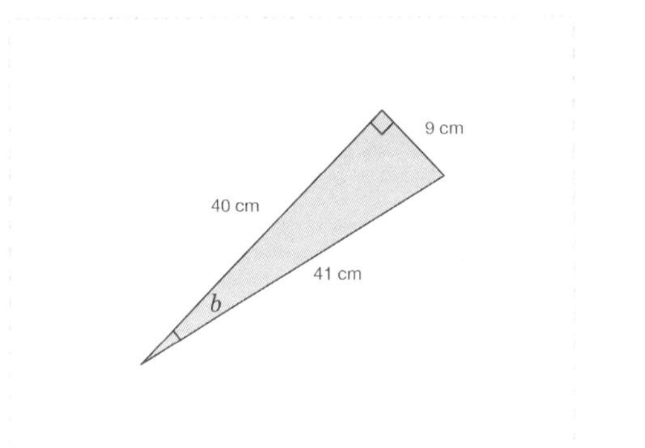

Solve Problems Using Pythagoras and Trigonometry

Objective: I can use Pythagoras' thoerem and trigonometry to solve problems.

Quick Search Ref: 10631

Level 1: Problem solving: Solve multi-step problems involving Pythagoras and trigonometry.

Required: 10/10 **Pupil Navigation:** on **Randomised:** off

1. A gardener designs the decking for a patio using square paving stones with a side length of 1.5 metres. Some of the paving stones are cut in half diagonally to make triangles. What is the perimeter of the patio?
Give your answer to 1 decimal place and include the units m (metres) in your answer.

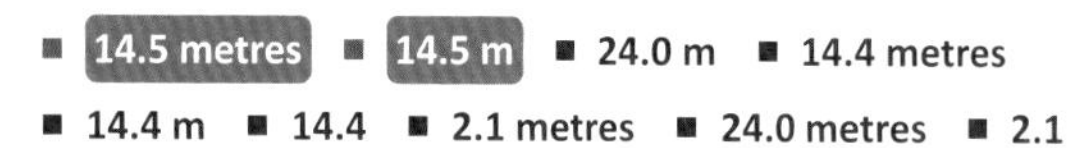

- 14.4 m ▪ 14.4 ▪ 2.1 metres ▪ 24.0 metres ▪ 2.1
- 24.0 ▪ 14.5 ▪ 2.1 m

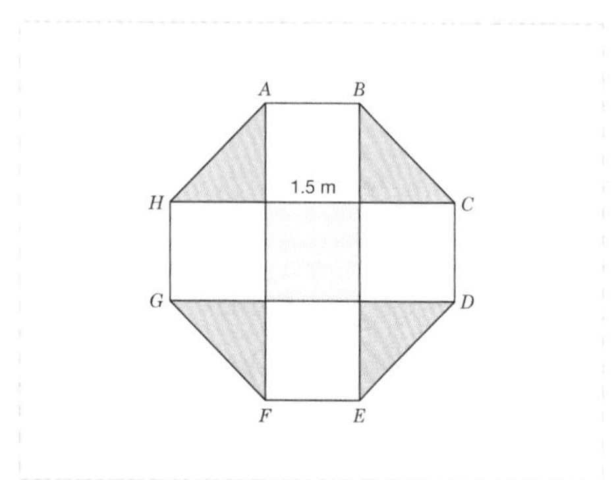

2. ABC is an equilateral triangle. M is the midpoint of BC. Calculate the perimeter of the equilateral triangle.
Give your answer to 1 decimal place and include the units m (metres) in your answer.

- 27.6 ▪ 9.2 ▪ 9.2 m ▪ **27.7 metres** ▪ **27.7 m**
- 27.7 ▪ 9.2 metres ▪ 27.6 metres ▪ 27.6 m

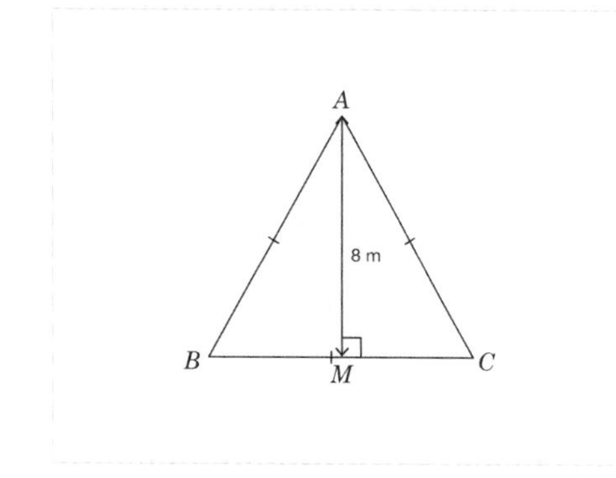

3. ABC is a straight line. BC = 5 metres. What is the length of AD?
Give your answer to 1 decimal place and include the units m (metres) in your answer.

- 14 metres ▪ 14.6 m ▪ 14.6 ▪ 9.8 m

- 14.7 ▪ 14.6 metres ▪ 9.8

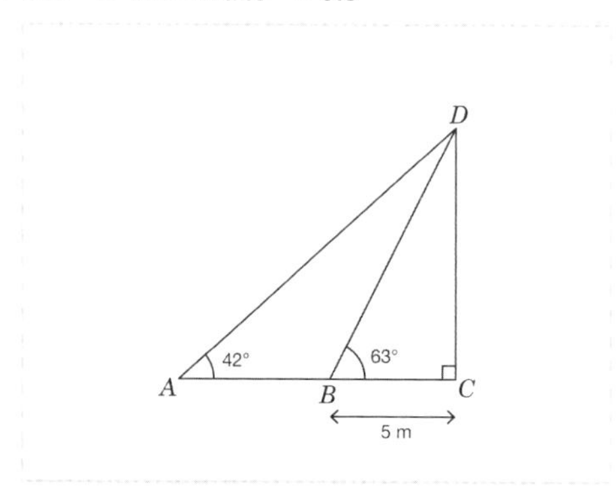

4. $ABCD$ is a kite. M is the point where the lines BD and AC intersect. BM = 0.86 metres. Calculate the perimeter of the kite.
Give your answer to 1 decimal place and include the units m (metres) in your answer.

- **5.4 metres** ▪ **5.4 m** ▪ 1.5 m ▪ 1.2 ▪ 1.2 m
- 1.5 ▪ 1.5 metres ▪ 1.2 metres ▪ 5.4

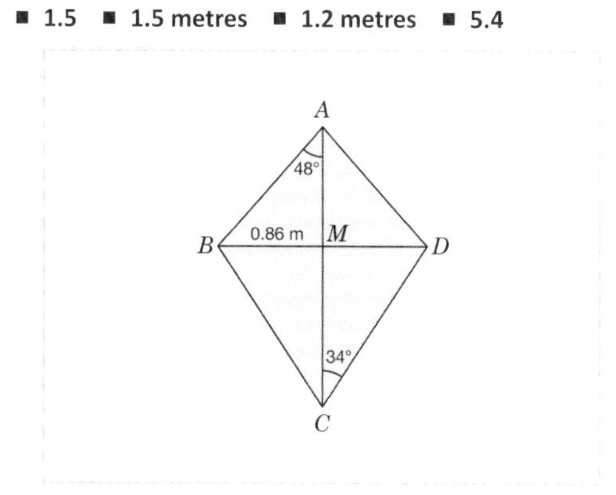

Level 1 *continued*

5. *ABC* is a straight line. What is the size of angle *y*?
Give your answer in degrees to one decimal place. Don't include the units in your answer.

- **24.2** ■ 17.1 ■ 1.3 ■ 41.4 ■ 52.2

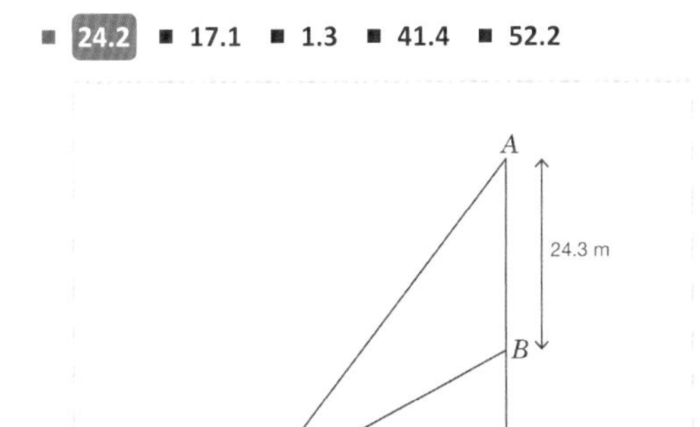

6. *ABCD* is a trapezium. *P* is the point such that *AP* is parallel to *BC*. Calculate the perimeter of the trapezium.
Give your answer to 1 decimal place and include the units m (metres) in your answer.

- **53.3 m** ■ **53.3 metres** ■ 13.0 m ■ 13.0 ■ 9.3 ■ 13.0 metres ■ 53.3 ■ 9.3 m ■ 9.3 metres

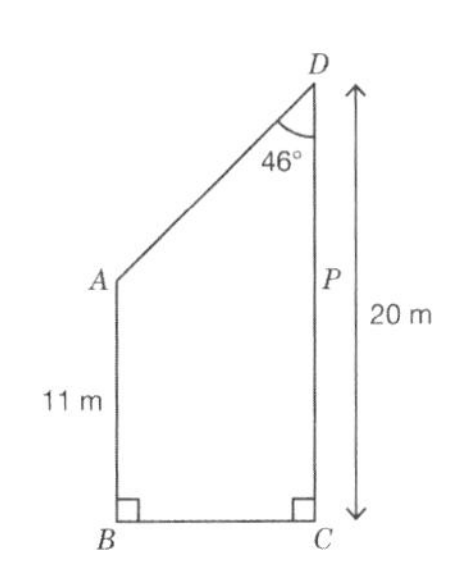

7. *AFE* and *CDE* are straight lines. What is the size of angle *m*?
Give your answer in degrees to 1 decimal place. Don't include the units in your answer.

- 1.1 ■ **15.7** ■ 16.1 ■ 0.3 ■ 20.2 ■ 4.1 ■ 14.5

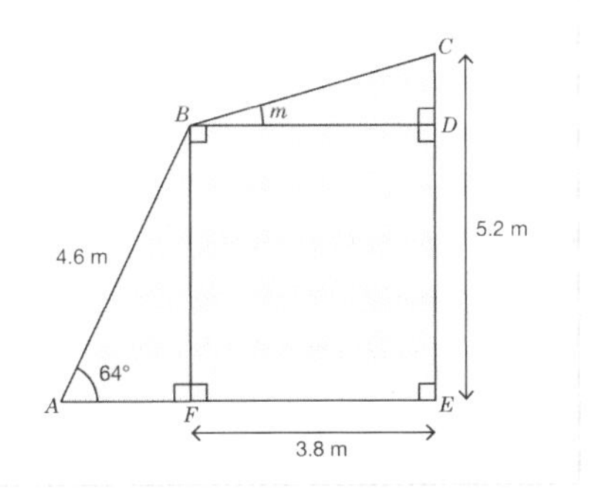

8. Calculate the area of the square *ABCD*.
Give your answer in square metres to 1 decimal place. Don't include the units in your answer.

- 22.2 ■ **754.9** ■ 109.9 ■ 27.4 ■ 751.6 ■ 750.8

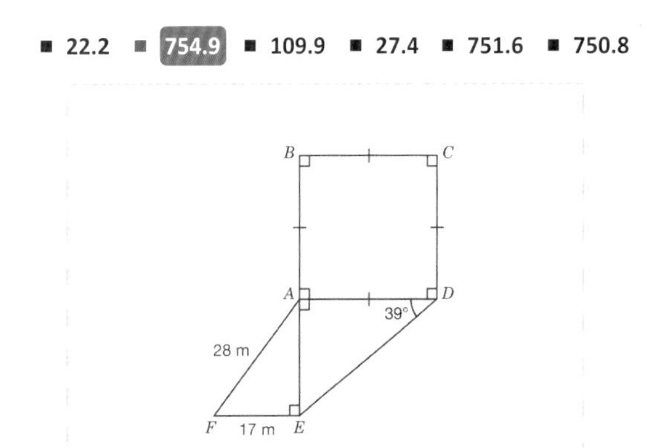

9. *ABC* is a straight line. *BD* is perpendicular to *AC*. Calculate the length of *AC*.
Give your answer to 1 decimal place and include the units cm (centimetres) in your answer.

- 49.6 ■ 38.8 centimetres ■ **104.1 cm** ■ **104.1 centimetres** ■ 49.6 cm ■ 65.3 ■ 104.1 ■ 49.6 centimetres ■ 65.3 centimetres ■ 38.8 cm ■ 38.8 ■ 65.3 cm

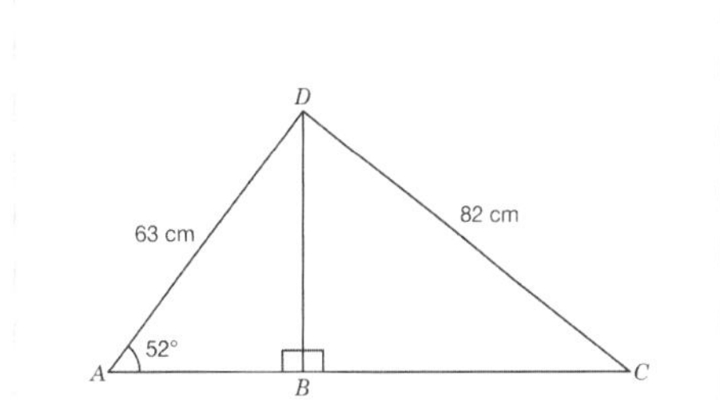

Level 1 *continued*

10. 
ABCDEF is a regular hexagon. *M* is the midpoint of *AB*. *O* is the centre of the regular hexagon. *OM* is perpendicular to *AB*. Calculate the perimeter of the regular hexagon.
Give your answer to 1 decimal place and include the units cm (centimetres) in your answer.

- **2.9 centimetres**
- **2.9 cm**
- **34.8 cm**
- **34.6 centimetres**
- **30**
- **2.9**
- **34.6 cm**
- **5.8 centimetres**
- **34.8**
- **34.8 centimetres**
- **30 cm**
- **5.8**
- **34.6**
- **30 centimetres**
- **5.8 cm**

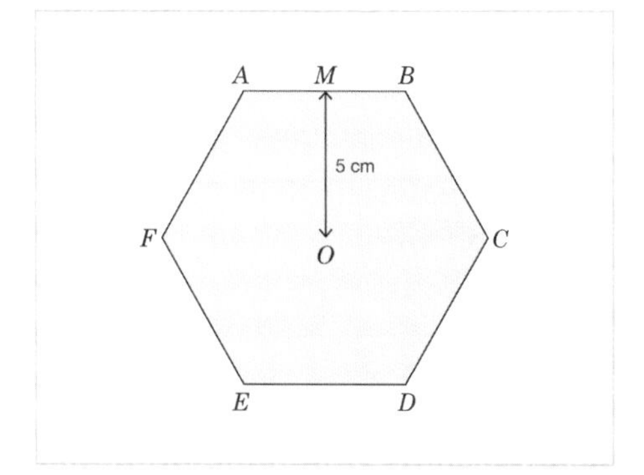

Use Pythagoras' Theorem to Find Missing Sides

Objective: I can use Pythagoras' theorem to find missing sides.

Quick Search Ref: 10707

Level 1: Understanding: Use Pythagoras' Theorem to find missing sides in right-angled triangles.

Required: 7/10 **Pupil Navigation:** on **Randomised:** off

1. Select the length of the hypotenuse of the right angled-triangle.

1/3

- 9 cm
- 12 cm
- **15 cm**

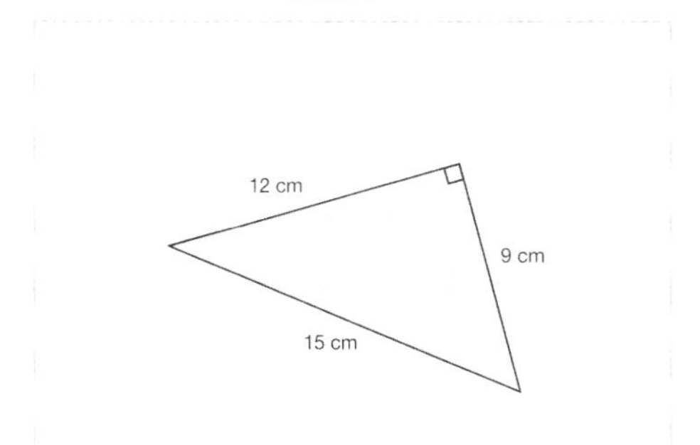

2. Look at the triangle in the diagram. Which of the following statements is true?

1/3

- $s^2 = r^2 - t^2$
- **$s^2 = r^2 + t^2$**
- $s^2 = t^2 - r^2$

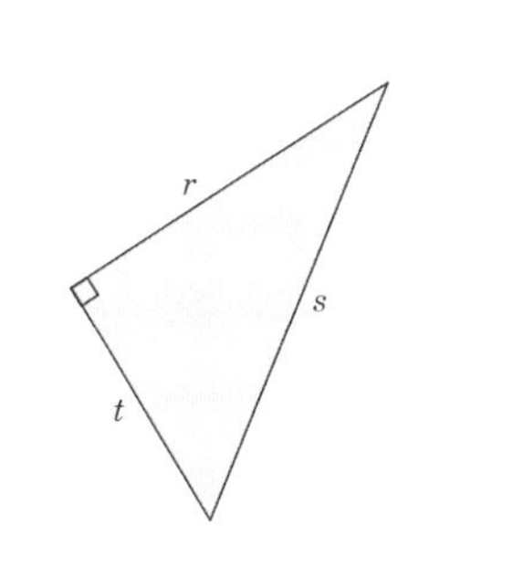

3. Look at the triangle in the diagram. Which of the following statements is true?

1/3

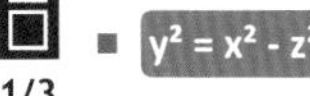
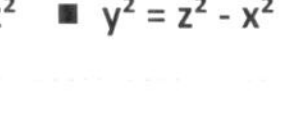

- **$y^2 = x^2 - z^2$**
- $y^2 = x^2 + z^2$
- $y^2 = z^2 - x^2$

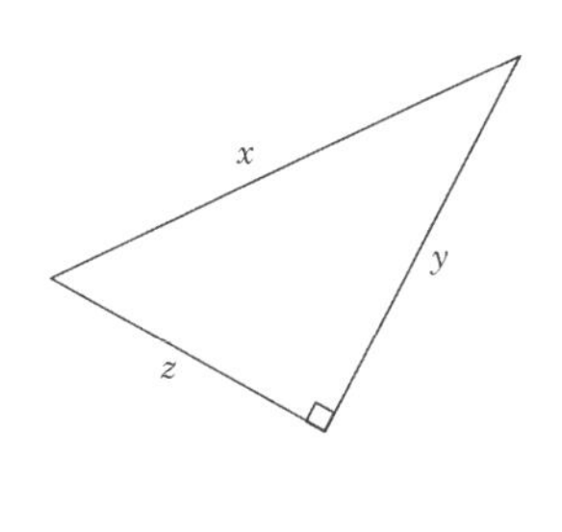

4. What is the length of the hypotenuse of the right-angled triangle?
Include the units cm (centimetres) in your answer.

- 10
- **10 centimetres**
- **10 cm**
- 100 centimetres
- 100
- 100 cm

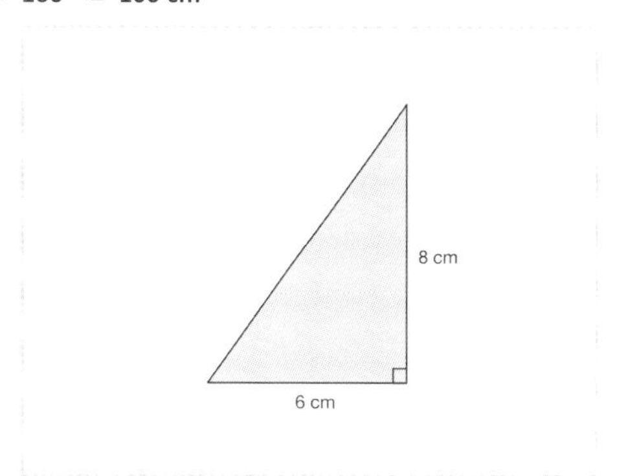

5. What is the value of *r*?
Give your answer to 2 decimal places and include the units m (metres) in your answer.

- **10.95 m**
- **10.95 metres**
- 120
- 21.40 m
- 21.40
- 10.95
- 120 m
- 120 metres
- 21.40 metres

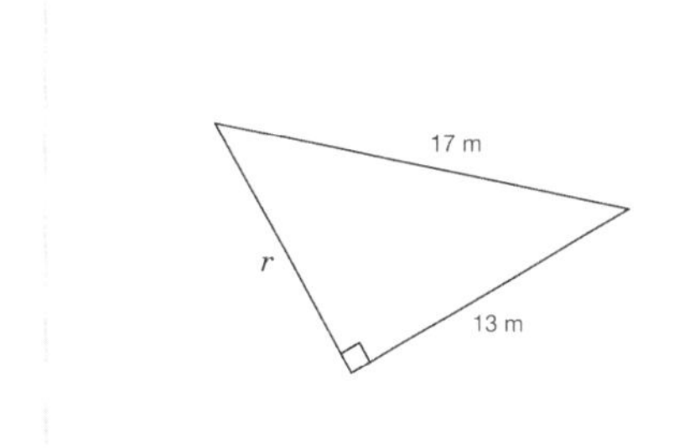

Level 1 *continued*

6. What is the length of AB?

Give your answer to 1 decimal place and include the units cm (centimetres) in your answer.

- 6.9 cm
- **10.5 centimetres**
- 6.9 centimetres
- **10.5 cm**
- 110.6 centimetres
- 110.6 cm
- 10.5
- 6.9
- 110.6

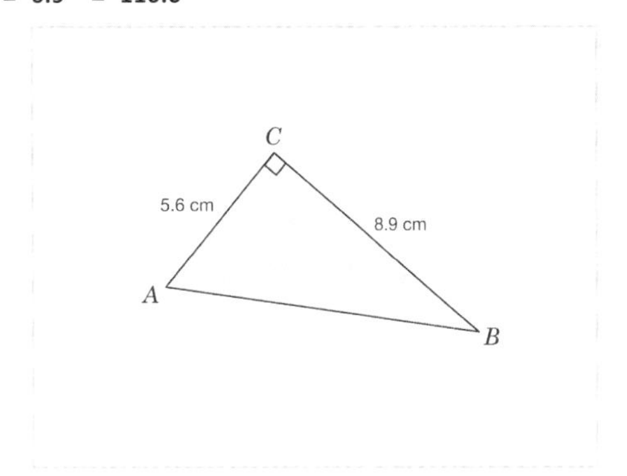

7. What is the value of d?

Give your answer to 3 significant figures and include the units mm (millimetres) in your answer.

- 35.3 mm
- **28.3 mm**
- 28.3
- 799 mm
- **28.3 millimetres**
- 35.3 millimetres
- 28.267 mm
- 28.267
- 35.3
- 799
- 799 millimetres
- 28.267 millimetres

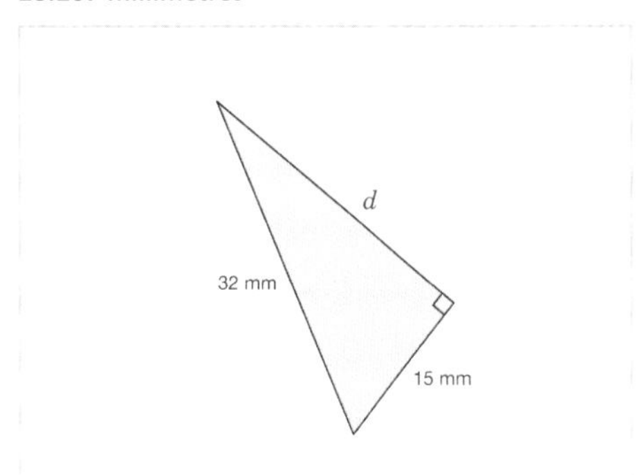

8. Look at the triangle in the diagram. Which of the following statements is true?

1/3

- $p^2 = r^2 - q^2$
- $p^2 = q^2 + r^2$
- **$p^2 = q^2 - r^2$**

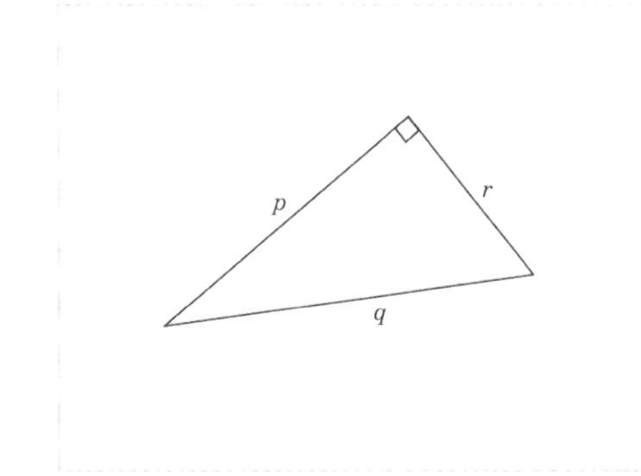

9. What is the value of z?

Give your answer to 2 decimal places and include the units cm (centimetres) in your answer.

- 28.62
- **28.62 cm**
- 819.13
- 819.13 cm
- **28.62 centimetres**
- 14.85 centimetres
- 819.13 centimetres
- 14.85
- 14.85 cm

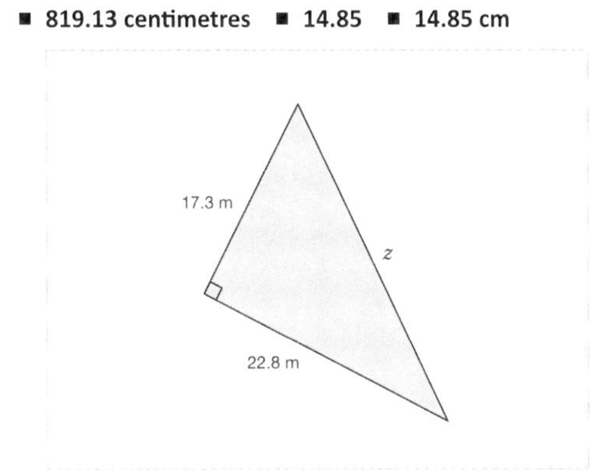

10. What is the length of QR?

Give your answer to 3 significant figures and include the units cm (centimetres) in your answer.

- 12.845 centimetres
- 23.6 centimetres
- 12.845
- **12.8 centimetres**
- **12.8 cm**
- 23.6
- 165
- 12.8
- 12.845 cm
- 165 cm
- 165 centimetres
- 23.6 cm

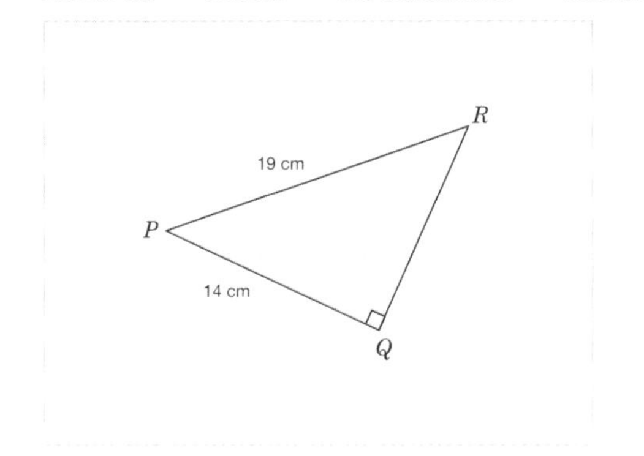

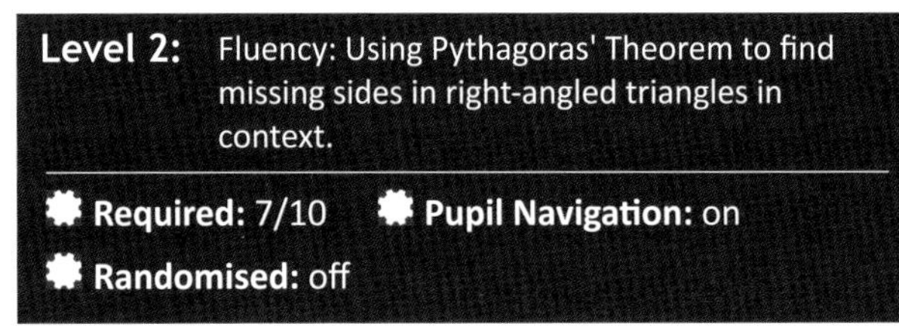

11. A sheet of A4 paper is 297 mm by 210 mm. What is the length of the diagonal of the paper?
Give your answer to 1 decimal place and include the units mm (millimetres) in your answer.

- **363.7 millimetres**
- **363.7 mm**
- 363.7
- 210.0 millimetres
- 210.0 mm
- 210.0

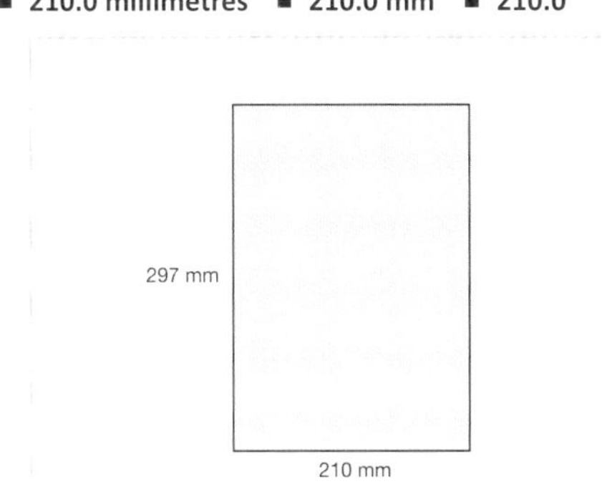

12. A 3 metre ladder is placed against a wall. The bottom of the ladder is 70 centimetres from the wall. How high up the wall will the ladder reach?
Give your answer to 3 decimal places and include the units m (metres) in your answer.

- **2.917 m**
- **2.917 metres**
- 3.081 metres
- 2.917
- 3.081
- 3.081 m

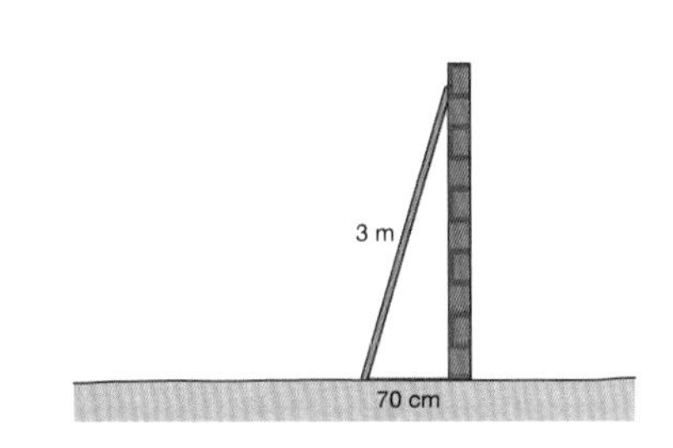

13. A plane flies 320 kilometres due north and then 470 kilometres due east. How far is the plane from its original starting point?
Give your answer to 4 significant figures and include the units km (kilometres) in your answer.

- 568.6
- **568.6 km**
- **568.6 kilometres**
- 344.2
- 344.2 kilometres
- 344.2 km
- 568.5947 kilometres
- 568.5947 km
- 568.5947

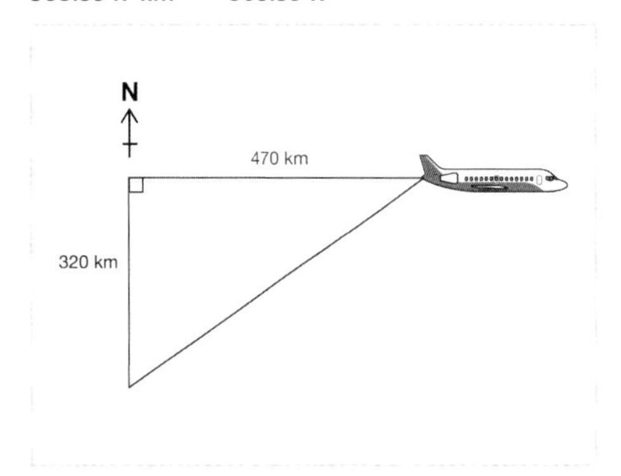

14. A coordinate grid is drawn on centimetre squared paper. What is the distance between the points (-2, -5) and (7, 1)?
Give your answer to 2 decimal places and include the units cm (centimetres) in your answer.

- 6.71 centimetres
- **10.82 centimetres**
- **10.82 cm**
- 6.71
- 6.71 cm
- 10.82

15. The diagonal of a rectangle is 14 metres (m). If the width of the rectangle is 9 m, what is the length of the rectangle?
Give your answer to 1 decimal place and include the units m (metres) in your answer.

- **10.7 metres**
- **10.7 m**
- 10.7
- 16.6 m
- 16.6 metres
- 16.6

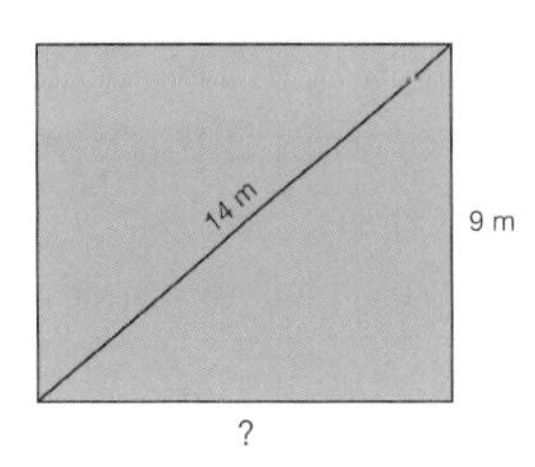

Level 2 *continued*

16. The penalty area on a full-size football pitch is 18 yards by 44 yards. What is the length of the diagonal of the penalty area?
Give your answer to 3 significant figures and include the units yd (yards) in your answer.

- 40.1
- **47.5 yards**
- **47.5 yd**
- 40.1 yd
- 47.539 yards
- 47.539 yd
- 40.1 yards
- 47.539
- 47.5

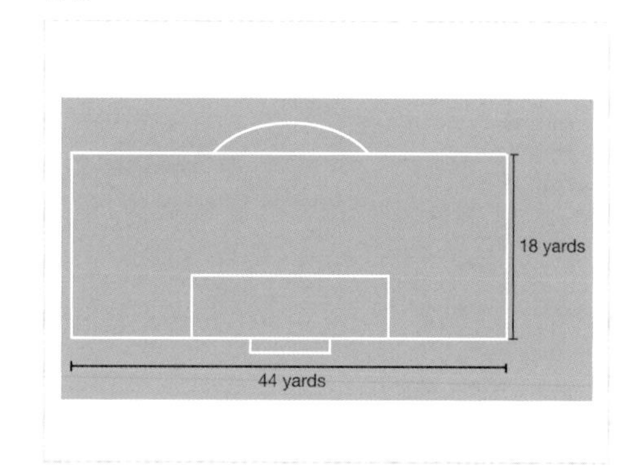

17. A 240 metre (m) zip-wire is attached between a 10 m tower and a 3 m tower. What is the distance between the two towers?
Give your answer to 1 decimal place and include the units m (metres) in your answer.

- 240.1 metres
- **239.9 metres**
- **239.9 m**
- 239.9
- 240.1
- 240.1 m

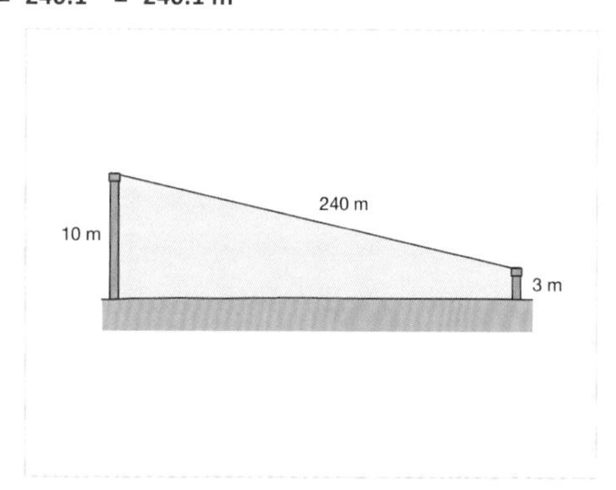

18. A coordinate grid is drawn on centimetre squared paper. What is the distance between the points (1, 3) and (6, 9)?
Give your answer to 2 decimal places and include the units cm (centimetres) in your answer.

- 3.32 cm
- 3.32 centimetres
- **7.81 centimetres**
- **7.81 cm**
- 3.32
- 7.81

19. Calculate the perpendicular height of an equilateral triangle with side lengths of 9 cm.
Give your answer to 1 decimal place and include the units cm (centimetres) in your answer.

- 10.1 cm
- 10.1 centimetres
- **7.8 centimetres**
- **7.8 cm**
- 7.8
- 10.1

9 cm

20. A ship sails 480 kilometres due south and then 260 kilometres due west. How far is the ship from its original starting point?
Give your answer to 2 decimal places and include the units km (kilometres) in your answer.

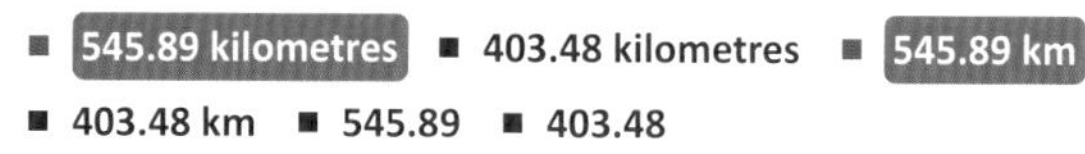

- **545.89 kilometres**
- 403.48 kilometres
- **545.89 km**
- 403.48 km
- 545.89
- 403.48

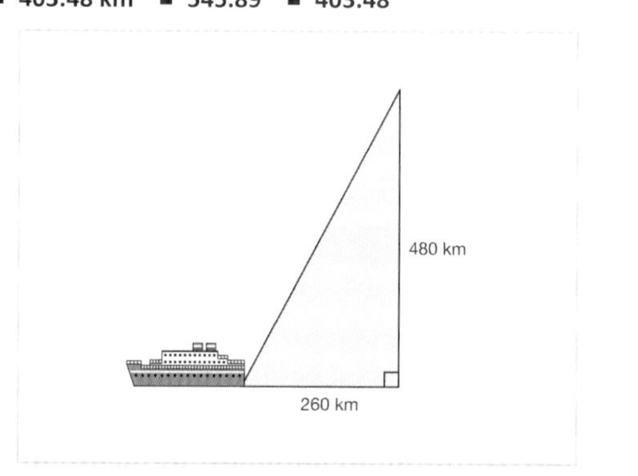

Level 3: Reasoning: Misconceptions and making inferences using Pythagoras' Theorem.

Required: 5/5 **Pupil Navigation:** on
Randomised: off

21. Fazel says that in a right-angled triangle, the hypotenuse is always the longest side. Is Fazel correct? Explain your answer.

- Open question, no set answer

Level 3 *continued*

22. Andrew has made a mistake with his maths homework. What is the correct answer?
Give your answer to 1 decimal place and include the units mm (millimetres) in your answer.

- 10.8 cm
- **108.2 millimetres**
- **108.2 mm**
- 108.2
- 10.8 centimetres
- 10.8

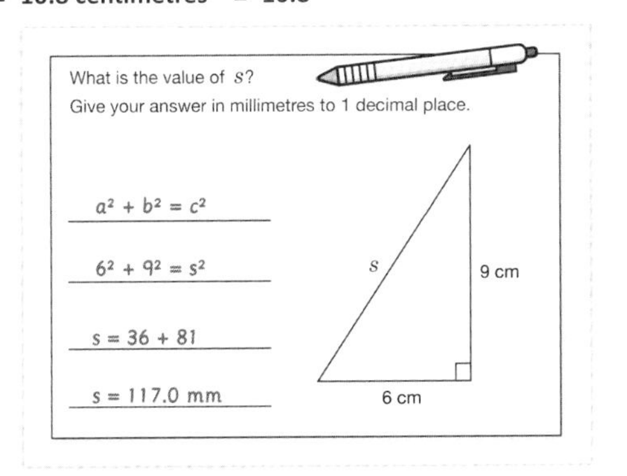

23. Select two statements that are true for the triangle shown.

2/4

- $x^2 - 13^2 = 8^2$
- $8^2 + 13^2 = x^2$
- **$x^2 + 8^2 = 13^2$**
- **$8^2 = 13^2 - x^2$**

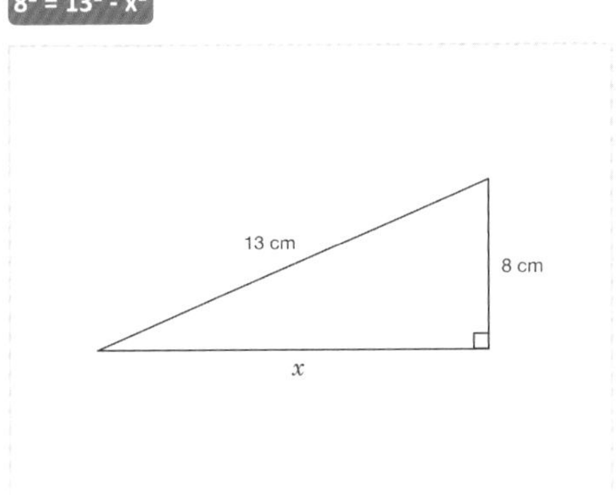

24. Is this a right-angled triangle? Explain your answer.

- Open question, no set answer

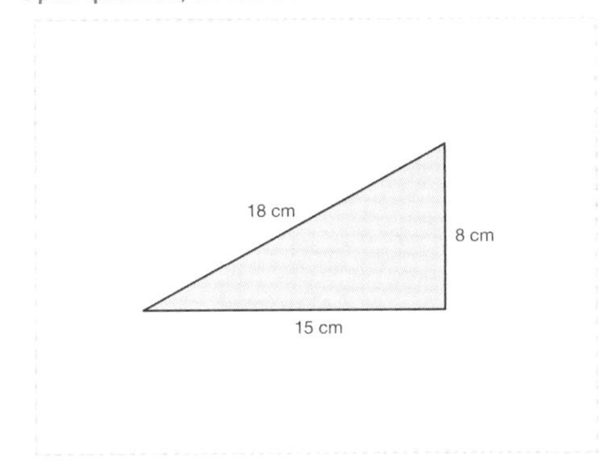

25. Sort the triangles according to the length of the missing side from smallest to largest.

- **triangle (ii)**
- **triangle (iii)**
- **triangle (i)**

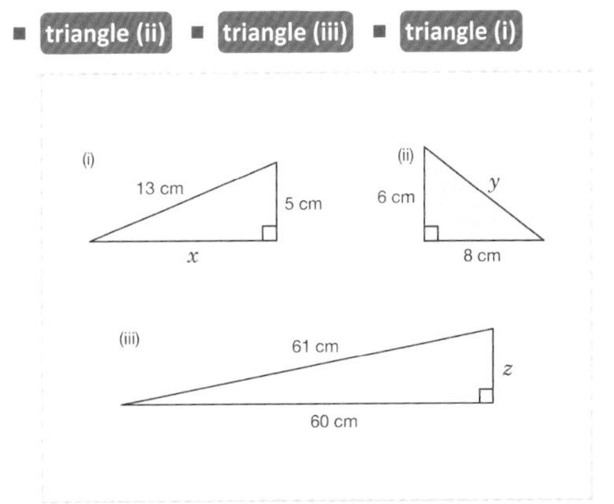

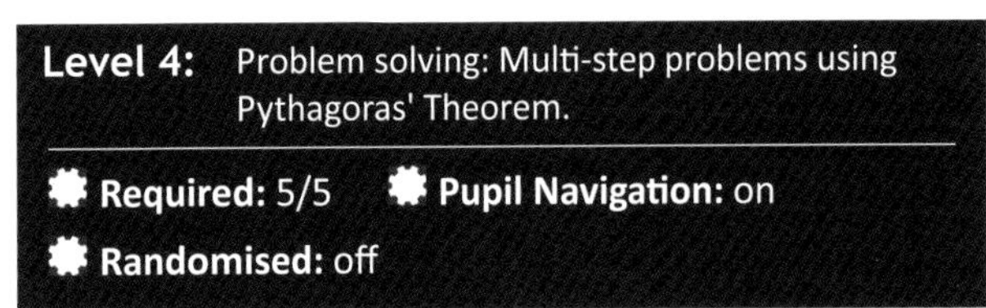

Level 4: Problem solving: Multi-step problems using Pythagoras' Theorem.

Required: 5/5 **Pupil Navigation:** on
Randomised: off

26. The diagonal of a square field is 320 metres. What is the perimeter of the field?
Give your answer to 1 decimal place and include the units m (metres) in your answer.

- **905.1 metres**
- **905.1 m**
- 905.1
- 226.3 m
- 226.3
- 226.3 metres

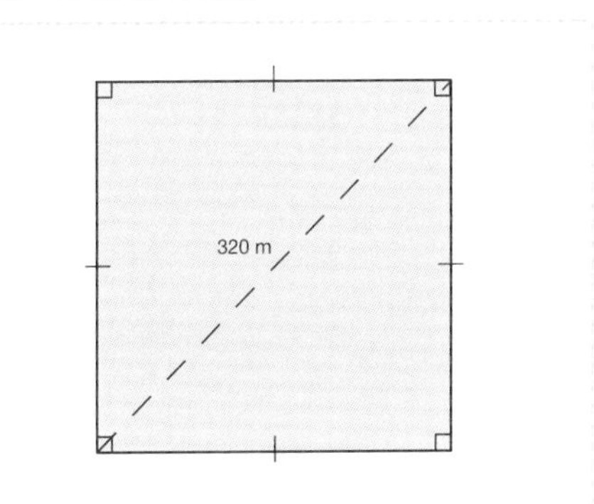

27. What is the length of AD?
Give your answer to 1 decimal place and include the units cm (centimetres) in your answer.

- **8.2 centimetres**
- **8.2 cm**
- 8.2

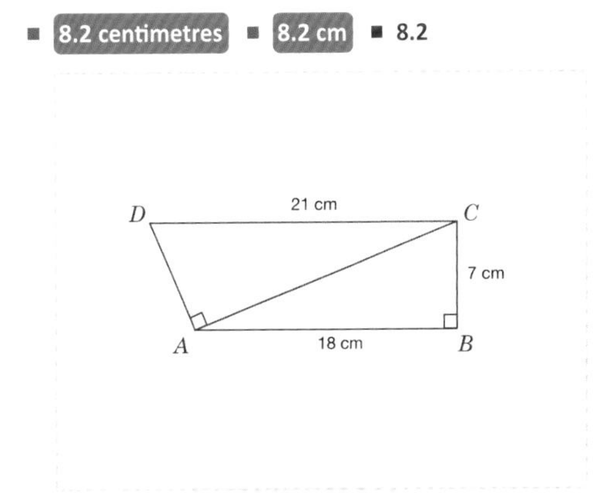

Level 4 *continued*

28. The diagonals of a rhombus are 5 cm (centimetres) and 14 cm. What is the perimeter of the rhombus?
Give your answer to 1 decimal place and include the units cm (centimetres) in your answer.

- 29.7 centimetres
- 7.4 cm
- 29.7 cm
- 7.4 centimetres
- 29.7
- 7.4

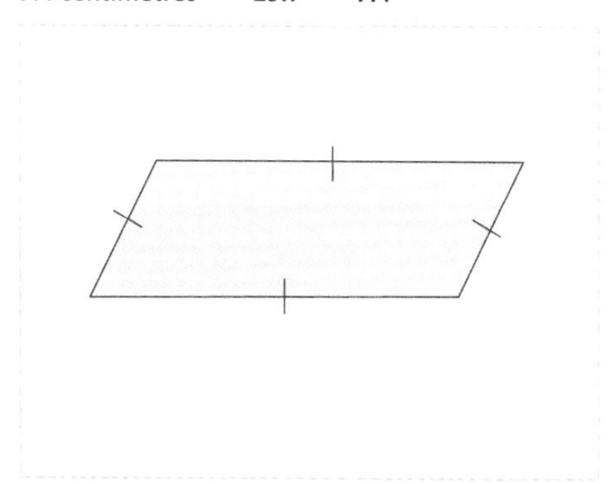

29. The box shown has a height of 3 centimetres (cm), a width of 4 cm and a length of 12 cm. Calculate the length of *AG*.
Include the units cm (centimetres) in your answer.

- 13 centimetres
- 13 cm
- 13

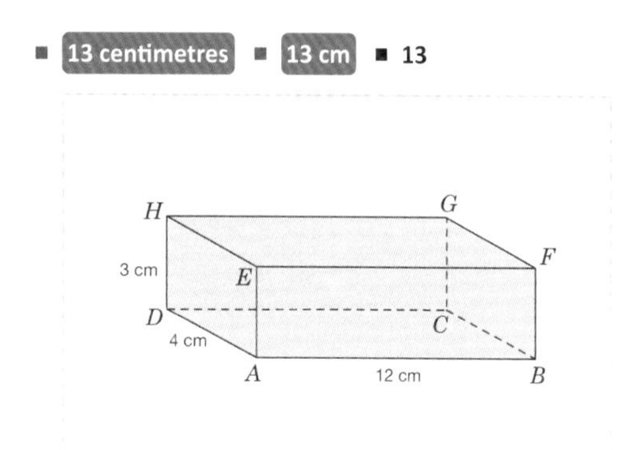

30. If the lengths of the three sides of a right-angled triangle are all integers, then these numbers form a Pythagorean Triple. One Pythagorean Triple is 3, 4 and 5 because $3^2 + 4^2 = 5^2$.
Select the 3 sets of numbers which are Pythagorean Triples.

3/5

- {9, 12, 15}
- {7, 9, 11}
- {5, 12, 13}
- {8, 15, 17}
- {3, 7, 8}

Use Pythagoras' Theorem to Find Shorter Sides

Objective: I can use Pythagoras' theorem to find shorter sides.

Quick Search Ref: 10624

Level 1: Understanding: Exploring the squares on the sides of a right-angled triangle.

Required: 7/10 **Pupil Navigation:** on **Randomised:** off

1. Select the **two** shorter sides of the right-angled triangle.

2/3

- a
- b
- c

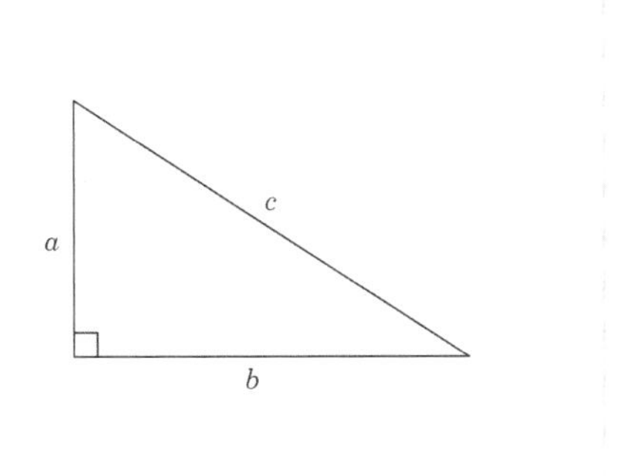

2. In the diagram, AC = 5 cm and BC = 4 cm. Select the calculation to find the area of the square on side AB in square centimetres.

1/4

- $5^2 - 3^2$
- $4^2 - 3^2$
- $4^2 - 5^2$
- $5^2 - 4^2$

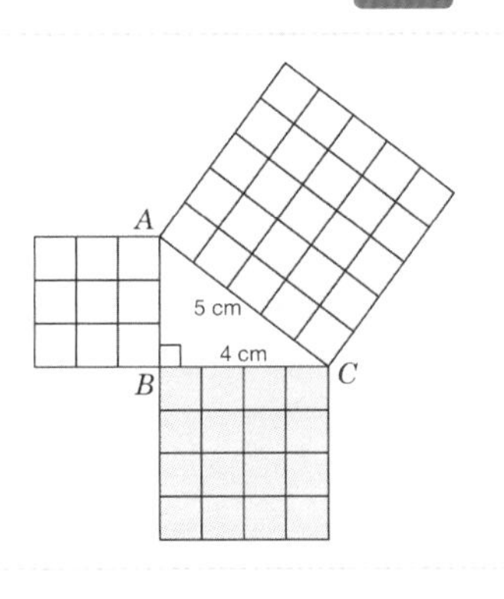

3. In the diagram, AC = 10 cm and BC = 8 cm. What is the area of the square on side AB in square centimetres?
Don't include the units in your answer.

- 6
- 36
- 64
- 100

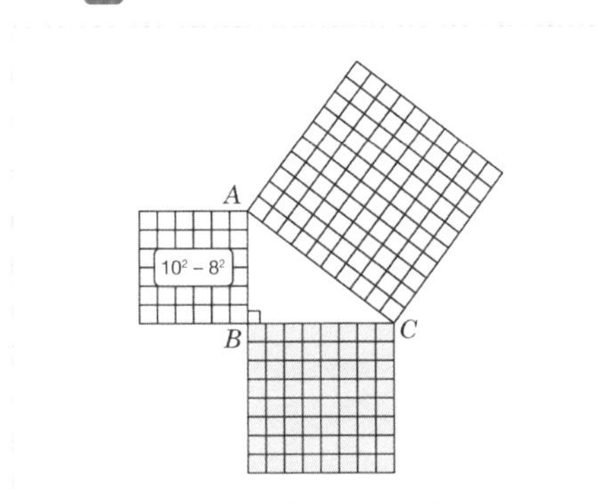

4. In the diagram, AC = 13 cm and BC = 5 cm. What is the area of the square on AB in square centimetres?
Don't include the units in your answer.

- 144
- 12
- 25
- 169

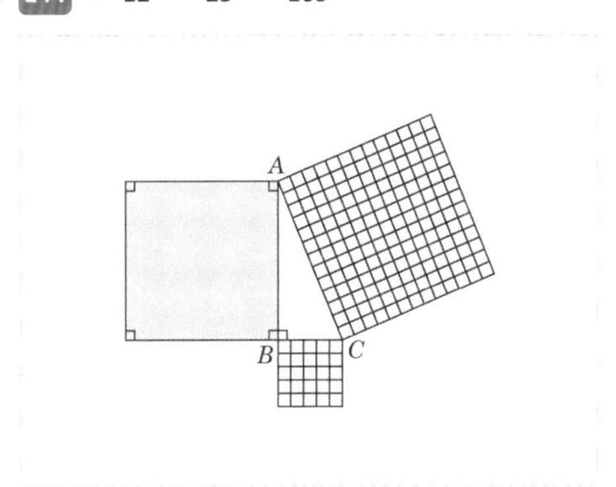

5. If the area of a square is 100 cm², what is the length of each of its side?
Include the units cm (centimetres) in your answer.

- 10 centimetres
- 10
- 10 cm
- 25 cm
- 25
- 25 centimetres

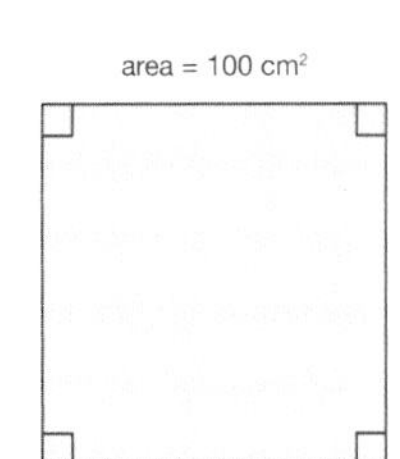

Level 1 *continued*

6. What is the length of side *AB*?
abc *Include the units cm (centimetres) in your answer.*

- 12 centimetres
- 12
- 36 cm
- 12 cm
- 36 centimetres
- 36

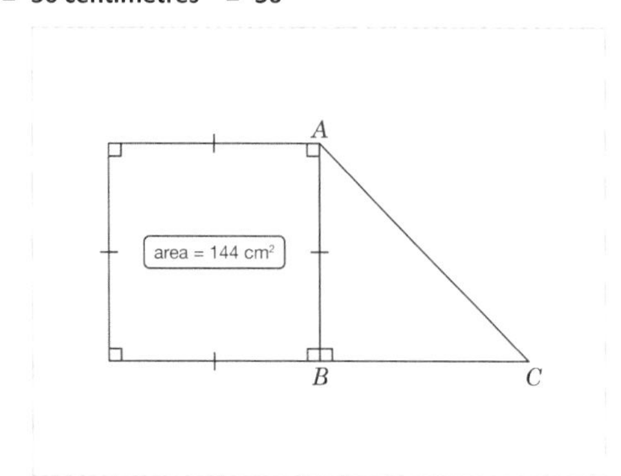

7. What is the length of side *AB*?
abc *Include the units cm (centimetres) in your answer.*

- 18.8
- 15 centimetres
- 15
- 15 cm
- 225 centimetres
- 225 cm
- 18.8 cm
- 225
- 18.8 centimetres

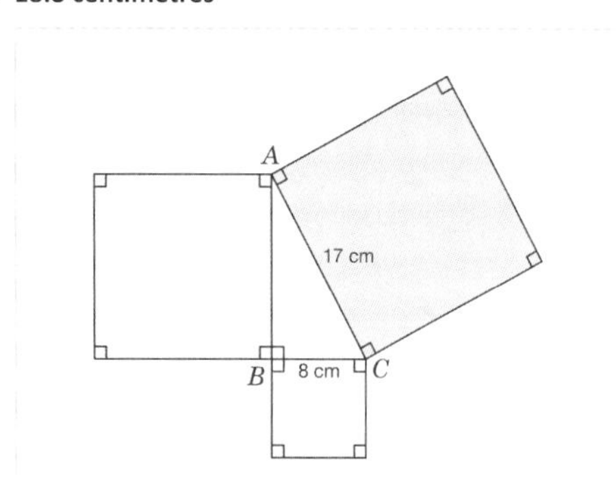

8. What is the length of side *AB*?
abc *Include the units cm (centimetres) in your answer.*

- 38.4 centimetres
- 18
- 324 centimetres
- 18 cm
- 18 centimetres
- 324
- 324 cm
- 38.4 cm
- 38.4

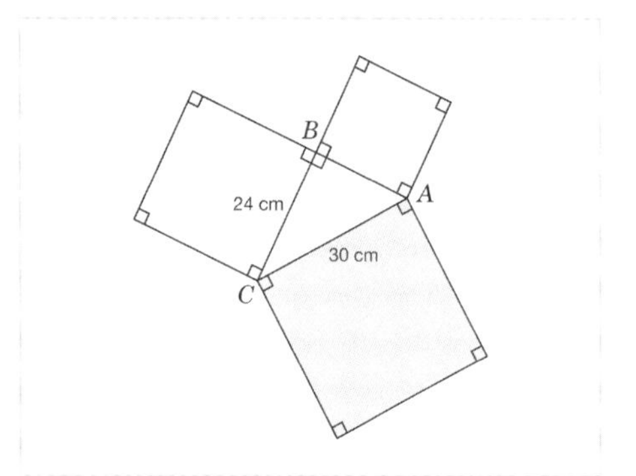

9. What is the length of side *AB*?
abc *Include the units cm (centimetres) in your answer.*

- 14 centimetres
- 14 cm
- 49 cm
- 49
- 49 centimetres
- 14

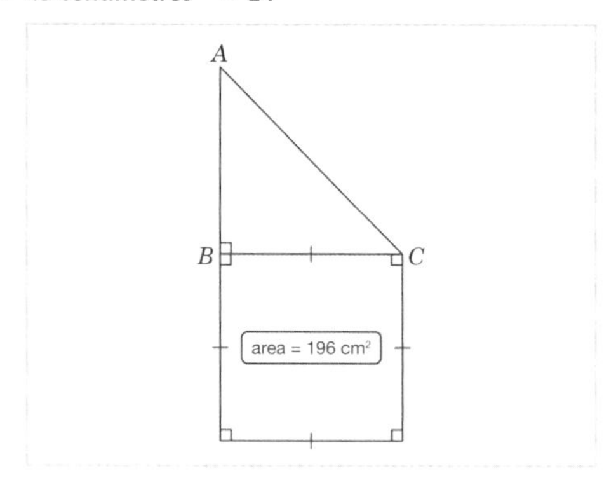

10. What is the length of side *AB*?
abc *Include the units cm (centimetres) in your answer.*

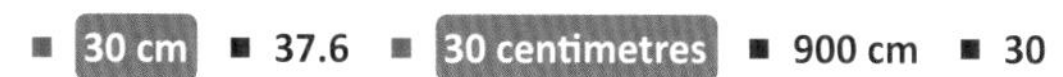

- 30 cm
- 37.6
- 30 centimetres
- 900 cm
- 30
- 37.6 centimetres
- 900
- 900 centimetres
- 37.6 cm

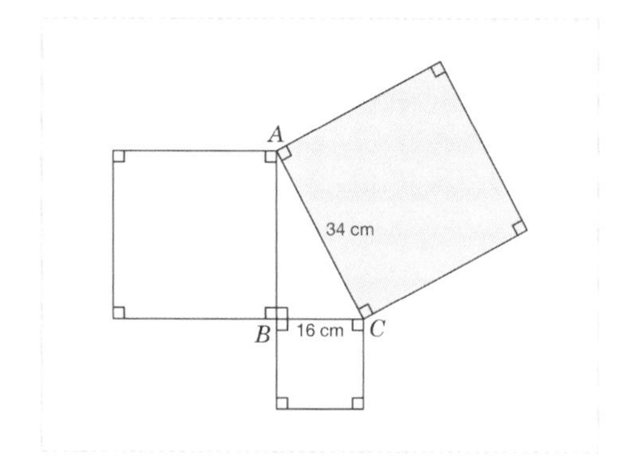

Level 2: Fluency: Using Pythagoras' theorem to find one of the shorter sides of a right angled triangle.

Required: 7/10 **Pupil Navigation:** on
Randomised: off

11. What is the value of b?
Give your answer to 1 decimal place and include the units cm (centimetres) in your answer.

- **24.2 centimetres**
- 46.4
- **24.2 cm**
- 585 centimetres
- 24.2
- 585
- 46.4 centimetres
- 585 cm
- 46.4 cm

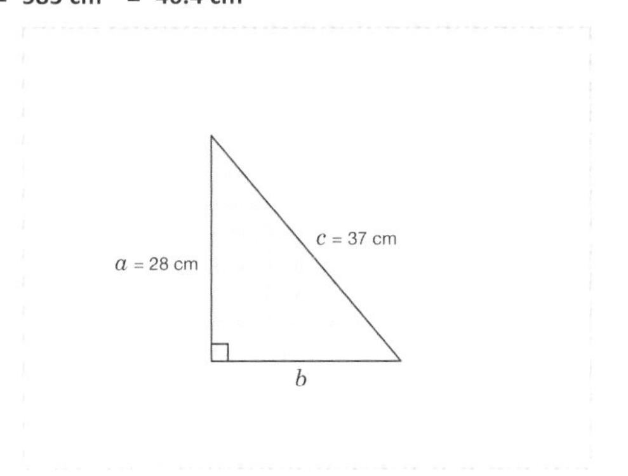

12. What is the value of b?
Give your answer to 1 decimal place and include the units cm (centimetres) in your answer.

- 73.1
- **30.4 centimetres**
- **30.4 cm**
- 73.1 centimetres
- 927 cm
- 927
- 30.4
- 927 centimetres
- 73.1 cm

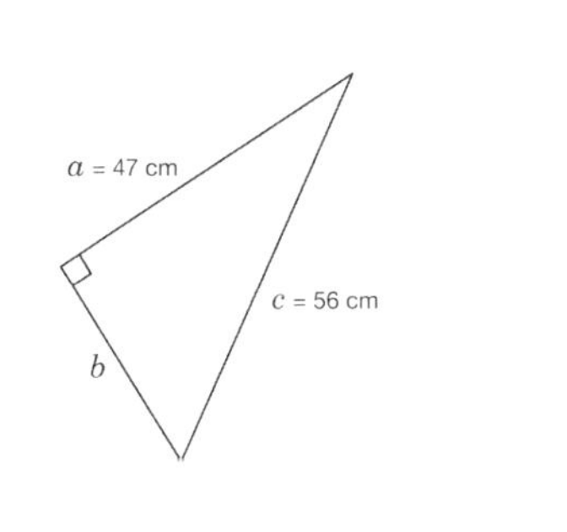

13. What is the value of x?
Give your answer to 1 decimal place and include the units cm (centimetres) in your answer.

- 25.6 cm
- **22.2 centimetres**
- 495 cm
- **22.2 cm**
- 25.6 centimetres
- 22.2
- 495
- 495 centimetres
- 25.6

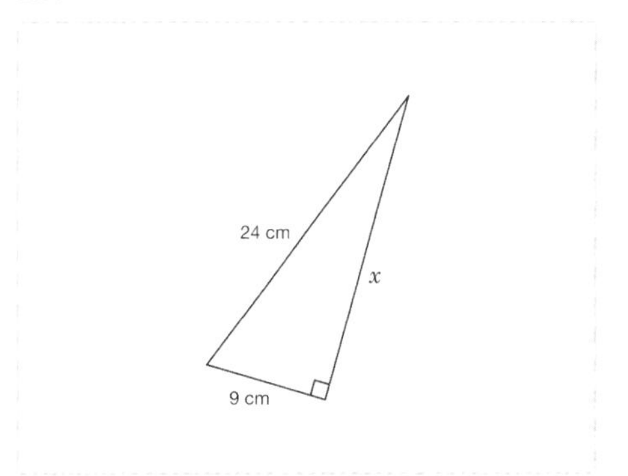

14. What is the value of x?
Give your answer to 1 decimal place and include the units cm (centimetres) in your answer.

- 413.3
- 36.1 cm
- **20.3 centimetres**
- 36.1 centimetres
- **20.3 cm**
- 36.1
- 20.3
- 413.3 centimetres
- 413.3 cm

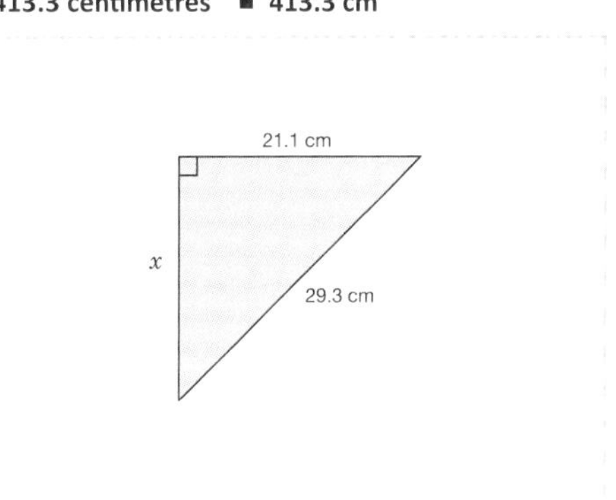

15. What is the value of x?
Give your answer to 1 decimal place and include the units cm (centimetres) in your answer.

- 119.3 centimetres
- 119.3 cm
- **10.9 cm**
- 22.4
- **10.9 centimetres**
- 22.4 cm
- 22.4 centimetres
- 119.3
- 10.9

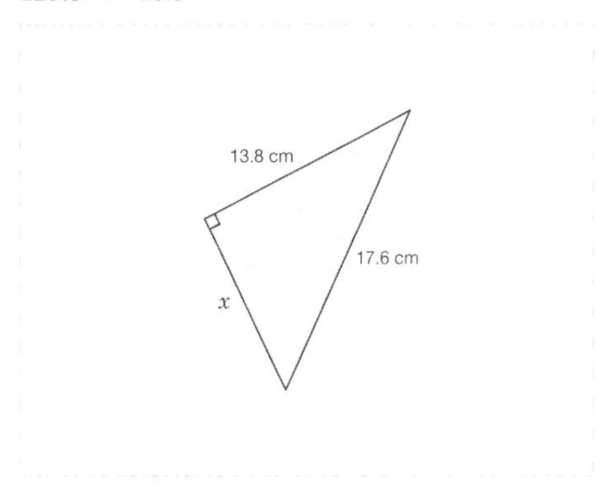

Level 2 *continued*

16. A ladder of length 4.9 metres reaches 4.2 metres up a vertical wall. What is the horizontal distance in metres between the foot of the ladder and the bottom of the wall?
Give your answer to 1 decimal place and include the units m (metres) in your answer.

- 41.7 m
- 6.4 m
- 2.5
- 2.5 metres
- 2.5 m
- 6.4 metres
- 41.7 metres
- 6.4
- 41.7

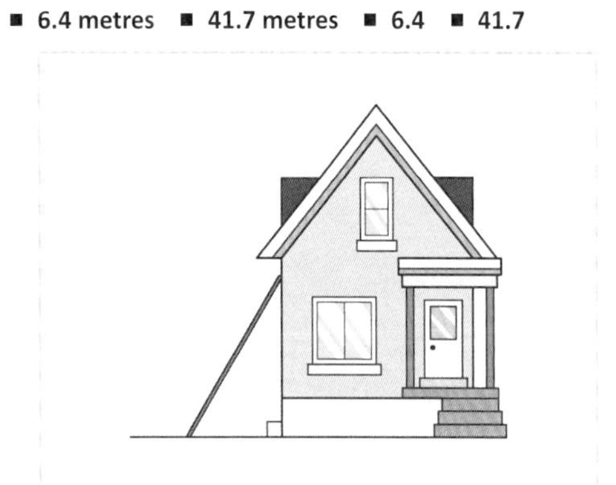

17. The horizontal distance from a point on the ground to the bottom of a tree is 28 metres. The vertical distance from the same point on the ground to the top of the tree is 35 metres. What is the height of the tree?
Include the units m (metres) in your answer.

- 21 metres
- 44.8 m
- 21
- 21 m
- 44.8
- 441
- 44.8 metres
- 441 m
- 441 metres

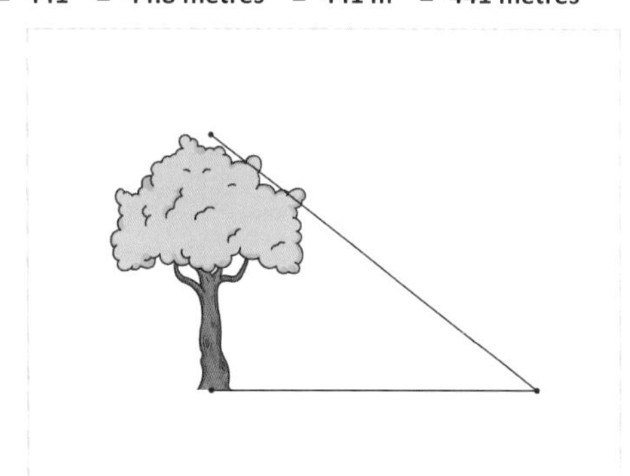

18. What is the value of x?
Give your answer to 1 decimal place and include the units cm (centimetres) in your answer.

- 37.4 centimetres
- 305.1 cm
- 17.5 cm
- 37.4 cm
- 17.5 centimetres
- 37.4
- 305.1
- 17.5
- 305.1 centimetres

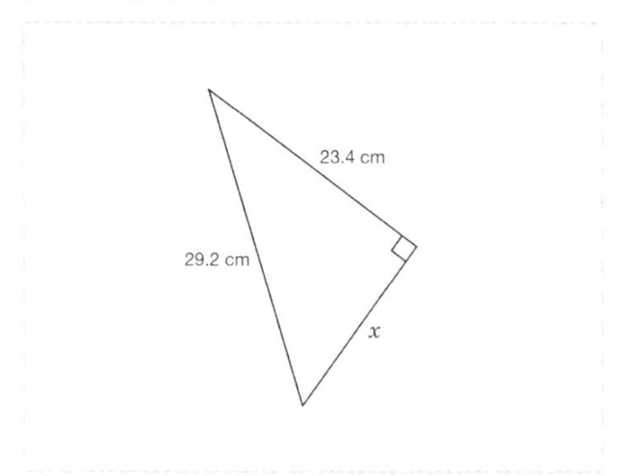

19. A ladder of length 5.3 metres reaches 5.1 metres up a vertical wall. What is the horizontal distance in metres between the foot of the ladder and the bottom of the wall?
Give your answer to 1 decimal place and include the units m (metres) in your answer.

- 1.4 m
- 7.4 metres
- 7.4
- 1.4 metres
- 2.1 m
- 7.4 m
- 1.4
- 2.1 metres
- 2.1

Level 2 *continued*

20. The horizontal distance from a point on the ground to the bottom of a tree is 22 metres. The vertical distance from the same point on the ground to the top of the tree is 32 metres. What is the height of the tree?
Include the units m (metres) in your answer.

- 540
- **23.2 metres**
- **23.2 m**
- 38.8 metres
- 23.2
- 38.8 m
- 540 metres
- 540 m
- 38.8

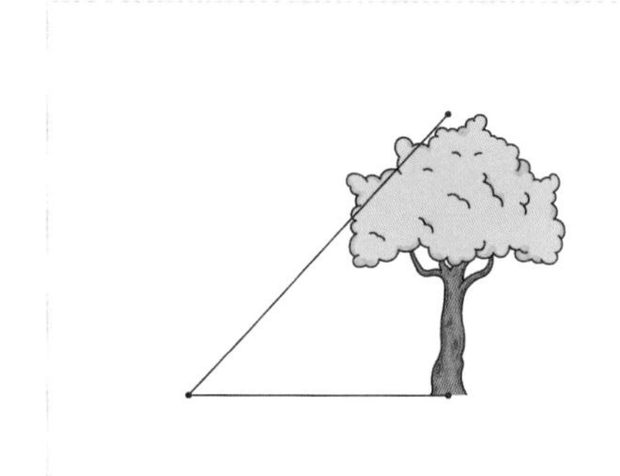

Level 3: Reasoning: Making inferences when using Pythagoras' theorem to find one of the shorter sides of a right angled triangle.

Required: 5/5 **Pupil Navigation:** on
Randomised: off

21. Simon says that for the following triangle, the value of its perimeter in centimetres is equal to the value of its area in square centimetres. Is he correct? Explain your answer.

- Open question, no set answer

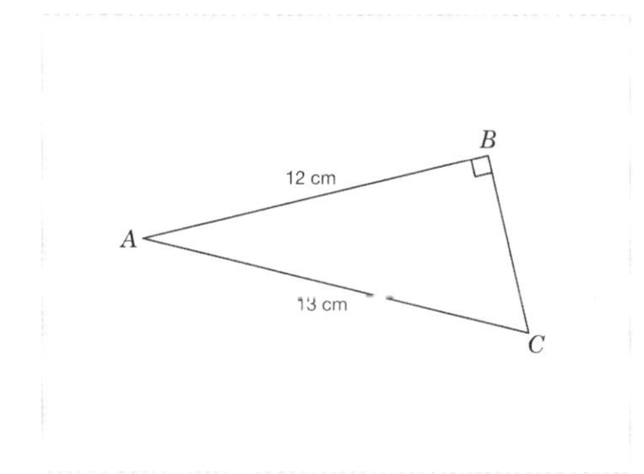

22. Select the **two** statements that are true for the following right-angled triangle.

2/4

- $x^2 - 11^2 = 15^2$
- **$15^2 - x^2 = 11^2$**
- $11^2 - x^2 = 15^2$
- **$15^2 - 11^2 = x^2$**

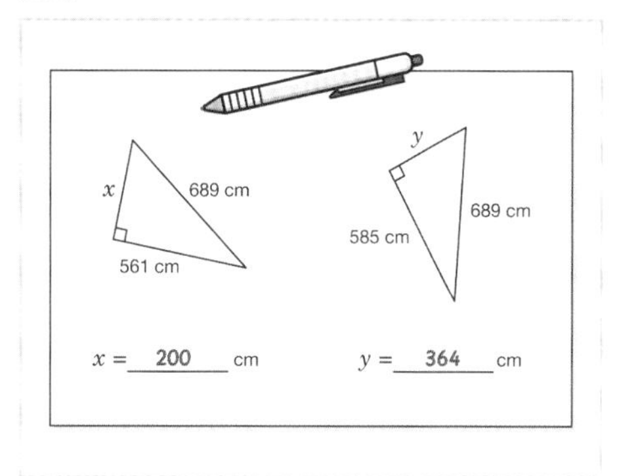

23. Poppy completes her homework but gets one of the questions wrong. What is the correct value for the question she has answered incorrectly?
Don't include the units in your answer.

- **400**
- 888.5
- 903.9

24. Raj says that the length of *BC* = 34,225 cm. Is he correct? Explain your answer.

- Open question, no set answer

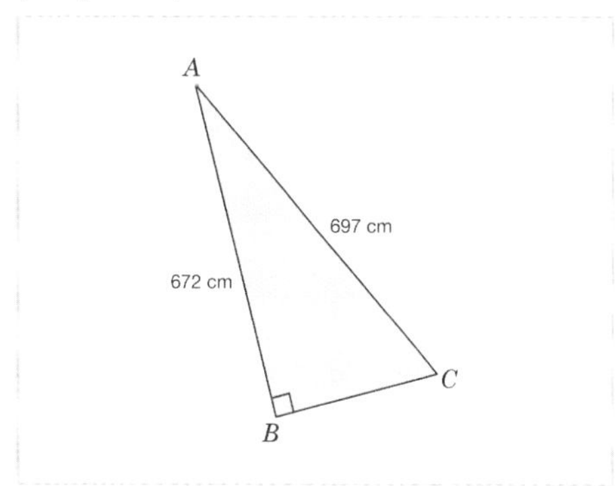

Level 3 *continued*

25. Sumbul says that the height *h* of the equilateral triangle is approximately 6.9 cm when rounded to one decimal place. Is she correct? Explain your answer.

- Open question, no set answer

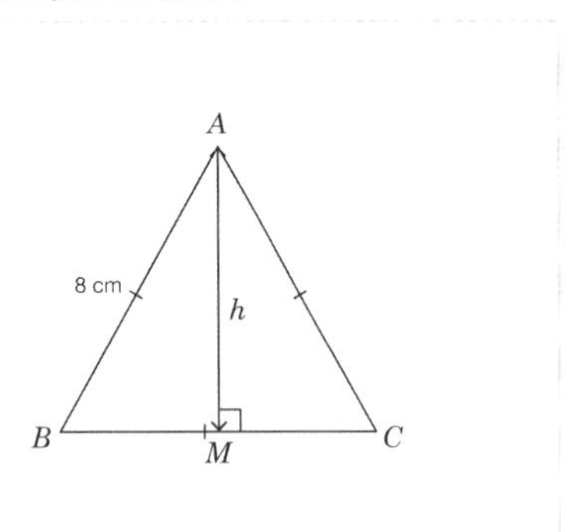

Level 4: Problem Solving: multi step problems using Pythagoras to find one of the shorter sides of a right-angled triangle.

Required: 5/5 **Pupil Navigation:** on
Randomised: off

26. Calculate the area of the equilateral triangle in square centimetres.
Give your answer to 1 decimal place and don't include the units in your answer.

- **43.3** ■ 8.7 ■ 43.5 ■ 55.9

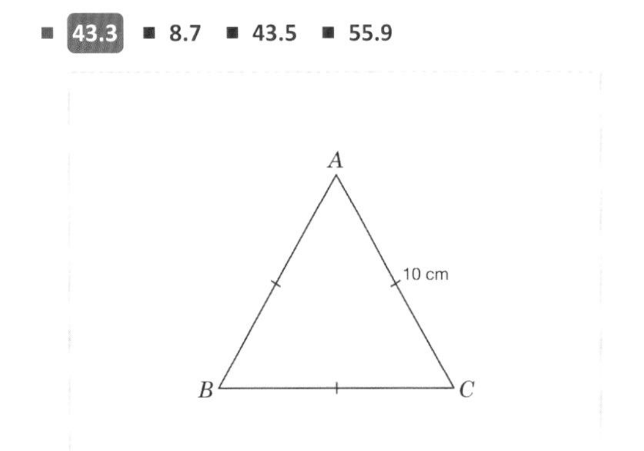

27. *ABCD* is a kite. *BD* = 27 cm, *AB* = 17 cm, *BC* = 43 cm. What is the length of the diagonal *AC*?
Give your answer to 1 decimal place and include the units cm (centimetres) in your answer.

■ 51.1 cm ■ **51.2 cm** ■ 51.1 ■ **51.2 centimetres**
■ 10.3 cm ■ 51.2 ■ 10.3 centimetres
■ 40.8 centimetres ■ 51.1 centimetres ■ 40.8 cm
■ 40.8 ■ 10.3

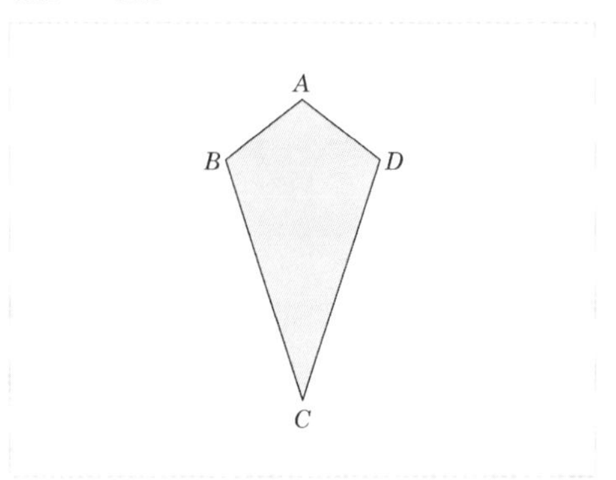

28. Calculate the length of *CD*.
Give your answer to 1 decimal place and include the units cm (centimetres) in your answer.

■ **22.4 centimetres** ■ 22.5 centimetres ■ **22.4 cm**
■ 22.5 ■ 22.4 ■ 22.5 cm

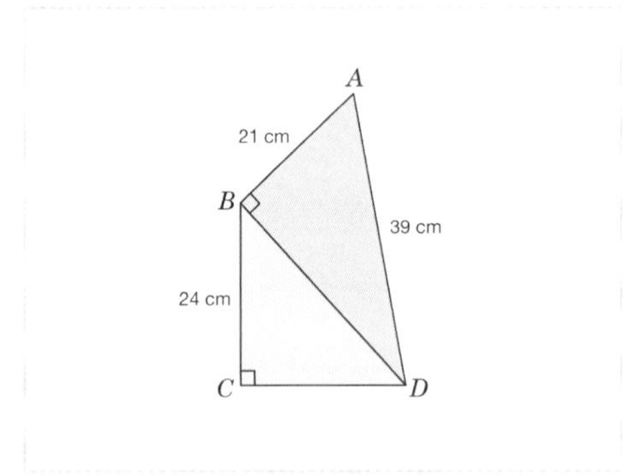

29. Calculate the area in square centimetres of a regular hexagon with perimeter 78 cm.
Give your answer to 1 decimal place and don't include the units in your answer.

■ 73.5 ■ **439.1** ■ 73.2 ■ 440.7 ■ 11.3

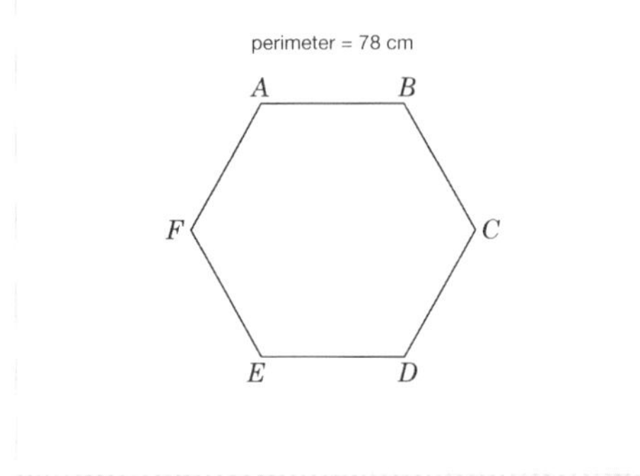

Level 4 *continued*

30. Calculate the length of *CD*.

Give your answer to 1 decimal place and include the units cm (centimetres) in your answer.

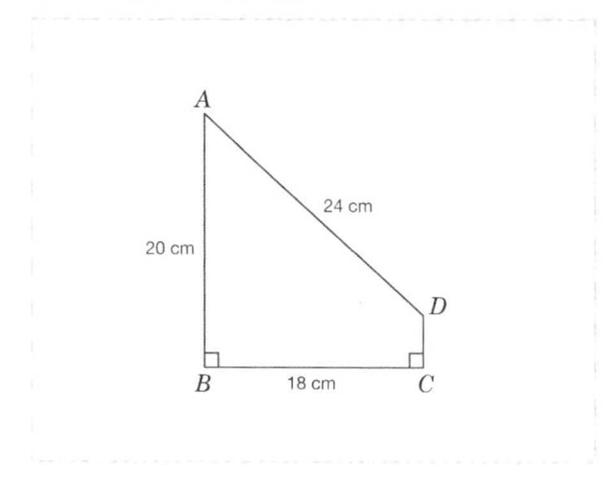

Use Pythagoras' Theorem to Find the Hypotenuse

Objective: I can use Pythagoras' theorem to find the hypotenuse.

Quick Search Ref: 10621

Level 1: Understanding: Exploring the squares on the sides of a right-angled triangle.

Required: 7/10 **Pupil Navigation:** on **Randomised:** off

1. Select the hypotenuse of the right-angled triangle.

1/3

- a
- **b**
- c

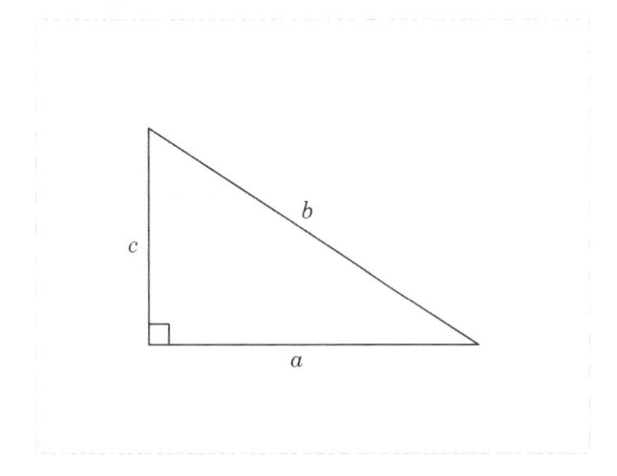

2. In the diagram, $AB = 3$ cm and $BC = 4$ cm. Select the calculation to find the area of the square on side AC in square centimetres.

1/4

- $4^2 + 5^2$
- **$3^2 + 4^2$**
- $6^2 + 3^2$
- $5^2 + 6^2$

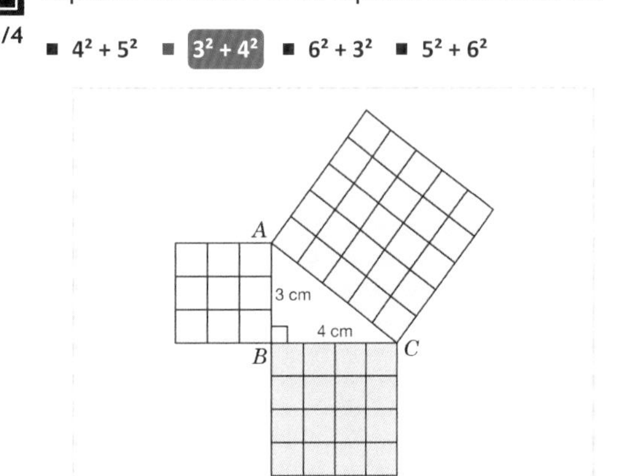

3. In the diagram, $AB = 6$ cm and $BC = 8$ cm. What is the area of the square on side AC in square centimetres?
Don't include the units in your answer.

- **100**
- 36
- 64
- 10

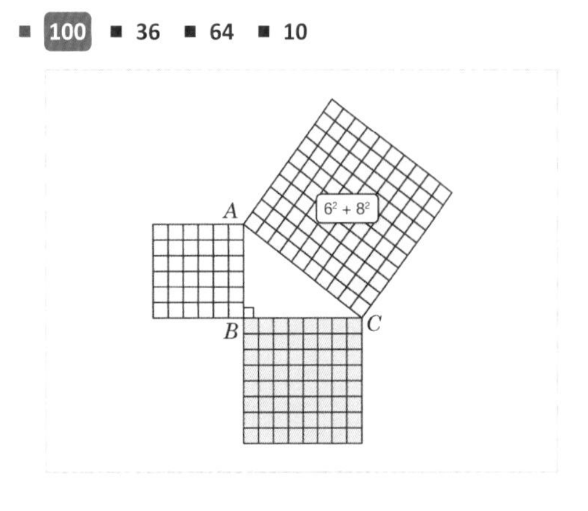

4. In the diagram, $AB = 5$ cm and $BC = 12$ cm. What is the area of the square on side AC in square centimetres?
Don't include the units in your answer.

- **169**
- 144
- 25

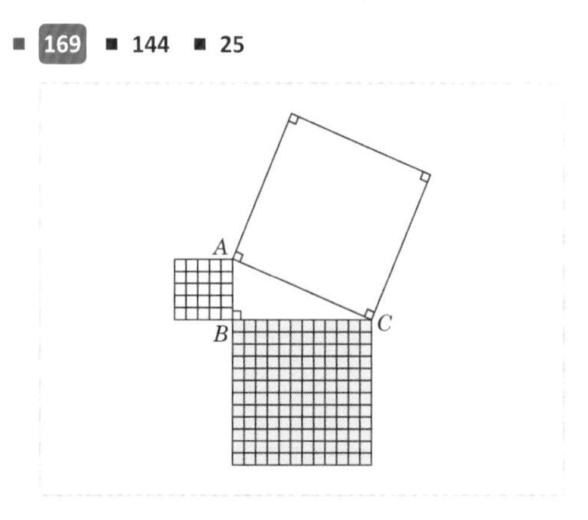

5. If the area of a square is 36 cm^2, what is the length of its side?
Include the units cm (centimetres) in your answer.

- 6
- **6 centimetres**
- 9
- **6 cm**
- 9 cm
- 9 centimetres

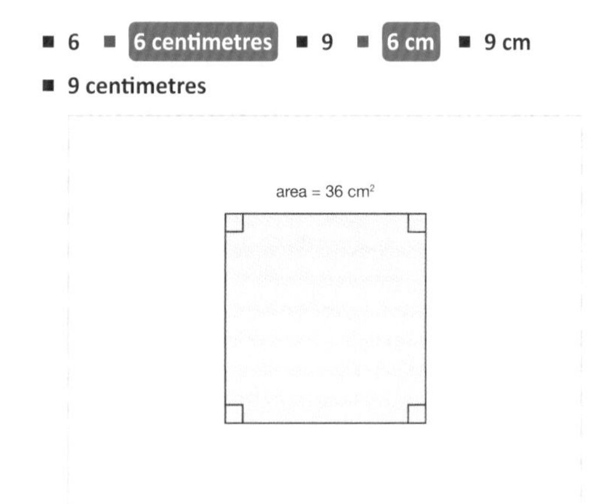

Level 1 *continued*

6. What is the length of the hypotenuse of the right-angled triangle?
Include the units cm (centimetres) in your answer.

- 8 cm
- 8 centimetres
- 16 centimetres.
- 16
- 8
- 16 cm

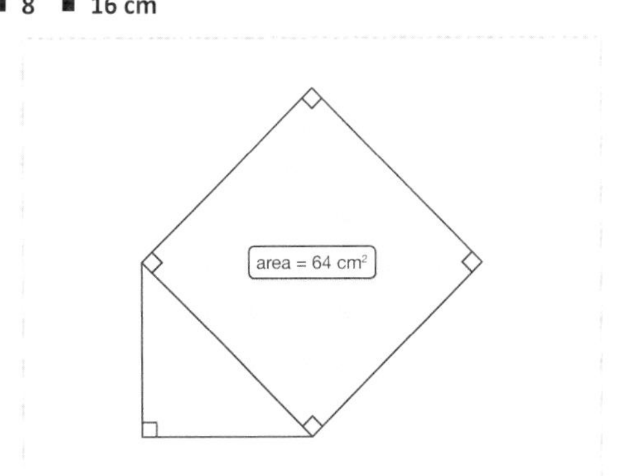

7. What is the length of the hypotenuse of the right-angled triangle?
Include the units cm (centimetres) in your answer.

- 289
- 17 cm
- 17 centimetres
- 289 centimetres
- 289 cm
- 17

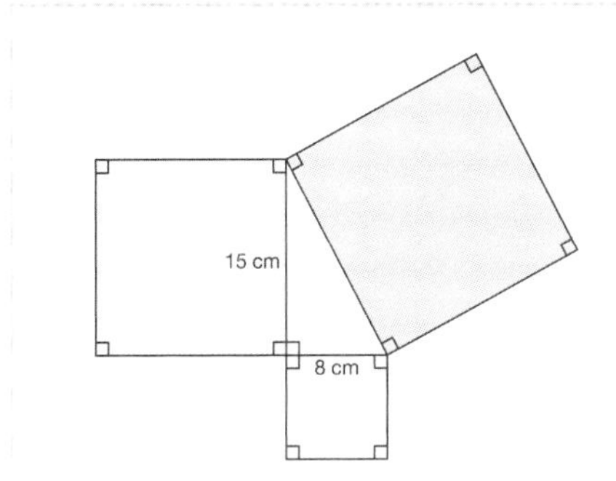

8. What is the length of the hypotenuse of the right-angled triangle?
Include the units cm (centimetres) in your answer.

- 841
- 29 centimetres
- 29 cm
- 841 centimetres
- 841 cm
- 29

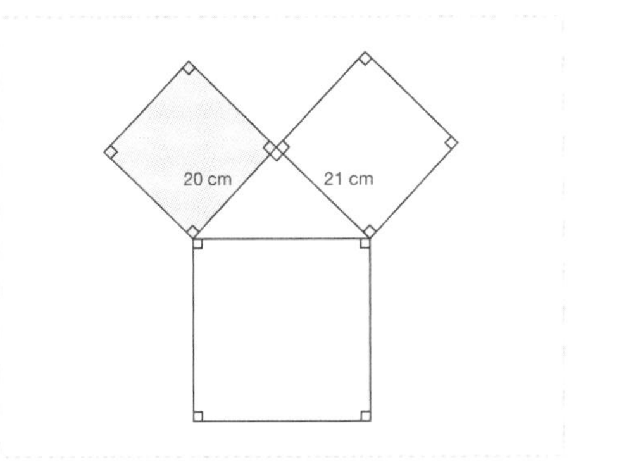

9. What is the length of the hypotenuse of the right-angled triangle?
Include the units cm (centimetres) in your answer.

- 1 cm
- 2 centimetres
- 1 centimetres.
- 2 cm
- 2
- 1

area = 4 cm²

10. What is the length of the hypotenuse of the right-angled triangle?
Include the units cm (centimetres) in your answer.

- 3,364 cm
- 3,364
- 3,364 centimetres
- 58 cm
- 58 centimetres
- 3364 cm
- 58
- 3364
- 3364 centimetres

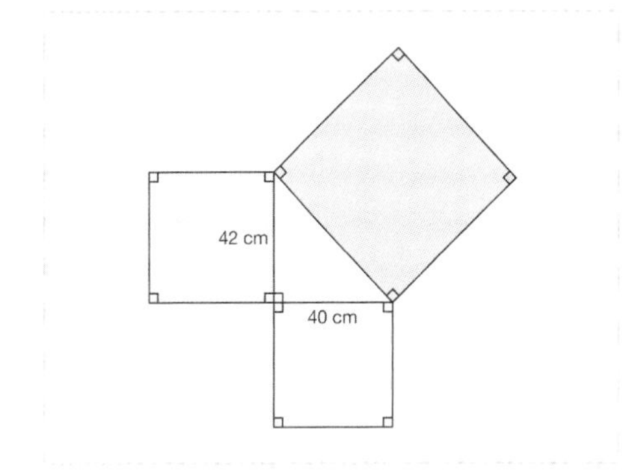

Level 2: Fluency: Calculate the hypotenuse of a right-angled triangle.

Required: 7/10 **Pupil Navigation:** on
Randomised: off

11. What is the length of the hypotenuse of the right-angled triangle?
Give your answer to one decimal place and include the units cm (centimetres) in your answer.

- **15.8 centimetres** ■ **15.8 cm** ■ 250 cm
- 250 centimetres ■ 250 ■ 15.8

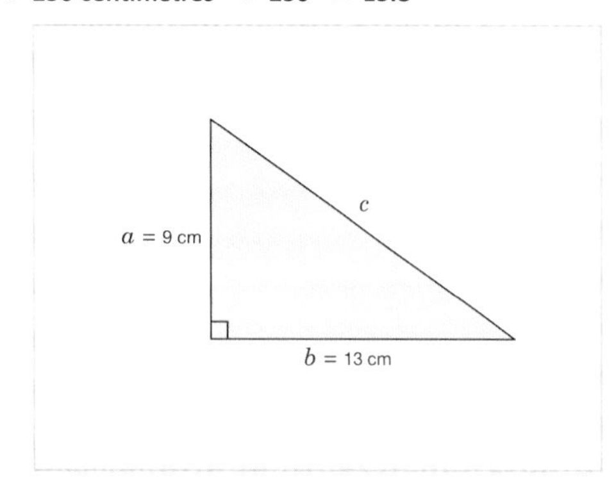

12. What is the length of the hypotenuse of the right-angled triangle?
Give your answer to one decimal place and include the units cm (centimetres) in your answer.

- **23.6 cm** ■ 557 centimetres ■ **23.6 centimetres**
- 557 cm ■ 557 ■ 23.6

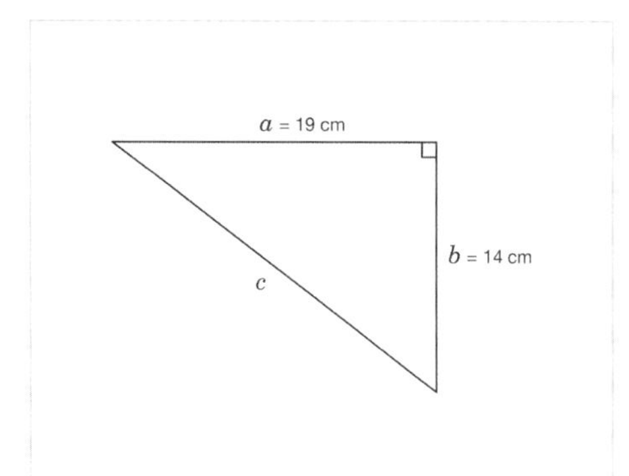

13. What is the length of the hypotenuse of the right-angled triangle?
Give your answer to one decimal place and include the units cm (centimetres) in your answer.

- **21.1 cm** ■ 445 centimetres ■ 445
- **21.1 centimetres** ■ 21.1 ■ 445 cm

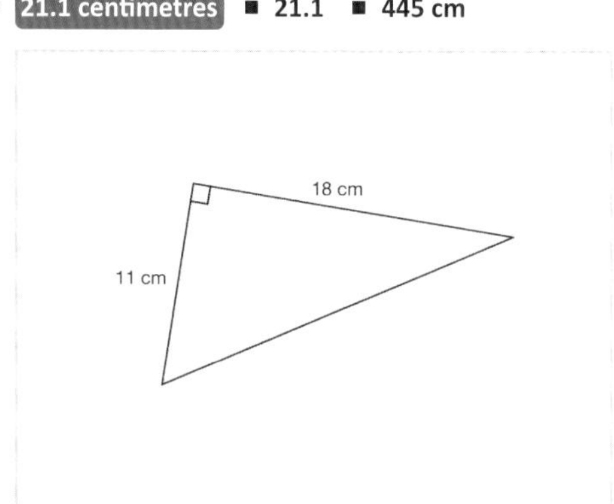

14. What is the length of the hypotenuse of the right-angled triangle?
Give your answer to one decimal place and include the units cm (centimetres) in your answer.

- 1129 ■ 1,129 ■ **33.6 centimetres** ■ **33.6 cm**
- 1129 cm ■ 1,129 centimetres ■ 33.6
- 1129 centimetres ■ 1,129 cm

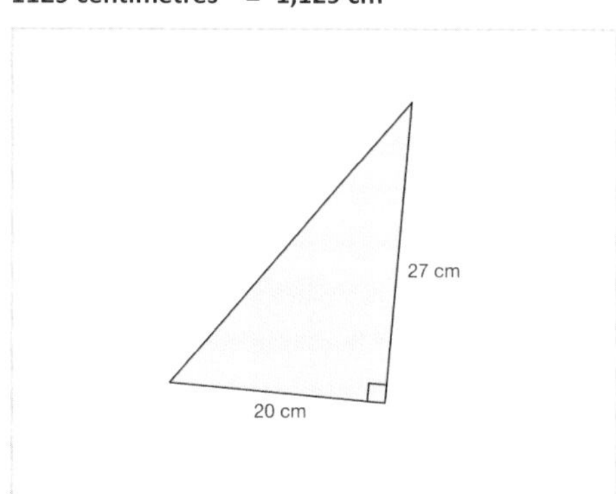

15. What is the length of the hypotenuse of the right-angled triangle?
Give your answer to one decimal place and include the units cm (centimetres) in your answer.

- **21.7 centimetres** ■ 470.6 cm ■ 470.6 centimetres
- **21.7 cm** ■ 470.6 ■ 21.7

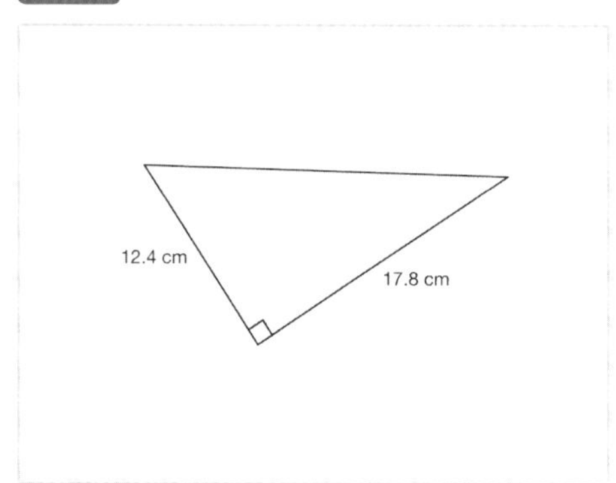

Level 2 *continued*

16. A ladder reaches 3.5 metres up a vertical wall. The foot of the ladder is 0.8 metres horizontally from the bottom of the wall. What is the length of the ladder?
Give your answer to one decimal place and include the units m (metres) in your answer.

- 3.6
- 3.6 m
- 3.6 metres
- 12.89
- 12.89 m
- 12.89 metres

17. A ship sails 32 kilometres due west. It then sails 27 kilometres due south. What is the shortest distance that the ship must travel to return to its starting point?
Give your answer to one decimal place and include the units km (kilometres) in your answer.

- 41.9 kilometres
- 1753 kilometres
- 41.9 km
- 1753 km
- 1753
- 41.9

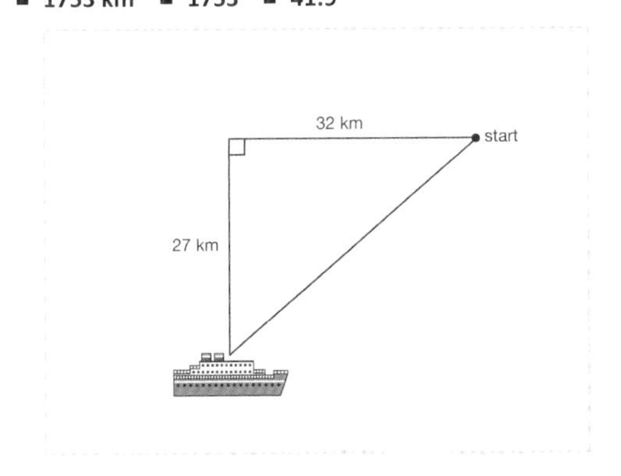

18. What is the length of the hypotenuse of the right-angled triangle?
Give your answer to one decimal place and include the units cm (centimetres) in your answer.

- 10.5 cm
- 10.5 centimetres
- 10.5
- 110.5 cm
- 110.5
- 110.5 centimetres

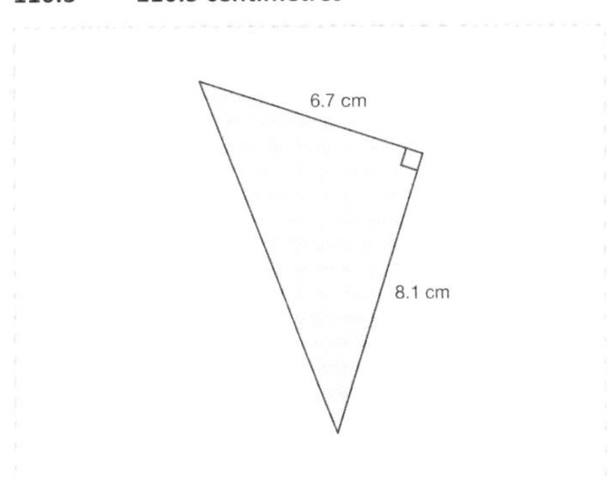

19. A ladder reaches 2.8 metres up a vertical wall. The foot of the ladder is 0.75 metres horizontally from the bottom of the wall. What is the length of the ladder?
Give your answer to one decimal place and include the units m (metres) in your answer.

- 2.9 metres
- 8.4 m
- 2.9 m
- 8.4
- 8.4 metres
- 2.9

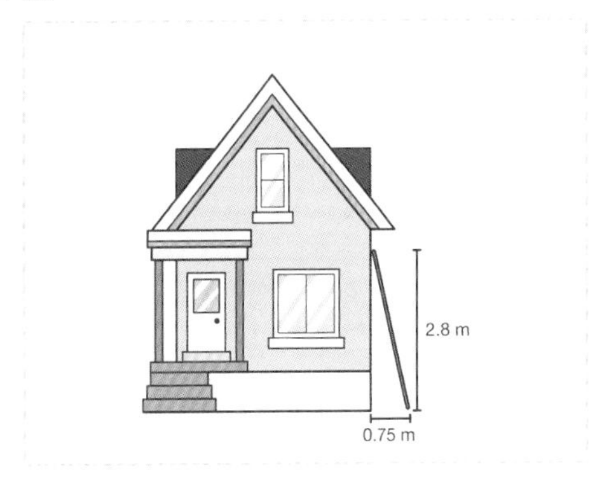

Level 2 *continued*

20. A ship sails 140 kilometres due north. It then sails 80 kilometres due east. What is the shortest distance that the ship must travel to return to its starting point?
Give your answer to one decimal place and include the units miles in your answer.

- **161.2 km**
- 161.2
- **161.2 kilometres**
- 26000 km
- 26000
- 26000 kilometres

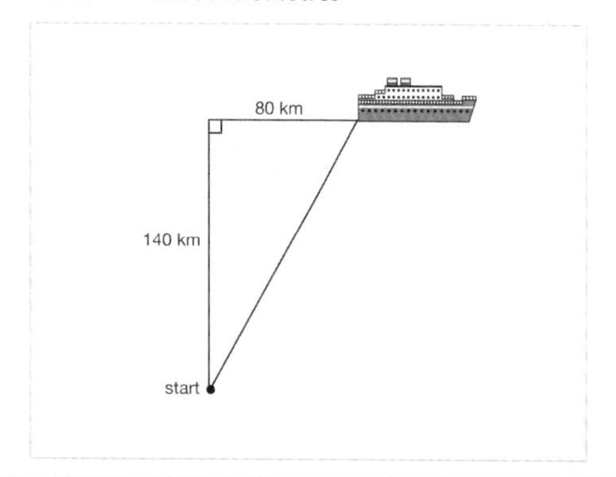

Level 3: Reasoning: Make inferences using Pythagoras' Theorem.

Required: 5/5 **Pupil Navigation:** on
Randomised: off

21. Is this a right-angled triangle? Explain your answer.

- Open question, no set answer

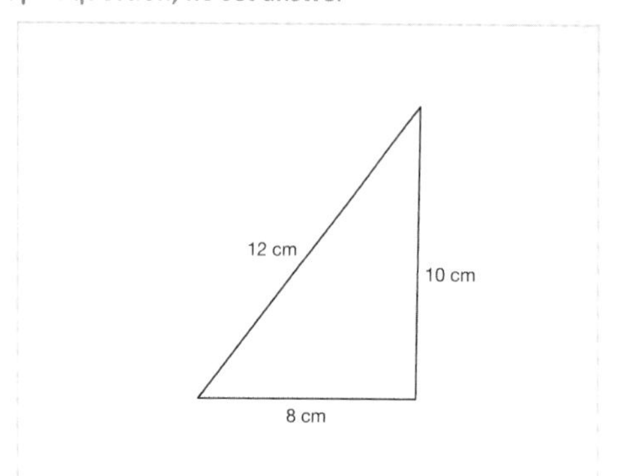

22. For this triangle, Sadie says that $a^2 + b^2 = c^2$. Is she correct? Explain your answer.

- Open question, no set answer

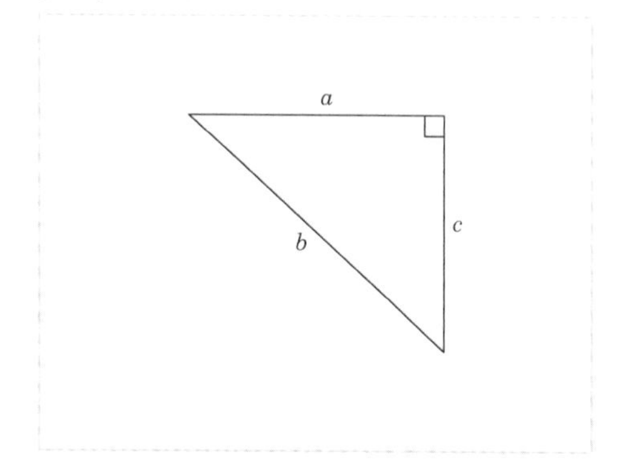

23. Work out the value of x.
Include the units cm (centimetres) in your answer.

- 576
- **24 cm**
- **24 centimetres**
- 24
- 26
- 26 cm
- 26 centimetres
- 576 cm
- 576 centimetres

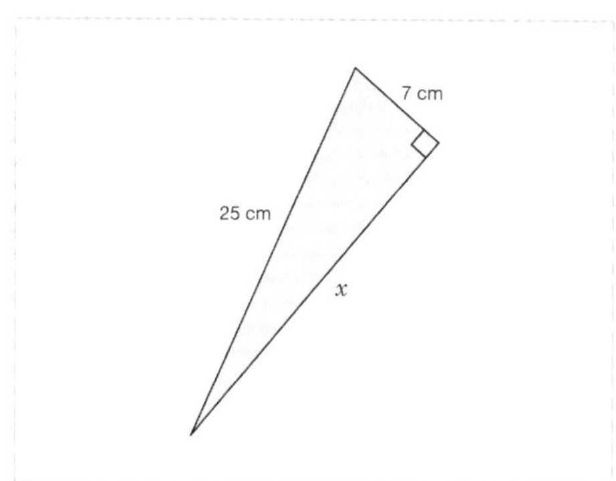

24. A shop sells pencils in four different lengths. What is the longest pencil that will fit inside the pencil case?

1/4

- 15 cm
- 16 cm
- **17 cm**
- 18 cm

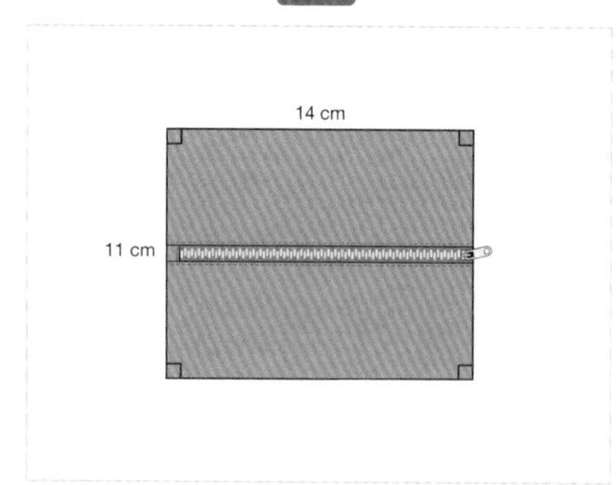

25. For this triangle, Sophie says that the value of x is 52 cm. Is she correct? Explain your answer.

- Open question, no set answer

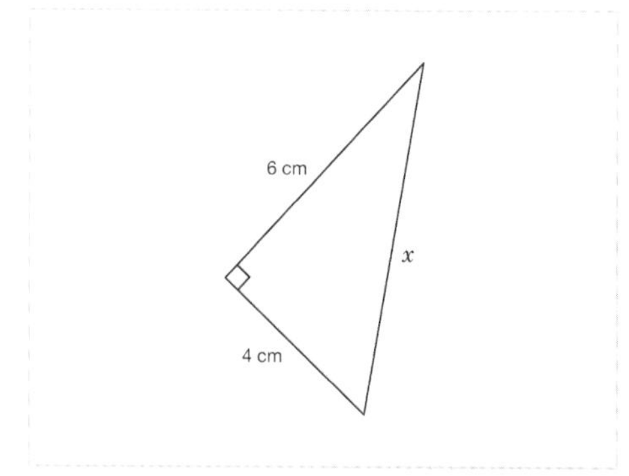

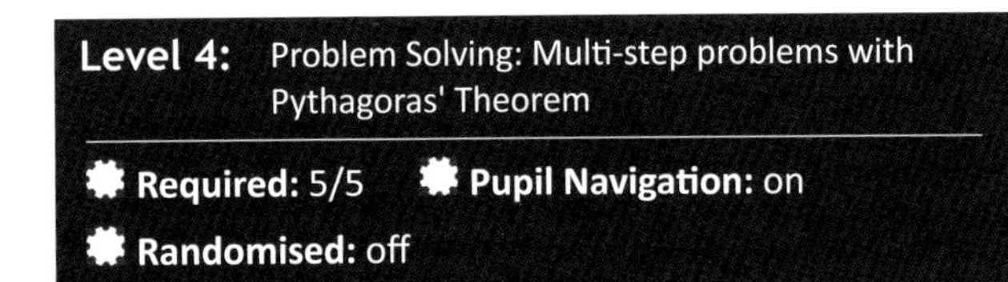

Level 4: Problem Solving: Multi-step problems with Pythagoras' Theorem

Required: 5/5 **Pupil Navigation:** on

Randomised: off

26. Calculate the length of *AC*.

abc *Give your answer to one decimal place and include the units cm (centimetres) in your answer.*

- 38.7 ■ 1,501 cm ■ 38.8 ■ **38.7 centimetres**
- 1501 ■ 1,501 centimetres ■ **38.7 cm**
- 38.8 centimetres ■ 1,501 ■ 38.8 cm ■ 1501 cm
- 1501 centimetres

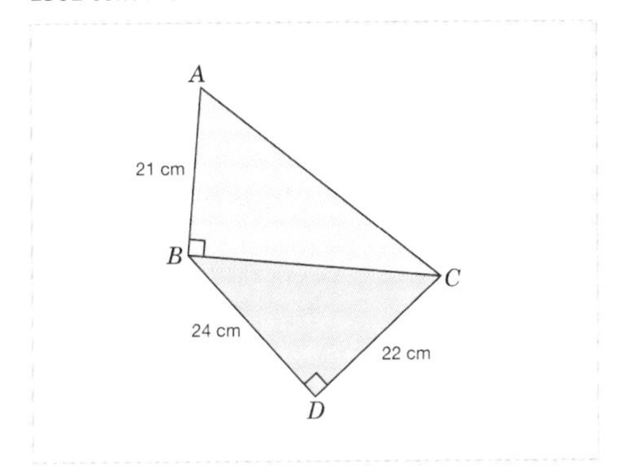

27. The diagram shows the design for a zip wire from *A* to *B*. What is the length of the zip wire?

abc *Give your answer to one decimal place and include the units m (metres) in your answer.*

- 15.3 ■ **15.3 m** ■ 234 metres ■ **15.3 metres**
- 234 m ■ 234

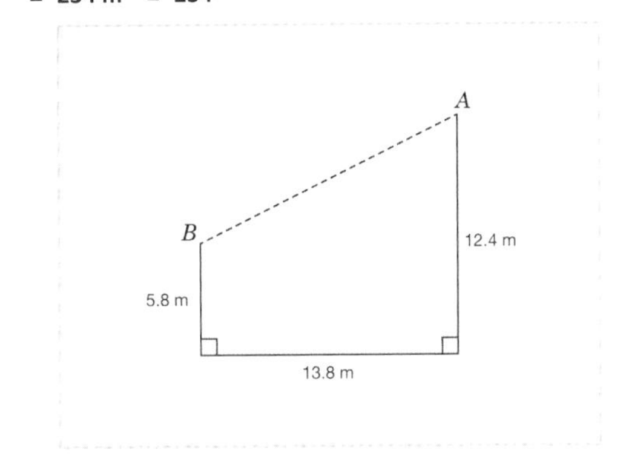

28. The points (-3, -1) and (4, 3) are shown on a 1 centimetre grid. What is the length of *AB*?

abc *Give your answer to one decimal place and include the units cm (centimetres) in your answer.*

- **8.1 centimetres** ■ **8.1 cm** ■ 8.1 ■ 65 ■ 65 cm
- 65 centimetres

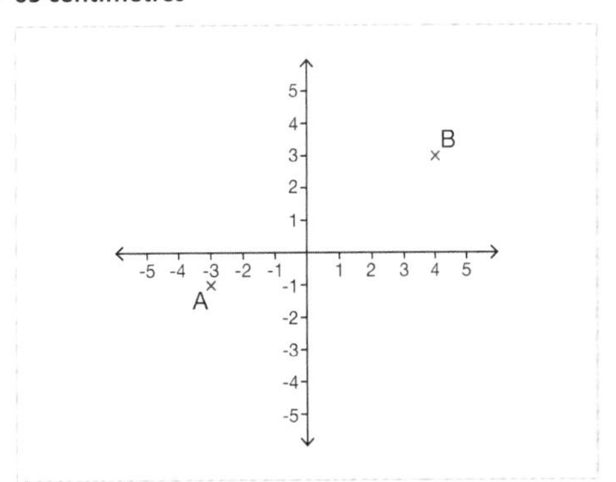

29. Select the triangle which does not contain a right angle.

1/4

- (i) ■ (ii) ■ (iii) ■ **(iv)**

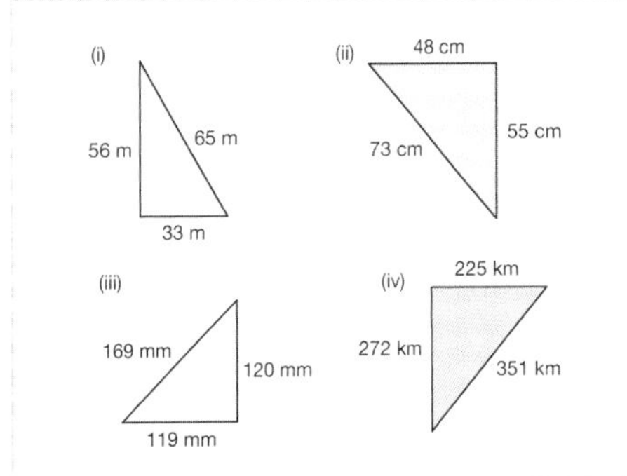

30. A tree is located in the corner of a rectangular field. The field is 9 metres wide and 7 metres long. Point *A* is at the top of the tree and is directly above point *B* which is at the foot of the tree. The tree is 6 metres high. Calculate the length of *AD*.

abc *Give your answer to one decimal place and include the units m (metres) in your answer.*

- 11.4 metres ■ 166 m ■ **12.9 m** ■ **12.9 metres**
- 11.4 ■ 12.9 ■ 166 ■ 11.4 m ■ 166 metres

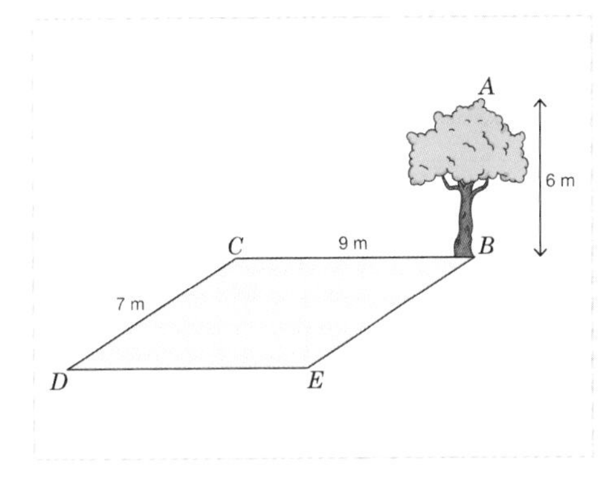

Use the Cosine Ratio

Objective: I can calculate the cosine of an angle in a triangle given the hypotenuse and adjacent sides. I can calculate missing sides in a triangle given the cosine of an angle.

Quick Search Ref: 10623

Level 1: Understanding: Calculate the cosine of an angle given the hypotenuse and adjacent side.

Required: 6/8 **Pupil Navigation:** on **Randomised:** off

1. Select the calculation which gives the cosine of an angle.

1/3

- i) opposite ÷ adjacent
- ii) opposite ÷ hypotenuse
- **iii) adjacent ÷ hypotenuse**

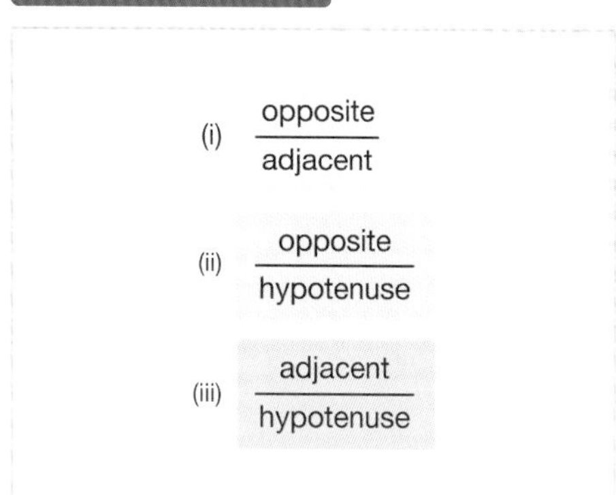

2. What is the value of cos x?

1/4

- 13/12
- 5/13
- 5/12
- **12/13**

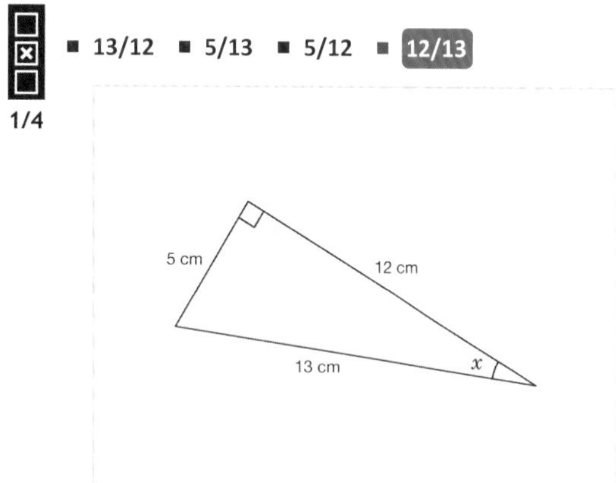

3. Give cos x as a fraction.

- **5/8**
- 8/5

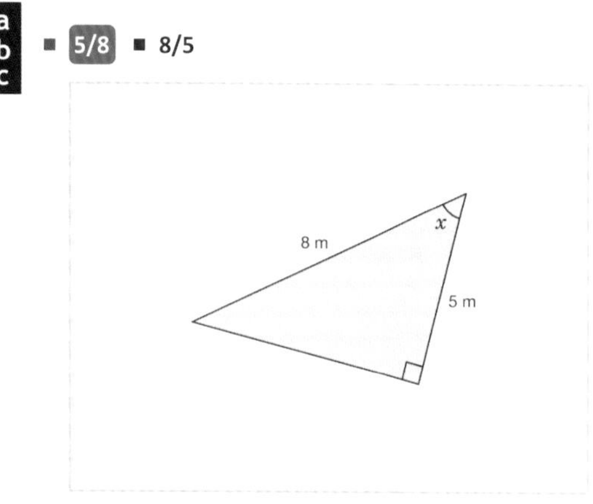

4. Give cos d as a fraction in its simplest form.

- 16/34
- **8/17**
- 17/8
- 15/17

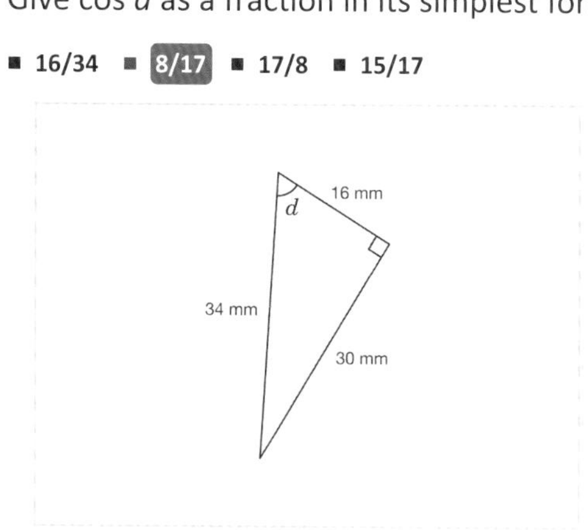

5. Give cos y as a decimal.

- **0.8**
- 16/20
- 4/5
- 0.6
- 8/10

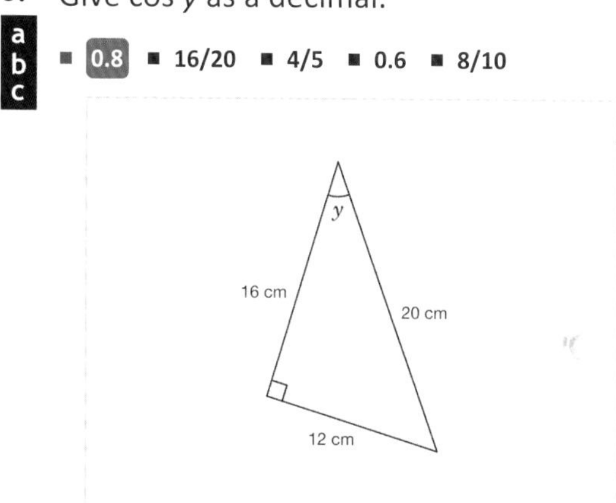

6. Give cos x as a fraction.

- **20/29**
- 29/20
- 21/29

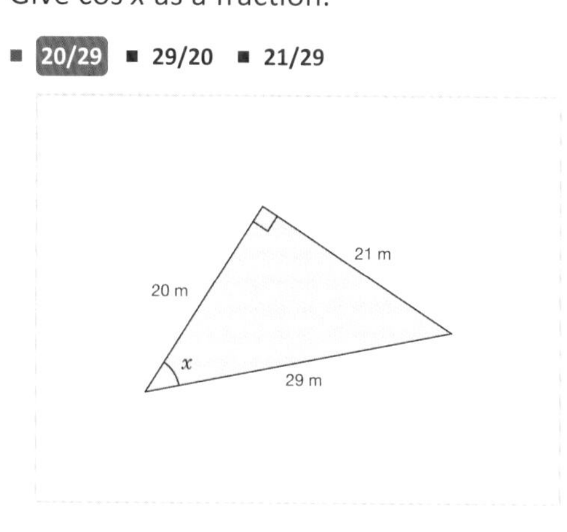

Level 1 *continued*

7. Give cos *c* as a fraction in its simplest form.

a b c

- 13/5
- **5/13**
- 10/26
- 12/13

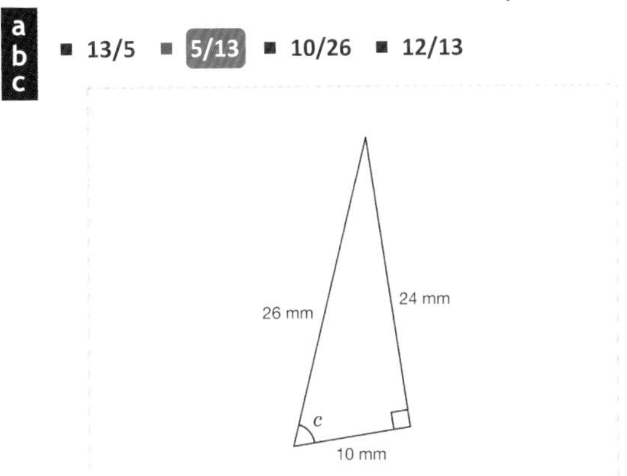

8. Give cos *z* as a decimal.

a b c

- **0.28**
- 14/50
- 0.96
- 7/25

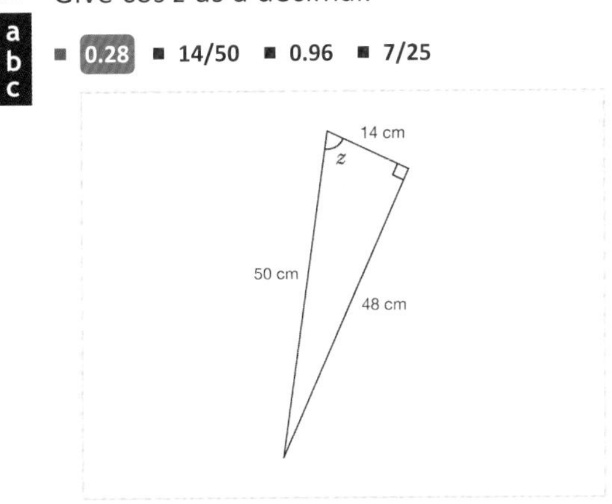

Level 2: Fluency: Calculate missing sides in triangles given the cosine of an angle.

Required: 6/8 **Pupil Navigation:** on

Randomised: off

9. If cos $x = 1/3$, find the value of *d*.
Include the units cm (centimetres) in your answer.

a b c

- 54 centimetres
- **6 centimetres**
- 54 cm
- **6 cm**
- 54
- 6

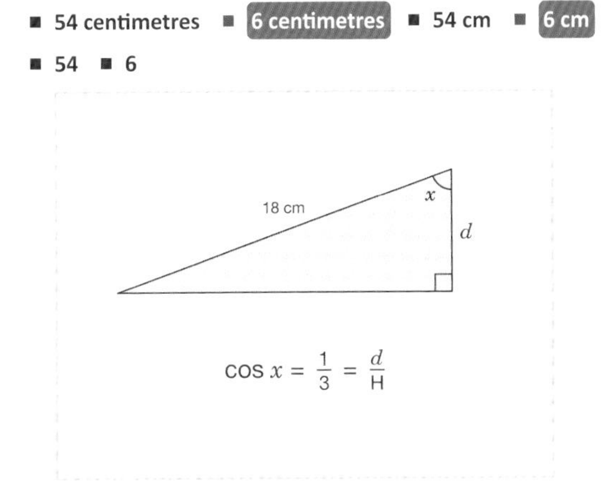

$\cos x = \frac{1}{3} = \frac{d}{H}$

10. If cos $x = 1/2$, find the value of *c*.
Include the units m (metres) in your answer.

a b c

- 7
- **28 m**
- **28 metres**
- 7 metres
- 28
- 7 m

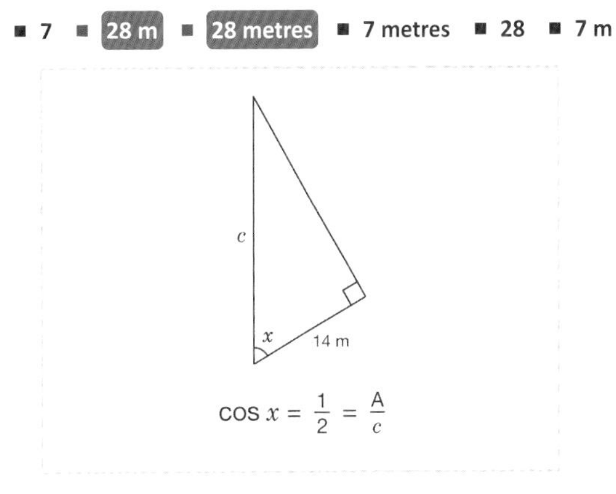

$\cos x = \frac{1}{2} = \frac{A}{c}$

11. If cos $x = 1/4$, find the value of *t*.
Include the units mm (millimetres) in your answer.

a b c

- **12 mm**
- 48 millimetres
- **12 millimetres**
- 48
- 12
- 48 mm

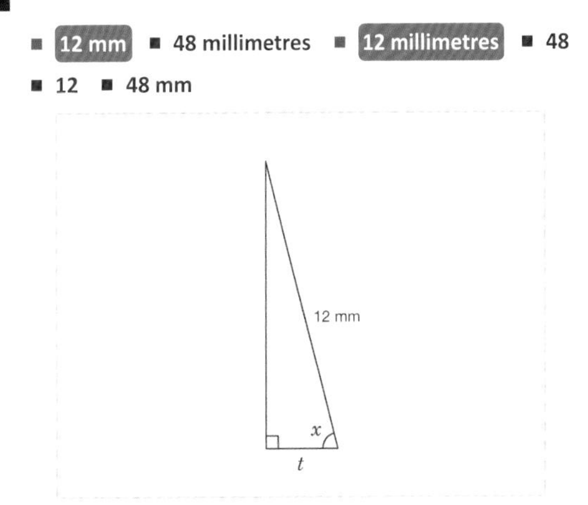

12. If cos $x = 0.5$, find the value of *s*.
Include the units cm (centimetres) in your answer.

a b c

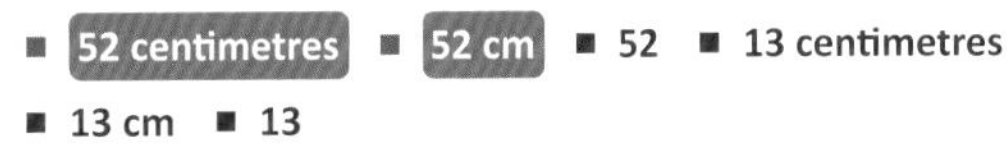

- **52 centimetres**
- **52 cm**
- 52
- 13 centimetres
- 13 cm
- 13

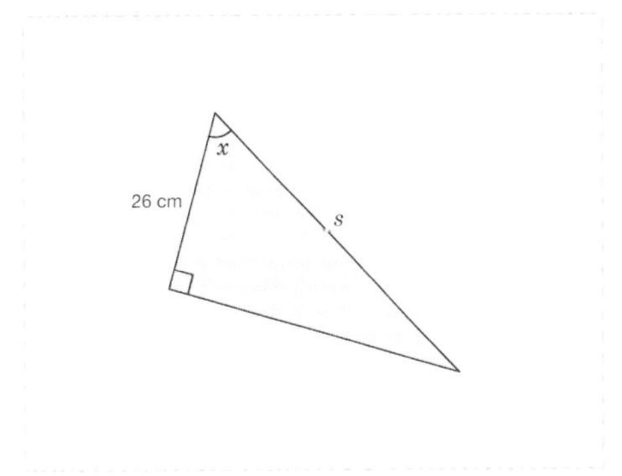

Level 2 *continued*

13. If cos $x = 2/5$, find the length of side *BC*.
Include the units m (metres) in your answer.

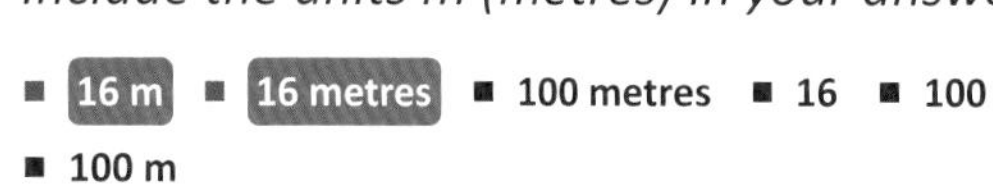
- 16 m
- 16 metres
- 100 metres
- 16
- 100
- 100 m

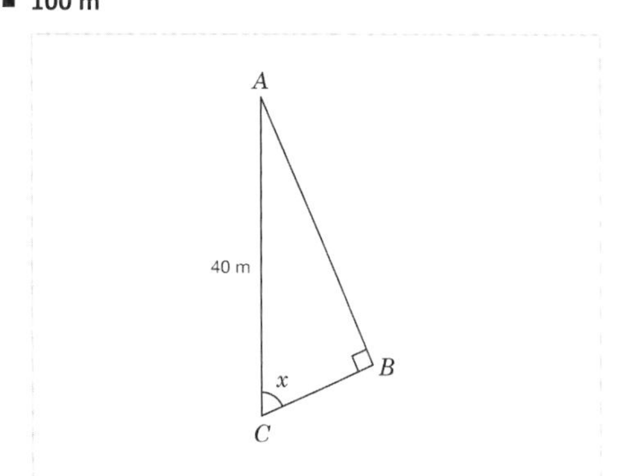

14. If cos $x = 3/4$, find the value of n.
Include the units mm (millimetres) in your answer.

- 32
- 16 mm
- 32 mm
- 32 millimetres
- 16
- 16 millimetres

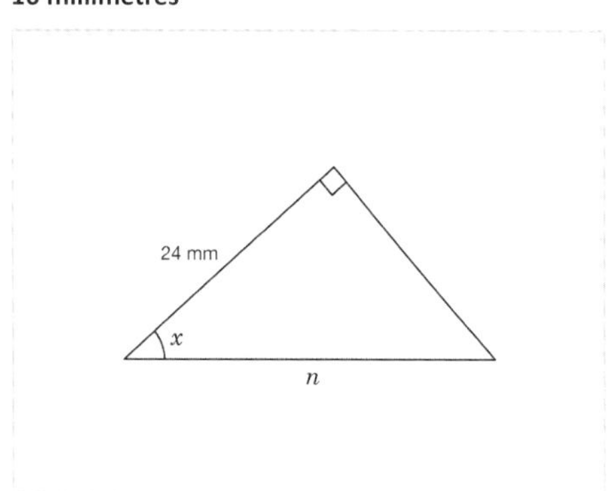

15. If cos $x = 2/3$, find the length of side *LM*.
Include the units cm (centimetres) in your answer.

- 54 cm
- 54 centimetres
- 54
- 24 centimetres
- 24
- 24 cm

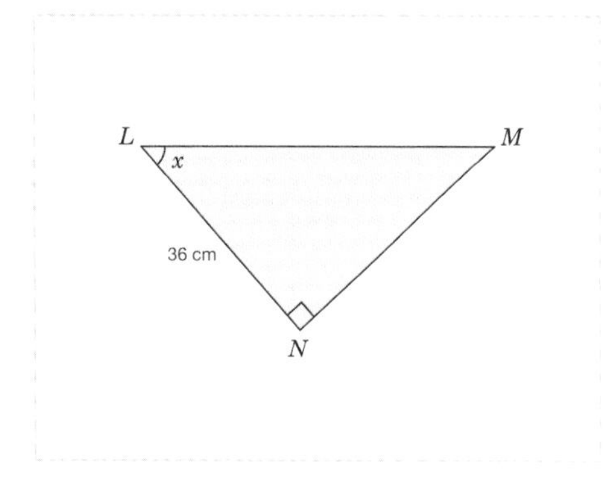

16. If cos $x = 0.24$, find the length of side s.
Include the units m (metres) in your answer.

- 312.5 metres
- 18 m
- 18 metres
- 312.5
- 18
- 312.5 m

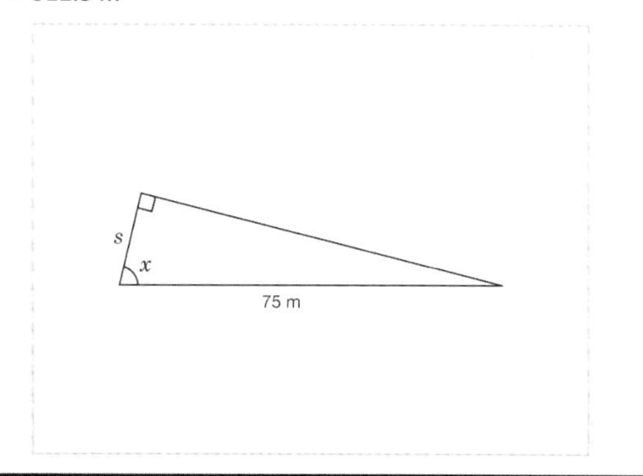

Level 3: Reasoning: Making inferences using the cosine ratio.

Required: 3/3 **Pupil Navigation:** on
Randomised: off

17. Calculate the value of cos 60°.

- 1/2
- 0.5

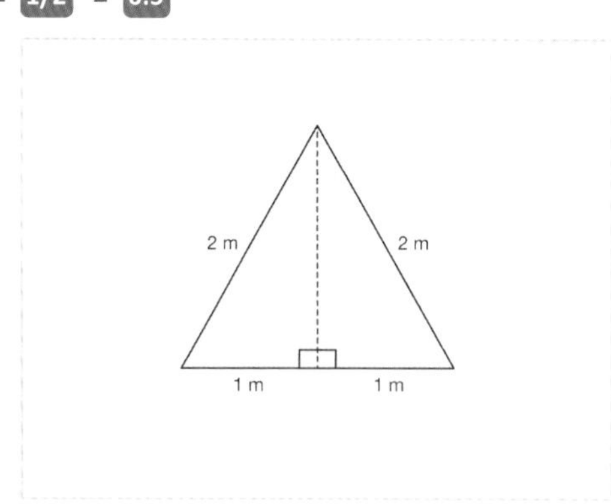

18. Angle *ACB* = x. Angle *CAB* = $90° - x$. Daniel says that cos x = sin(90– x).
Is Daniel correct? Explain your answer.

- Open question, no set answer

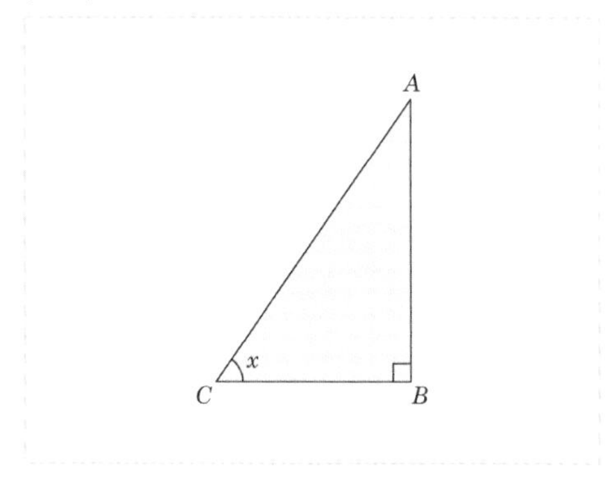

Level 3 *continued*

19. Using the diagrams, give the value of cos 40° correct to two decimal places.

1/4

- 0.72
- 0.64
- **0.77**
- 0.67

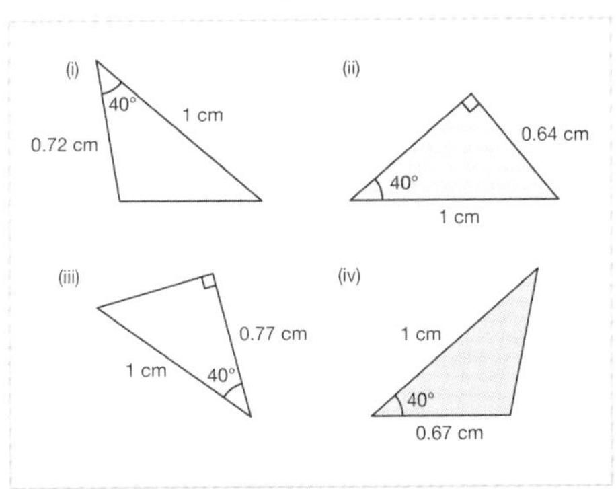

Level 4: Problem solving: Multi-step problems using the cosine ratio.

Required: 3/3 **Pupil Navigation:** on

Randomised: off

20. Select the correct value of cos 30°.

1/4

- (i)
- (ii)
- (iii)
- **(iv)**

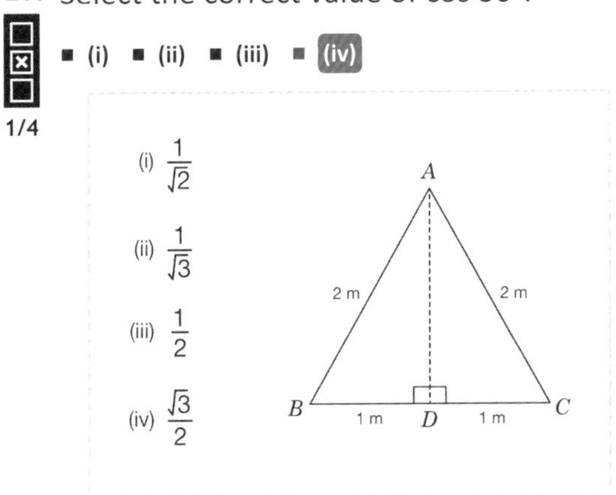

21. Give the value of cos *x* as a fraction.

- **13/15**

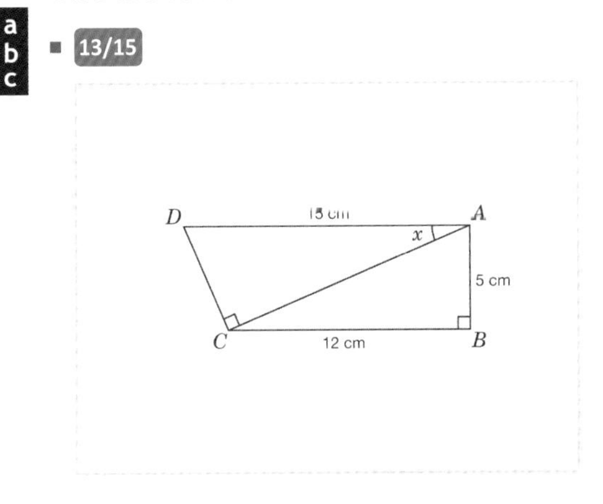

22. Give the value of $(\cos x)^2 + (\sin x)^2$.
Give your answer as an integer.

- **1**
- 25/25

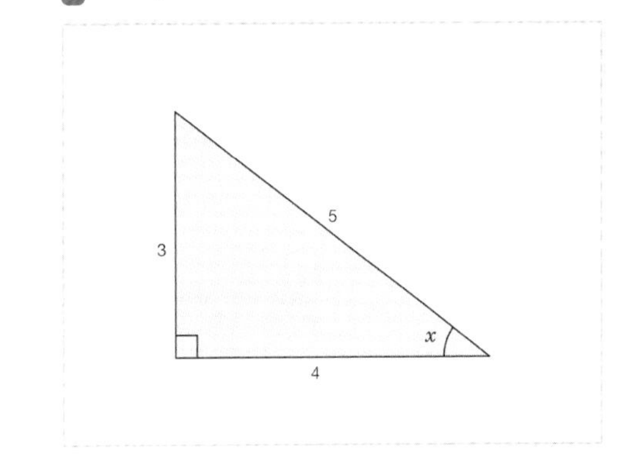

Use the Sine Ratio

Objective: I can calculate the sine of an angle in a triangle given the hypotenuse and opposite sides. I can calculate missing sides in a triangle given the sine of an angle.

Quick Search Ref: 10619

Level 1: Understanding: Calculate the sine of an angle given the hypotenuse and opposite side.

Required: 6/8 **Pupil Navigation:** on **Randomised:** off

1. Select the calculation which gives the sine of an angle.

1/3

- i) opposite ÷ adjacent
- ii) opposite ÷ hypotenuse
- iii) adjacent ÷ hypotenuse

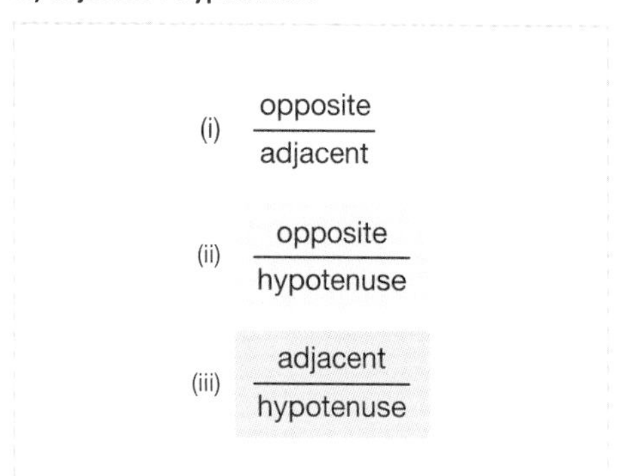

2. What is the value of sin x?

1/4

- 5/3
- 3/4
- 4/5
- 3/5

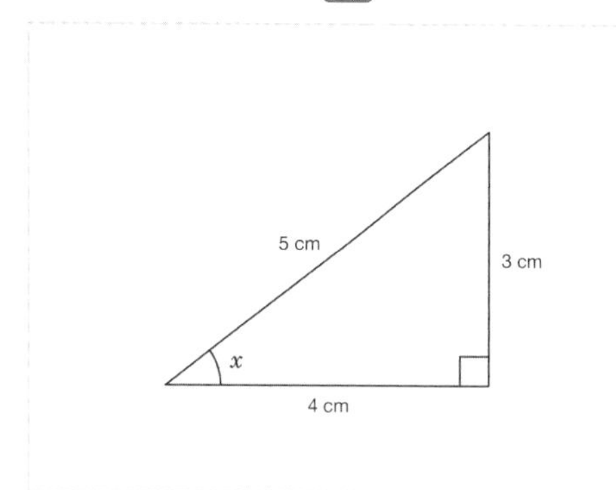

3. Give sin x as a fraction.

- 4/9
- 9/4

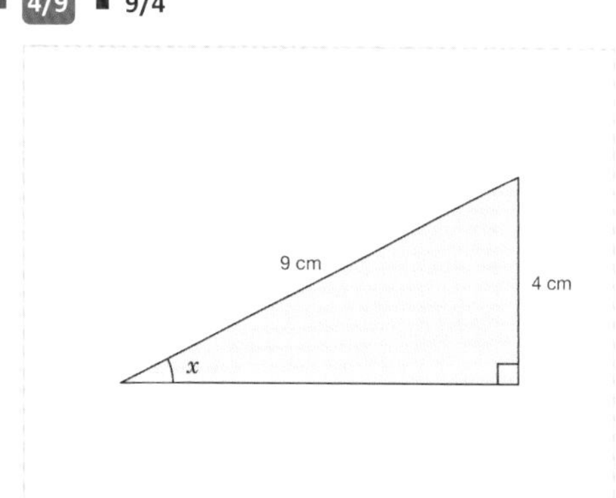

4. Give sin a as a fraction in its simplest form.

- 12/13
- 13/12
- 24/26
- 5/13

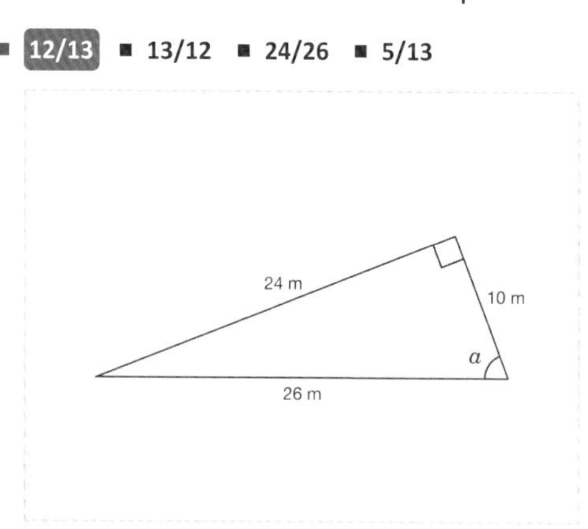

5. Give sin y as a decimal.

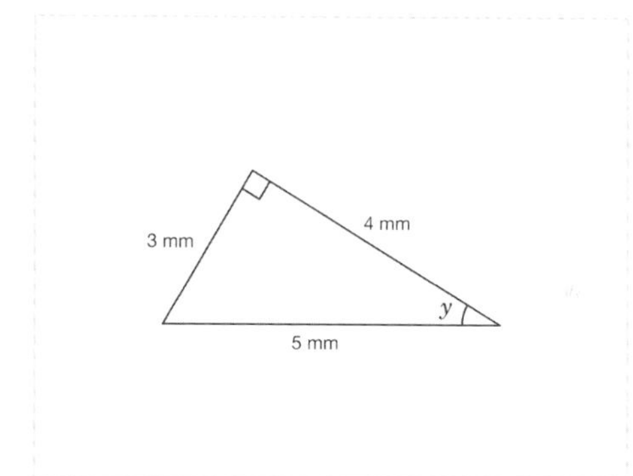

6. Give sin c as a fraction in its simplest form.

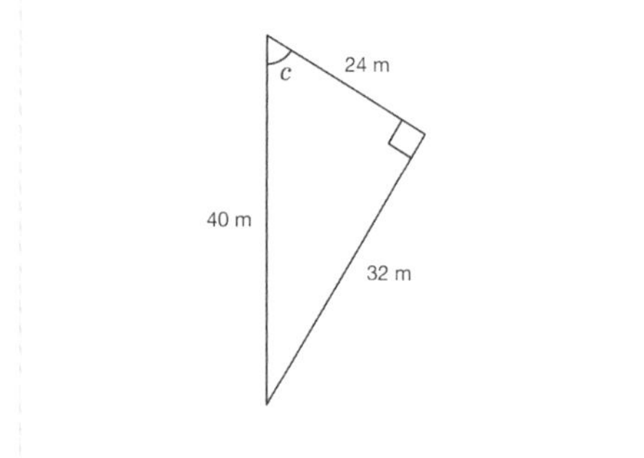

Level 1 *continued*

7. Give sin *x* as a fraction.

a b c

- **15/17** (highlighted)
- **8/17**
- **17/15**

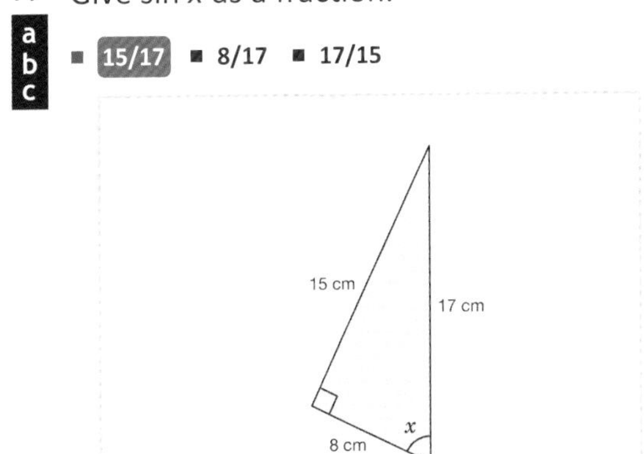

8. Give sin *z* as a decimal.

a b c

- **0.96** (highlighted)
- **24/25**
- **0.36**

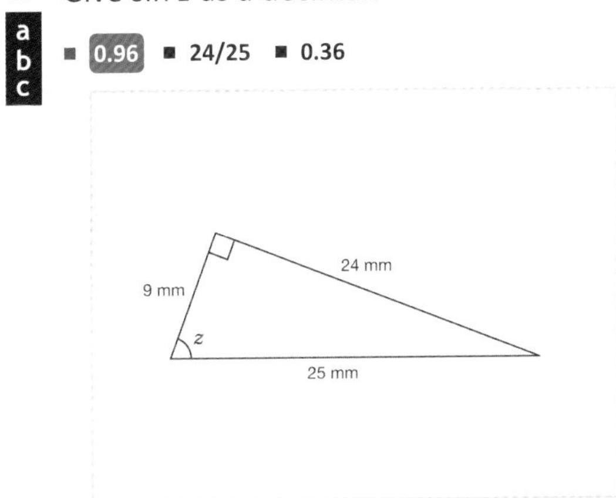

Level 2: Fluency: Calculate missing sides in triangles given the sine of an angle.

Required: 6/8 **Pupil Navigation:** on

Randomised: off

9. If sin $x = 1/2$, find the value of *d*.

a b c *Include the units cm (centimetres) in your answer.*

- **6 centimetres** (highlighted)
- **6 cm** (highlighted)
- **24 centimetres**
- **6**
- **24**
- **24 cm**

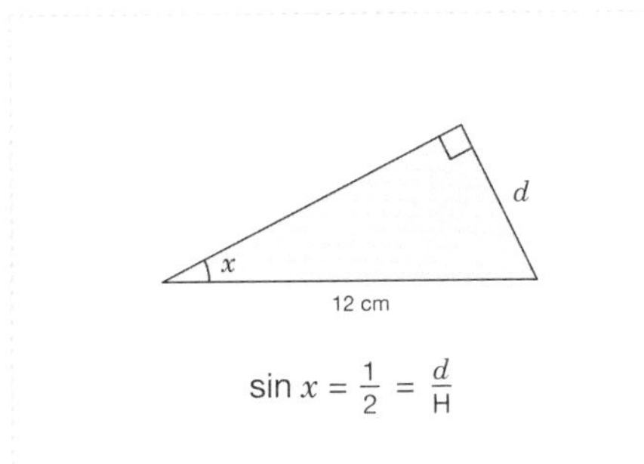

$\sin x = \frac{1}{2} = \frac{d}{H}$

10. If sin $x = 1/4$, find the value of *c*.

a b c *Include the units m (metres) in your answer.*

- **32 metres** (highlighted)
- **32**
- **32 m** (highlighted)
- **2 m**
- **2**
- **2 metres**

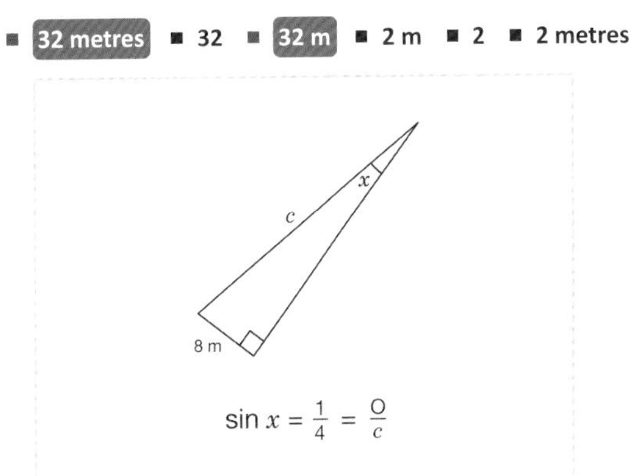

$\sin x = \frac{1}{4} = \frac{O}{c}$

11. If sin $x = 1/3$, find the value of *t*.

a b c *Include the units m (metres) in your answer.*

- **90 metres** (highlighted)
- **90**
- **90 m** (highlighted)
- **10 m**
- **10**
- **10 metres**

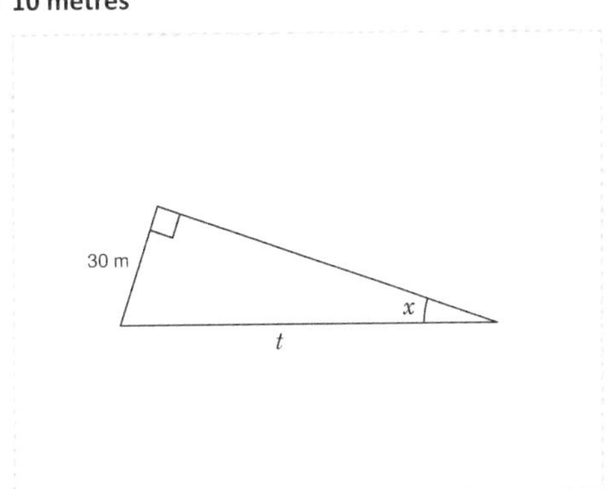

12. If sin $x = 0.5$, find the value of *s*.

a b c *Include the units mm (millimetres) in your answer.*

- **7 mm** (highlighted)
- **7 millimetres** (highlighted)
- **28 mm**
- **7**
- **28**
- **28 millimetres**

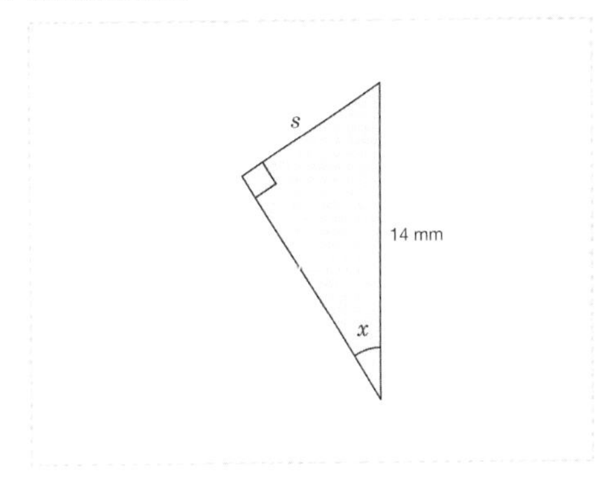

Level 2 *continued*

13. If sin x = 2/3, find the length of side *BC*.
Include the units cm (centimetres) in your answer.

- 12 cm
- 12
- 12 centimetres
- 27 cm
- 27 centimetres
- 27

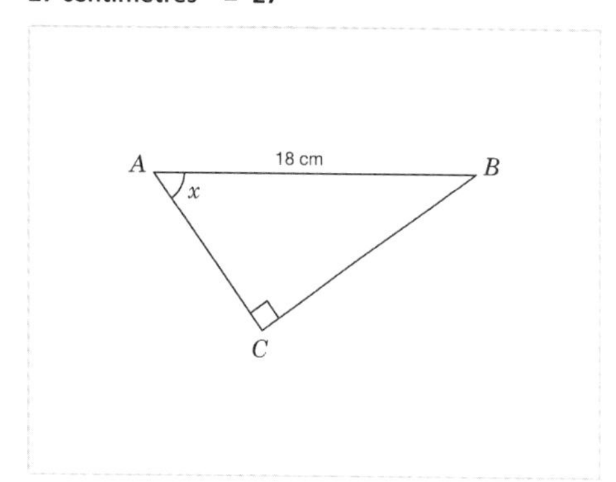

14. If sin x = 0.3, find the length of side *s*.
Include the units m (metres) in your answer.

- 20 m
- 20 metres
- 20
- 1.8 m
- 1.8 metres
- 1.8

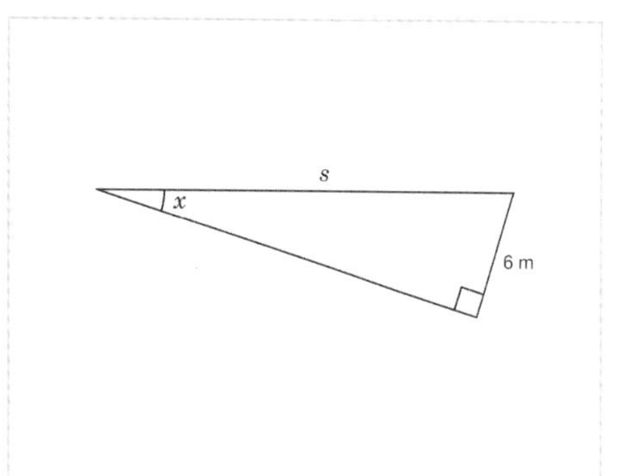

15. If sin x = 3/4, find the length of side *LM*.
Include the units mm (millimetres) in your answer.

- 27 mm
- 48 millimetres
- 48 mm
- 27
- 27 millimetres
- 48

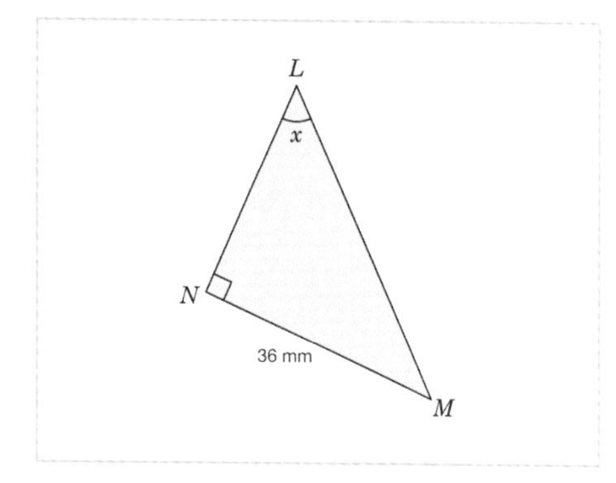

16. If sin x = 4/5, find the value of *a*.
Include the units cm (centimetres) in your answer.

- 16 cm
- 25 centimetres
- 16 centimetres
- 25 cm
- 16
- 25

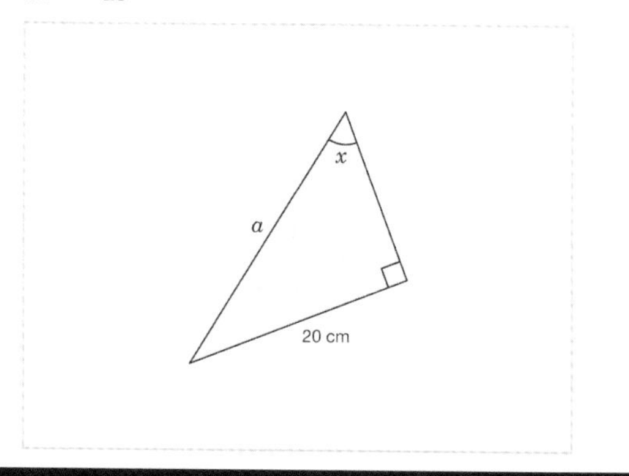

Level 3: Reasoning: Making inferences using the sine ratio.

Required: 3/3 **Pupil Navigation:** on
Randomised: off

17. Calculate the value of sin 30°.

- 0.5
- 1/2

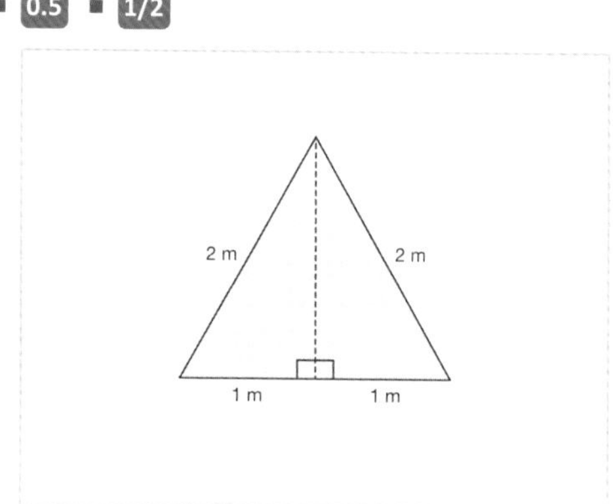

18. William has made one mistake on his homework. What is the correct answer to the question he got wrong?

- 15/17

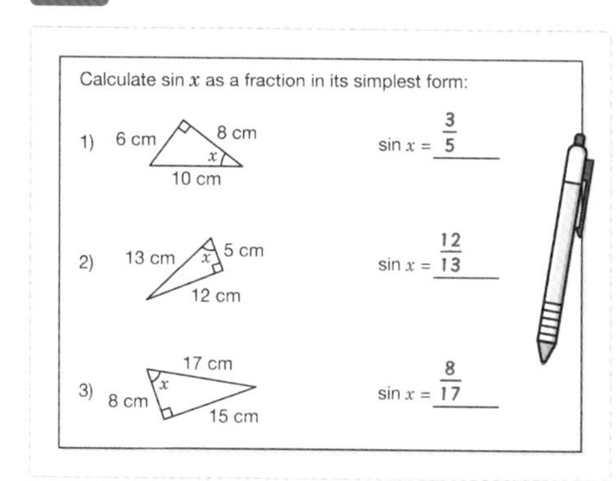

Calculate sin x as a fraction in its simplest form:

1) 6 cm, 8 cm, 10 cm — sin x = $\frac{3}{5}$

2) 13 cm, 5 cm, 12 cm — sin x = $\frac{12}{13}$

3) 17 cm, 8 cm, 15 cm — sin x = $\frac{8}{17}$

Level 3 *continued*

19. Give the value of sin 40° correct to two decimal places.

1/4

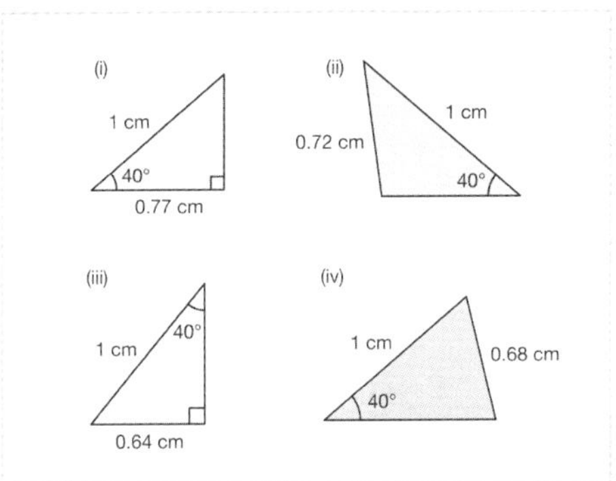

Level 4: Problem solving: Multi-step problems using the sine ratio.

Required: 3/3 **Pupil Navigation:** on

Randomised: off

20. Select the correct value of sin 60°.

- (i)
- (ii)
- (iii)
- (iv)

1/4

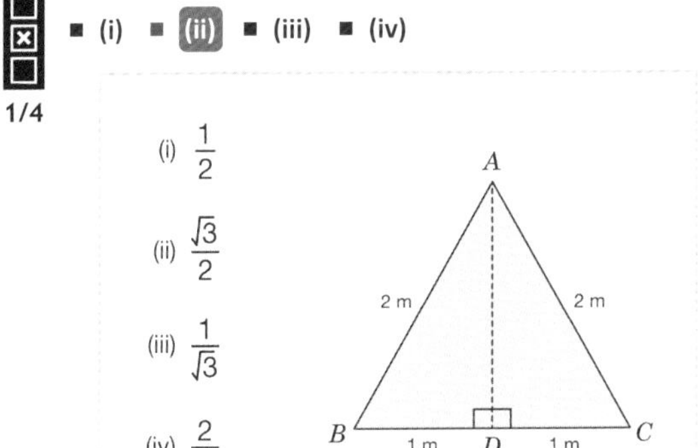

21. Give the value of y.

Give your answer as a decimal and include the units cm (centimetres).

- 8.25 cm
- 8.25 centimetres
- 8.25

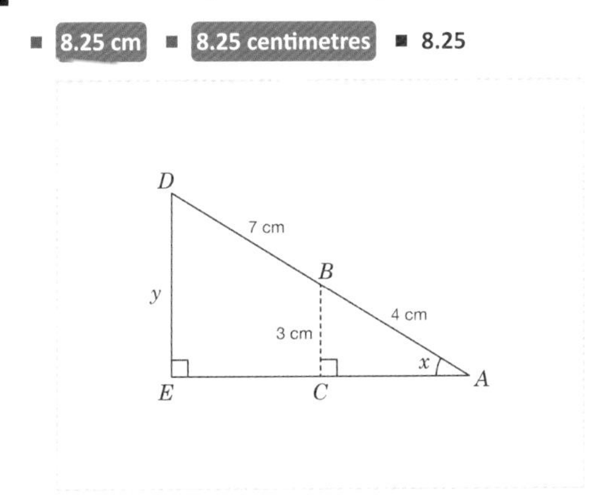

22. The perimeter of the triangle is 120 centimetres and sin x = 8/17. Calculate the value of t.

- 51 centimetres
- 51 cm
- 51

t

x

45 cm

Use the Tangent Ratio

Objective: I can calculate the tangent of an angle in a triangle given the opposite and adjacent sides. I can calculate missing sides in a triangle given the tangent of an angle.

Quick Search Ref: 10622

Level 1: Understanding: Calculate the tangent of an angle given the hypotenuse and adjacent side.

Required: 6/8 **Pupil Navigation:** on **Randomised:** off

1. Select the calculation which gives the tangent of an angle.

1/3

- i) opposite ÷ adjacent
- ii) opposite ÷ hypotenuse
- iii) adjacent ÷ hypotenuse

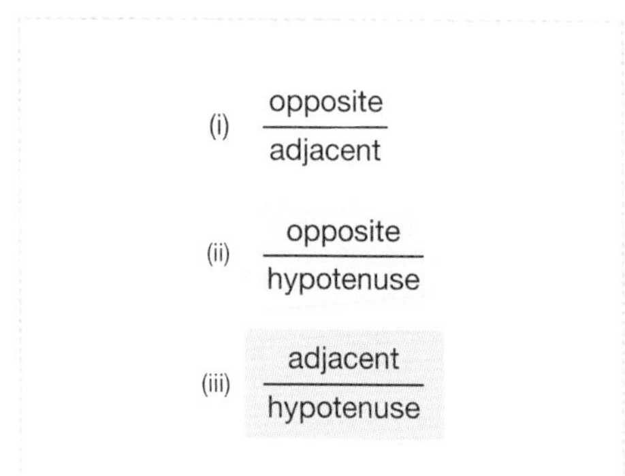

2. What is the value of tan x?

1/4

- 15/8
- 8/17
- 15/17
- 8/15

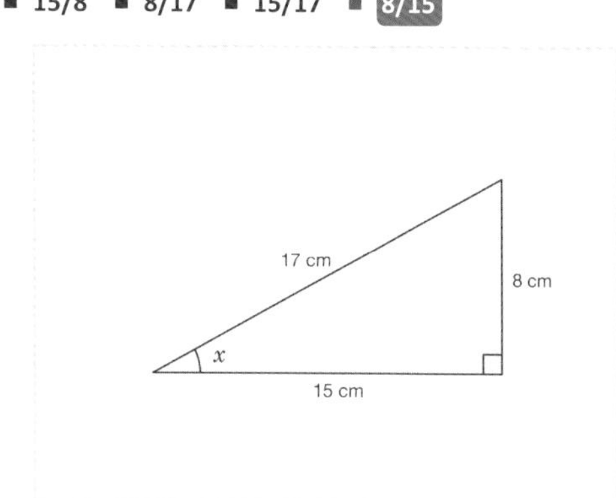

3. Give tan x as a fraction.

- 7/5
- 1 2/5
- 5/7

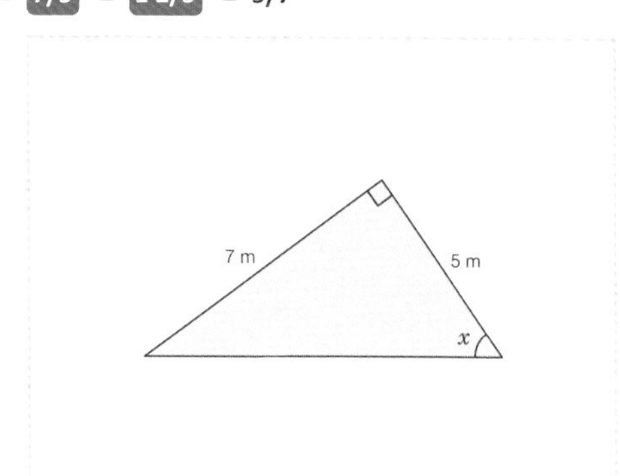

4. Give tan y as a fraction in its simplest form.

- 21/72
- 24/25
- 7/25
- 7/24
- 3 3/7
- 24/7
- 72/75
- 21/75
- 72/21

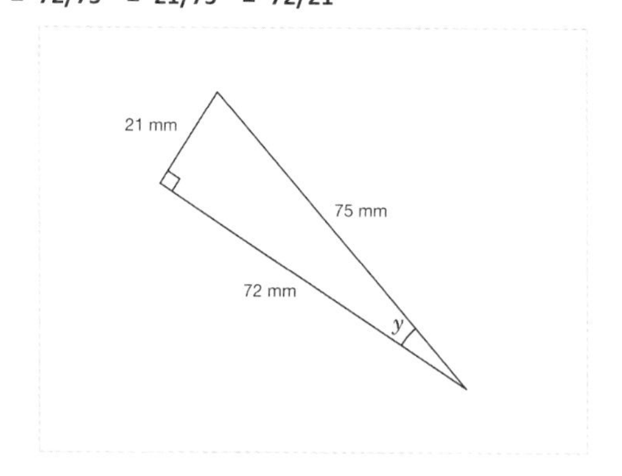

5. Give tan n as a decimal.

- 0.75
- 3/4
- 0.8
- 0.6

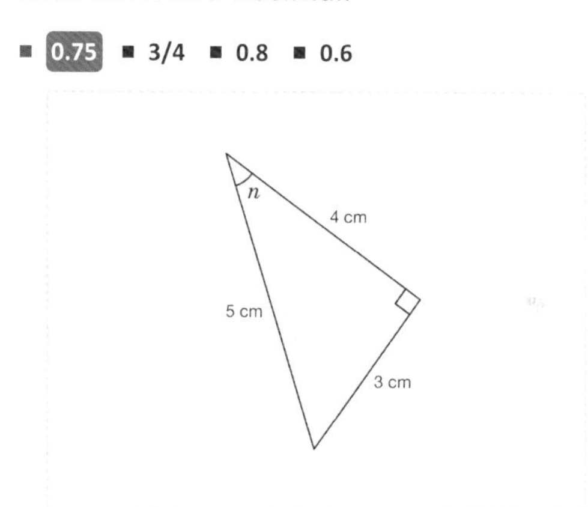

6. Give tan z as a fraction.

- 2 2/5
- 12/5
- 12/13
- 5/12
- 5/13

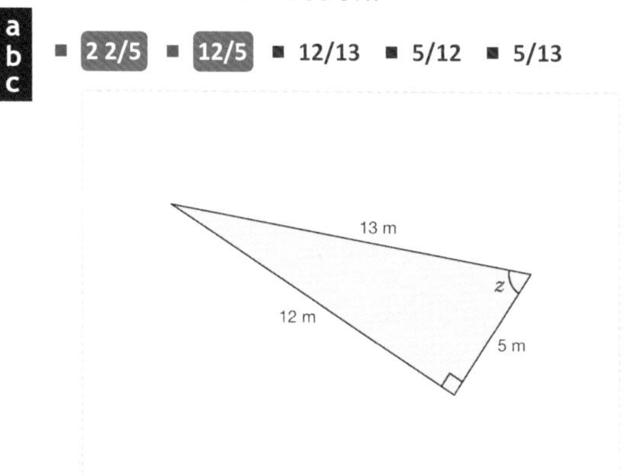

Level 1 *continued*

7. Give tan *c* as a decimal.

abc

- 2.4
- 12/5
- 24/10
- 48/20

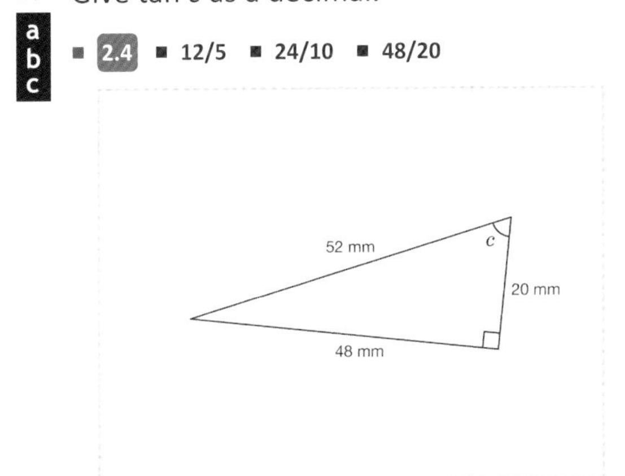

8. Give tan *c* as a fraction in its simplest form.

abc

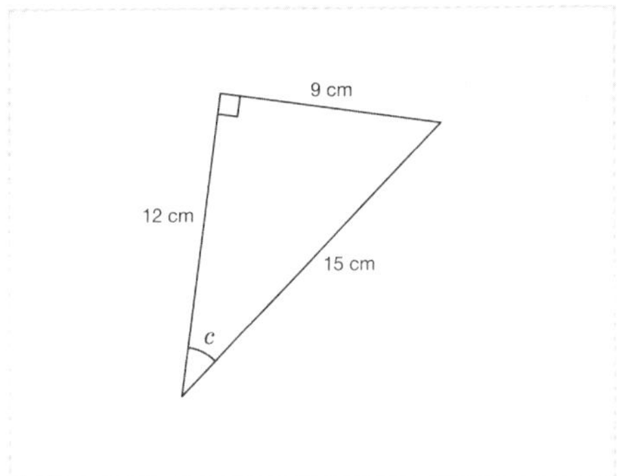

Level 2: Fluency: Calculate missing sides in triangles given the tangent of an angle.

Required: 6/8 **Pupil Navigation:** on

Randomised: off

9. If tan x = 1/4, find the value of r.

abc *Include the units cm (centimetres) in your answer.*

- 64 centimetres
- 4 centimetres
- 4 cm
- 64
- 64 cm
- 4

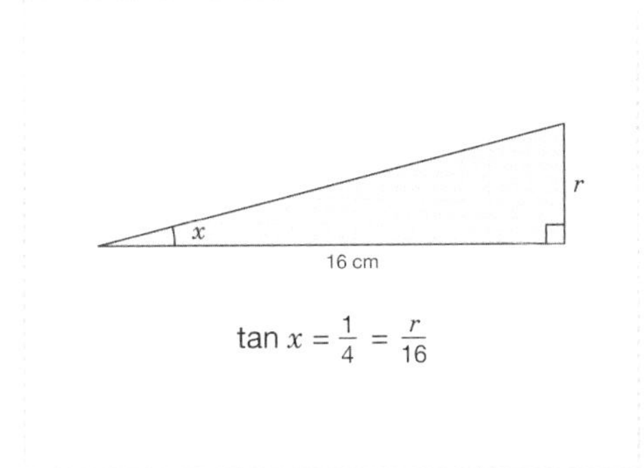

10. If tan x = 2, find the value of t.

abc *Include the units m (metres) in your answer.*

- 21

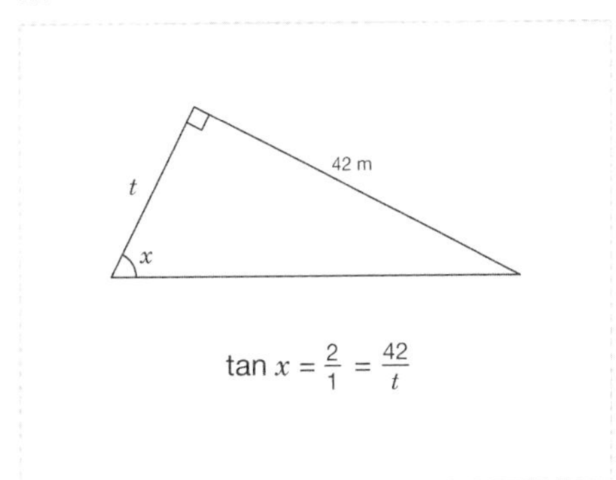

11. If tan x = 1/3, find the value of z.

abc *Include the units mm (millimetres) in your answer.*

- 72 mm

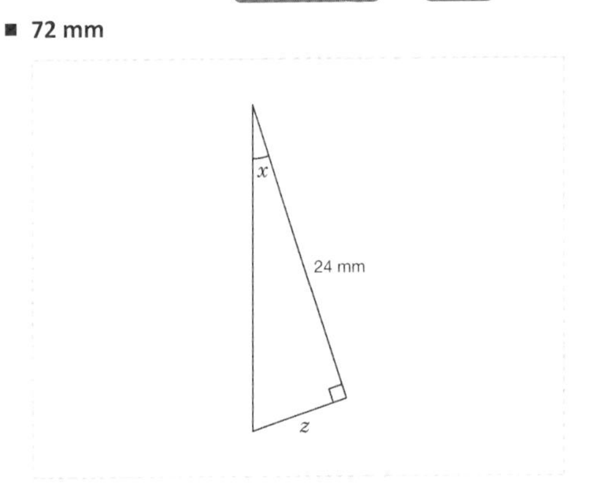

12. If tan x = 0.6, find the value of s.

abc *Include the units cm (centimetres) in your answer.*

- 50
- 18

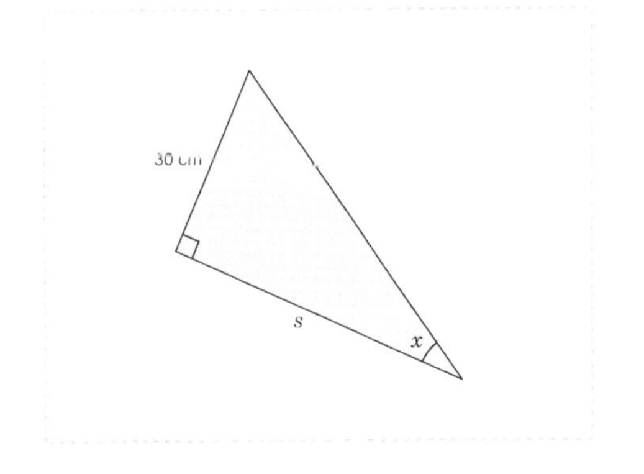

Level 2 *continued*

13. If tan $x = 2/7$, find the length of side AC.
Include the units m (metres) in your answer.

- **147 m** ■ 12 metres ■ **147 metres** ■ 12 ■ 12 m ■ 147

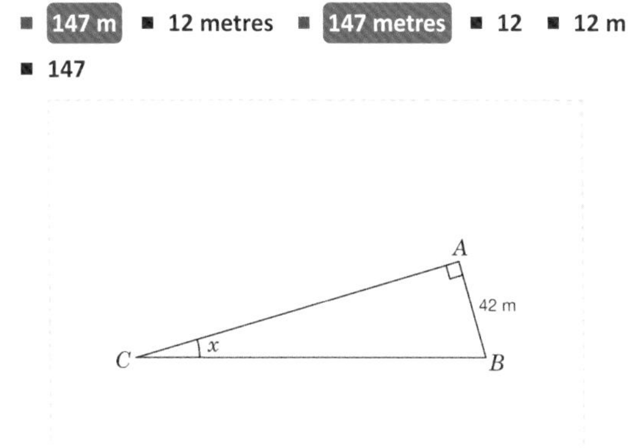

14. If tan $x = 2.4$, find the value of n.
Include the units mm (millimetres) in your answer.

- **144 mm** ■ 25 mm ■ **144 millimetres** ■ 144 ■ 25 millimetres ■ 25

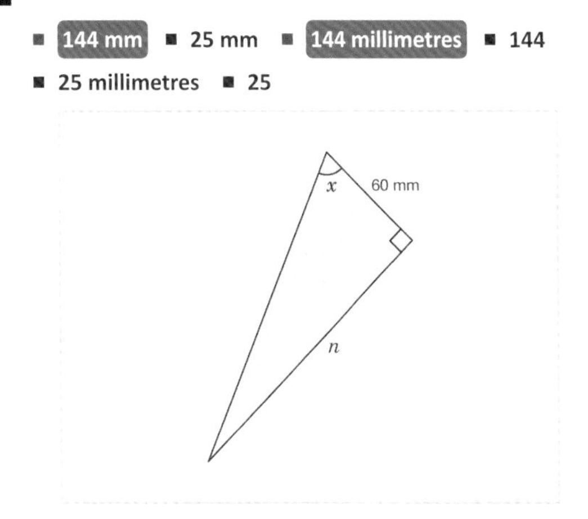

15. If tan $x = 3/4$, find the value of r.
Include the units cm (centimetres) in your answer.

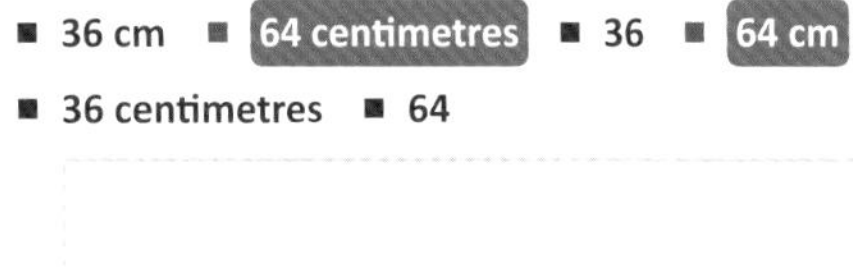

- 36 cm ■ **64 centimetres** ■ 36 ■ **64 cm** ■ 36 centimetres ■ 64

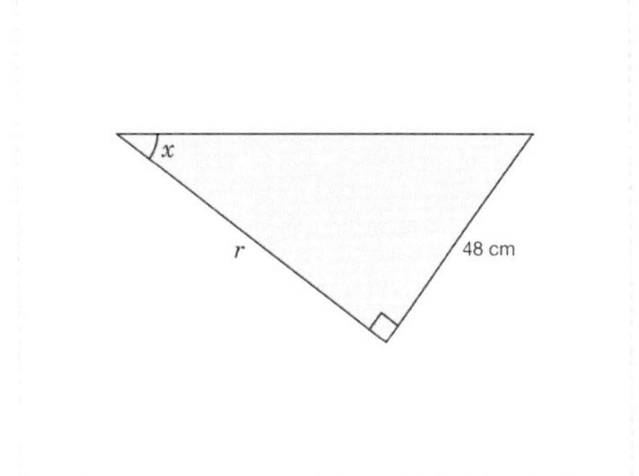

16. If tan $x = 1.2$, find the length of side ST.
Include the units mm (millimetres) in your answer.

- **36 mm** ■ 36 ■ **36 millimetres** ■ 25 millimetres ■ 25 m ■ 25

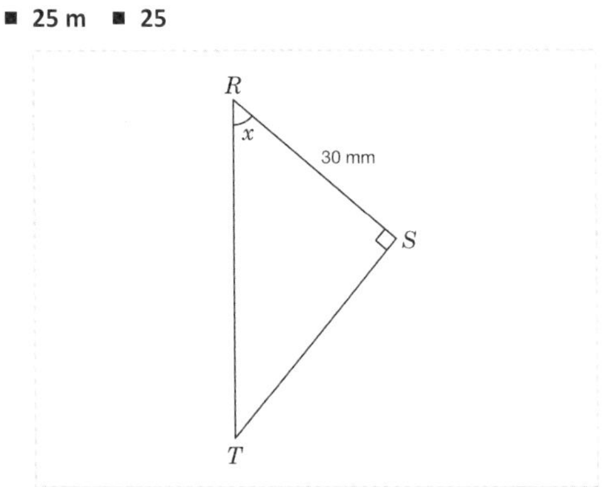

Level 3: Reasoning: Making inferences using the tangent ratio.

Required: 3/3 **Pupil Navigation:** on
Randomised: off

17. Calculate the value of tan 45°.

- **1**

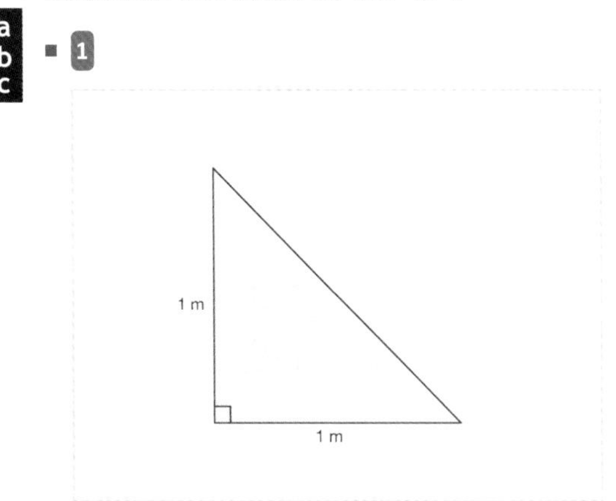

18. Owais says that sin x divided by cos x equals tan x. Is he correct? Explain your answer.

- Open question, no set answer

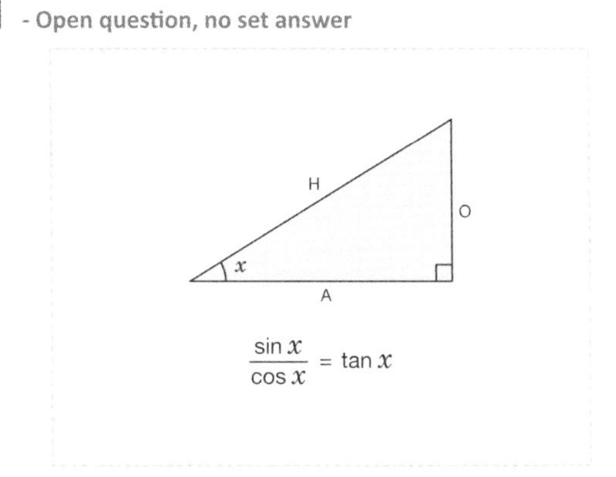

Level 3 continued

19. Using the diagrams, give the value of tan 43° correct to two decimal places.

1/4

- 1.07
- 1.02
- 0.89
- **0.93**

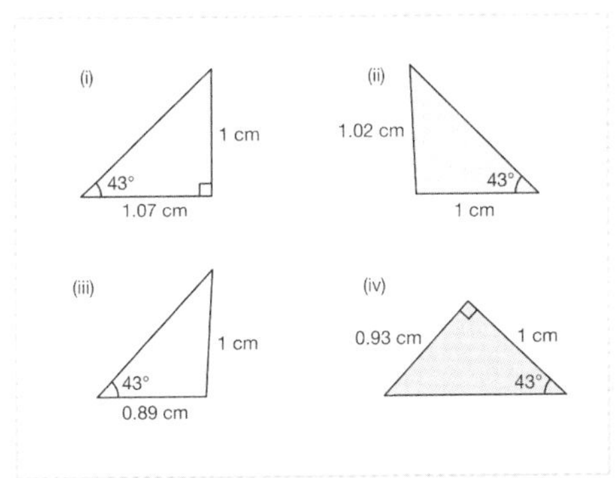

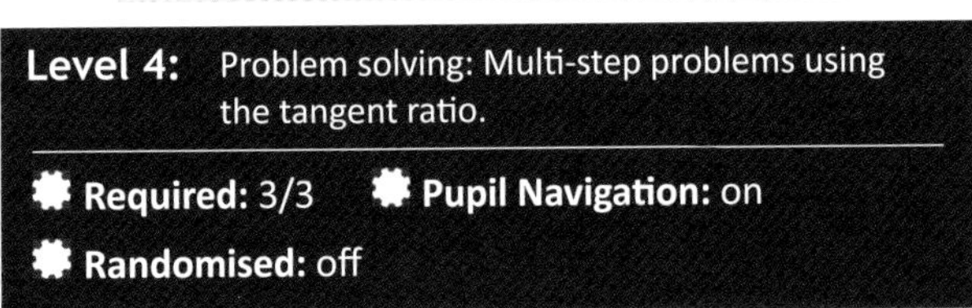

Level 4: Problem solving: Multi-step problems using the tangent ratio.

Required: 3/3 **Pupil Navigation:** on

Randomised: off

20. Select the correct value of tan 30°.

1/4

- (i)
- (ii)
- **(iii)**
- (iv)

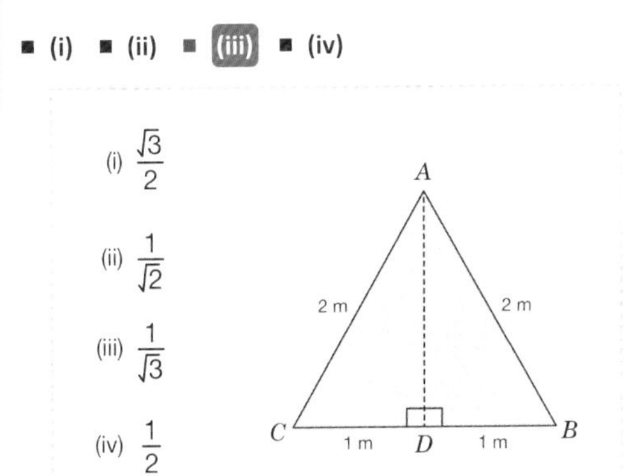

21. A sailor uses a clinometer to measure the angle between the sea level and the top of a lighthouse on a cliff.
If the top of the lighthouse is 90 metres above sea level and the tangent of the angle measured is 0.3, how far is the ship from the cliff?
Include the units m (metres) in your answer.

- **300 metres**
- **300 m**
- 300

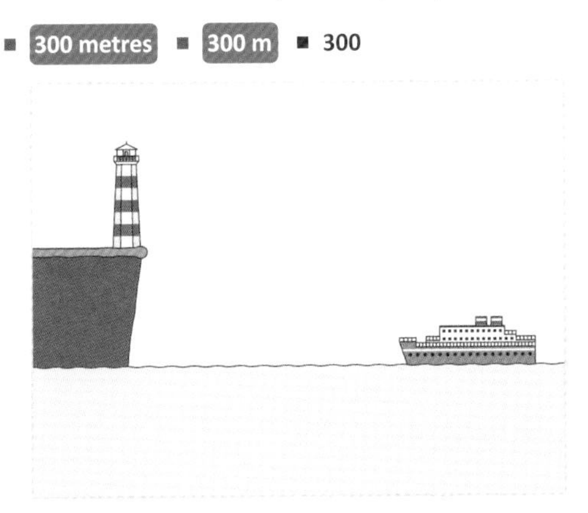

22. *ABCD* is a rhombus. Tan x = 0.4 and the length of *OA* is 2 cm. What is the area of the rhombus?
Don't include the units in your answer.

- **20**
- 5

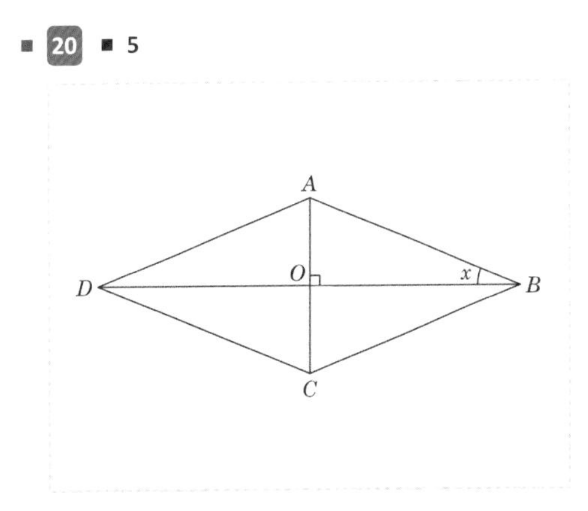

Use Trigonometry to Find a Missing Angle

Objective: I can use trigonometry to find a missing angle in a triangle.

Quick Search Ref: 10701

Level 1: Understanding: Find missing angles when given the correct trig ratio.

Required: 10/10 **Pupil Navigation:** on **Randomised:** off

1. Sort the trigonometric ratios into the same order as the corresponding calculations.

- sin
- tan
- cos

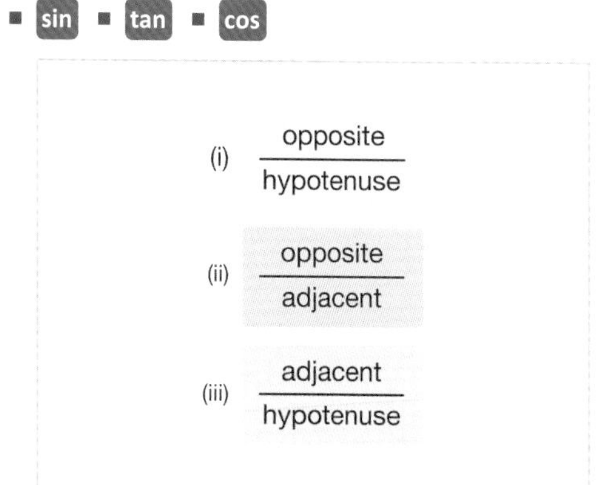

2. Use $\sin^{-1} x$ to calculate x such that $\sin x = 3/5$.
Give your answer in degrees to 2 decimal places. Don't include the units in your answer.

- 0.01
- 36.87
- 0.64

3. Use $\tan^{-1} x$ to calculate x such that $\tan x = 1.35$.
Give your answer in degrees to 1 decimal place. Don't include the units in your answer.

- 53.5
- 0
- 0.9

4. Use $\cos x$ = A/H to calculate the size of angle x.
Give your answer in degrees to 1 decimal place. Don't include the units in your answer.

- 41.4
- 0.7
- 1

12 cm

x

9 cm

5. Use $\tan z$ = O/A to calculate the size of angle z.
Give your answer in degrees to 1 decimal place. Don't include the units in your answer.

- 41.6
- 0
- 0.7

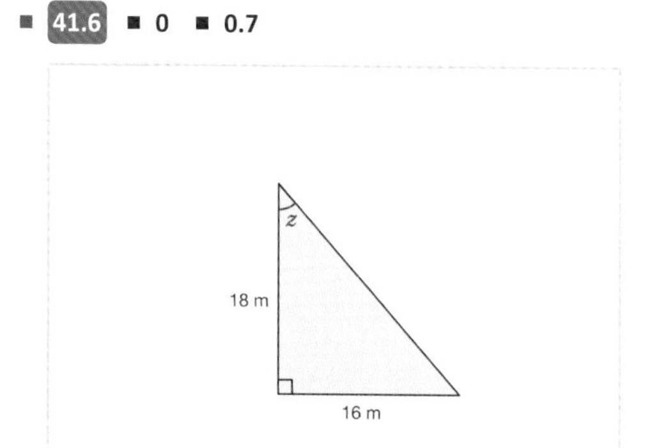

6. Use $\cos r$ = A/H to calculate the size of angle r.
Give your answer in degrees to 2 decimal places. Don't include the units in your answer.

- 23.90
- 0.42
- 1
- 23.9
- 1.00

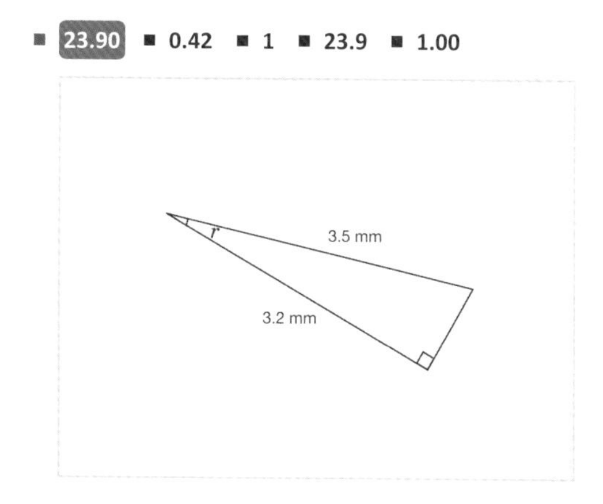

Level 1 *continued*

7. Use $\sin d$ = O/H to calculate the size of angle d.
Give your answer in degrees to 3 significant figures. Don't include the units in your answer.

- 0.0127
- **46.7**
- 0.814
- 46.658

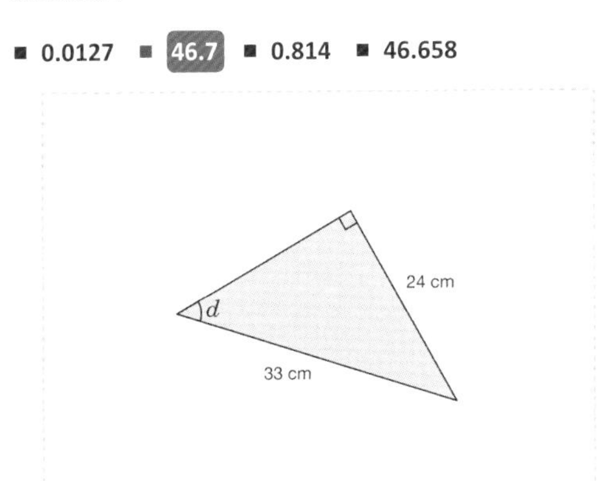

8. Use $\cos^{-1} x$ to calculate x such that $\cos x$ = 0.7.
Give your answer in degrees to 1 decimal place. Don't include the units in your answer.

- **45.6**
- 1
- 0.8

9. Use $\tan m$ = O/A to calculate the size of angle m.
Give your answer in degrees to 2 decimal places. Don't include the units in your answer.

- 51.6
- **51.60**
- 0.02
- 0.90
- 0.9

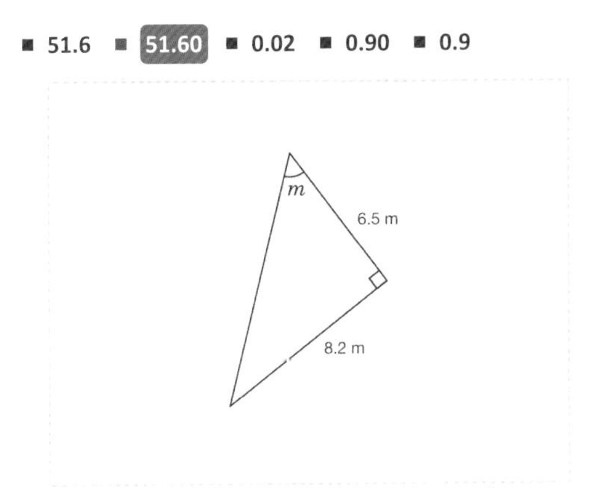

10. Use $\sin q$ = O/H to calculate the size of angle q.
Give your answer in degrees to 4 significant figures. Don't include the units in your answer.

- 0.8070
- **46.24**
- 46.2383
- 0.01260

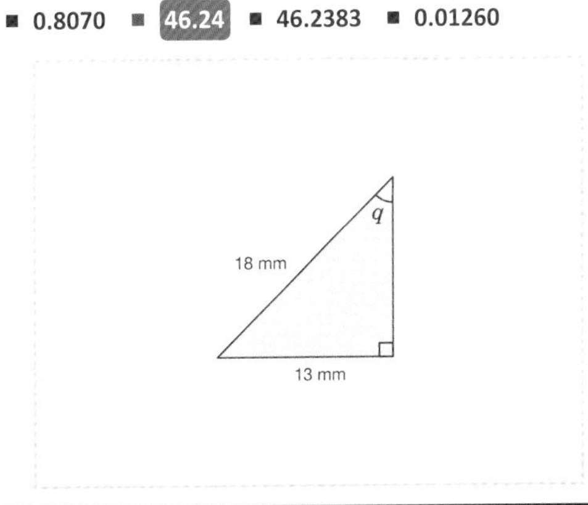

Level 2: Fluency: Selecting the correct trig ratio and finding missing angles.

Required: 9/10 **Pupil Navigation:** on
Randomised: off

11. Which trigonometric ratio would be used to calculate the value of t?

1/3

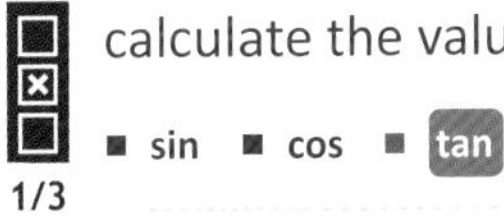

- sin
- cos
- **tan**

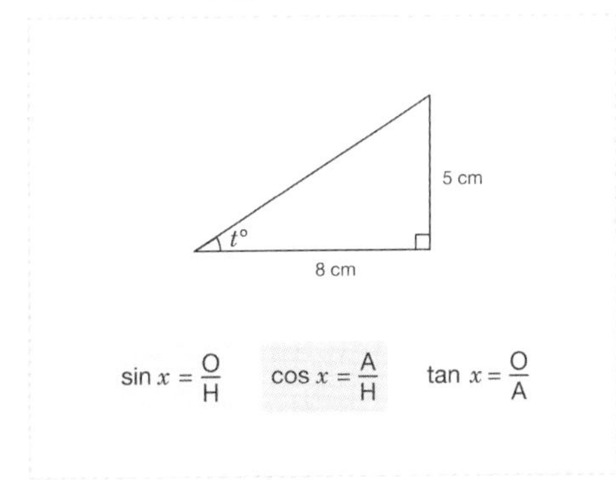

12. Which trigonometric ratio would be used to calculate the value of v?

1/3

- **sin**
- cos
- tan

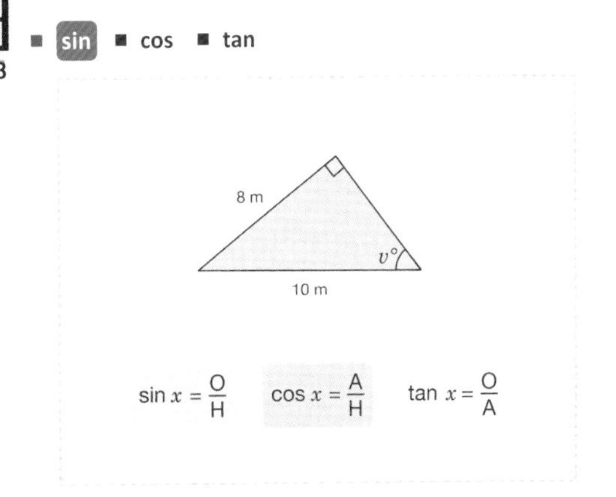

Level 2 *continued*

13. Calculate the value of x.

Give your answer to 1 decimal place. Don't include the units in your answer.

■ 57.3

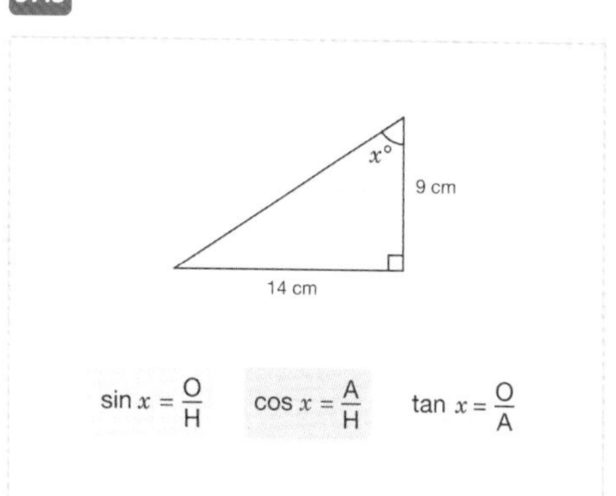

14. Calculate the value of e.

Give your answer in degrees to 2 decimal places. Don't include the units in your answer.

■ 49.29

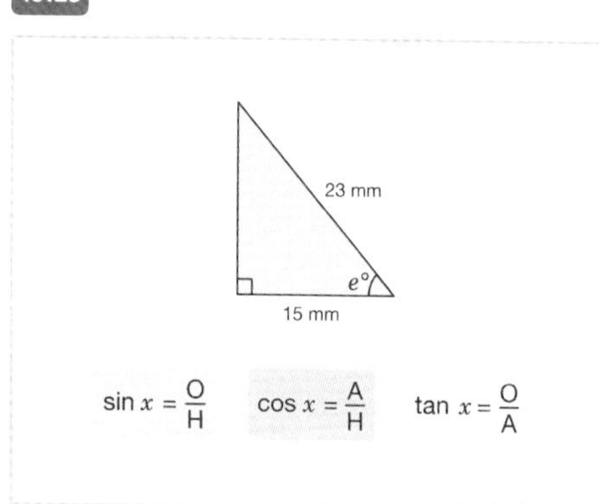

15. Calculate the size of $\angle LMN$.

Give your answer in degrees to 1 decimal place. Don't include the units in your answer.

■ 37.7

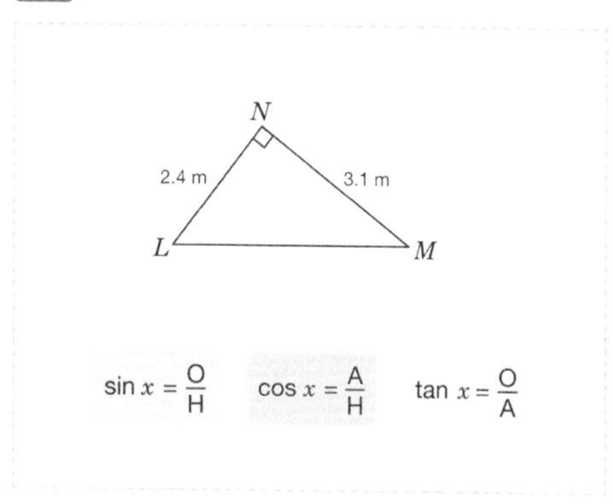

16. Calculate the value of s.

Give your answer to 3 significant figures. Don't include the units in your answer.

■ 63.8 ■ 63.823

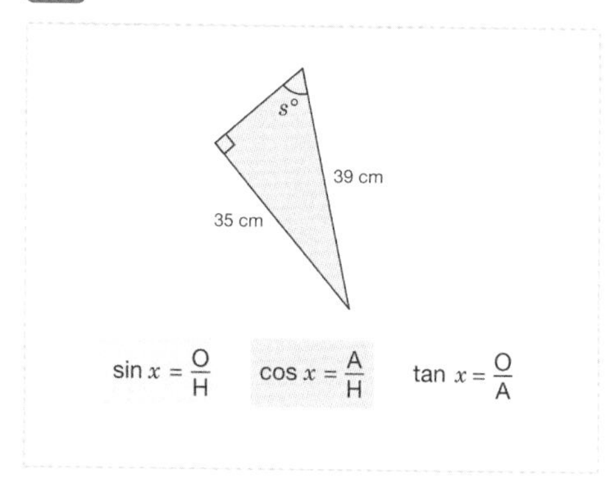

17. Calculate the size of $\angle XYZ$.

Give your answer in degrees to 2 decimal places. Don't include the units in your answer.

■ 60.39

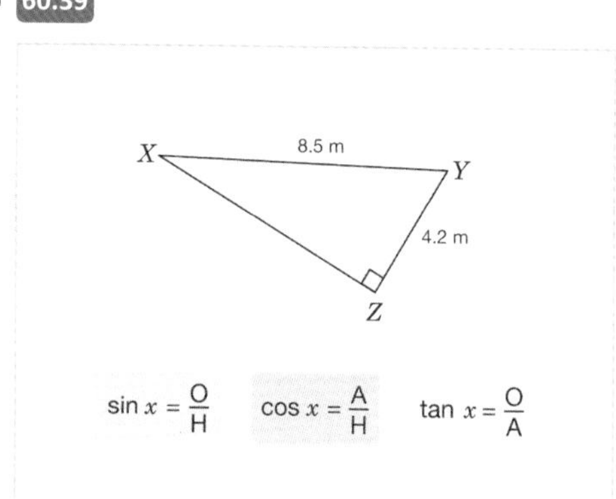

18. Which trigonometric ratio would be used to calculate the value of d?

1/3

■ sin ■ cos ■ tan

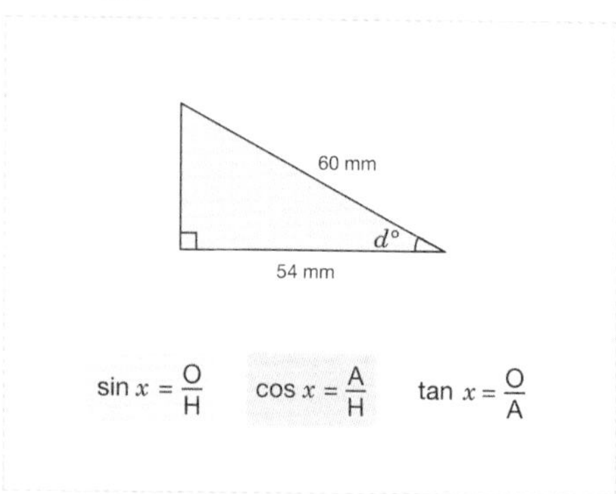

Level 2 continued

19. Calculate the value of z.
Give your answer to 1 decimal place. Don't include the units in your answer.

- 56.3

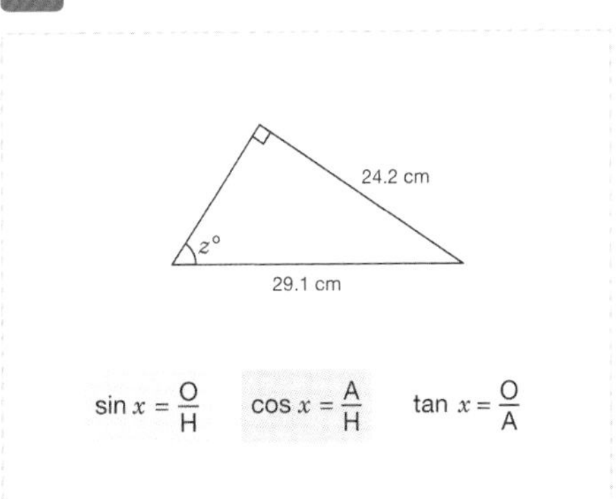

20. Calculate the value of ∠PQR.
Give your answer in degrees to 4 significant figures. Don't include the units in your answer.

- 71.53
- 71.514

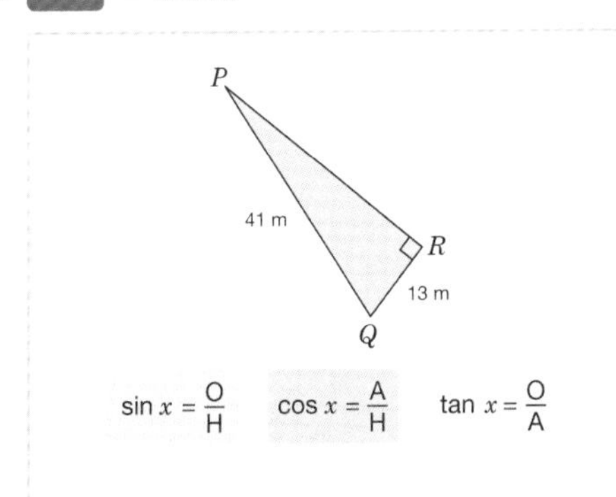

Level 3: Reasoning: Misconceptions and making inferences using trig ratios.

Required: 5/5 **Pupil Navigation:** on
Randomised: off

21. George has made a mistake with a trigonometry question. What is the correct size of ∠XYZ?
Give your answer in degrees to 1 decimal place. Don't include the units in your answer.

- 32.7

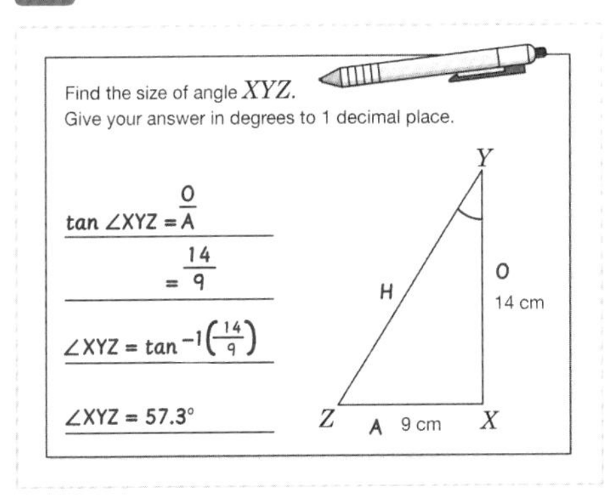

22. Use the diagrams to identify which of these angles has a sine of 1/√2.

1/3

- 30 degrees
- 45 degrees
- 60 degrees

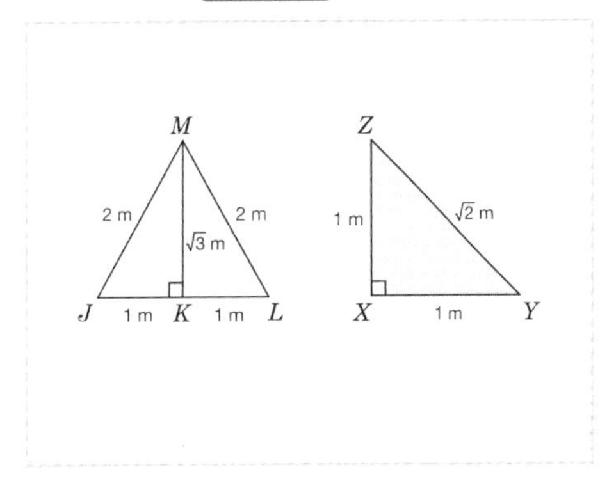

23. Ella says the size of angle x is 56.5°. Is Ella correct? Explain your answer.

- Open question, no set answer

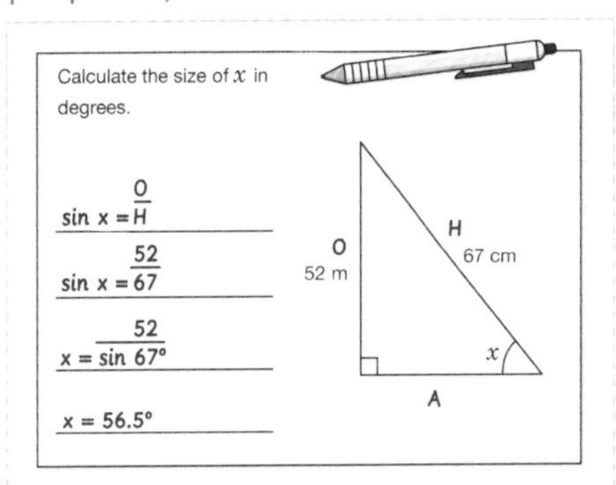

24. Arrange angles x, y and z by size, from smallest to largest.

- z
- x
- y

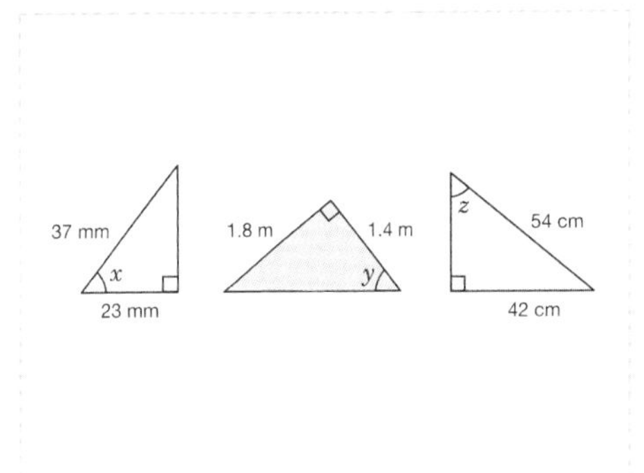

Level 3 *continued*

25. Use the diagrams to identify which of these angles has a tangent of 1/√3.

1/3

- **30 degrees**
- **45 degrees**
- **60 degrees**

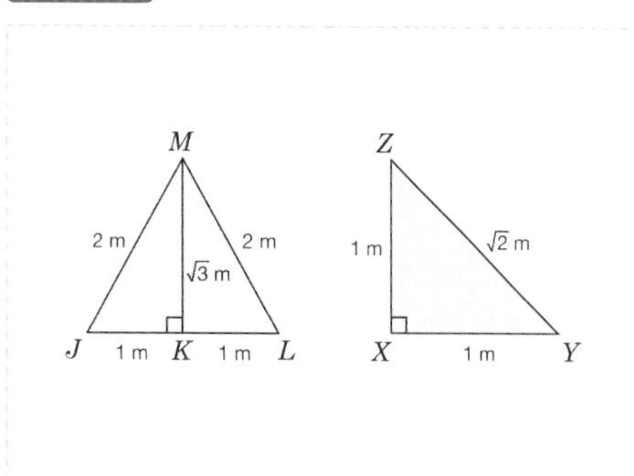

Level 4: Problem solving: Multi-step problems finding missing angles.

Required: 5/5 **Pupil Navigation:** on
Randomised: off

26. Calculate the size of ∠ADC.
Give your answer in degrees to 1 decimal place. Don't include the units in your answer.

- **73.3**

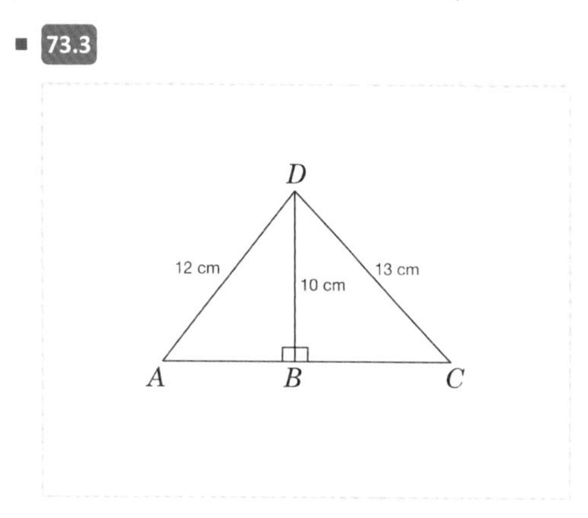

27. A kite is tied to the ground using a piece of string that is 50 metres long. If the kite is 36 metres above ground level, what angle does the string make with the ground?
Give your answer in degrees to 1 decimal place. Don't include the units in your answer.

- **46.1**

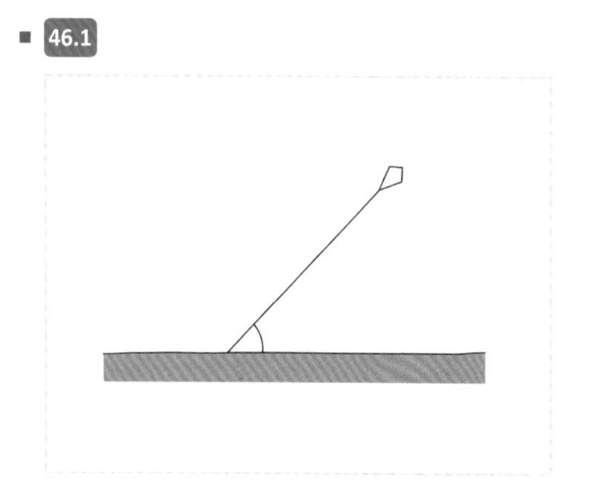

28. Calculate the size of ∠XZY.
Give your answer in degrees to 2 decimal places. Don't include the units in your answer.

- **40.97**

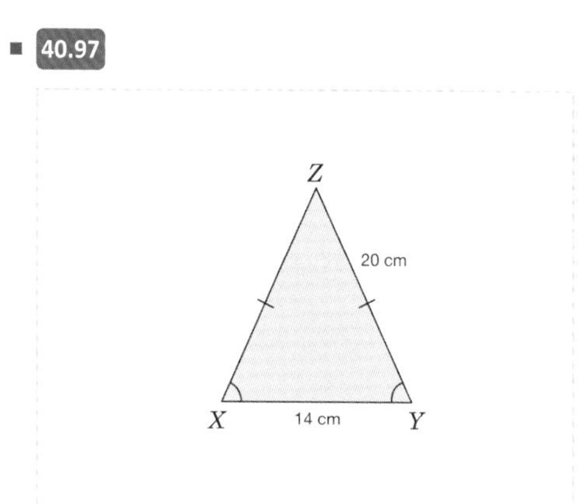

29. Match the trigonometric functions to their graphs and sort the functions into the same order as the graphs.

- **sin x**
- **tan x**
- **cos x**

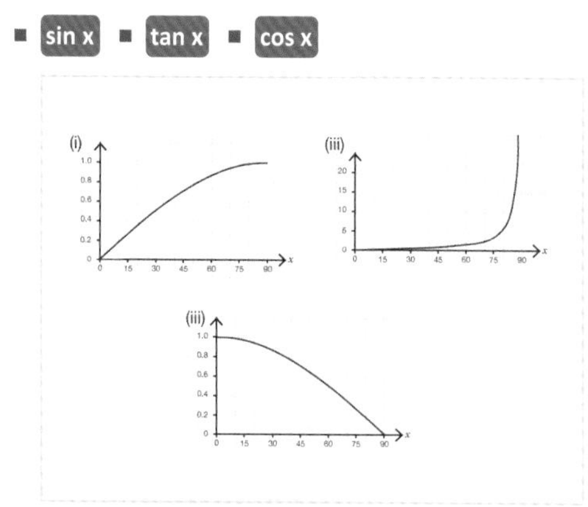

30. *ABCD* is a rhombus such that the length of *AC* is three times the length of *DB*. Calculate the size of ∠ADC.
Give your answer in degrees to 1 decimal place. Don't include the units in your answer.

- **143.1**

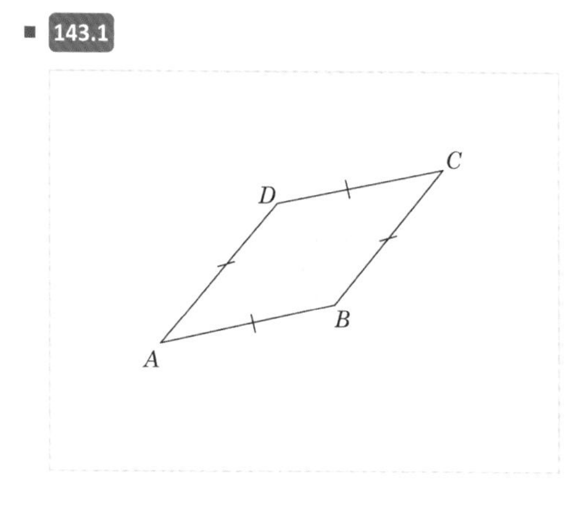

Use Trigonometry to Find a Missing Side 1

Objective: I can use trigonometry to find a missing side in a triangle.

Quick Search Ref: 10650

Level 1: Understanding: Find missing sides when given the correct trig ratio.

Required: 7/10 **Pupil Navigation:** on **Randomised:** off

1. Sort the side lengths into this order:

1. hypotenuse
2. opposite side to x
3. adjacent side to x

■ 13 cm ■ 5 cm ■ 12 cm

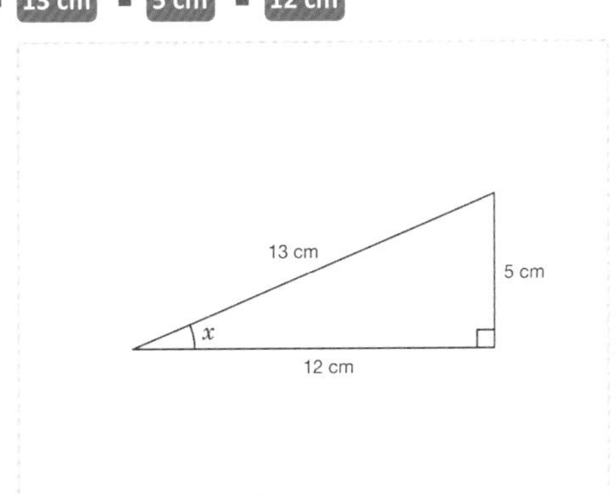

2. Sort the trigonometric ratios into the same order as the corresponding calculations.

■ cos ■ tan ■ sin

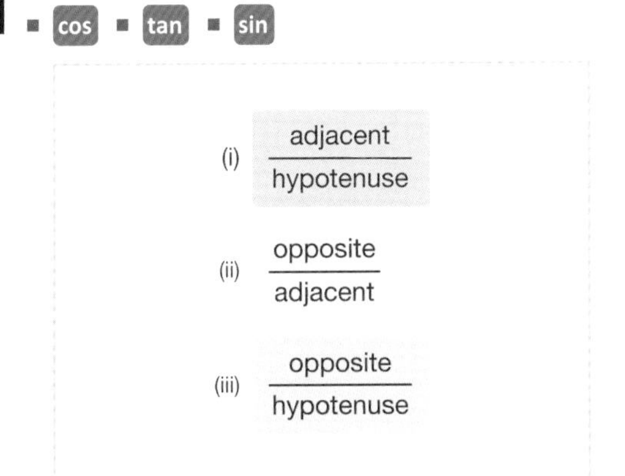

3. Use your calculator to find the value of sin 20°.
Give your answer to 2 decimal places.

■ 0.34

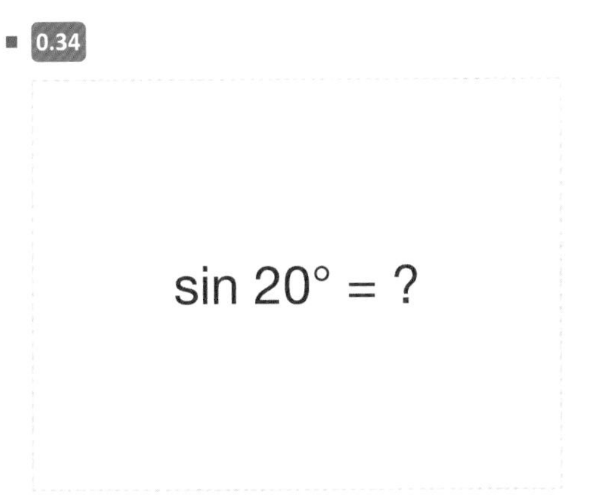

4. Use cos x = A/H to calculate the value of n.
Give your answer to 2 decimal places and include the units cm (centimetres) in your answer.

■ 9.19 centimetres ■ 9.12 cm ■ 9.24 centimetres
■ 9.24 cm ■ 9.19 cm ■ 9.12 centimetres ■ 9.24
■ 9.12 ■ 9.19

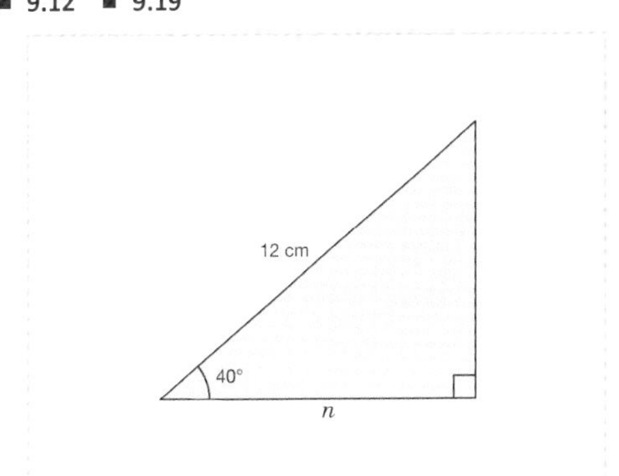

5. Use tan x = O/A to calculate the value of n.
Give your answer to 1 decimal place and include the units m (metres) in your answer.

■ 11.9 m ■ 11.9 metres ■ 11.9 ■ 11.7 ■ 11.7 m
■ 11.7 metres

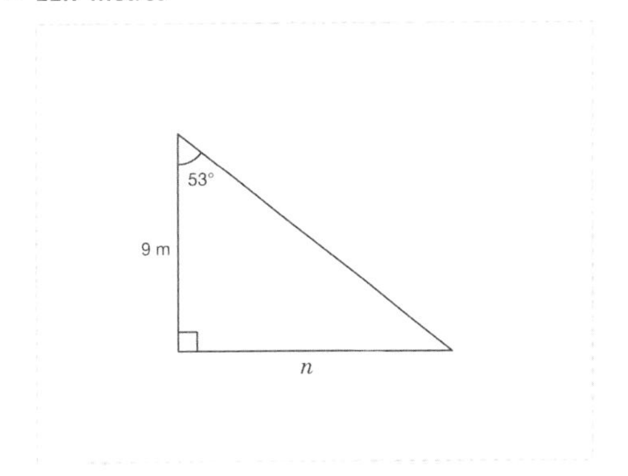

Level 1 *continued*

6. Use sin x = O/H to calculate the value of r.
a b c *Give your answer to 1 decimal place and include the units mm (millimetres) in your answer.*

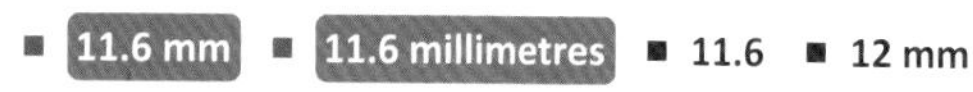

- **11.6 mm** ■ **11.6 millimetres** ■ 11.6 ■ 12 mm
- 12 millimetres ■ 12

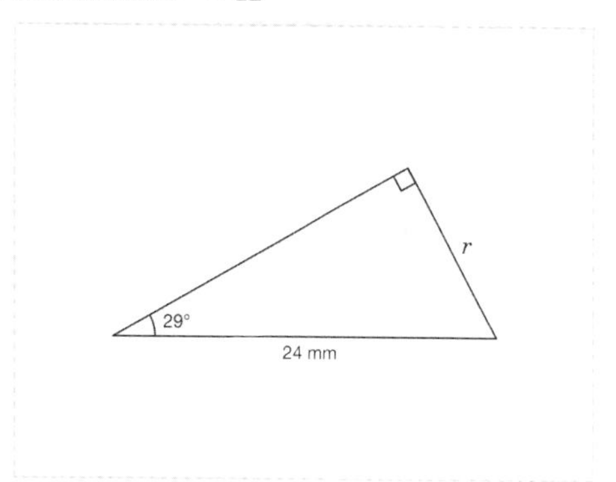

7. Use cos 62° = A/H to calculate the value of e.
a b c *Give your answer to 3 significant figures and include the units cm (centimetres) in your answer.*

- **3.38 centimetres** ■ **3.38 cm** ■ 3.38

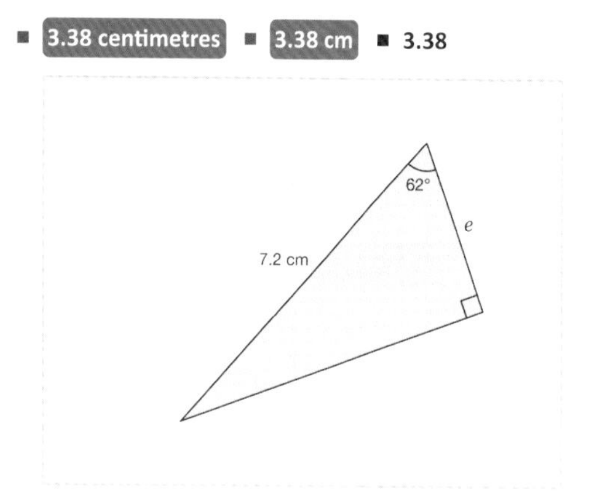

8. Use your calculator to find the value of cos 45°.
1 2 3 *Give your answer to 2 decimal places.*

- **0.71**

cos 45° = ?

9. Use sin 19° = O/H to calculate the value of y.
a b c *Give your answer to 2 decimal places and include the units m (metres) in your answer.*

- 0.98 ■ **0.98 m** ■ **0.98 metres** ■ 0.99 m
- 0.99 metres ■ 0.99

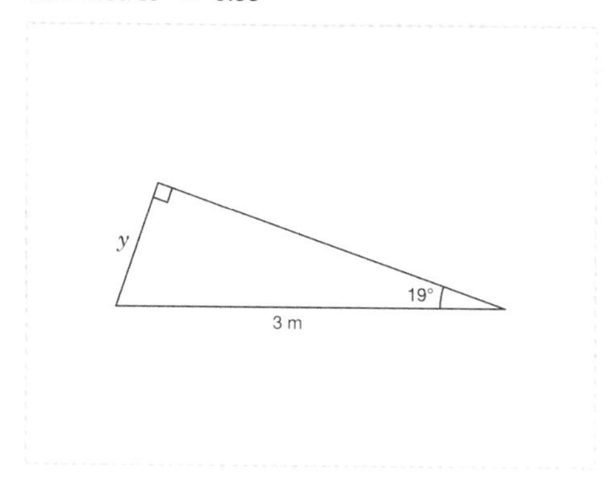

10. Use tan x = O/A to calculate the value of z.
a b c *Give your answer to 1 decimal place and include the units mm (millimetres) in your answer.*

- **13.6 mm** ■ **13.6 millimetres** ■ 12.8 millimetres
- 13.6 ■ 12.8 ■ 12.8 mm

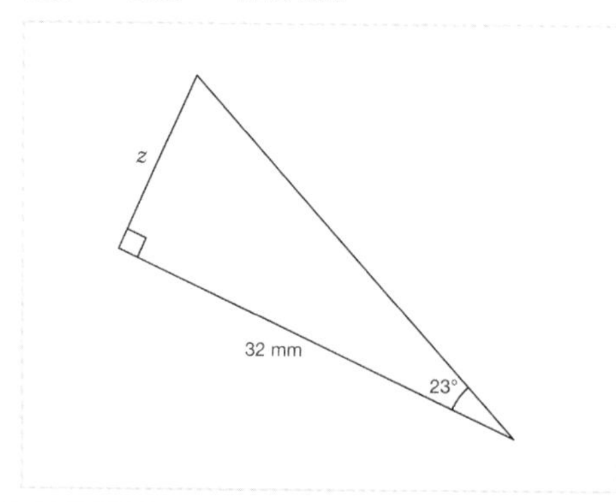

Level 2: Fluency: Select the correct trig ratio and find missing sides.

Required: 7/10 **Pupil Navigation:** on
Randomised: off

11. Which trigonometric ratio would you use to calculate the value of t?
1/3

- sin ■ cos ■ **tan**

30°
7 cm
t

$\sin x = \frac{O}{H}$ $\cos x = \frac{A}{H}$ $\tan x = \frac{O}{A}$

Level 2 *continued*

12. Which trigonometric ratio would you use to calculate the value of *v*?

1/3

- sin
- cos
- tan

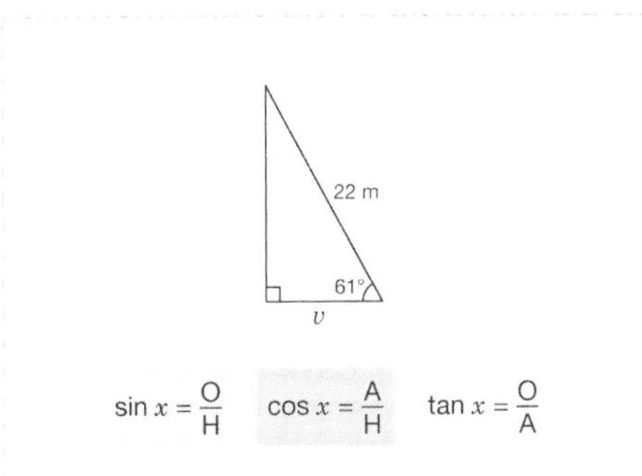

13. Calculate the value of *d*.

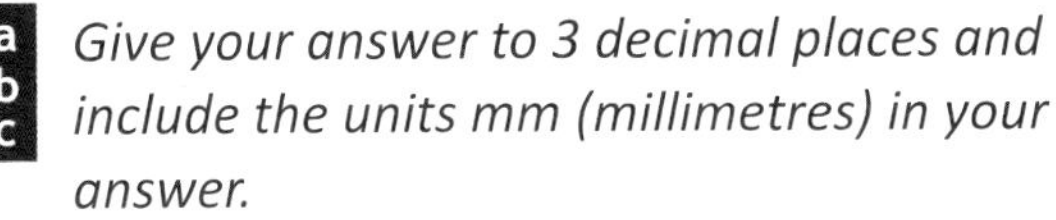

Give your answer to 3 decimal places and include the units mm (millimetres) in your answer.

- 4.653 mm
- 4.649 millimetres
- 4.649 mm
- 4.649
- 4.653 millimetres
- 4.653

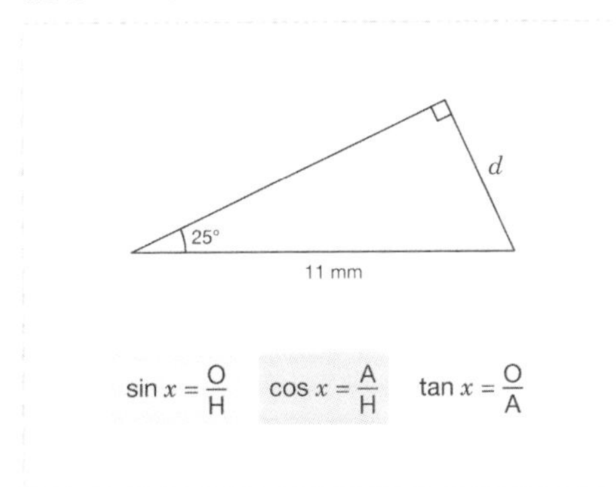

14. Calculate the value of *r*.

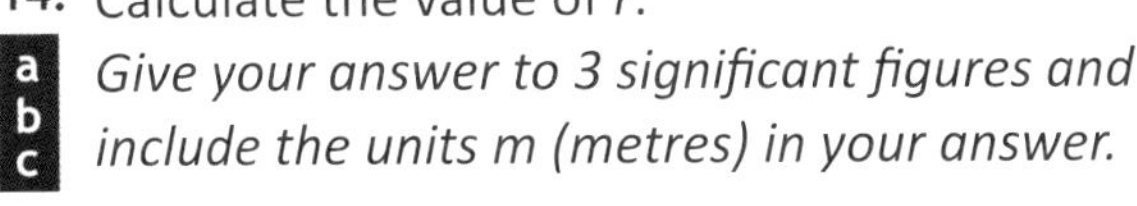

Give your answer to 3 significant figures and include the units m (metres) in your answer.

- 16.8
- 16.8 m
- 16.8 metres
- 16.794
- 16.794 m
- 16.794 metres

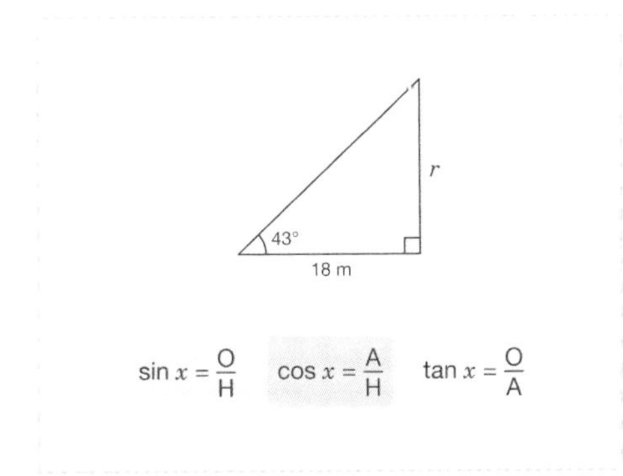

15. Calculate the value of *k*.

Give your answer to 2 decimal places and include the units cm (centimetres) in your answer.

- 25.44 cm
- 25.44 centimetres
- 25.44

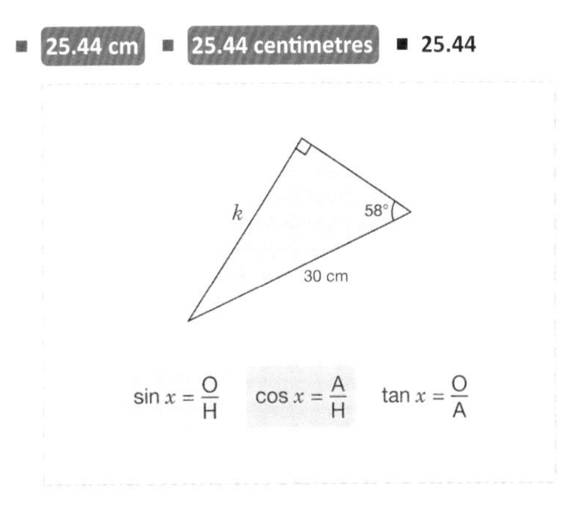

16. Calculate the value of *z*.

Give your answer to 2 decimal places and include the units mm (millimetres) in your answer.

- 14.18 millimetres
- 14.25
- 14.18 mm
- 14.25 millimetres
- 14.18
- 14.25 mm

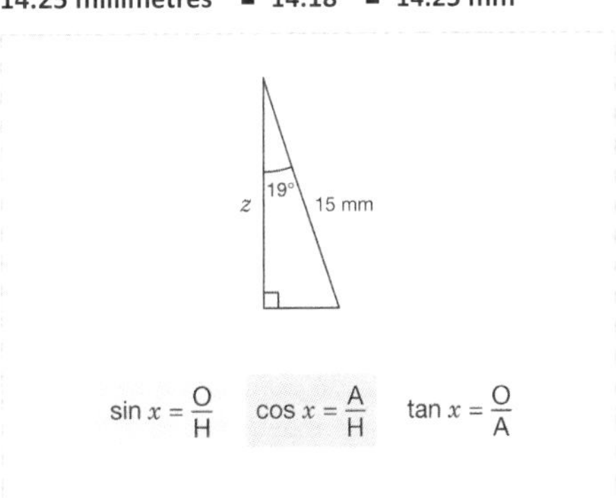

17. Calculate the value of *n*.

Give your answer to 1 decimal place and include the units cm (centimetres) in your answer.

- 43.2 cm
- 45.4 cm
- 45.4 centimetres
- 43.2 centimetres
- 43.2
- 45.4

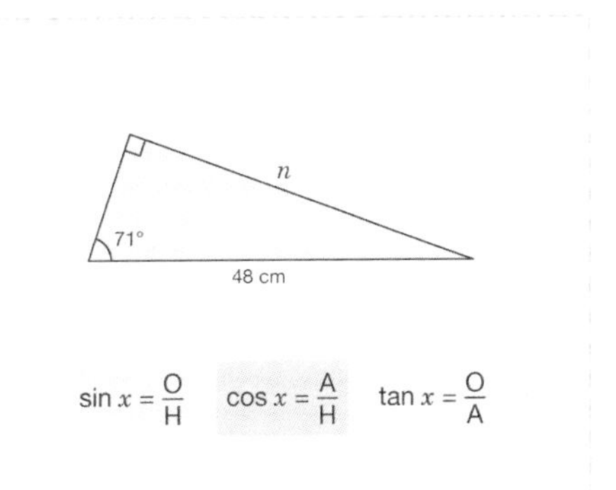

Level 2 *continued*

18. Which trigonometric ratio would you use to calculate the value of w?

1/3

- sin
- cos
- tan

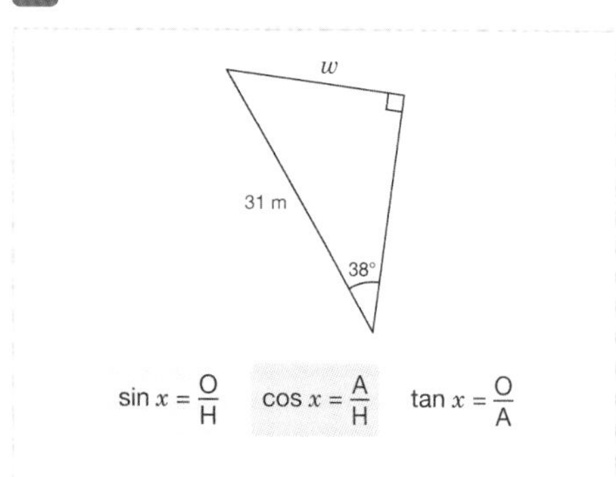

19. Calculate the value of s.
Give your answer to 3 decimal places and include the units mm (millimetres) in your answer.

- 6.464 millimetres
- 6.464 mm
- 6.464

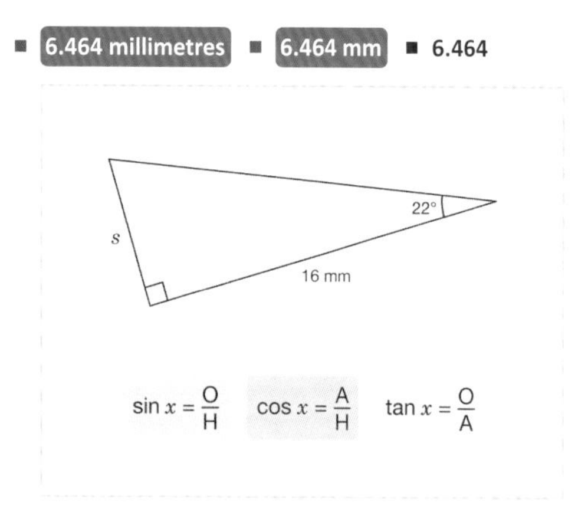

20. Calculate the value of p.
Give your answer to 4 significant figures and include the units cm (centimetres) in your answer.

- 6.138 cm
- 6.137 cm
- 6.138
- 6.137 centimetres
- 6.137
- 6.138 centimetres

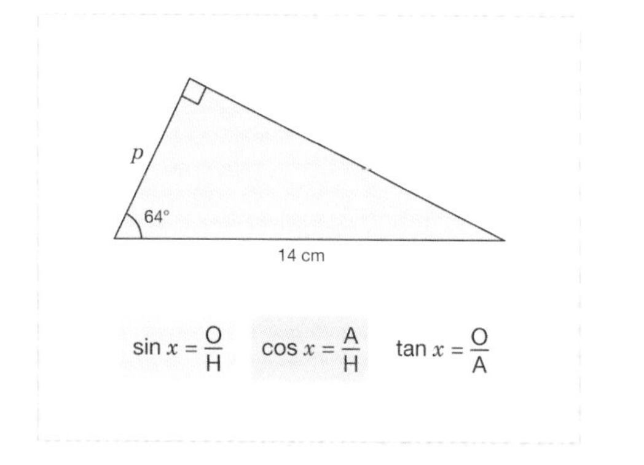

Level 3: Reasoning: Rearrange trig ratios and avoid misconceptions.

Required: 5/5 **Pupil Navigation:** on
Randomised: off

21. Mueez is rearranging sin x = O/H to make O the subject. He says that O = H × sin x. Is he correct? Explain your answer.

- Open question, no set answer

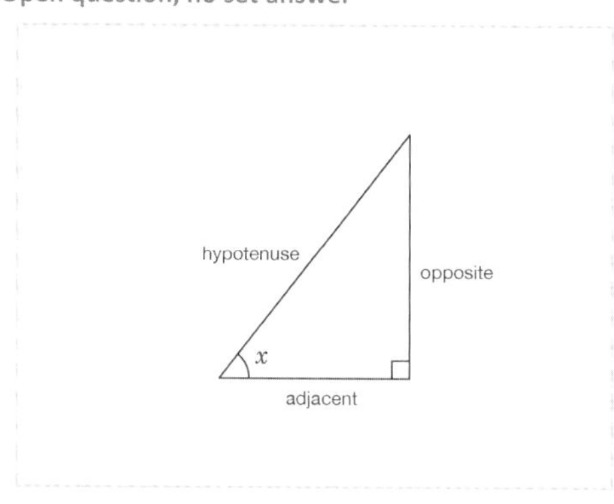

22. Sean has made a mistake with a trigonometry question. What is the correct length of line YZ?

- 13.8 mm
- 13.8 millimetres
- 13.8

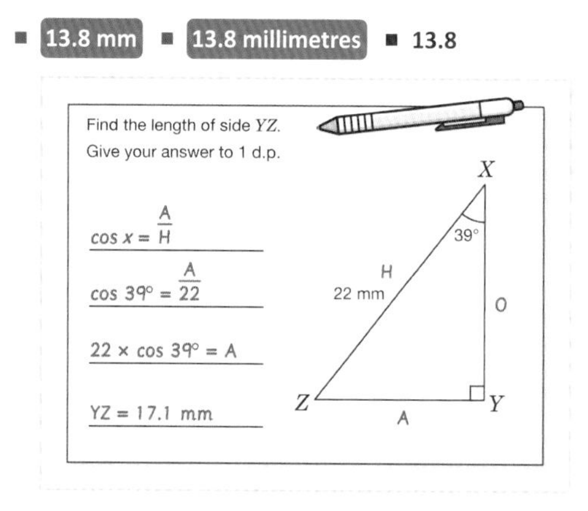

23. Which pair of values could you **not** use to find the value of a?

1/4

- sides b and c
- side c and angle x
- side b and angle y
- angles x and y

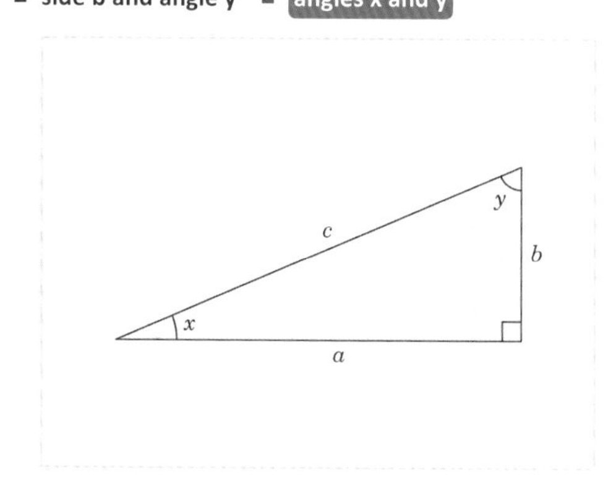

Level 3 *continued*

24. Which calculation can be used to find the value of d?

1/3

- $15 \div \tan 48°$
- **$15 \times \tan 48°$**
- $\tan 48° \div 15$

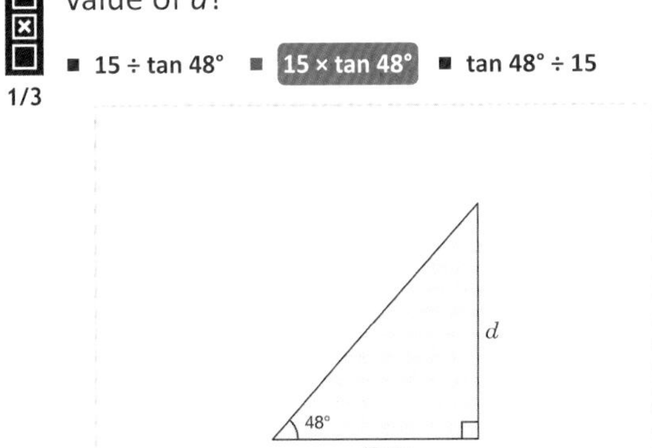

25. An equilateral triangle has a side length of 10 centimetres. What is the perpendicular height of the triangle?
Give your answer to 2 decimal places and include the units cm (centimetres) in your answer.

- **8.66 centimetres**
- **8.66 cm**
- 8.66

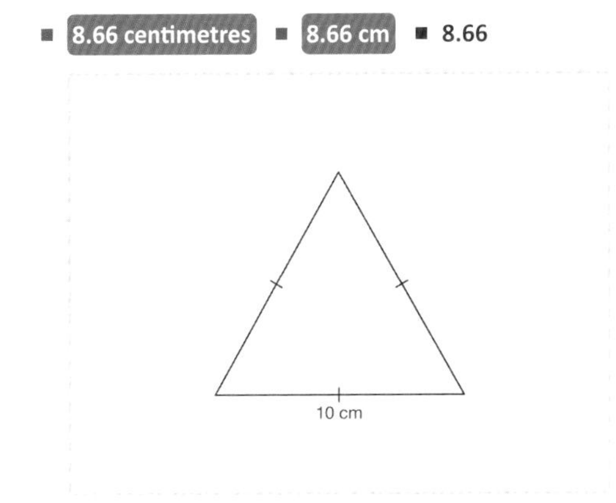

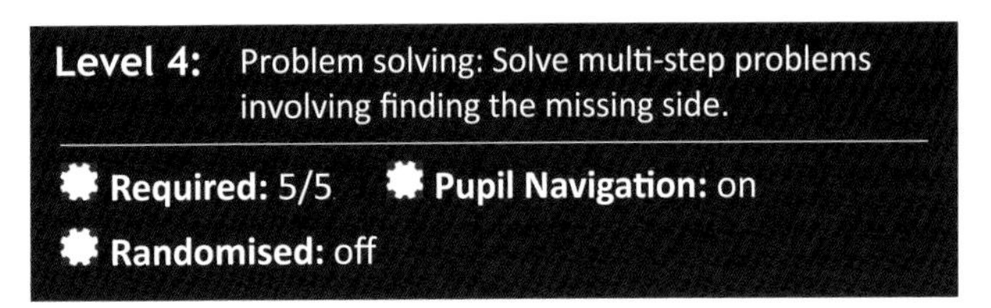

Level 4: Problem solving: Solve multi-step problems involving finding the missing side.

Required: 5/5 **Pupil Navigation:** on
Randomised: off

26. Calculate the length of line CD.
Give your answer to 2 decimal places and include the units cm (centimetres) in your answer.

- **7.05 cm**
- **7.05 centimetres**
- 7.05

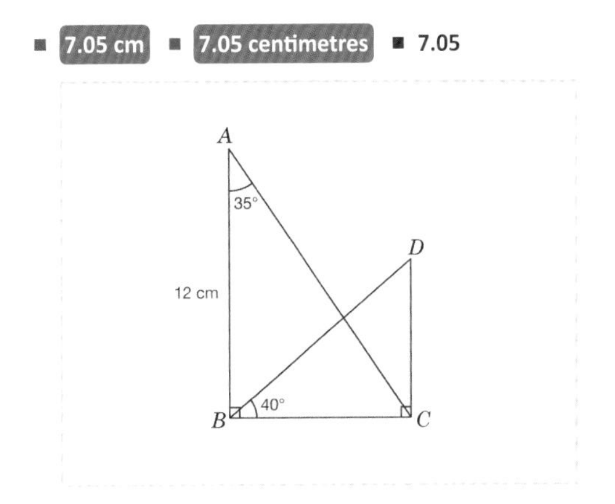

27. The rhombus shown has a side length of 12 cm. Calculate its area.
Give your answer in square centimetres to 1 decimal place, but don't include the units.

- **139.7**

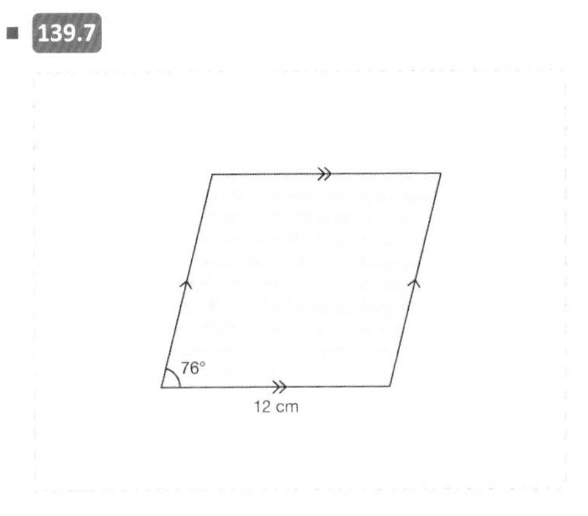

28. Calculate the length of line LM.
Give your answer to 2 decimal places and include the units cm (centimetres) in your answer.

- **10.91 cm**
- **10.91 centimetres**
- 10.91

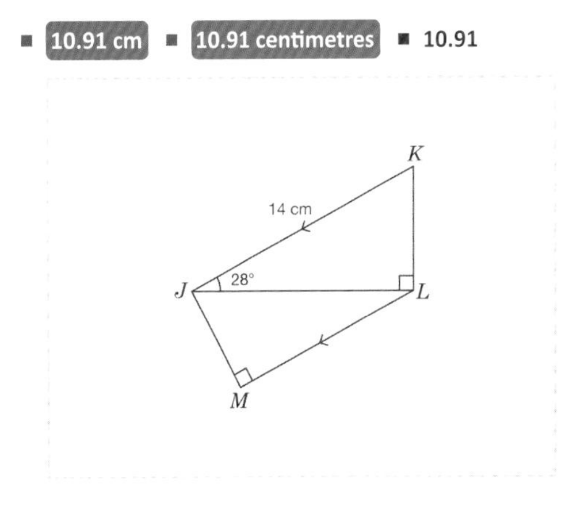

Level 4 ***continued***

29. Calculate the length of line *PR*.

Give your answer to 2 decimal places and include the units cm (centimetres) in your answer.

- 18.12 cm
- 24.04
- 24.04 centimetres
- 24.04 cm
- 5.92 cm
- 5.92
- 5.92 centimetres
- 18.12
- 18.12 centimetres

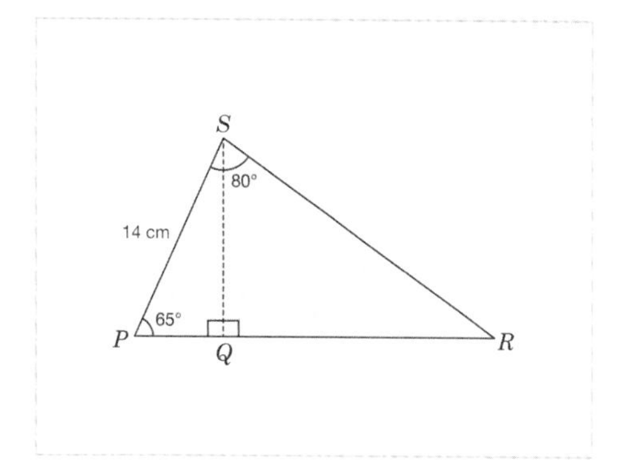

30. Calculate the area of triangle *XYZ*.

Give your answer in square centimetres to 1 decimal place, but don't include the units.

- 145.6

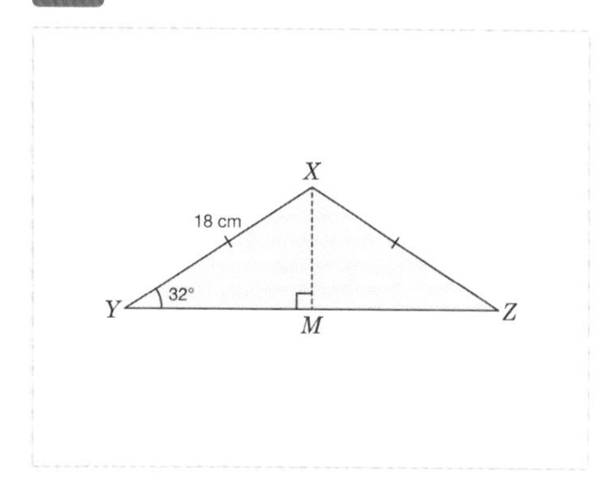

Use Trigonometry to Find a Missing Side 2

Objective: I can use trigonometry to find a missing side in a triangle.

Quick Search Ref: 10665

Level 1: Understanding: Find missing sides when given the correct trig ratio.

Required: 7/10 **Pupil Navigation:** on **Randomised:** off

1. Sort the side lengths into this order:

1. hypotenuse
2. opposite side to x
3. adjacent side to x

- 25 cm
- 7 cm
- 24 cm

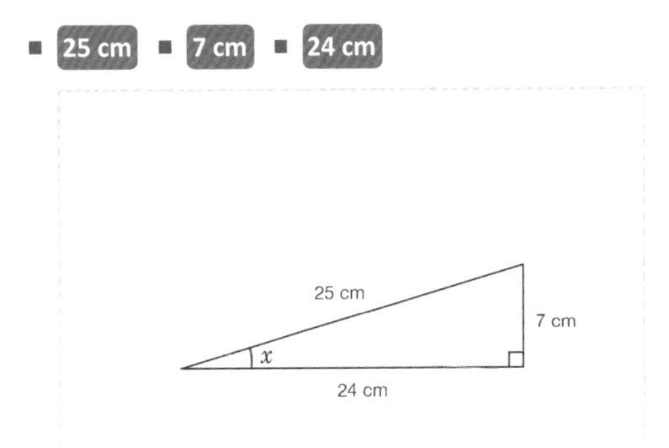

2. Sort the trigonometric ratios into the same order as the corresponding calculations.

- tan
- cos
- sin

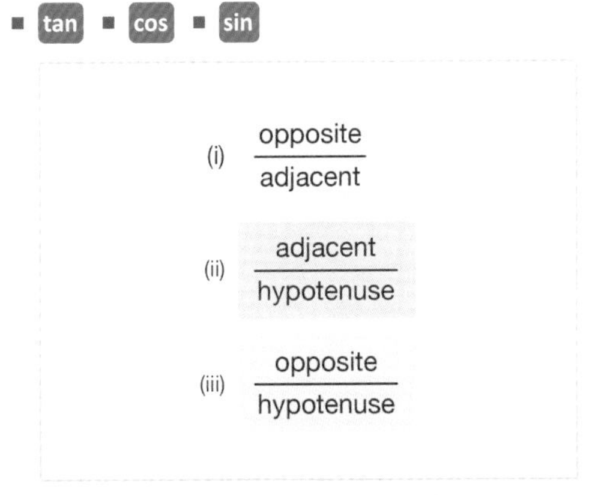

3. Rearrange cos x = A/H to make H the subject of the formula.
Select the equation that matches your answer.

1/3

- (i)
- (ii)
- (iii)

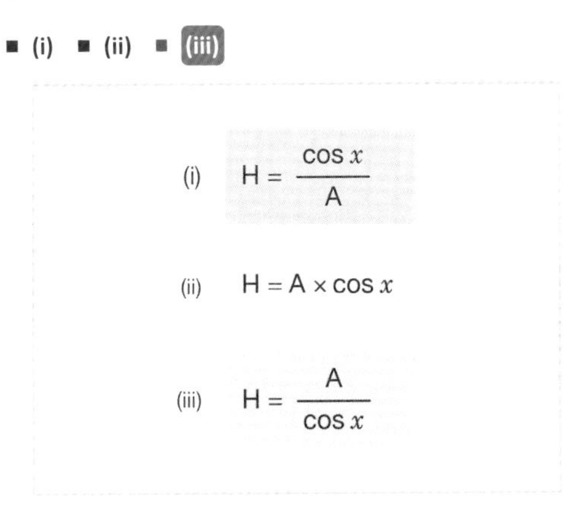

4. Use tan x = O/A to calculate the value of n.
Give your answer to 2 decimal places and include the units cm (centimetres) in your answer.

- 5.46 cm
- 5.47 centimetres
- 5.45 cm
- 5.46 centimetres
- 5.46
- 5.47
- 5.45 centimetres
- 5.45
- 5.47 cm

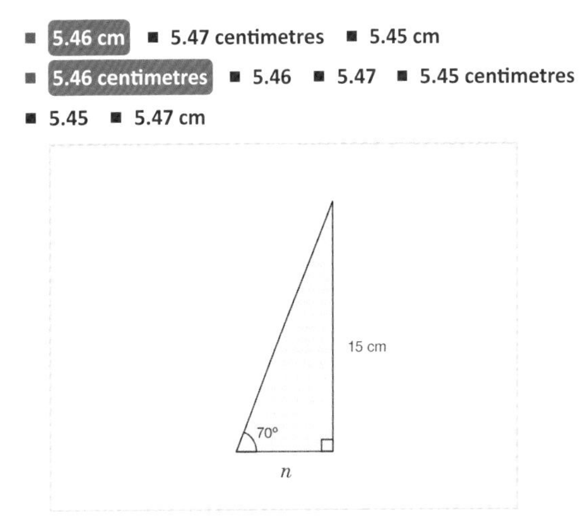

5. Use sin x = O/H to calculate the value of r.
Give your answer to 1 decimal place and include the units m (metres) in your answer.

- 21.3 metres
- 22.5 metres
- 21.3 m
- 22.5 m
- 21.3
- 22.5

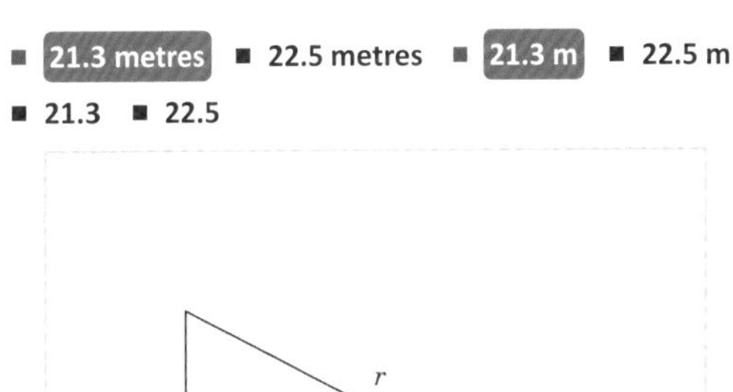

Level 1 *continued*

6. Use tan x = O/A to calculate the length of side *XY*.
a b c
Give your answer to 3 decimal places and include the units mm (millimetres) in your answer.

- 1.277
- **1.276 mm**
- **1.276 millimetres**
- 1.277 millimetres
- 1.277 mm
- 1.276

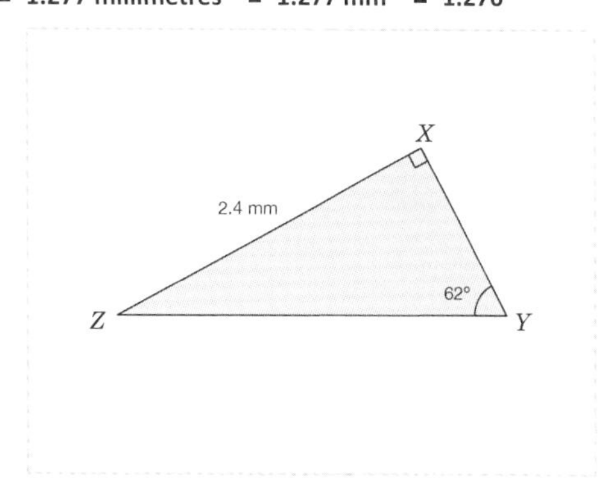

7. Use cos x = A/H to calculate the value of f.
a b c
Give your answer to 4 significant figures and include the units cm (centimetres) in your answer.

- 56.9756 cm
- **56.98 centimetres**
- **56.98 cm**
- 56.98
- 56.9756 centimetres
- 56.9756

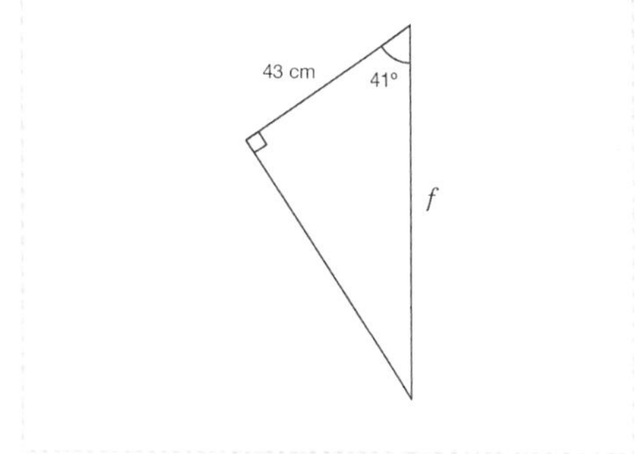

8. Rearrange tan x = O/A to make A the subject of the formula.
Select the equation that matches your answer.
1/3

- **(i)**
- (ii)
- (iii)

(i) $A = \dfrac{O}{\tan x}$

(ii) $A = O \times \tan x$

(iii) $A = \dfrac{\tan x}{O}$

9. Use cos x = A/H to calculate the length of side *MN*.
a b c
Give your answer to 2 decimal places and include the units m (metres) in your answer.

- 8.97 m
- 8.97
- **8.85 metres**
- **8.85 m**
- 8.85
- 8.81
- 8.81 m
- 8.97 metres
- 8.81 metres

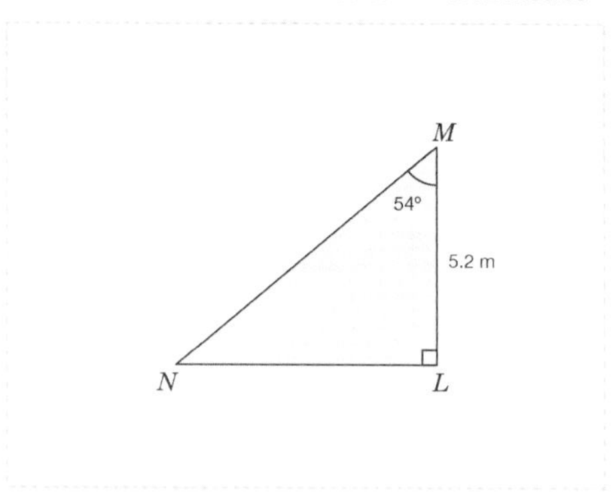

10. Use sin x = O/H to calculate the value of q.
a b c
Give your answer to 3 significant figures and include the units mm (millimetres) in your answer.

- **32.2 mm**
- 32.189
- **32.2 millimetres**
- 32.189 millimetres
- 32.2
- 32.189 mm

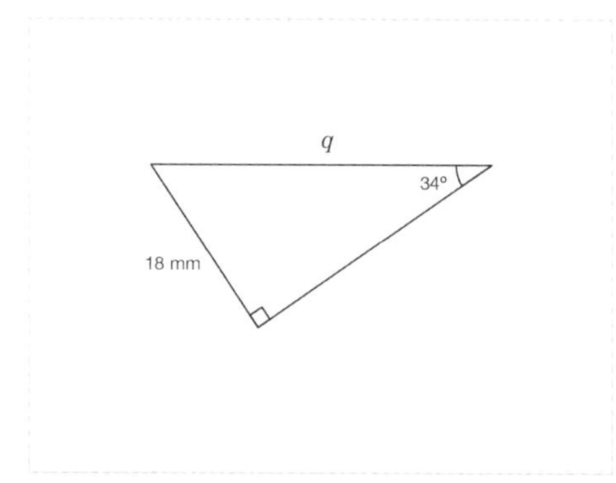

Level 2: Fluency: Select the correct trig ratio and finding missing sides.

Required: 7/10 **Pupil Navigation:** on
Randomised: off

11. Which trigonometric ratio would be used to calculate the value of s?
1/3

- sin
- **cos**
- tan

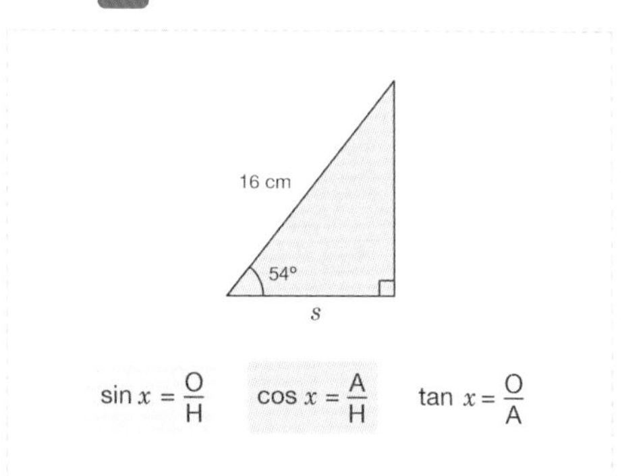

Level 2 *continued*

12. Which trigonometric ratio would be used to calculate the value of w?

1/3

- sin
- cos
- tan

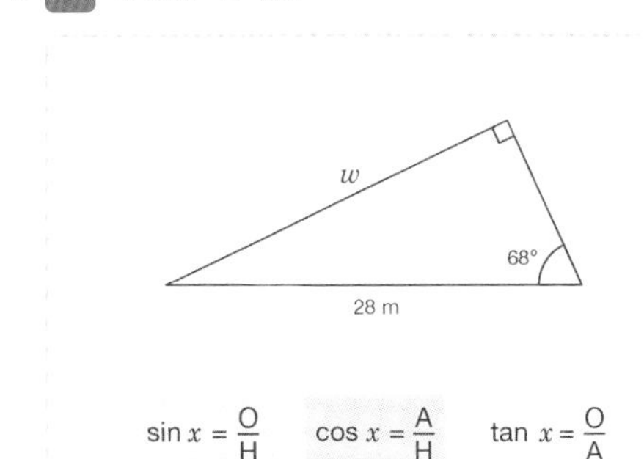

13. Calculate the value of p.

Give your answer to 2 decimal places and include the units cm (centimetres) in your answer.

- 9.14 cm
- 9.14 centimetres
- 9.14
- 9.09 centimetres
- 9.09 cm
- 9.21 cm
- 9.09
- 9.21 centimetres
- 9.21

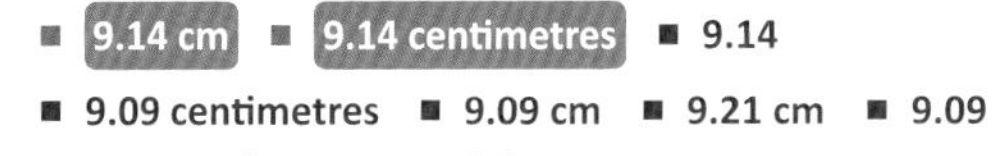

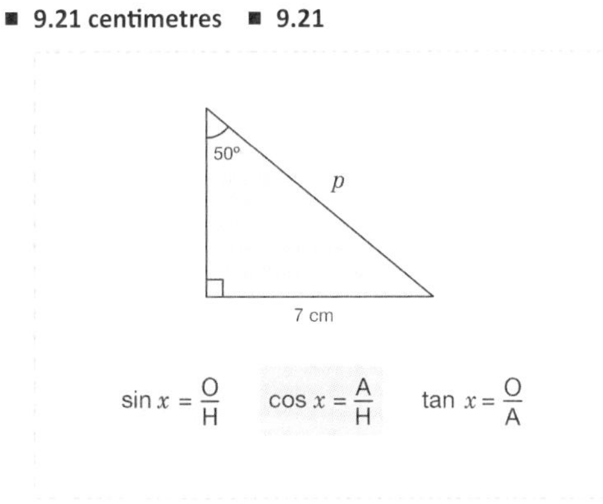

14. Calculate the value of d.

Give your answer to 3 decimal places and include the units m (metres) in your answer.

- 30.733 m
- 30.733
- 30.761 metres
- 30.761 m
- 30.806
- 30.761
- 30.806 m
- 30.806 metres
- 30.733 metres

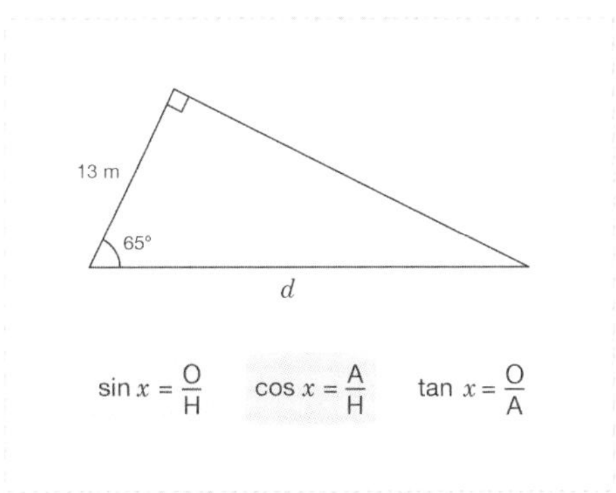

15. Calculate the length of side PQ.

Give your answer to 2 decimal places and include the units mm (millimetres) in your answer.

- 19.84
- 19.68 millimetres
- 19.68 mm
- 19.68
- 19.84 mm
- 19.84 millimetres

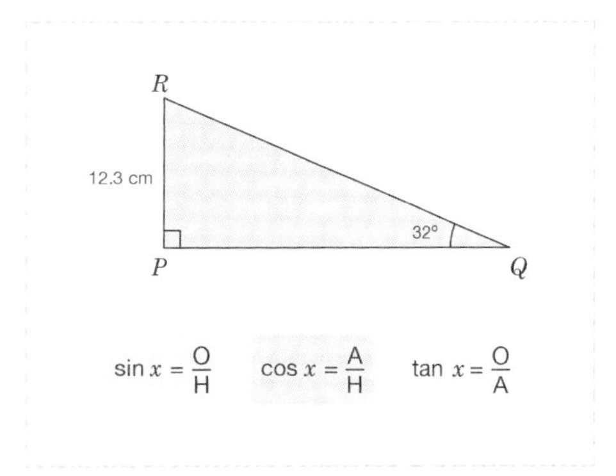

16. Calculate the value of z.

Give your answer to 3 significant figures and include the units cm (centimetres) in your answer.

- 8.95 cm
- 8.96 cm
- 8.95 centimetres
- 8.96 centimetres
- 8.946
- 8.96
- 8.946 cm
- 8.95
- 8.946 centimetres

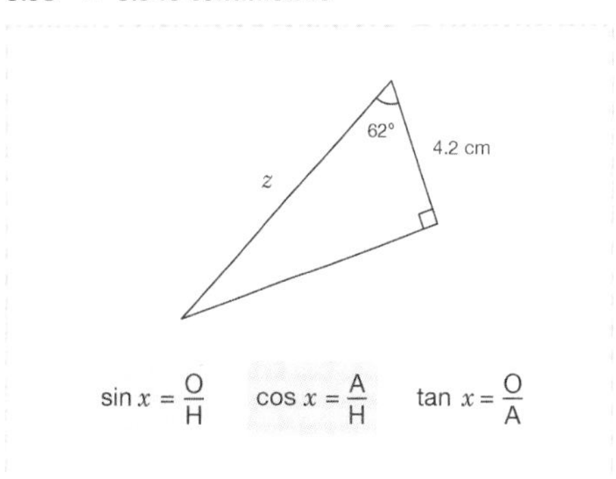

17. Calculate the value of r.

Give your answer to 1 decimal place and include the units m (metres) in your answer.

- 85.2 metres
- 91.3 m
- 91.3
- 85.2 m
- 81.1
- 81.1 metres
- 81.1 m
- 91.3 metres
- 85.2

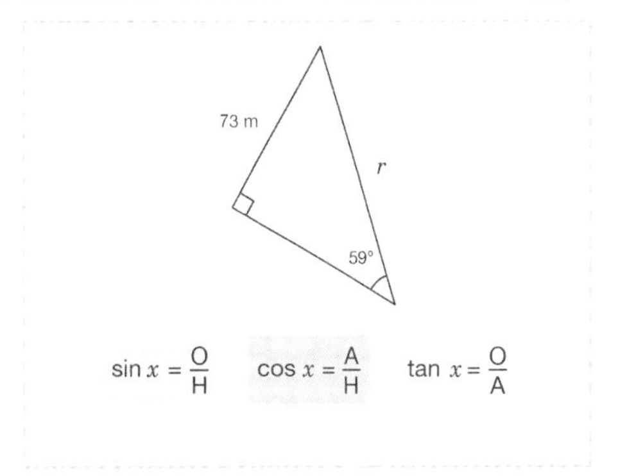

Level 2 *continued*

18. Which trigonometric ratio would you use to calculate the length of side *YZ*?

1/3

- sin
- cos
- **tan**

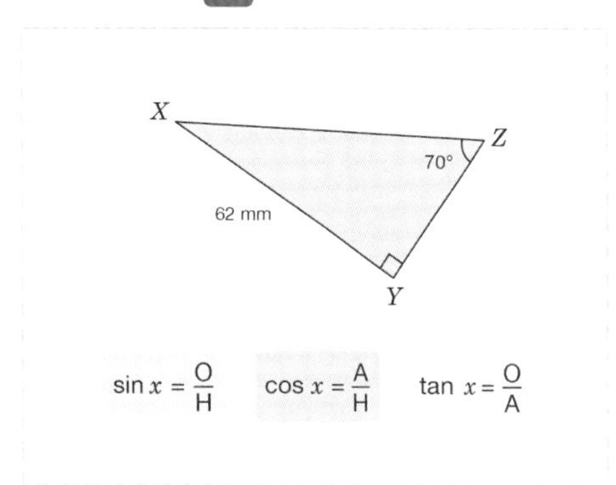

19. Calculate the length of side *RT*.
Give your answer to 2 decimal places and include the units mm (millimetres) in your answer.

- 57.5
- **56.63 mm**
- **56.63 millimetres**
- 57.5 mm
- 56.63
- 57.5 millimetres

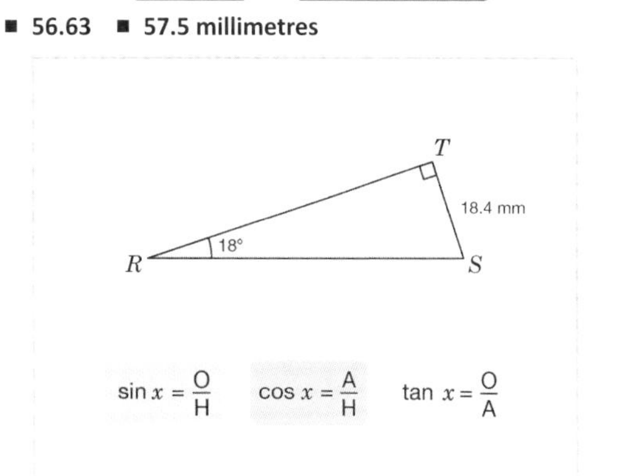

20. Calculate the value of *v*.
Give your answer to 3 significant figures and include the units cm (centimetres) in your answer.

- 81.3
- 81.134 centimetres
- **81.1 cm**
- 81.134
- **81.1 centimetres**
- 81.1
- 81.3 cm
- 81.134 cm
- 81.3 centimetres

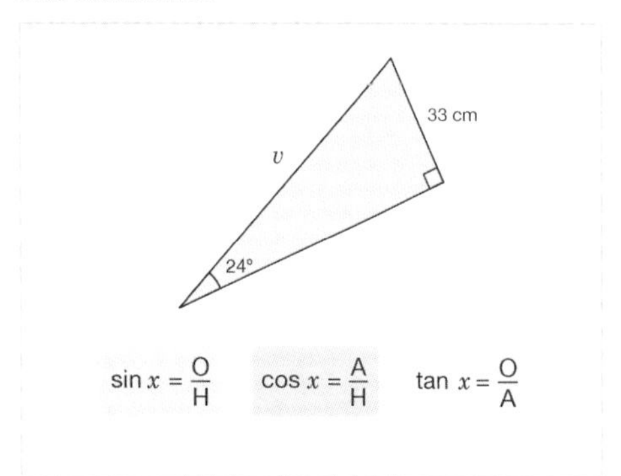

Level 3: Reasoning: Make inferences using trig ratios and identify misconceptions.

Required: 5/5 **Pupil Navigation:** on
Randomised: off

21. Darcey is rearranging tan *x* = O/A to make A the subject. Darcey says that A = O ÷ tan *x*. Is she correct? Explain your answer.

- Open question, no set answer

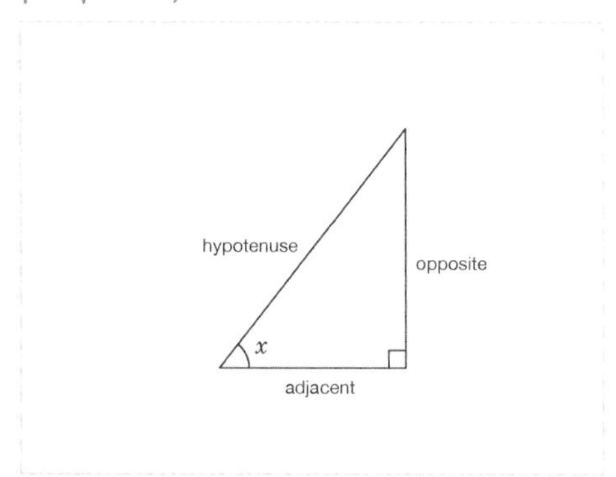

22. James has made a mistake with a trigonometry question. What is the correct length of side *XZ*?
Give your answer to 2 decimal places and include the units m (metres) in your answer.

- **31.57 metres**
- **31.57 m**
- 31.57

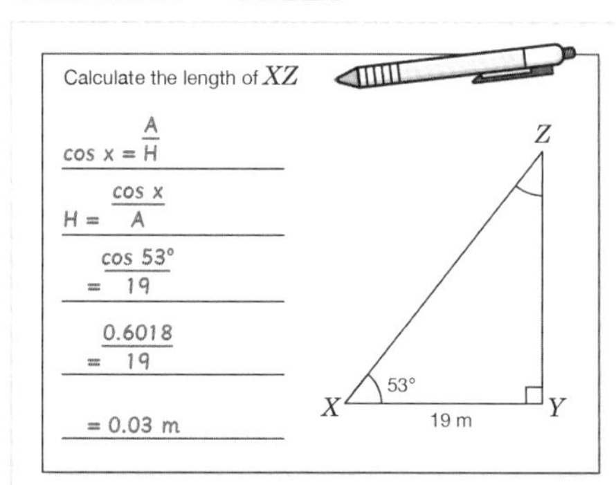

23. An equilateral triangle has a perpendicular height of 12 cm. Calculate the side length of the triangle.
Give your answer to 2 decimal places and include the units cm (centimetres) in your answer.

- **13.86 centimetres**
- **13.86 cm**
- 13.86

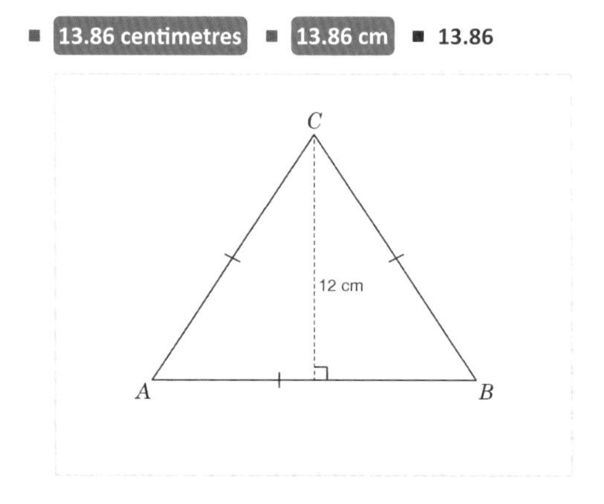

Level 3 *continued*

24. Which calculation would you use to find the value of r?

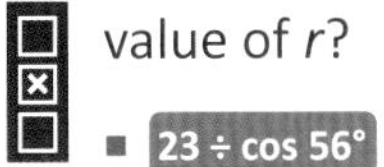

1/3

- 23 ÷ cos 56°
- 23 × cos 56°
- cos 56° ÷ 23

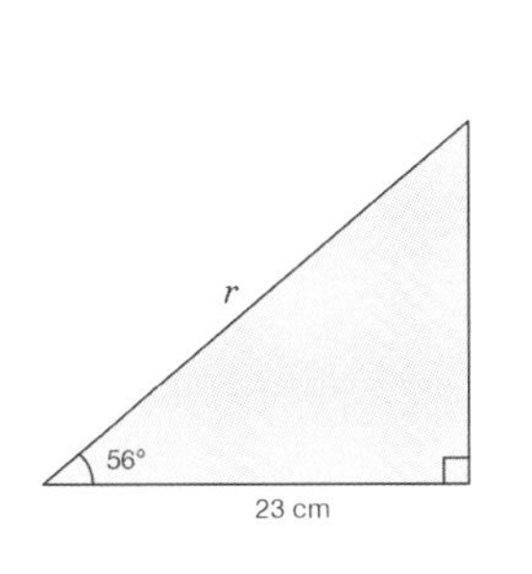

25. The cos button is broken on Danny's calculator. Explain in words how Danny can calculate the length of side *LN*.

- Open question, no set answer

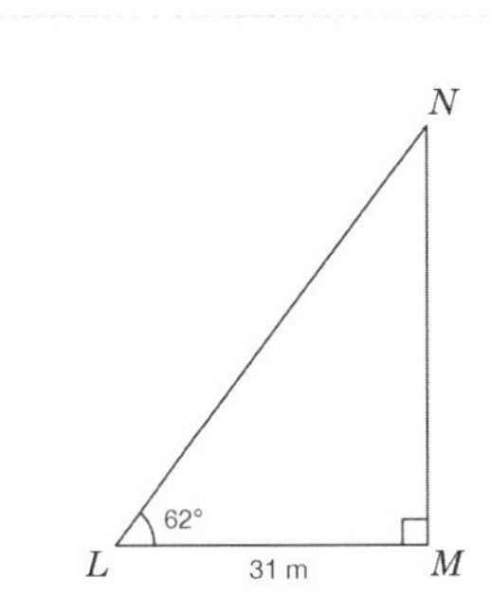

Level 4: Problem solving: Solve multi-step problems involving finding missing sides.

Required: 5/5 **Pupil Navigation:** on

Randomised: off

26. A skateboard ramp is at an angle of 42° to the ground. At its highest point, it is 150 centimetres above ground level. What is the length of the slope of the ramp in metres?
Give your answer to 3 decimal places and include the units m (metres) in your answer.

- 2.241 metres
- 2.241 m
- 224.171 centimetres
- 2.241
- 224.171 cm
- 224.171

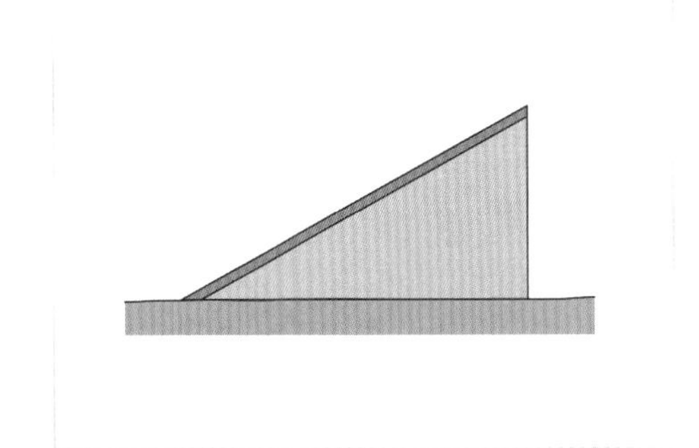

27. Triangle *RST* is an isosceles triangle. Calculate the length of side *RT*.
Give your answer to 1 decimal place and include the units cm (centimetres) in your answer.

- 28.9 cm
- 28.9
- 28.9 centimetres
- 14.9
- 14.9 cm
- 14.9 centimetres

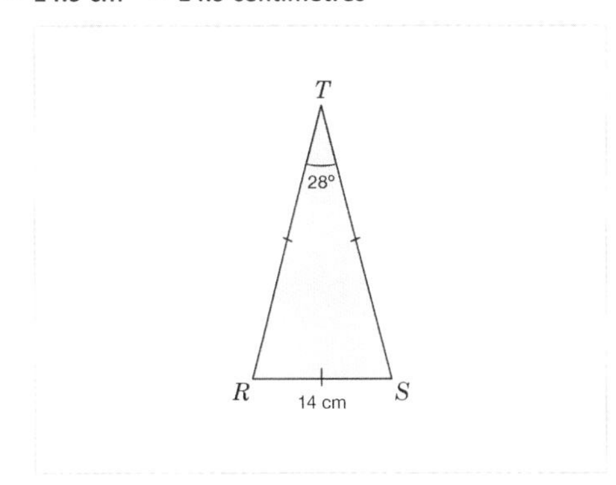

28. Calculate the length of side *XZ*.
Give your answer to 1 decimal place and include the units m (metres) in your answer.

- 38.8 metres
- 38.8 m
- 38.8

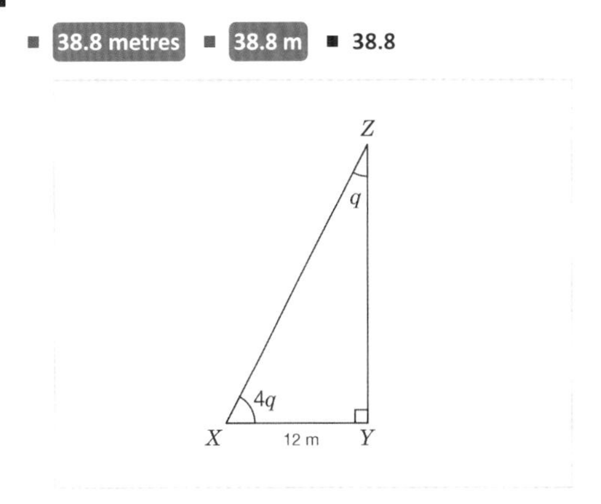

29. Calculate the length of side *AD*.
Give your answer to 2 decimal places and include the units cm (centimetres) in your answer.

- 27.35 cm
- 27.35 centimetres
- 27.35

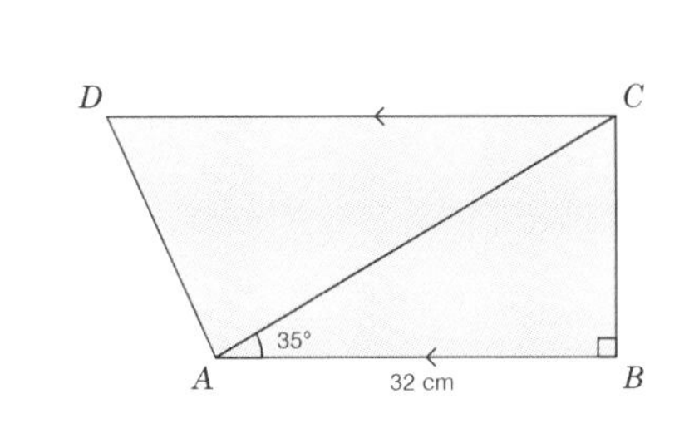

Level 4 *continued*

30. A regular hexagon has a perpendicular height of 10 cm when measured from the centre of one side of the shape to the centre of the opposite side. What is the side length of the hexagon?
Give your answer to 2 decimal places and include the units cm (centimetres) in your answer.

- 11.55 cm
- 11.55 centimetres
- 11.55

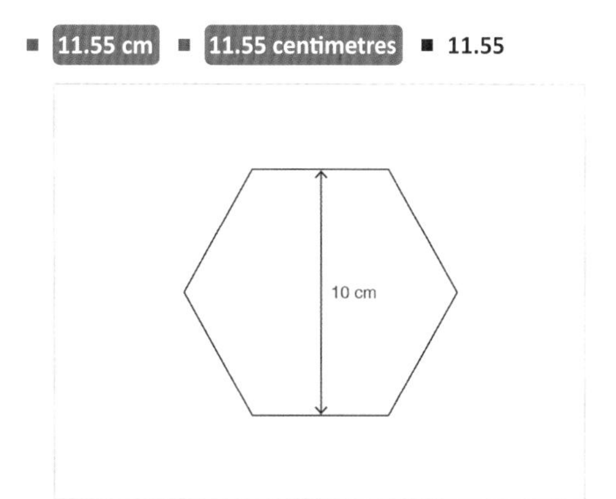

Pythagoras Topic Review

Objective: I can use Pythagorus' theorem to answer questions.

Quick Search Ref: 10768

Level 1: Understanding

Required: 10/10 **Pupil Navigation:** on **Randomised:** off

1. Select the hypotenuse of the right-angled triangle.

1/3

- a
- **b**
- c

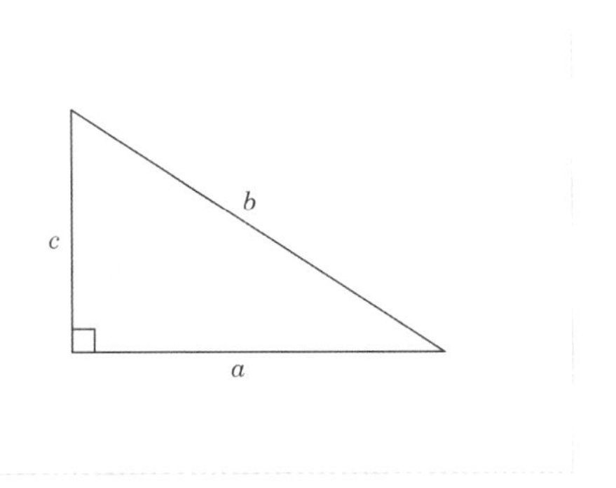

2. In the diagram, $AB = 6$ cm and $BC = 8$ cm. What is the area of the square on side AC in square centimetres?
Don't include the units in your answer.

- **100**
- 64
- 36
- 10

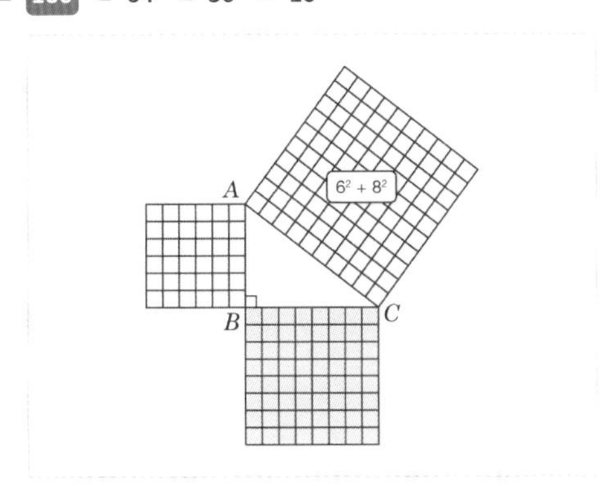

3. In the diagram, $AC = 10$ cm and $BC = 8$ cm. What is the area of the square on side AB in square centimetres?
Don't include the units in your answer.

- **36**
- 100
- 6
- 64

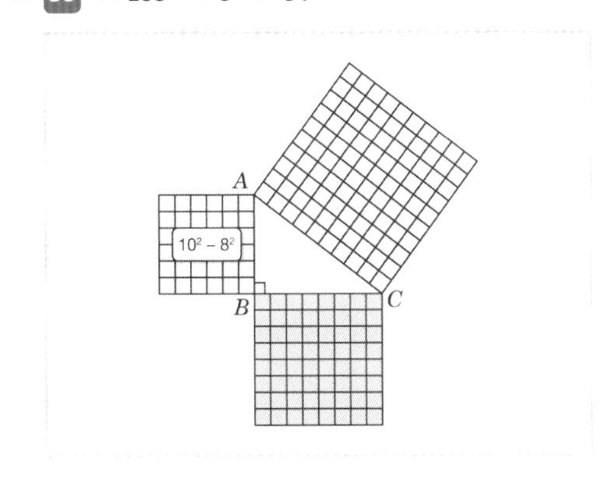

4. Look at the triangle in the diagram. Which of the following statements is true?

1/3

- $p^2 = r^2 - q^2$
- $p^2 = q^2 + r^2$
- **$p^2 = q^2 - r^2$**

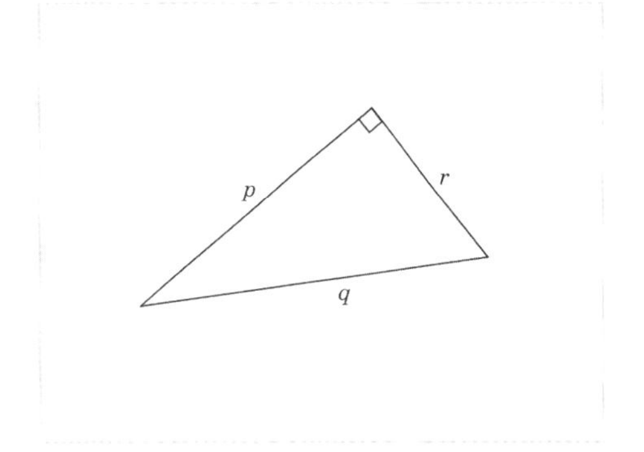

5. In the diagram, $AB = 5$ cm and $BC = 12$ cm. What is the area of the square on side AC in square centimetres?
Don't include the units in your answer.

- **169**
- 25
- 144

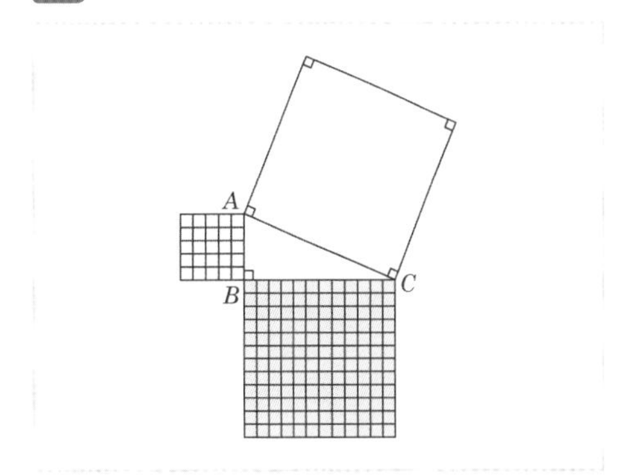

6. In the diagram, $AC = 13$ cm and $BC = 5$ cm. What is the area of the square on AB in square centimetres?
Don't include the units in your answer.

- **144**
- 25
- 12
- 169

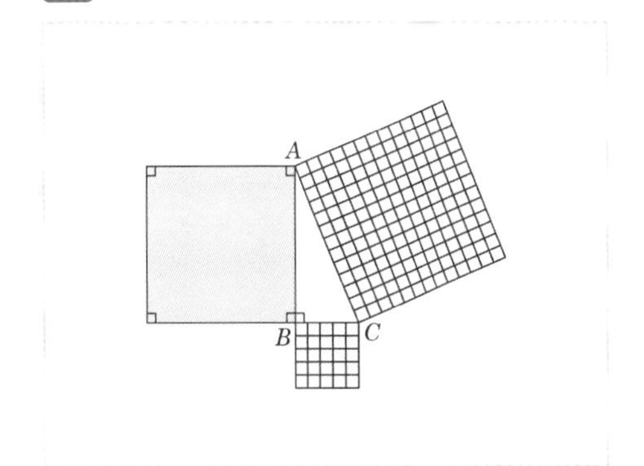

Level 1 *continued*

7. What is the length of the hypotenuse of the right-angled triangle?
Include the units cm (centimetres) in your answer.

- **29 centimetres** ■ **29 cm** ■ **841 cm** ■ **841**
- **841 centimetres** ■ **29**

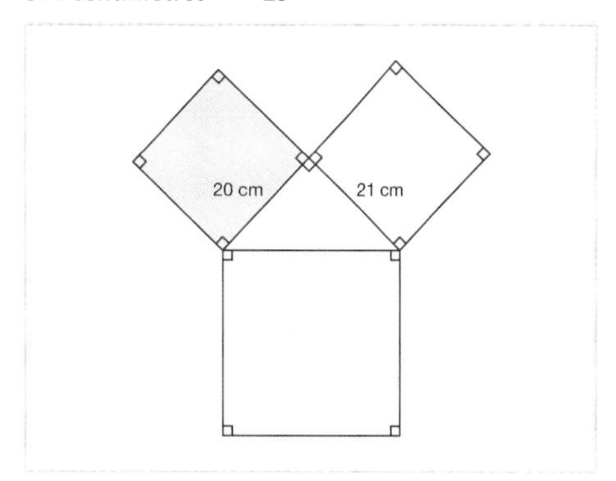

8. What is the length of the hypotenuse of the right-angled triangle?
Include the units cm (centimetres) in your answer.

- **17 cm** ■ **17 centimetres** ■ **289 cm** ■ **289**
- **289 centimetres** ■ **17**

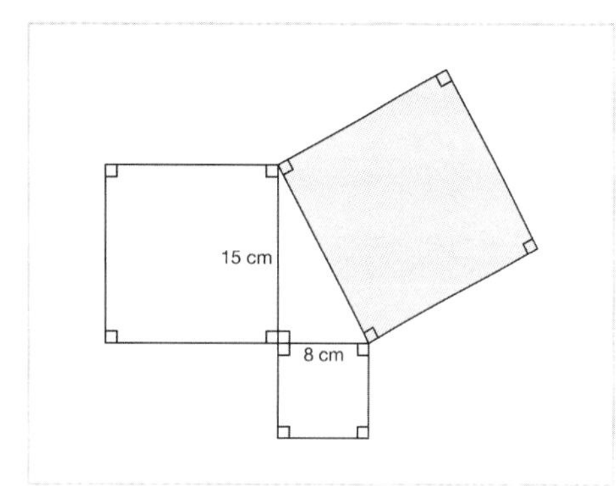

9. What is the length of side AB?
Include the units cm (centimetres) in your answer.

- **18** ■ **18 cm** ■ **18 centimetres** ■ **324 cm** ■ **38.4**
- **38.4 centimetres** ■ **324 centimetres** ■ **324**
- **38.4 cm**

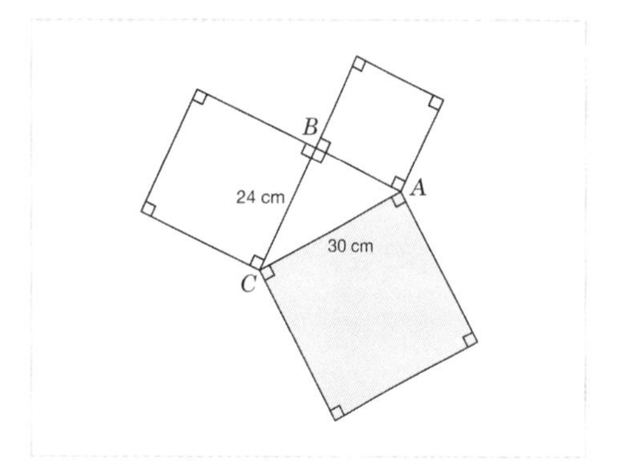

10. What is the length of the hypotenuse of the right-angled triangle?
Include the units cm (centimetres) in your answer.

- **10 centimetres** ■ **10 cm** ■ **100** ■ **10**
- **100 centimetres** ■ **100 cm**

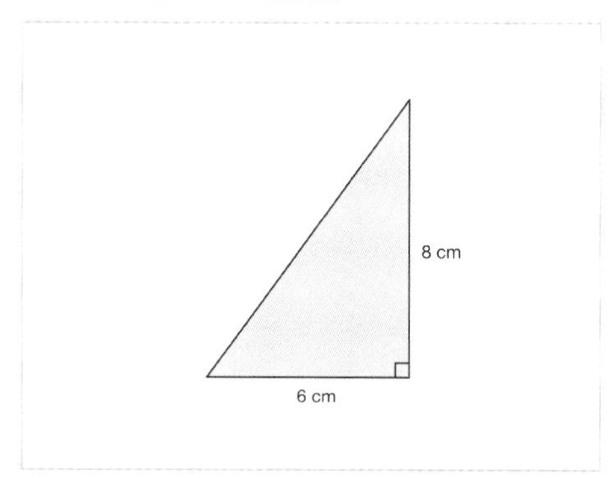

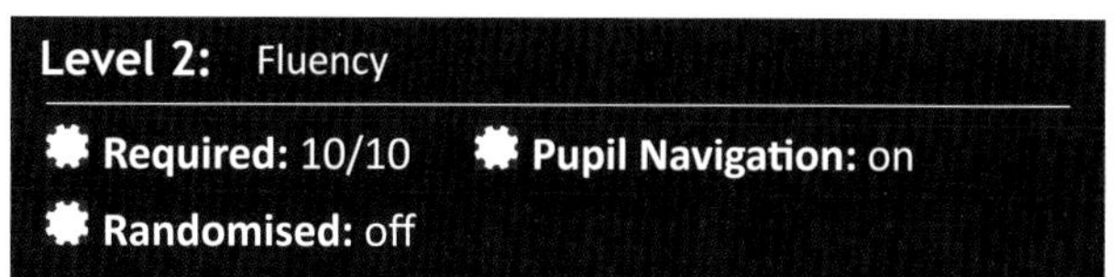

Level 2: Fluency

Required: 10/10 **Pupil Navigation:** on
Randomised: off

11. What is the length of the hypotenuse of the right-angled triangle?
Give your answer to one decimal place and include the units cm (centimetres) in your answer.

- **15.8 centimetres** ■ **15.8 cm** ■ **250 centimetres**
- **15.8** ■ **250 cm** ■ **250**

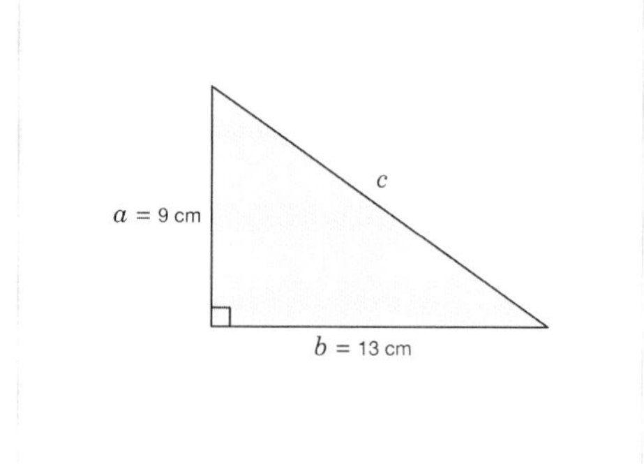

Level 2 *continued*

12. What is the value of b?

Give your answer to 1 decimal place and include the units cm (centimetres) in your answer.

- 30.4 centimetres ■ 30.4 cm ■ 927 cm ■ 30.4
- 73.1 cm ■ 73.1 ■ 73.1 centimetres ■ 927
- 927 centimetres

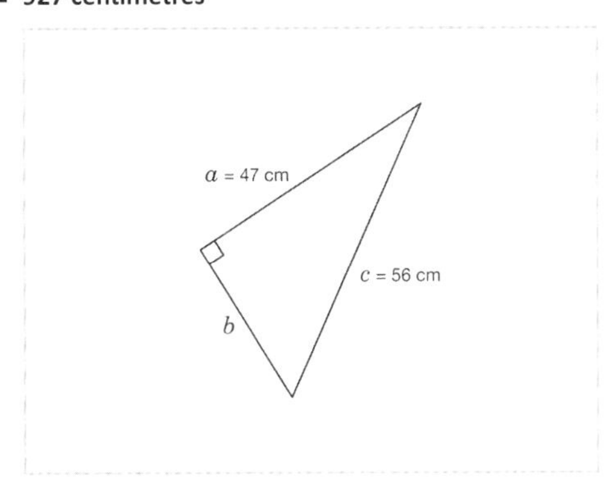

13. What is the length of QR?

Give your answer to 3 significant figures and include the units cm (centimetres) in your answer.

- 23.6 centimetres ■ 12.8 centimetres ■ 12.8 cm
- 165 ■ 12.845 cm ■ 165 centimetres
- 12.845 centimetres ■ 12.845 ■ 23.6 ■ 12.8
- 165 cm ■ 23.6 cm

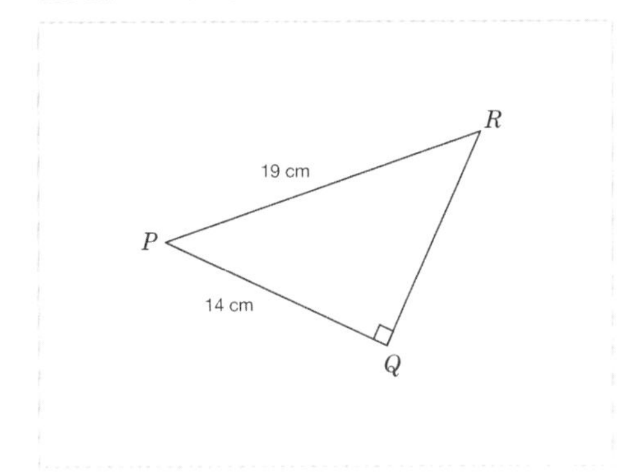

14. A ladder of length 5.3 metres reaches 5.1 metres up a vertical wall. What is the horizontal distance in metres between the foot of the ladder and the bottom of the wall?

Give your answer to 1 decimal place and include the units m (metres) in your answer.

- 1.4 m ■ 7.4 ■ 1.4 metres ■ 7.4 m ■ 2.1 metres
- 7.4 metres ■ 2.1 m ■ 1.4 ■ 2.1

15. The horizontal distance from a point on the ground to the bottom of a tree is 22 metres. The vertical distance from the same point on the ground to the top of the tree is 32 metres. What is the height of the tree?

Include the units m (metres) in your answer.

- 23.2 metres ■ 23.2 m ■ 23.2 ■ 540 metres
- 38.8 ■ 540 ■ 38.8 metres ■ 38.8 m ■ 540 m

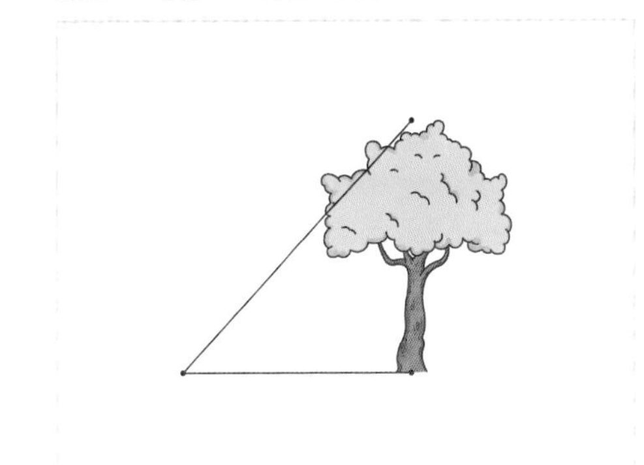

Level 2 *continued*

16. A ship sails 480 kilometres due south and then 260 kilometres due west. How far is the ship from its original starting point?
Give your answer to 2 decimal places and include the units km (kilometres) in your answer.

- 545.89 kilometres
- 545.89 km
- 545.89
- 403.48 kilometres
- 403.48 km
- 403.48

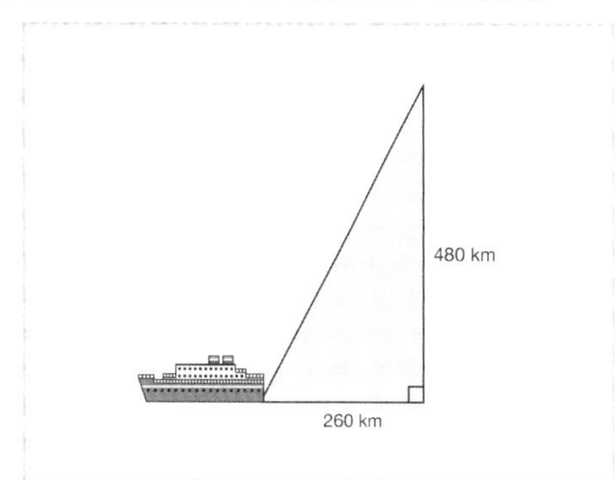

17. A coordinate grid is drawn on centimetre squared paper. What is the distance between the points (1, 3) and (6, 9)?
Give your answer to 2 decimal places and include the units cm (centimetres) in your answer.

- 3.32 centimetres
- 7.81 centimetres
- 7.81 cm
- 7.81
- 3.32 cm
- 3.32

18. Calculate the perpendicular height of an equilateral triangle with side lengths of 9 cm.
Give your answer to 1 decimal place and include the units cm (centimetres) in your answer.

- 10.1 centimetres
- 7.8 centimetres
- 7.8 cm
- 10.1
- 10.1 cm
- 7.8

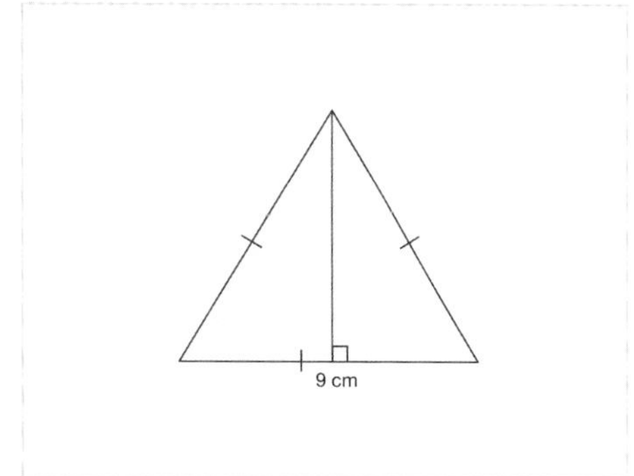

19. What is the value of d?
Give your answer to 3 significant figures and include the units mm (millimetres) in your answer.

- 28.3 mm
- 799 mm
- 28.3 millimetres
- 28.267 mm
- 35.3
- 799 millimetres
- 35.3 mm
- 28.3
- 35.3 millimetres
- 28.267
- 799
- 28.267 millimetres

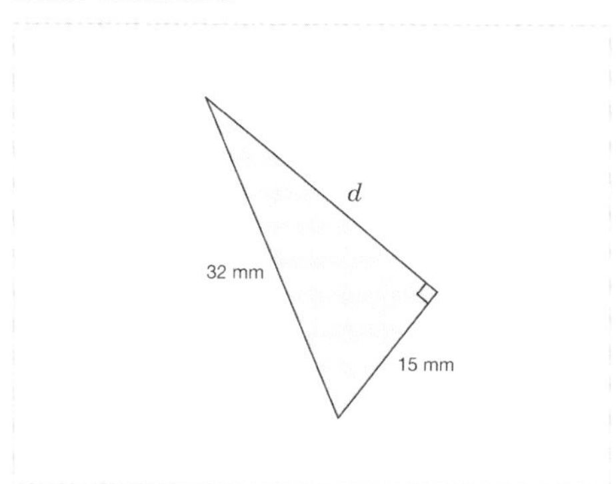

20. A plane flies 320 kilometres due north and then 470 kilometres due east. How far is the plane from its original starting point?
Give your answer to 4 significant figures and include the units km (kilometres) in your answer.

- 568.6 km
- 568.6 kilometres
- 344.2 kilometres
- 568.5947 kilometres
- 568.5947
- 568.6
- 344.2
- 344.2 km
- 568.5947 km

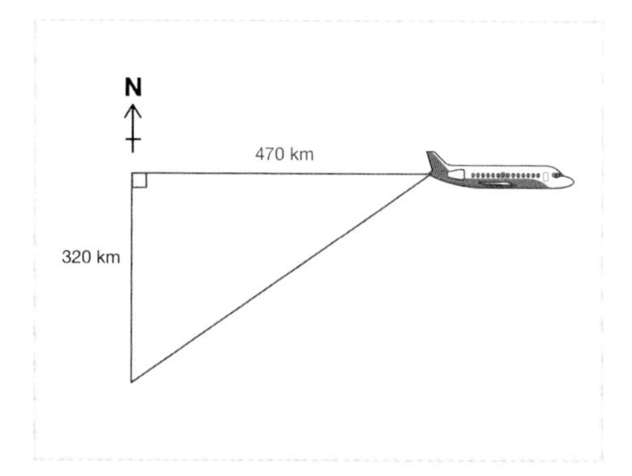

21. Select the **two** statements that are true for the following right-angled triangle.

2/4

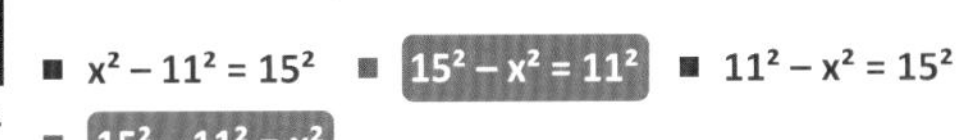

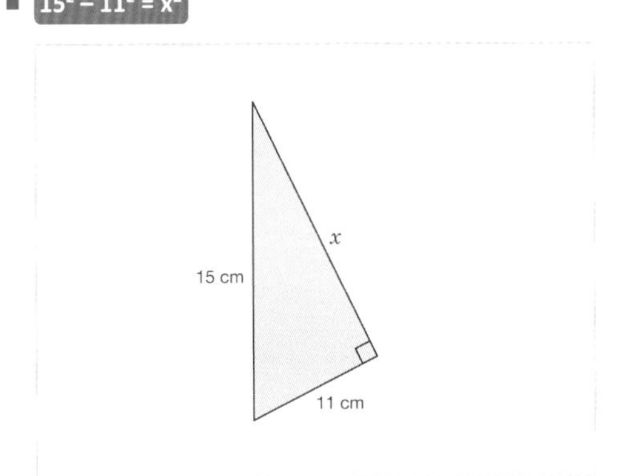

22. Is this a right-angled triangle? Explain your answer.

- Open question, no set answer

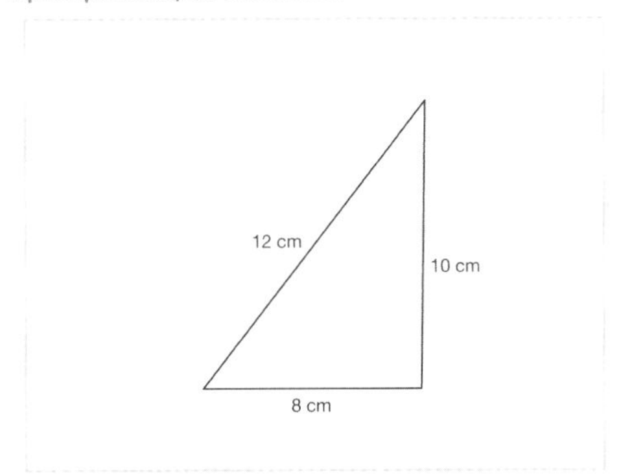

23. Poppy completes her homework but gets one of the questions wrong. What is the correct value for the question she has answered incorrectly?
Don't include the units in your answer.

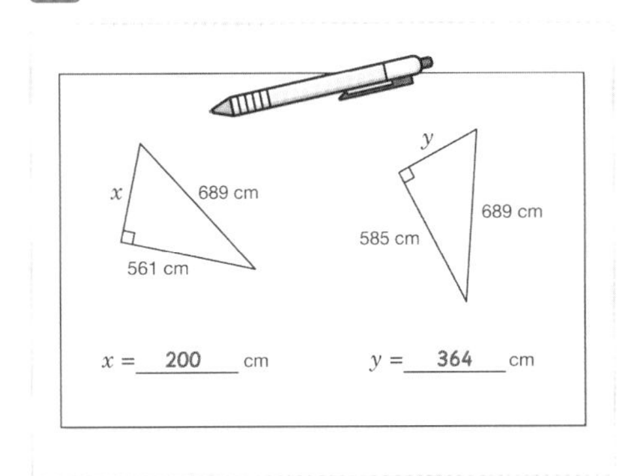

24. Fazel says that in a right-angled triangle, the hypotenuse is always the longest side. Is Fazel correct? Explain your answer.

- Open question, no set answer

25. A shop sells pencils in four different lengths. What is the longest pencil that will fit inside the pencil case?

1/4

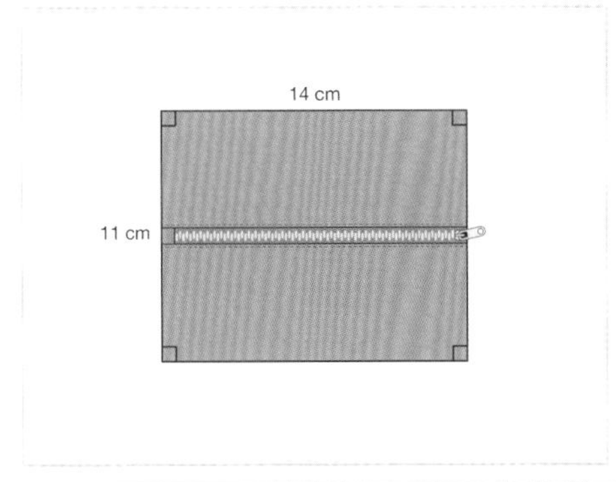

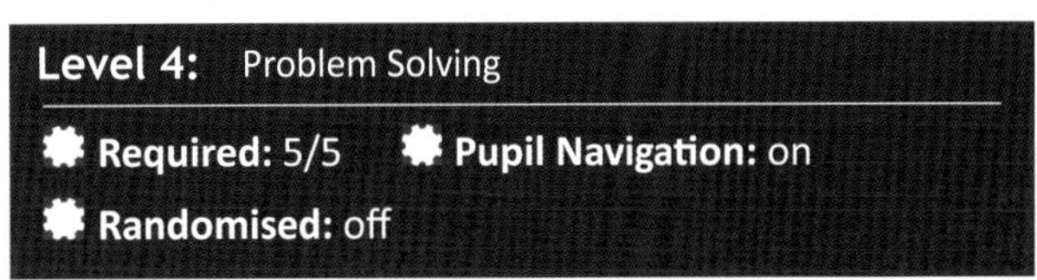

26. The diagram shows the design for a zip wire from *A* to *B*. What is the length of the zip wire?
Give your answer to one decimal place and include the units m (metres) in your answer.

- 234 metres
- 234 m

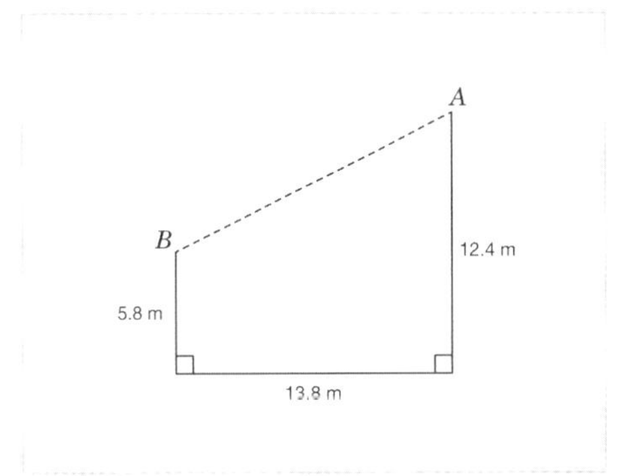

Level 4 *continued*

27. The diagonal of a square field is 320 metres. What is the perimeter of the field?
Give your answer to 1 decimal place and include the units m (metres) in your answer.

■ 905.1 metres ■ 905.1 m ■ 226.3 m

■ 226.3 metres ■ 905.1 ■ 226.3

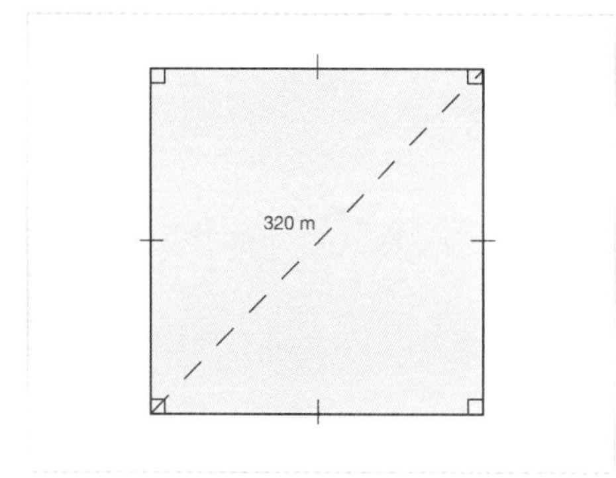

28. What is the length of *AD*?
Give your answer to 1 decimal place and include the units cm (centimetres) in your answer.

■ 8.2 centimetres ■ 8.2 cm ■ 8.2

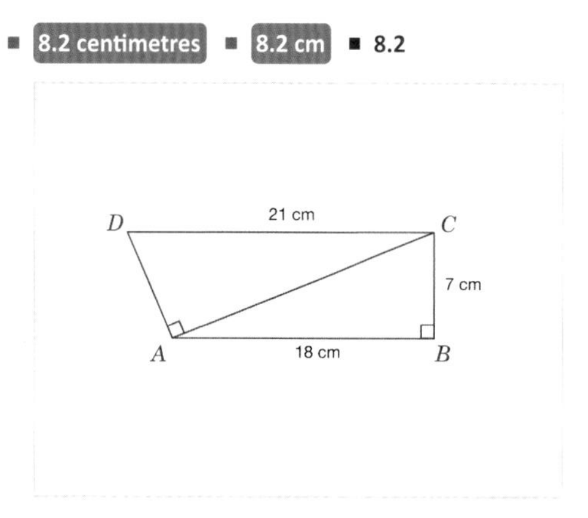

29. The box shown has a height of 3 centimetres (cm), a width of 4 cm and a length of 12 cm. Calculate the length of *AG*.
Include the units cm (centimetres) in your answer.

■ 13 centimetres ■ 13 cm ■ 13

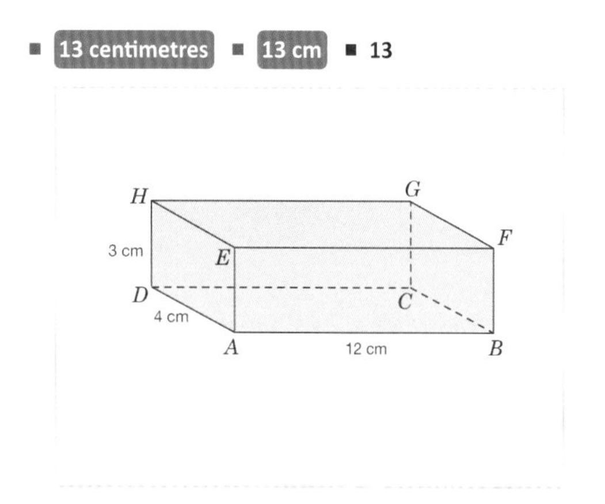

30. A gardener designs the decking for a patio using square paving stones with a side length of 1.5 metres. Some of the paving stones are cut in half diagonally to make triangles. What is the perimeter of the patio?
Give your answer to 1 decimal place and include the units m (metres) in your answer.

■ 14.5 metres ■ 14.5 m ■ 14.4 metres ■ 14.4
■ 24.0 metres ■ 24.0 ■ 2.1 m ■ 24.0 m ■ 14.4 m
■ 2.1 metres ■ 2.1 ■ 14.5

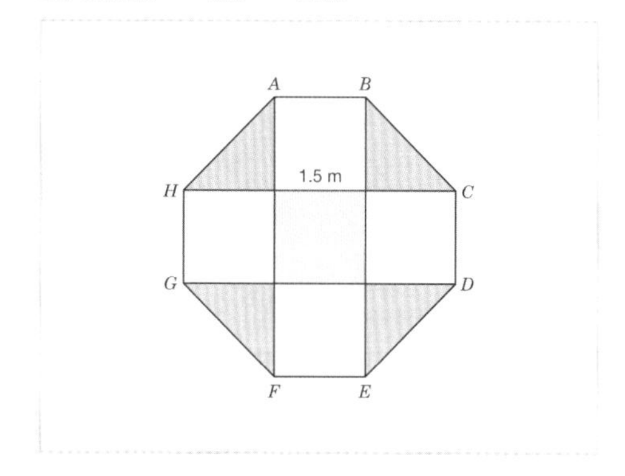

Transformations Topic Review

Objective: I can answer questions about transformations of shapes.

Quick Search Ref: 10757

Level 1: Understanding

Required: 10/10 **Pupil Navigation:** on **Randomised:** off

1. The shape is rotated 180° about the origin (0, 0). What are the coordinates that point L is rotated to?
Give your answer in the form (x, y).

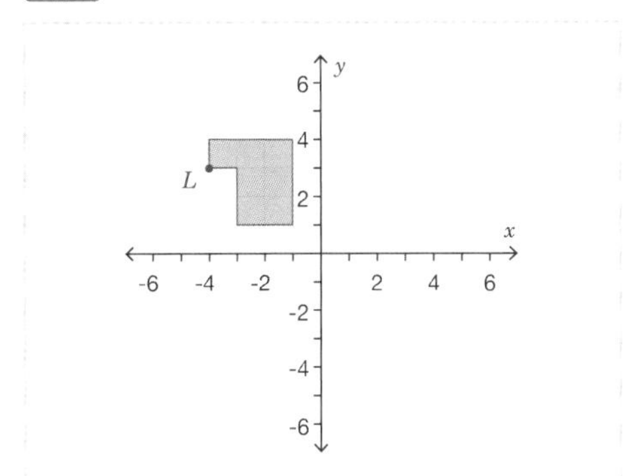

2. The shape is reflected in the line $y = x$. What are the coordinates of the point that G is mapped to?
Give your answer in the form (x, y).

■ (7, 1) ■ (7 1) ■ 7, 1 ■ (1, 7)

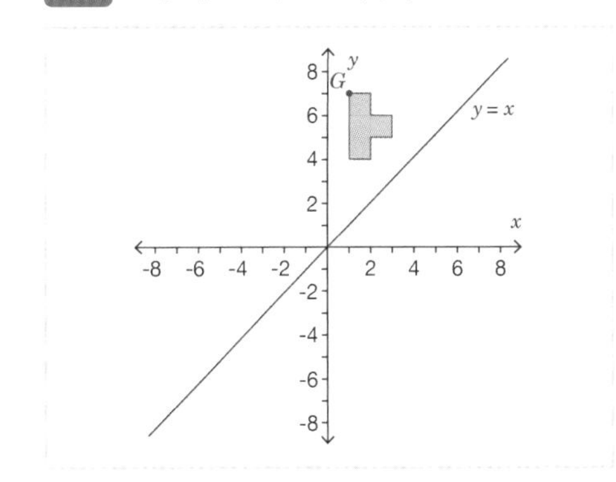

3. The shape is enlarged by a scale factor of 2 using centre of enlargement (-2, 1). What are the coordinates of the image of point Z?
Give your answer in the form (x, y).

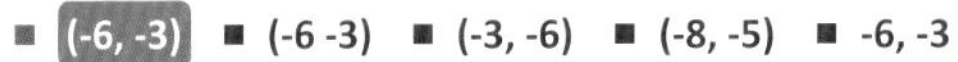

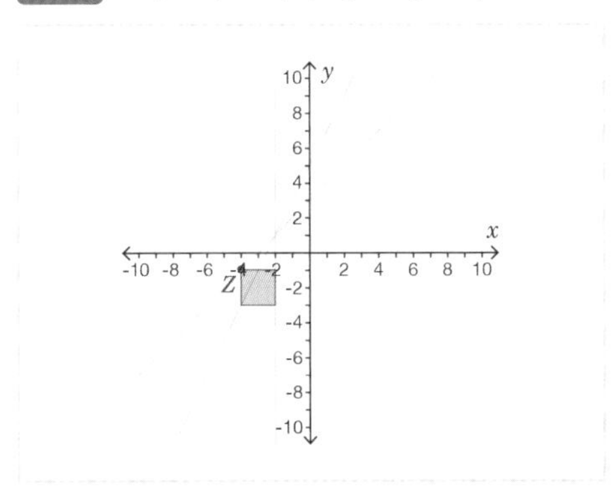

4. Shape A is translated by the column vector shown in the diagram. What are the coordinates of the point that P is translated to?
Give your answer in the form (x, y).

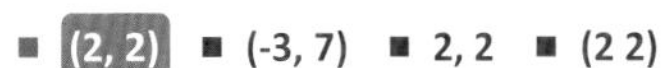

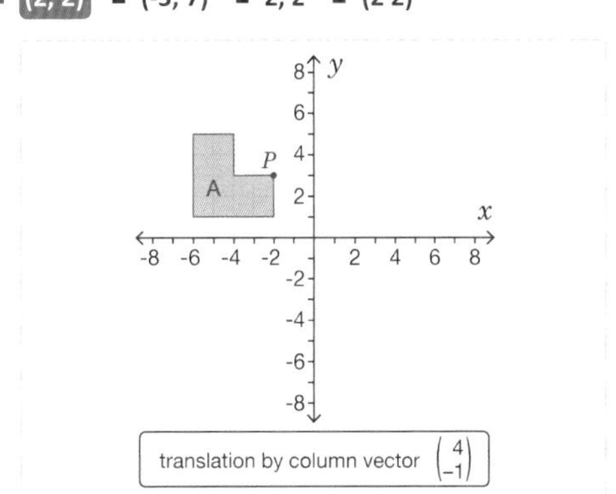

5. The shape is rotated 90° clockwise about the origin (0, 0). What are the coordinates that point C is rotated to?
Give your answer in the form (x, y).

■ (-5, -4) ■ -5, -4 ■ (-4, -5) ■ (-5 -4)

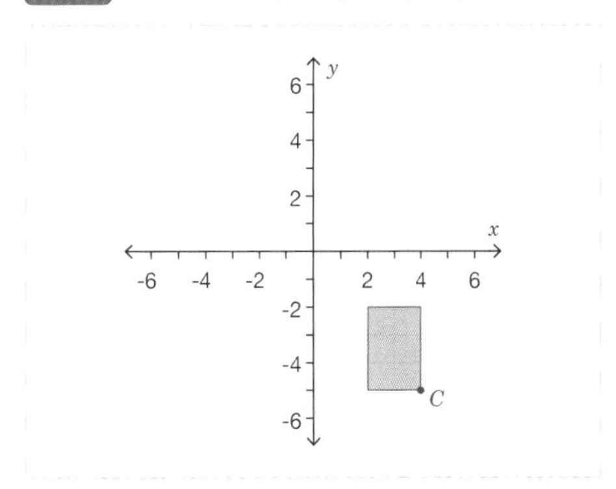

Level 1 continued

6. The shape is reflected in the line $y = 3$. What are the coordinates of the point that E is reflected to?
Give your answer in the form (x, y).

■ **(2, 6)** ■ **(4, 0)**

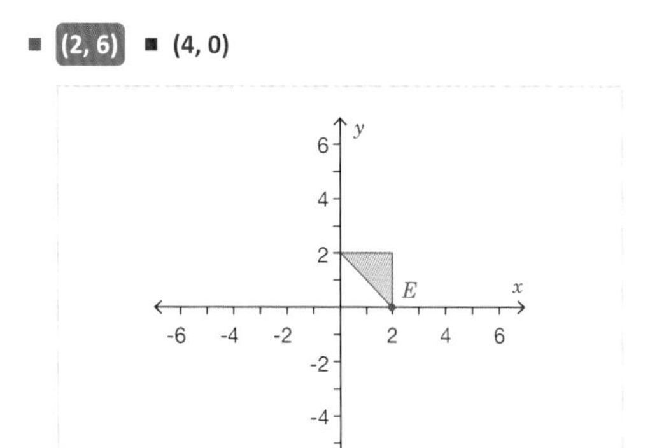

7. The shape is enlarged by a scale factor of 2 using centre of enlargement (4, -6). What are the coordinates of the image of point W?
Give your answer in the form (x, y).

■ **(0, 0)** ■ **(-2, 3)**

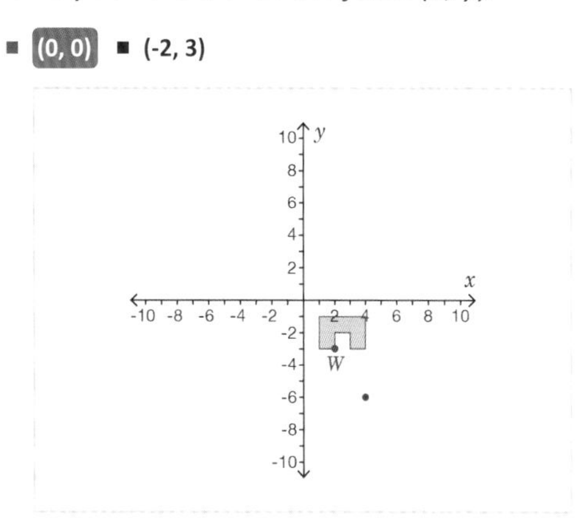

8. Shape S is mapped onto shape T using the column vector shown in the diagram. Which image shows the correct position of T?

1/4 ■ (i) ■ (ii) ■ (iii) ■ **(iv)**

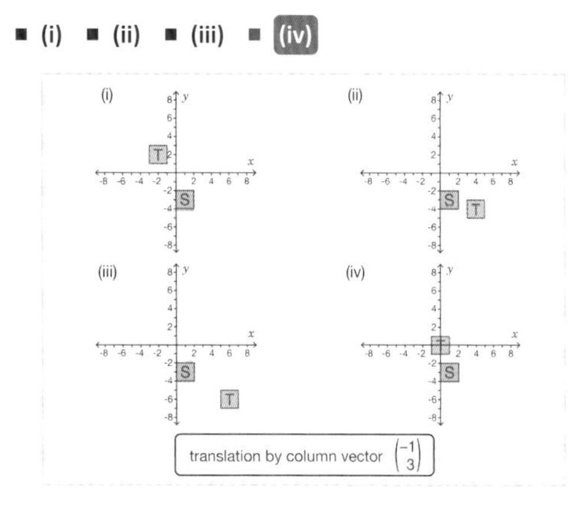

9. Point Q is reflected in the line $y = -4$. What are the coordinates of the point that Q is mapped to?
Give your answer in the form (x, y).

■ **(1, -2)** ■ **(-2, 1)** ■ **1, -2** ■ **(1 -2)**

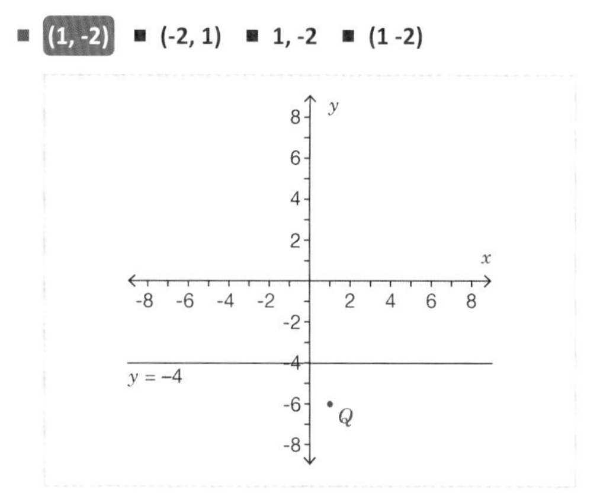

10. The shape is enlarged by a scale factor of 3 using centre of enlargement (5, -5). What are the coordinates of the image of point E?
Give your answer in the form (x, y).

■ **(-1, 7)** ■ **(7, -1)** ■ **-1, 7** ■ **(-1 7)** ■ **(-3, 11)**

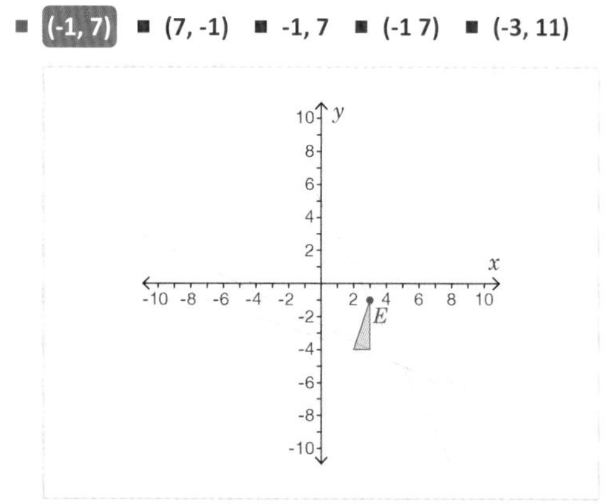

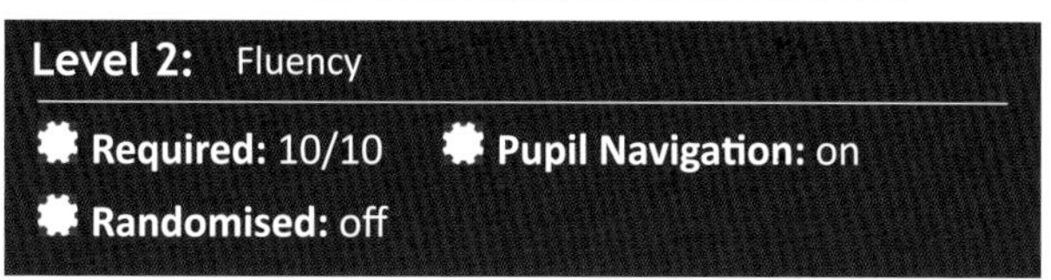

Level 2: Fluency

Required: 10/10 **Pupil Navigation:** on

Randomised: off

11. What number is missing from the bottom of the column vector that maps shape U onto shape V?

■ **-4** ■ **5** ■ **4**

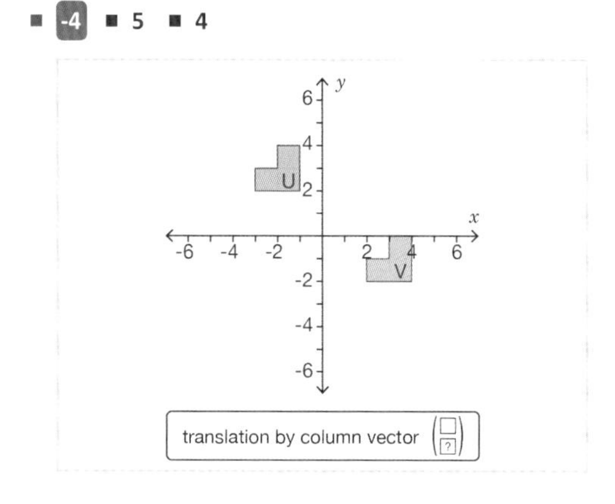

Level 2 *continued*

12. Shape M is reflected onto shape N. What is the equation of the line of reflection?

a b c

- x = -1
- y = -1

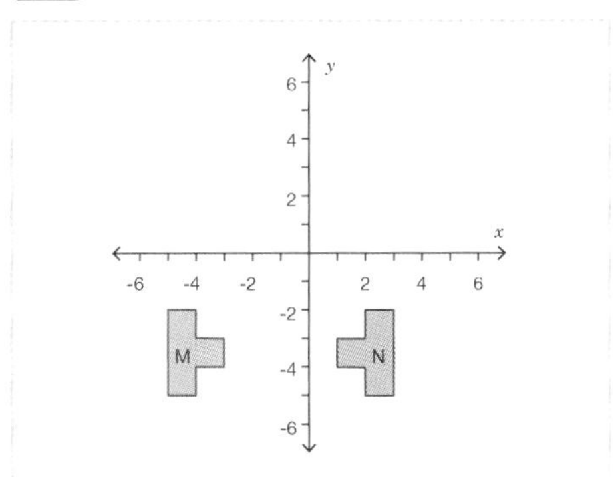

13. Select the **two statements** that describe the rotation that maps shape S onto shape T.

2/6

- centre of rotation (0, 1)
- rotation of 90° clockwise
- centre of rotation (1, 1)
- rotation of 180°
- rotation of 90° anti-clockwise
- centre of rotation (0, 0)

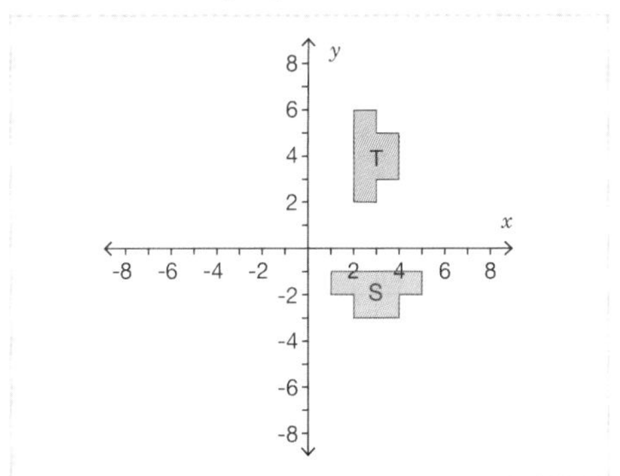

14. Select the transformation that would map shape S onto shape T.

1/4

- a translation by a column vector with -5 as the top number and 0 as the bottom number
- a reflection through the line x = -1
- a translation by a column vector with 0 as the top number and -2 as the bottom number
- a rotation of 180° about the point (-2, -1)

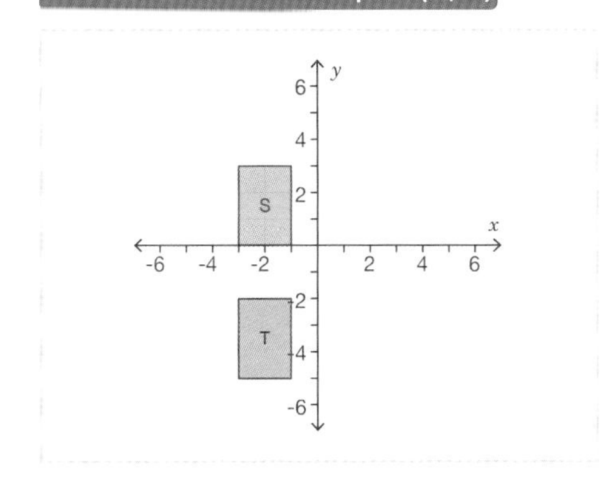

15. Give the coordinates of the centre of enlargement that maps triangle T onto triangle U.
Give your answer in the form (x, y).

a b c

- (-2, 3)
- (-5, 4)
- (-5 4)
- (-4, 3)
- (-3, 1)
- -5, 4
- (4, -5)

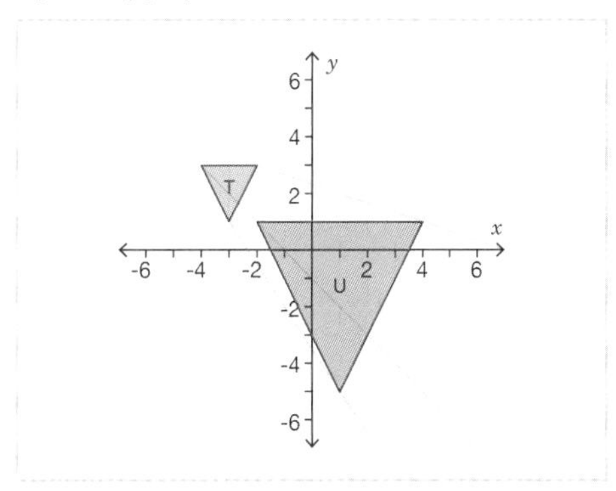

16. Which diagram shows the reflection of shape J onto shape K in the line $y = x$?

1/4

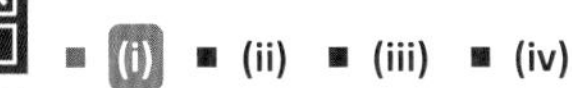

- (i)
- (ii)
- (iii)
- (iv)

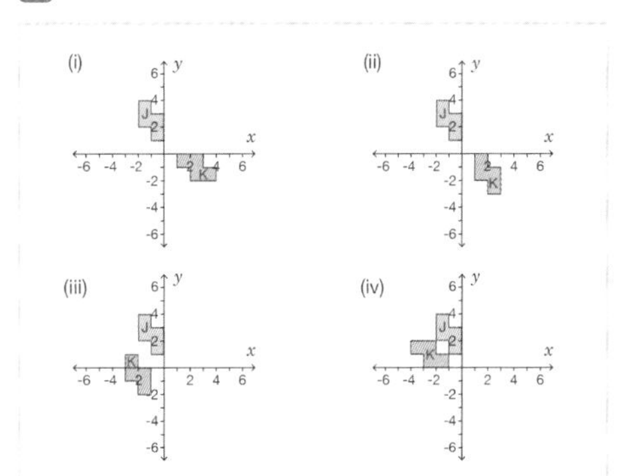

17. Describe fully the single transformation that maps shape G onto shape H.

a b c

- Open question, no set answer

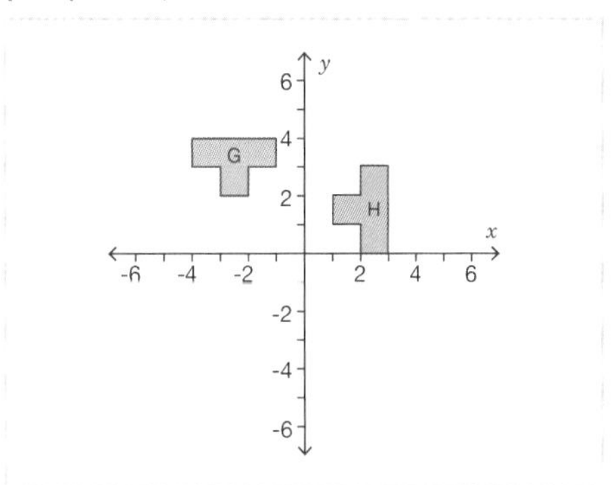

Level 2 *continued*

18. Shape C is rotated to map onto shape D. What are the coordinates of the centre of rotation?

a b c

Give your answer in the form (x, y).

- (0, -1)

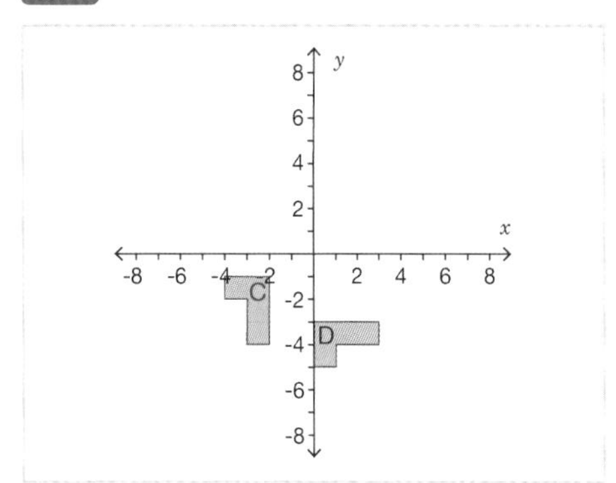

19. Shape P is reflected onto shape Q. What is the equation of the line of reflection?

a b c

- y = -2
- x = -2

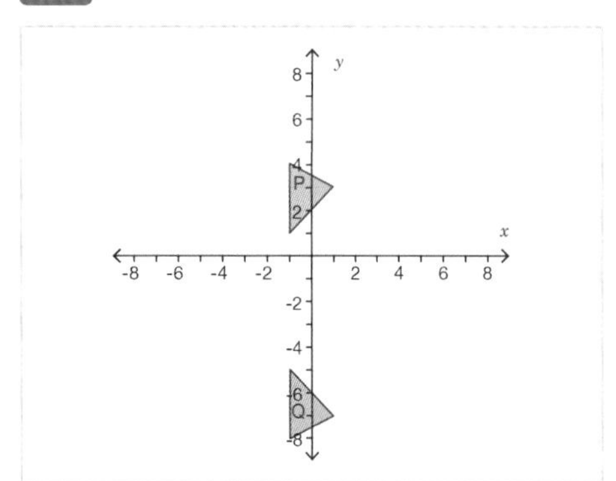

20. Each diagram shows a different transformation that maps shape T onto shape U. Sort the descriptions into the order that the transformations are shown, from (i) to (iv).

↑↓

- a translation by a column vector with -5 as the top number and 6 as the bottom number
- a reflection in the line y = x
- a rotation of 180° about the point (0, 0)
- a reflection in the line y = 1

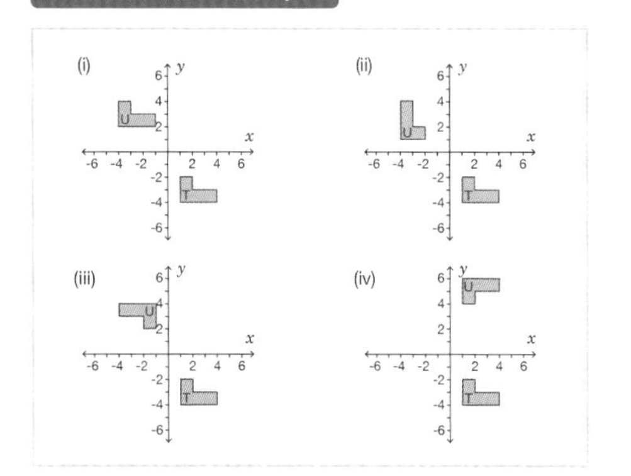

Level 3: Reasoning

Required: 5/5 **Pupil Navigation:** on

Randomised: off

21. The column vector that maps shape A onto shape B is shown in the diagram. Select the column vector that maps shape B onto shape A.

1/4

- (i)
- (ii)
- (iii)
- (iv)

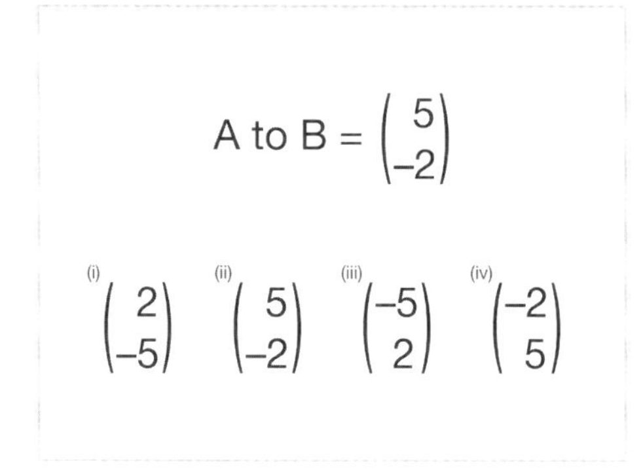

22. What is the scale factor of enlargement that maps shape D onto shape E?

a b c

Give your answer as a decimal.

- 0.5
- 1/2
- 2

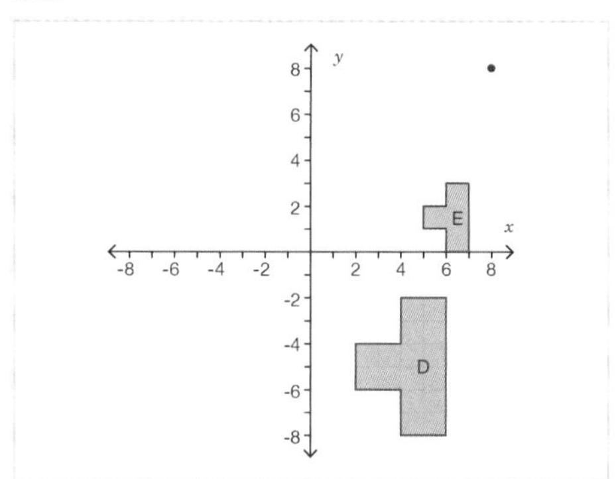

23. Bailey says the answer to the question in the image is 90° clockwise about the point (0, 1). Is Bailey correct? Explain your answer.

a b c

- Open question, no set answer

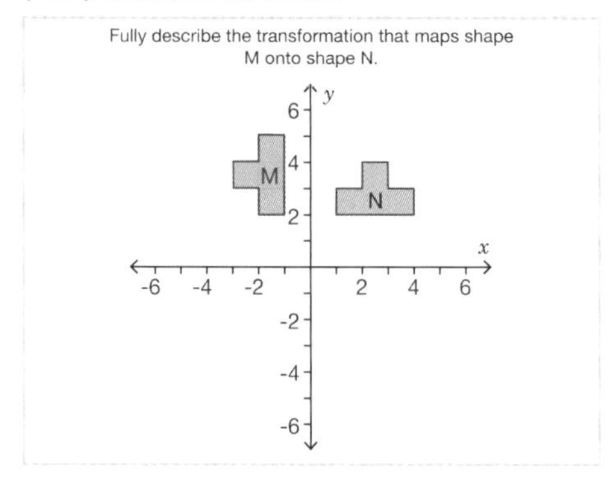

Level 3 *continued*

24. The point (0, 4) is rotated by 90° anti-clockwise about the origin. What are the coordinates of the new point?
Give your answer in the form (x, y).

- (-4, 0)
- (4, 0)

25. Which of the four images does **not** show a reflection of the shape on the left?

1/4

- (i)
- (ii)
- (iii)
- (iv)

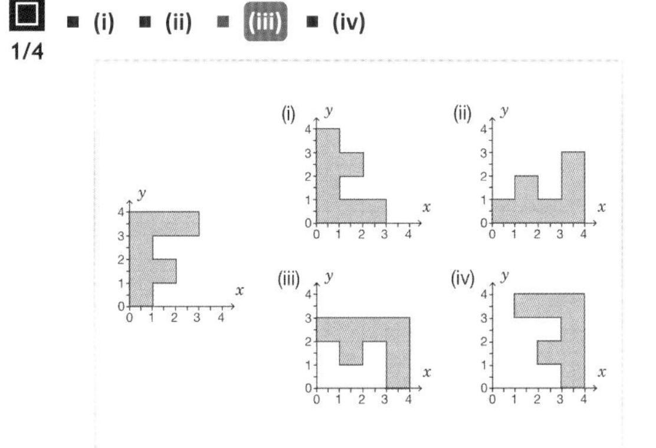

Level 4: Problem Solving

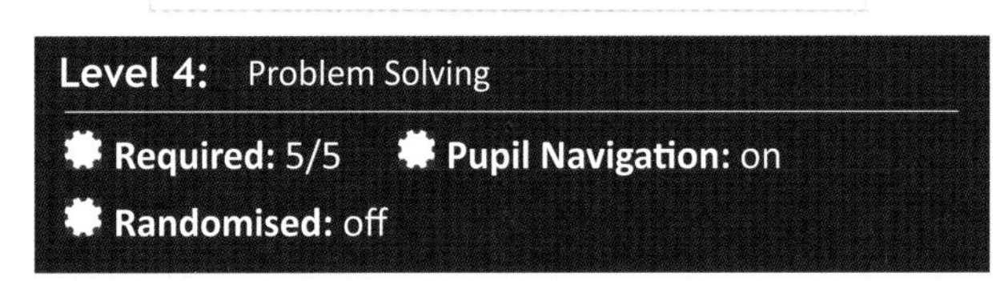

26. The column vectors that map shape R onto shapes S and T are shown in the diagram. What number is missing from the bottom of the column vector that maps shape S onto shape T?

- -5
- 5

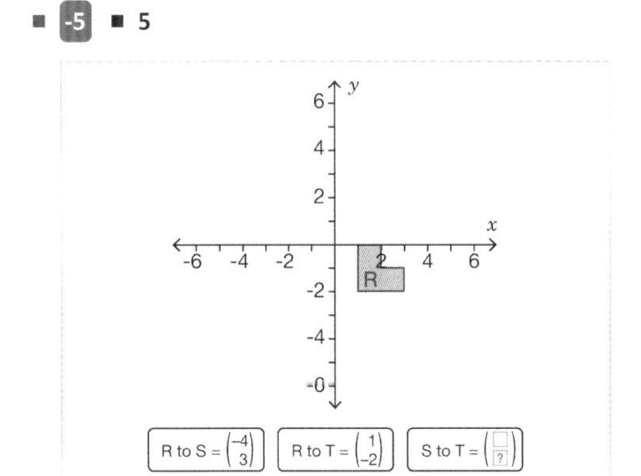

27. Rectangle A can be rotated clockwise 90° about the origin (0, 0) to map onto rectangle B. Rectangle A can also be rotated onto rectangle B using an anti-clockwise rotation of 90°. What are the coordinates of the centre of rotation for the anti-clockwise rotation?
Give your answer in the form (x, y).

- (7, 1)

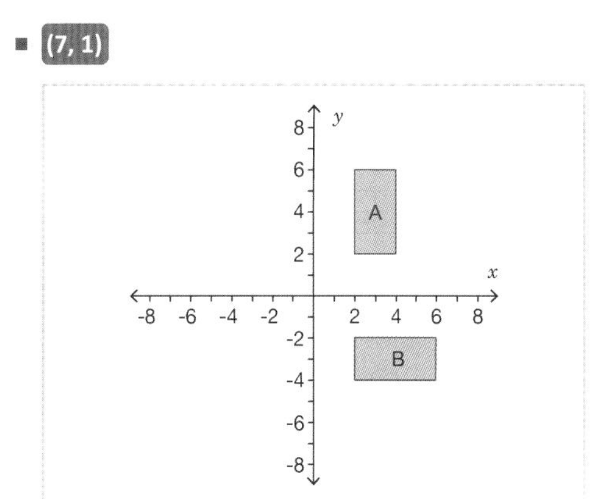

28. Katie is completing an enlargement task but has lost part of the question. She knows that triangle *ABC* is enlarged by a scale factor of 3. What are the coordinates of the centre of enlargement?
Give your answer in the form (x, y).

- (4, 9)

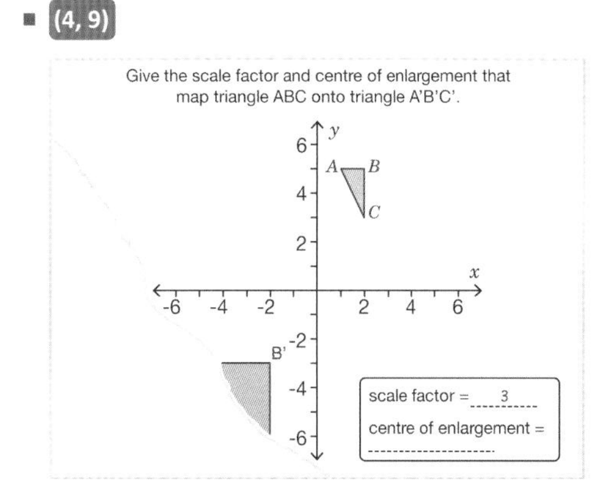

Level 4 *continued*

29. Shape R is translated onto shape S by a column vector with 0 as the top number and 8 as the bottom number. Shape S is then reflected in the line $y = -1$ onto shape T. Give the equation of the line of reflection that maps shape R onto shape T.

■ 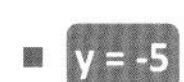

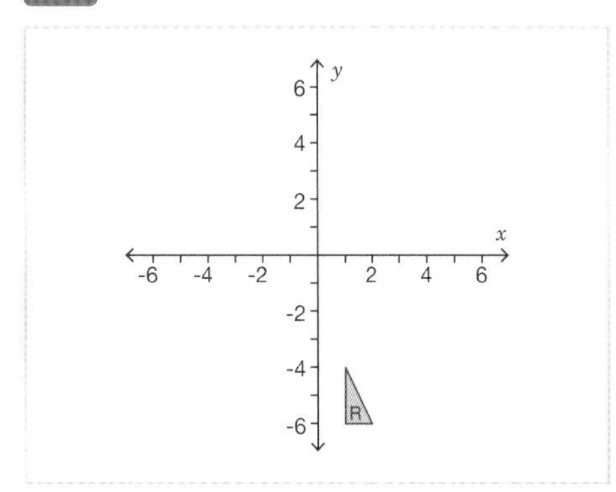

30. Shape S is reflected onto shape T. What is the equation of the line of reflection?
Give your answer in the form $y = mx + c$.

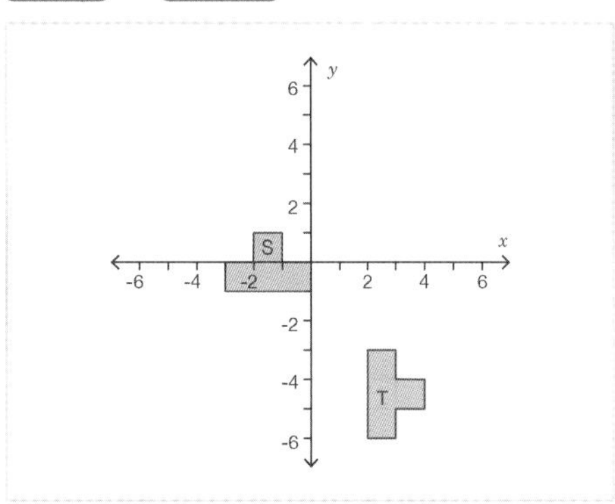

Trigonometry Topic Review

Objective: I can use trigonometry to answer questions.

Quick Search Ref: 10764

Level 1: Understanding

Required: 10/10 **Pupil Navigation:** on **Randomised:** off

1. Select the side opposite angle x in the triangle.

1/3

- AB
- BC
- AC

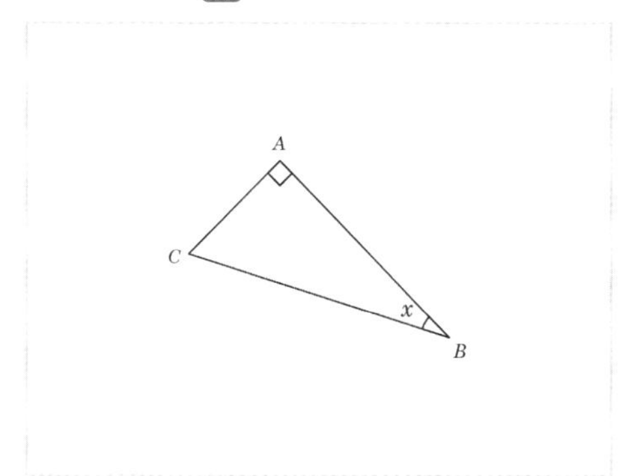

2. Select the calculation which gives the cosine of an angle.

1/3

- i) opposite ÷ adjacent
- ii) opposite ÷ hypotenuse
- iii) adjacent ÷ hypotenuse

(i) $\frac{\text{opposite}}{\text{adjacent}}$

(ii) $\frac{\text{opposite}}{\text{hypotenuse}}$

(iii) $\frac{\text{adjacent}}{\text{hypotenuse}}$

3. Give sin x as a fraction.

- 15/17
- 17/15
- 8/17

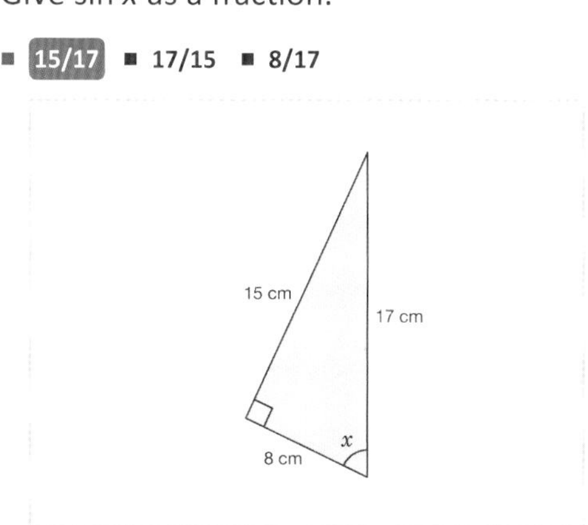

4. Give tan c as a fraction in its simplest form.

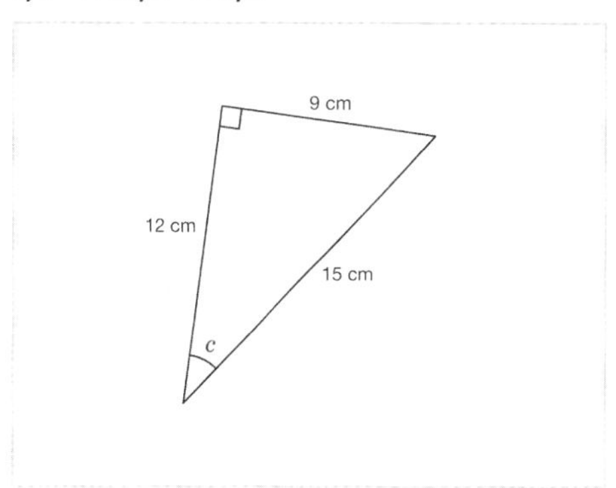

5. If cos x = 2/3, find the length of side LM. *Include the units cm (centimetres) in your answer.*

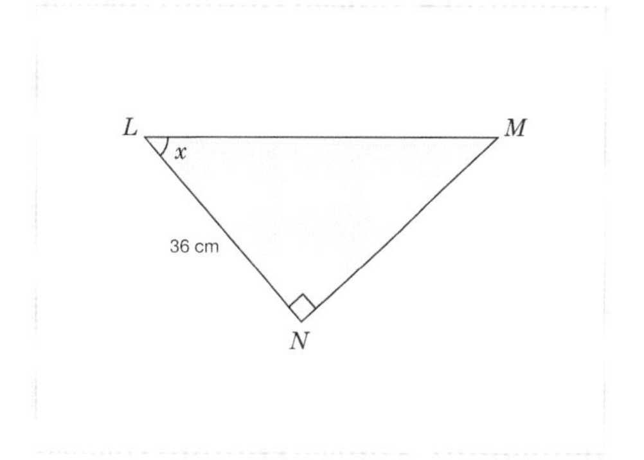

6. Use your calculator to find the value of sin 20°. *Give your answer to 2 decimal places.*

- 0.34

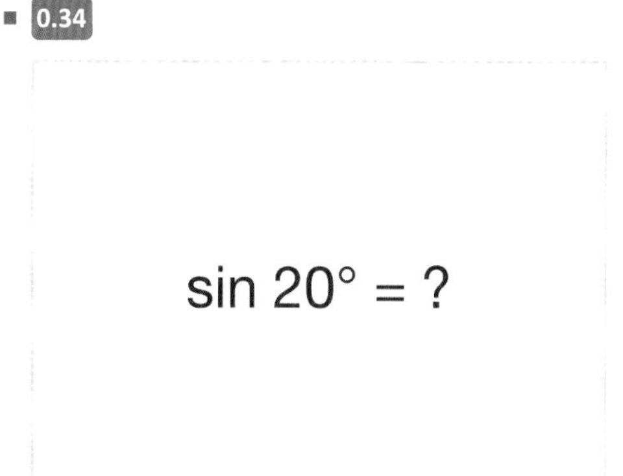

Level 1 *continued*

7. Use tan x = O/A to calculate the value of n.
Give your answer to 1 decimal place and include the units m (metres) in your answer.

- 11.9 m ■ 11.9 metres ■ 11.7 ■ 11.7 metres
- 11.9 ■ 11.7 m

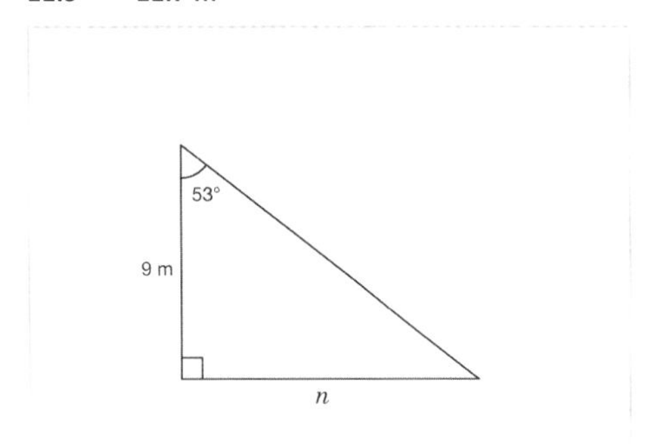

8. Use your calculator to find the value of cos 45°.
Give your answer to 2 decimal places.

- 0.71

cos 45° = ?

9. Give sin z as a decimal.

- 0.96 ■ 0.36 ■ 24/25

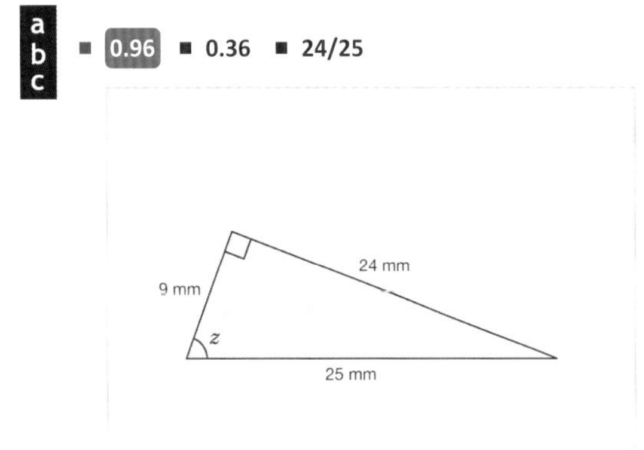

10. If tan x = 3/4, find the value of r.
Include the units cm (centimetres) in your answer.

- 64 centimetres ■ 64 cm ■ 64 ■ 36 cm ■ 36
- 36 centimetres

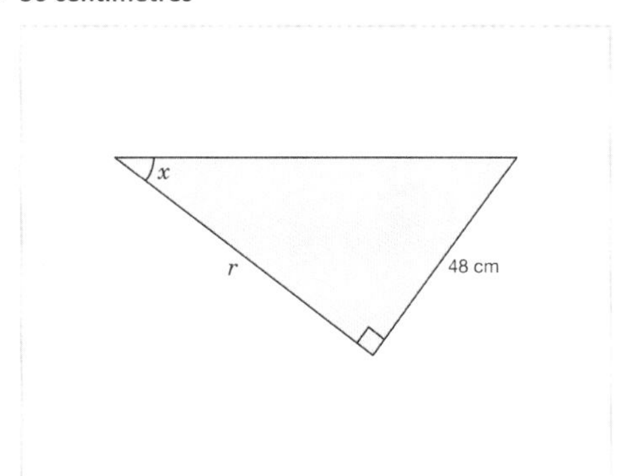

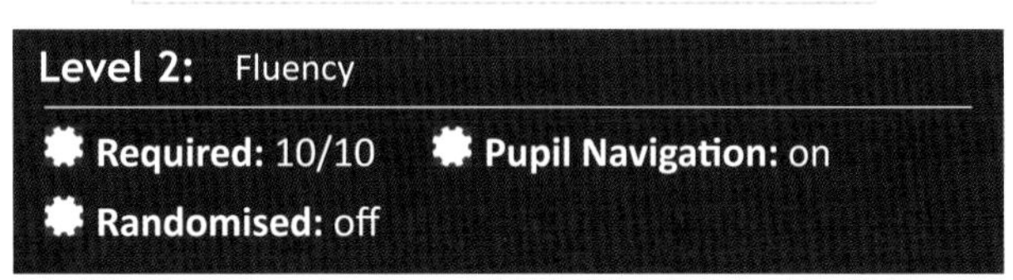

Level 2: Fluency

Required: 10/10 **Pupil Navigation:** on
Randomised: off

11. Which trigonometric ratio would you use to calculate the length of side YZ?

1/3

- sin ■ cos ■ tan

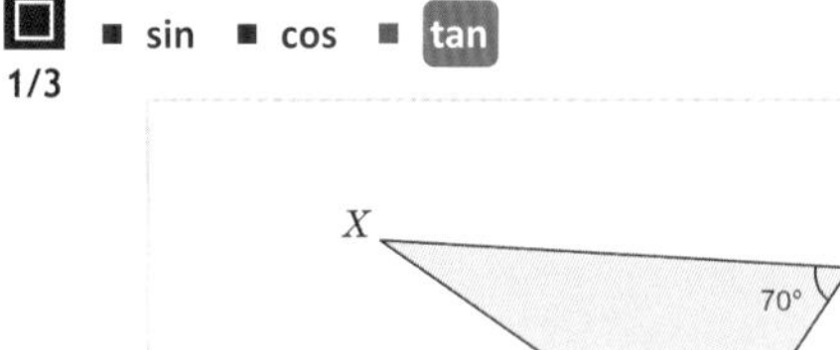
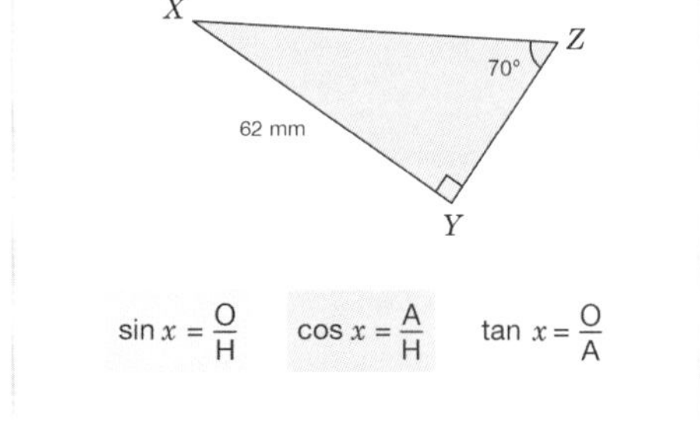

12. Calculate the value of s.
Give your answer to 3 decimal places and include the units mm (millimetres) in your answer.

- 6.464 millimetres ■ 6.464 mm ■ 6.464

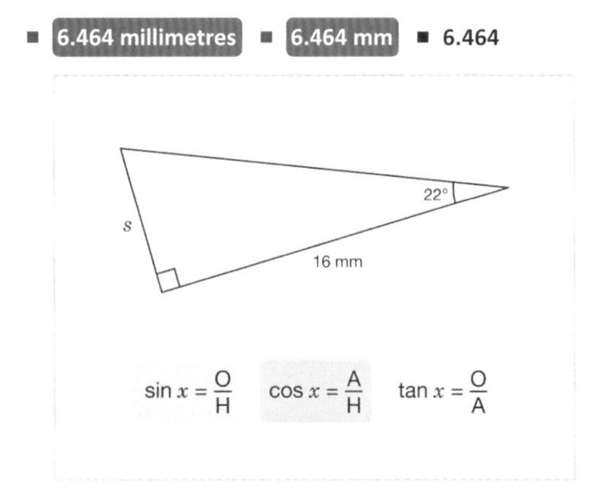

Level 2 continued

13. Calculate the length of side RT.

a b c *Give your answer to 2 decimal places and include the units mm (millimetres) in your answer.*

- 56.63 mm
- 56.63 millimetres
- 56.63
- 57.5
- 57.5 mm
- 57.5 millimetres

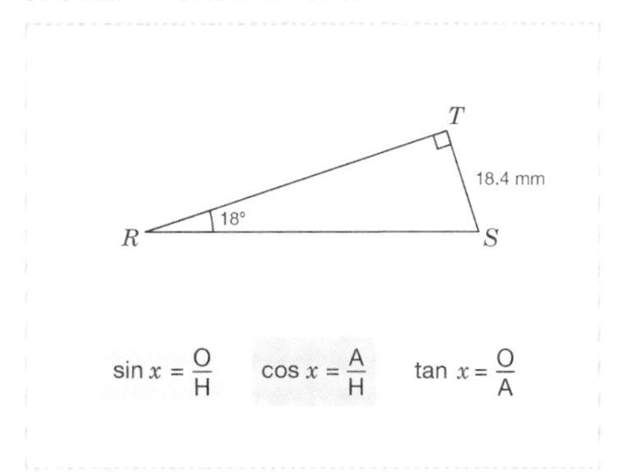

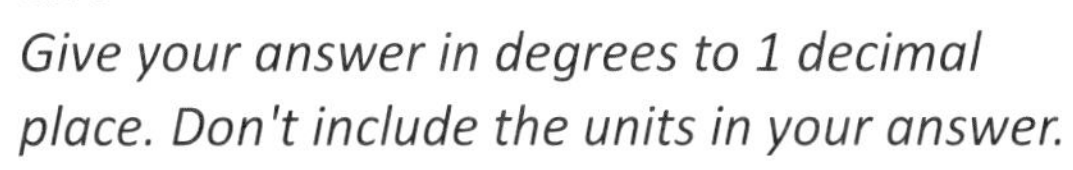

14. Use $\cos^{-1} x$ to calculate x such that $\cos x = 0.7$.

1 2 3 *Give your answer in degrees to 1 decimal place. Don't include the units in your answer.*

- 45.6
- 0.8
- 1

15. Calculate the value of z.

1 2 3 *Give your answer to 1 decimal place. Don't include the units in your answer.*

- 56.3

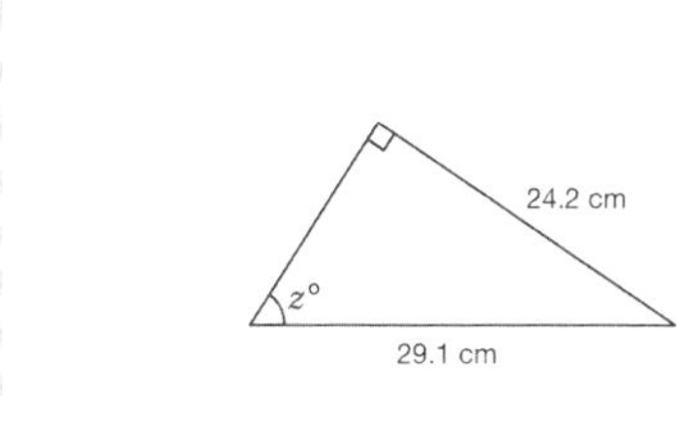

16. Calculate the value of p.

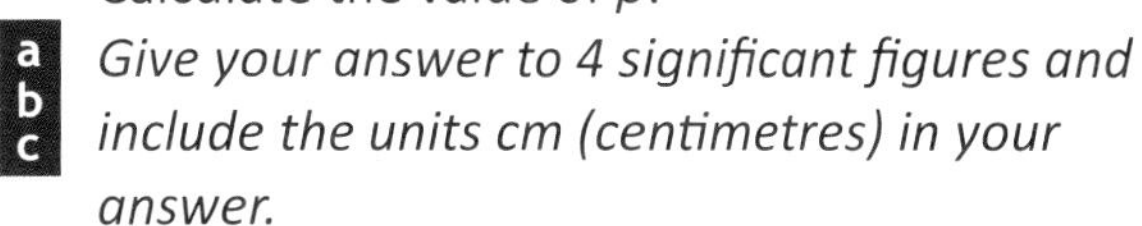

a b c *Give your answer to 4 significant figures and include the units cm (centimetres) in your answer.*

- 6.137 cm
- 6.137 centimetres
- 6.138 centimetres
- 6.138 cm
- 6.138
- 6.137

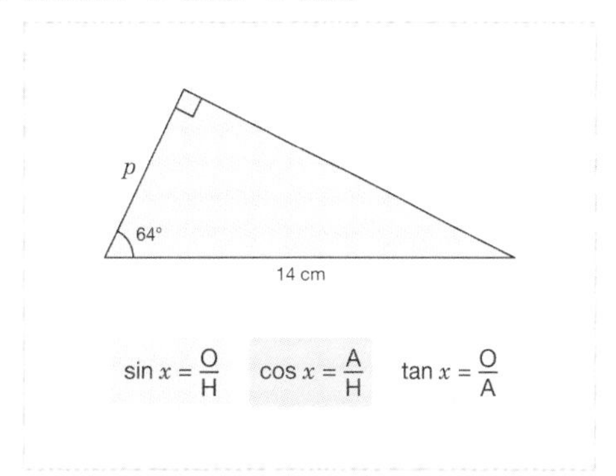

17. Calculate the value of $\angle PQR$.

1 2 3 *Give your answer in degrees to 4 significant figures. Don't include the units in your answer.*

- 71.53
- 71.514

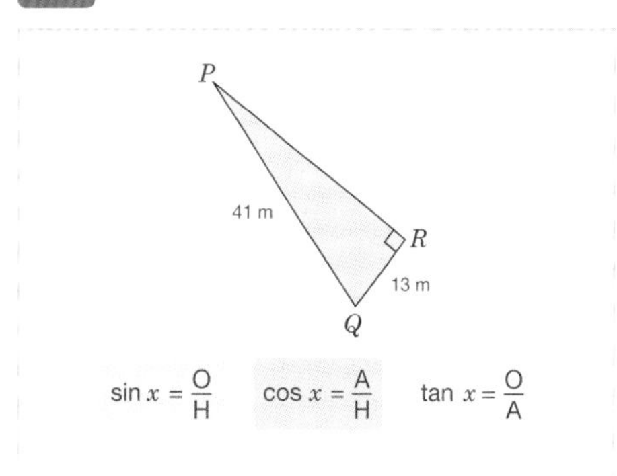

18. Calculate the value of r.

a b c *Give your answer to 1 decimal place and include the units m (metres) in your answer.*

- 85.2 metres
- 91.3
- 85.2 m
- 81.1 metres
- 91.3 metres
- 91.3 m
- 81.1
- 81.1 m
- 85.2

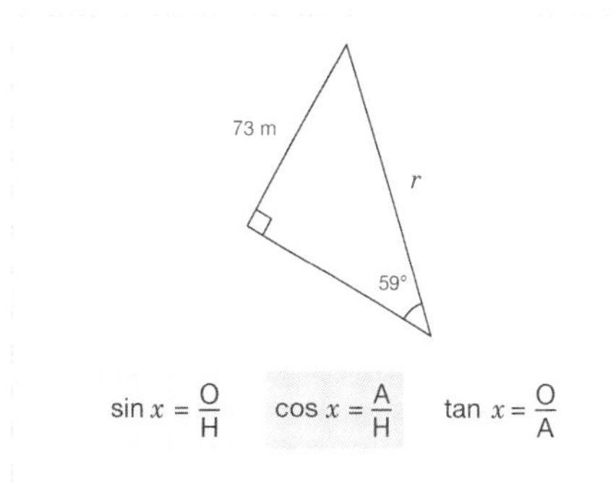

Level 2 *continued*

19. Calculate the value of *n*.

Give your answer to 1 decimal place and include the units cm (centimetres) in your answer.

- 43.2 centimetres
- 45.4

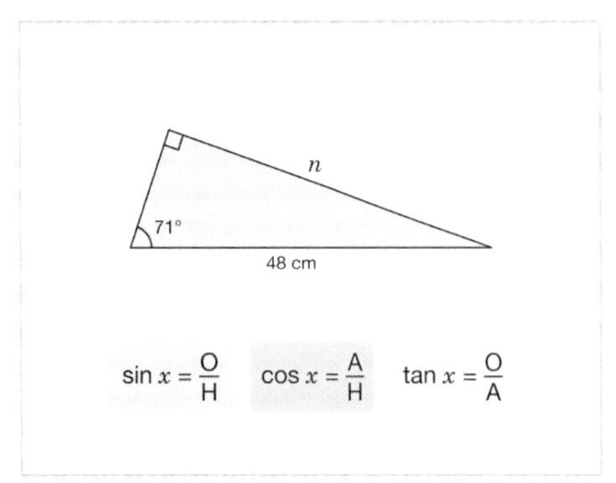

20. Calculate the size of $\angle XYZ$.

Give your answer in degrees to 2 decimal places. Don't include the units in your answer.

- 60.39

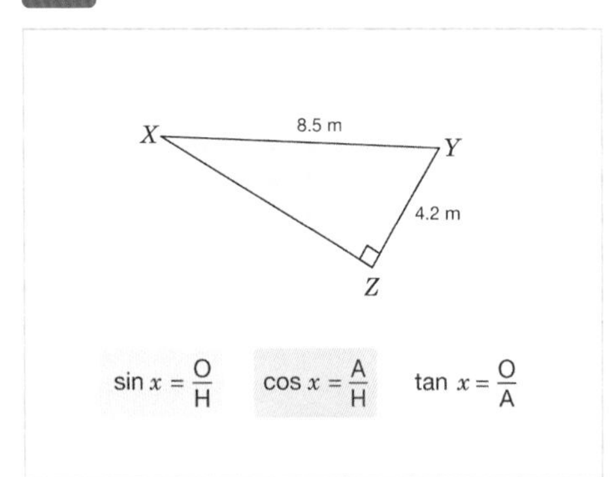

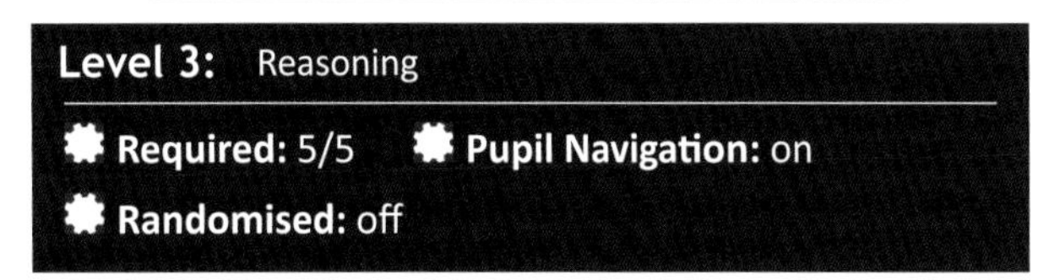

21. Calculate the value of sin 30°.

- 0.5
- 1/2

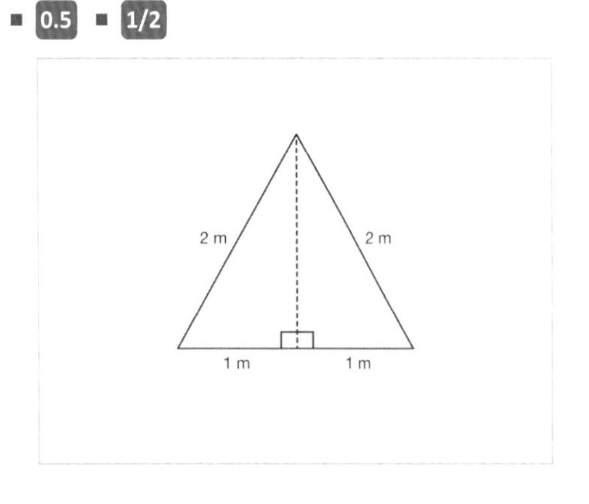

22. Angle *ACB* = *x*. Angle *CAB* = 90° – *x*. Daniel says that cos *x* = sin(90– *x*).

Is Daniel correct? Explain your answer.

- Open question, no set answer

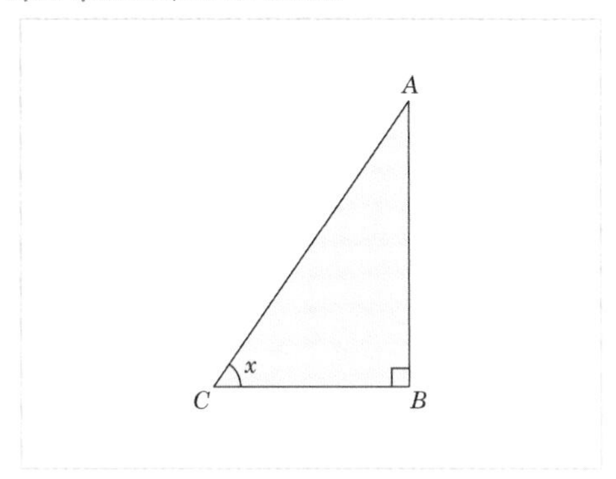

23. Which calculation would you use to find the value of *r*?

1/3

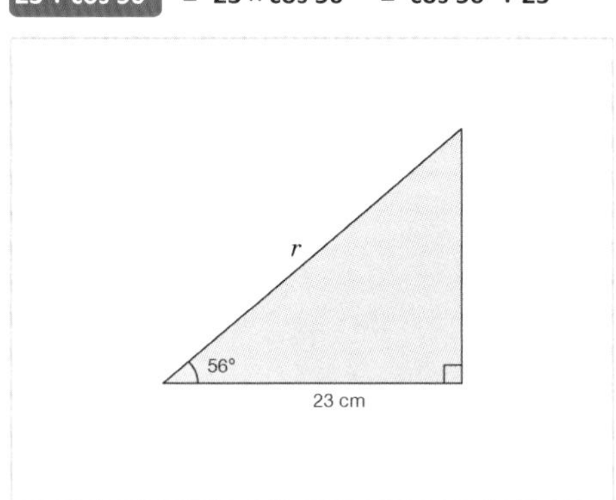

24. George has made a mistake with a trigonometry question. What is the correct size of $\angle XYZ$?

Give your answer in degrees to 1 decimal place. Don't include the units in your answer.

- 32.7

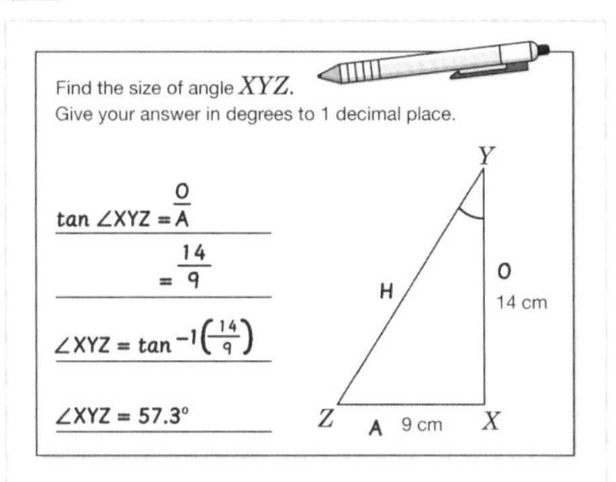

Level 3 continued

25. Owais says that sin *x* divided by cos *x* equals tan x. Is he correct? Explain your answer.

- Open question, no set answer

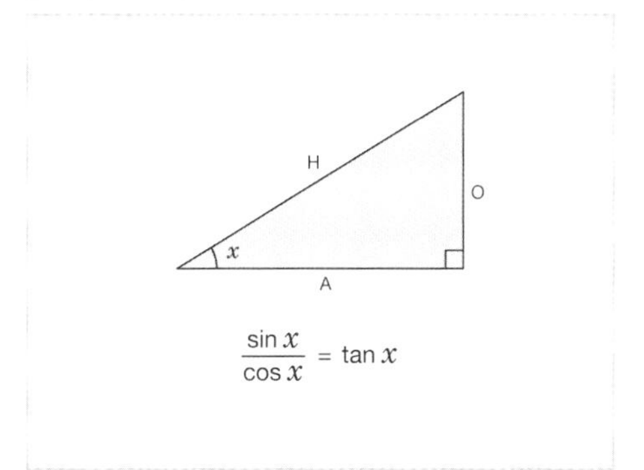

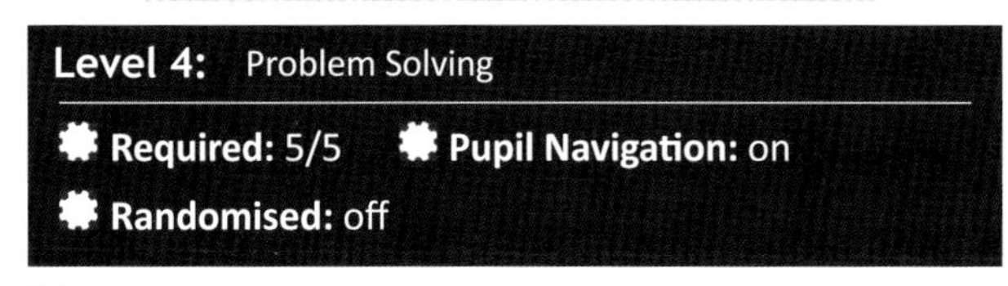

26. The perimeter of the triangle is 120 centimetres and sin x = 8/17. Calculate the value of t.

- 51 centimetres
- 51 cm
- 51

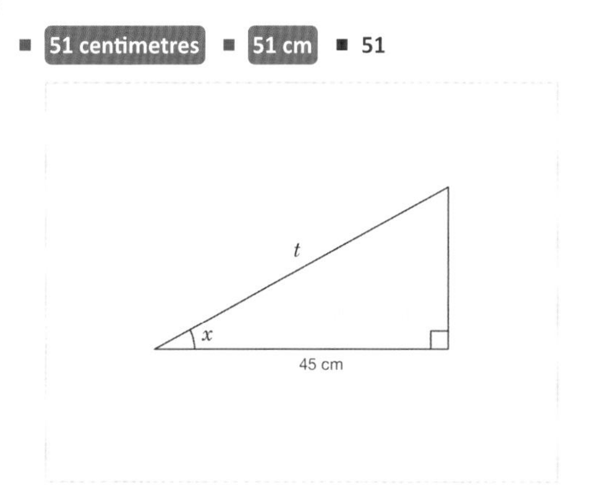

27. Calculate the length of line *CD*.
Give your answer to 2 decimal places and include the units cm (centimetres) in your answer.

- 7.05 cm
- 7.05 centimetres
- 7.05

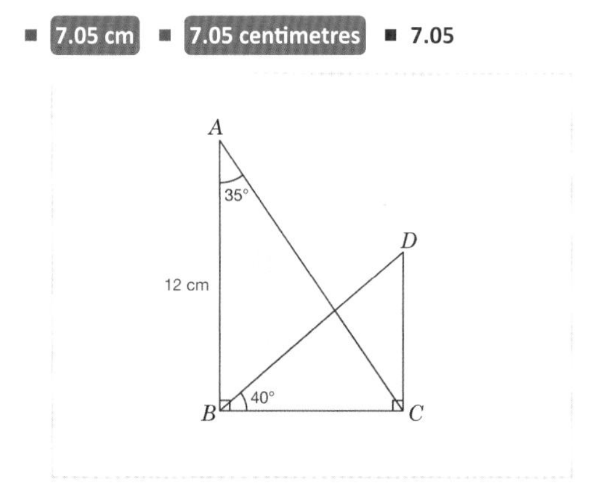

28. Match the trigonometric functions to their graphs and sort the functions into the same order as the graphs.

- sin x
- tan x
- cos x

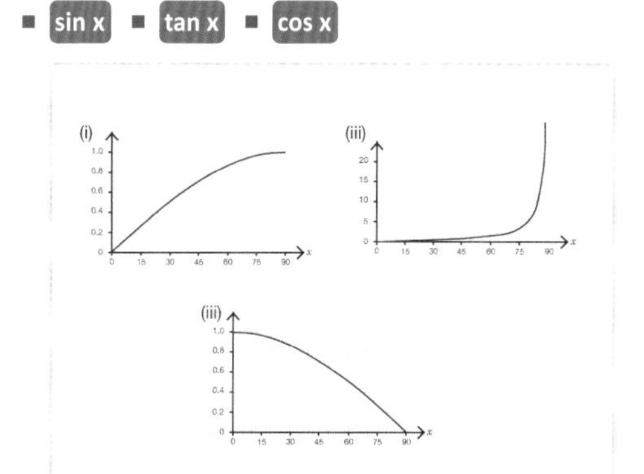

29. A regular hexagon has a perpendicular height of 10 cm when measured from the centre of one side of the shape to the centre of the opposite side. What is the side length of the hexagon?
Give your answer to 2 decimal places and include the units cm (centimetres) in your answer.

- 11.55 cm
- 11.55 centimetres
- 11.55

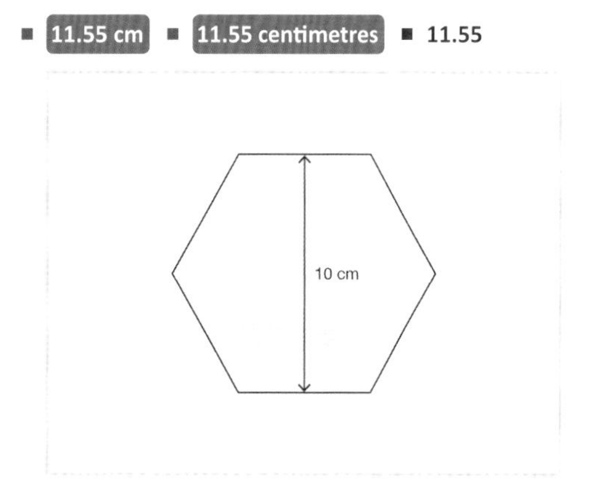

Level 4 *continued*

30. A sailor uses a clinometer to measure the angle between the sea level and the top of a lighthouse on a cliff.
If the top of the lighthouse is 90 metres above sea level and the tangent of the angle measured is 0.3, how far is the ship from the cliff?
Include the units m (metres) in your answer.

- 300 metres
- 300 m
- 300

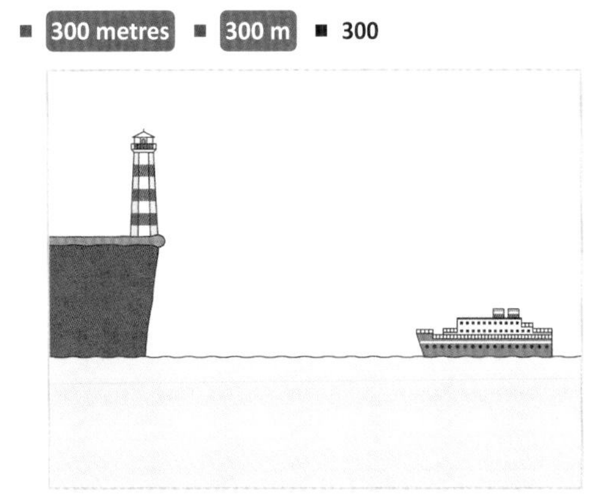

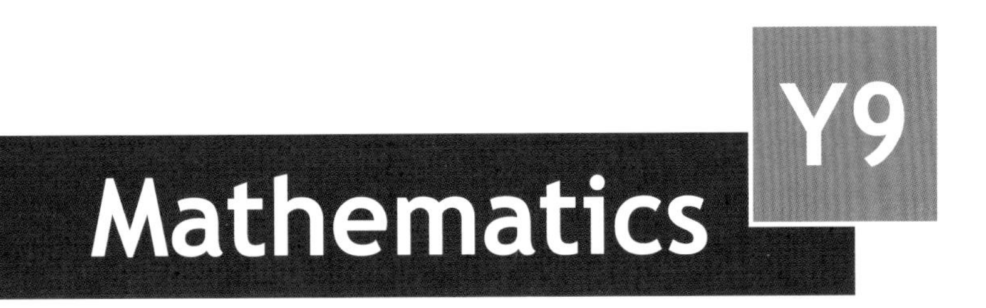

Statistics

Scatter Graphs
Averages and Range
Frequency Polygons

Use and Interpret Scatter Graphs

Objective: I can use and interpret scatter graphs.

Quick Search Ref: 10433

Level 1: Understanding: Plot points and describe relationships and correlation.

Required: 7/10 **Pupil Navigation:** on **Randomised:** off

1. Which set of data would you represent in a scatter graph?

1/3

- monthly rainfall in a city
- **English and maths results of ten students**
- temperature readings throughout the day

2. The scatter graph shows the marks that a maths class received for two test papers. Mary scored 33 marks on the non-calculator paper. What was her score on the calculator paper?

-  **37**
- 26

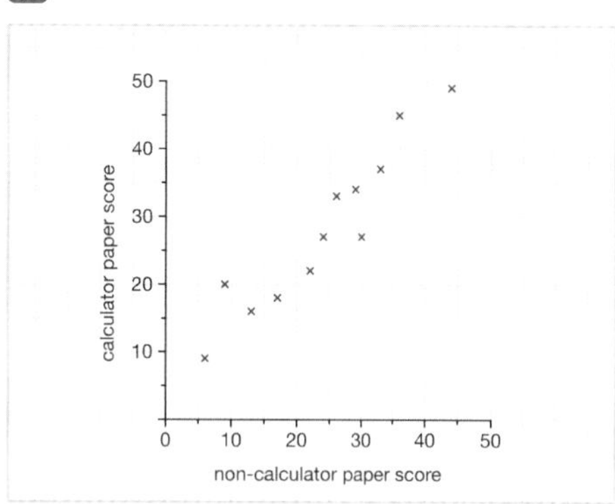

3. The scatter graph shows information about the distance of ten flats from London and their weekly rental cost. How many kilometres from London is the flat that costs £625 per week to rent?
Give your answer in km (kilometres).

- **9 km**
- **9 kilometres**
- 9

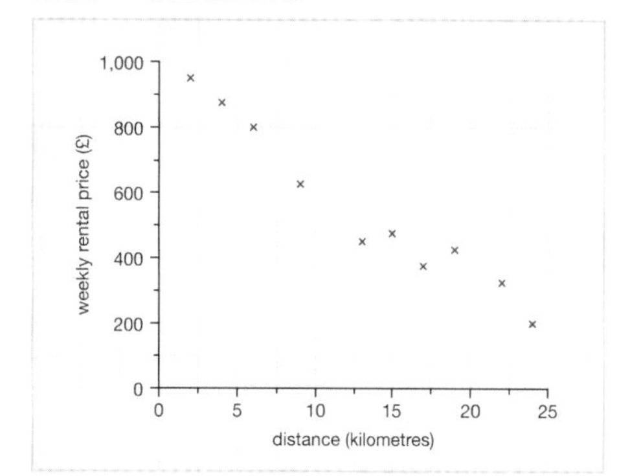

4. The scatter graph shows the marks that a maths class received for two test papers. What type of correlation does the graph show?

1/5

- negative
- The students with lower marks on the non-calculator paper received higher marks on the calculator paper.
- none
- **positive**
- The students with higher marks on the non-calculator paper received higher marks on the calculator paper.

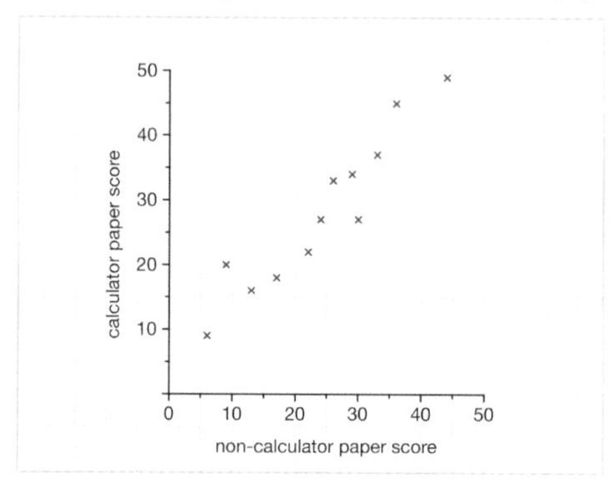

5. The scatter graph shows the marks that a maths class received for two test papers. How many students are in the set?

-  **12**

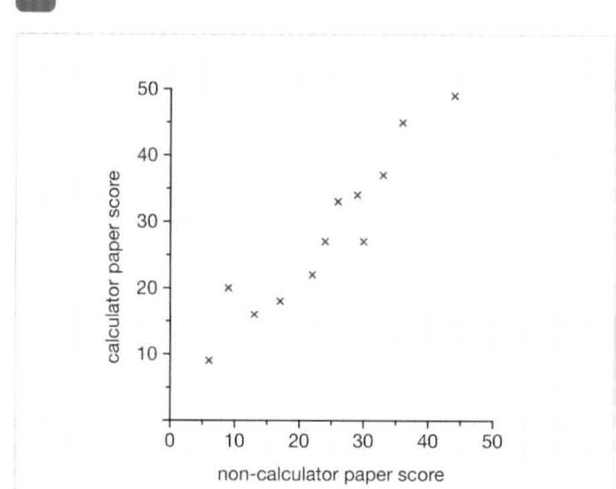

Level 1 *continued*

6. The scatter graph shows information about the distance of ten flats from London and their weekly cost to rent. What is the relationship between the distance of a flat from London and its weekly rental price?

1/5

- **negative**
- **As the distance from London decreases, the weekly rent decreases.**
- **As the distance from London increases, the weekly rent decreases.**
- **positive**
- **As the distance from London increases, the weekly rent increases.**

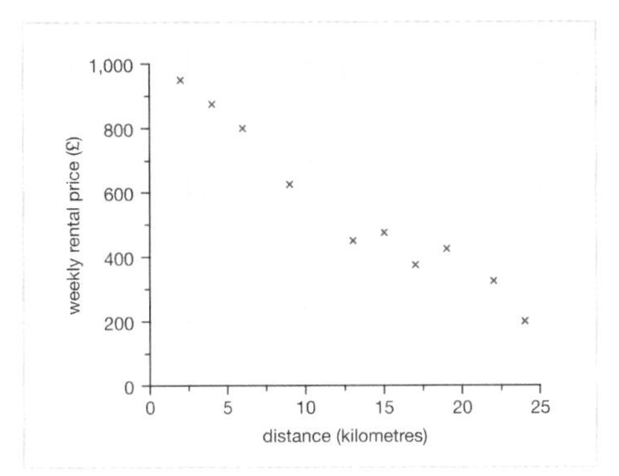

7. The scatter graph shows the marks that a maths class received for two test papers. One student received the same mark in both tests. What was their mark?

- **22**

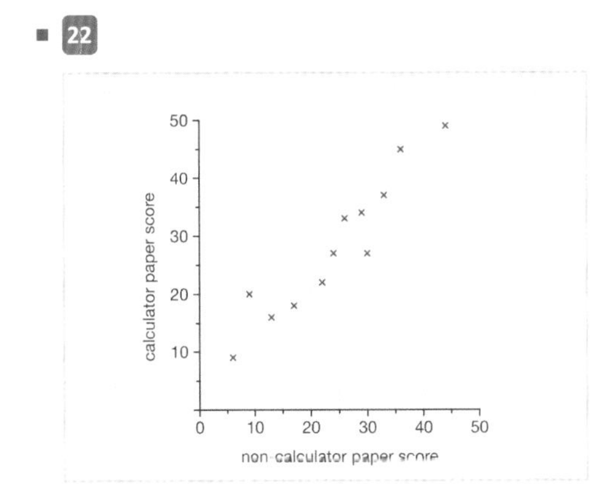

8. The scatter graph shows the marks that a maths class received for two test papers. Joel scored 9 marks on the calculator paper. What was his score on the non-calculator paper?

- **6**

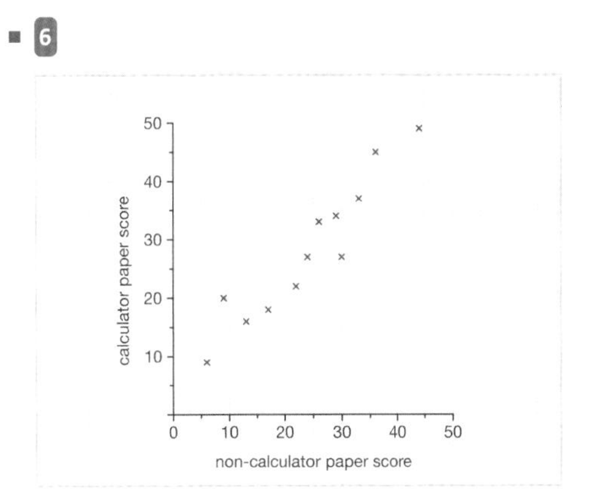

9. The scatter graph shows information about the distance of ten flats from London and their weekly rental cost. What type of correlation does the graph show?

1/3

- **negative**
- **none**
- **positive**

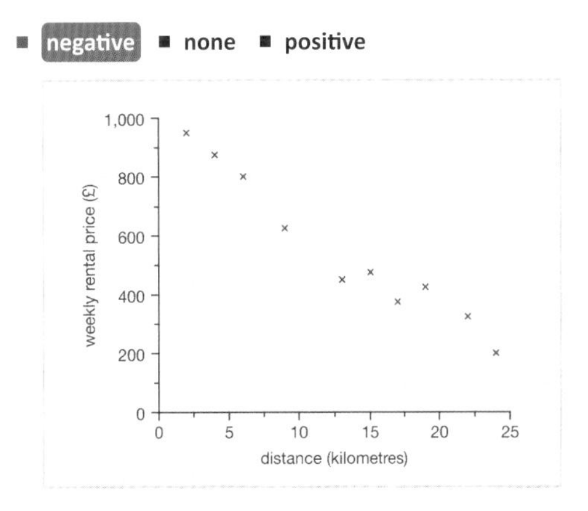

10. The scatter graph shows the marks that a maths class received for two test papers. What is the combined score out of 100 for both papers of the student with the highest marks?

- **93**

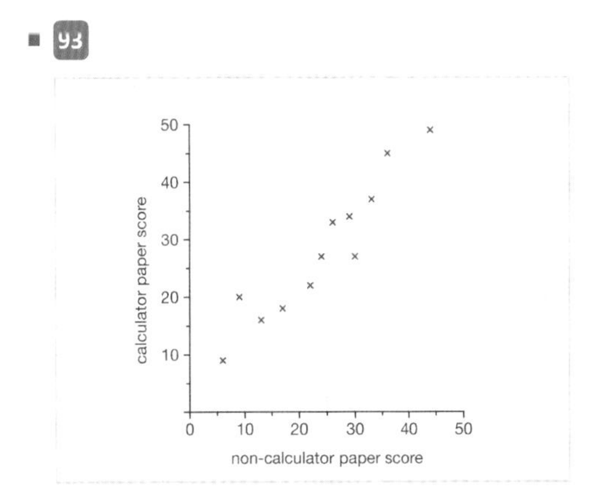

Level 2: Fluency: Using lines of best fit to estimate missing values.

Required: 8/10 **Pupil Navigation:** on
Randomised: off

11. Which is a good line of best fit?

1/4

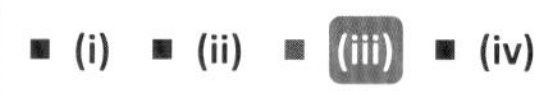

■ (i) ■ (ii) ■ (iii) ■ (iv)

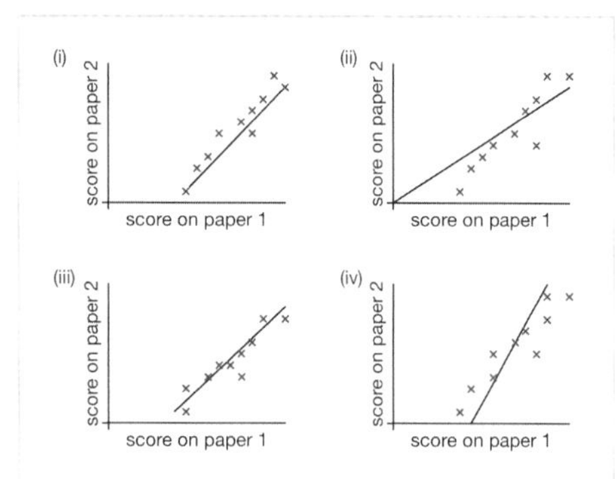

12. An ice cream van records the midday temperature and the number of ice creams sold each day. If tomorrow's midday temperature is forecast to be 15°C, how many ice creams can they expect to sell?

■ 40

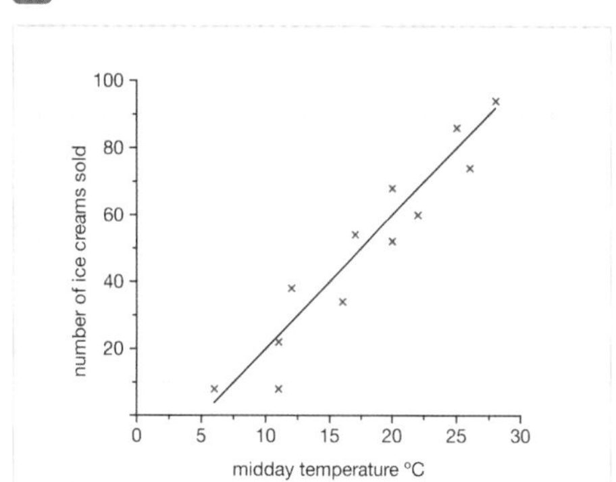

13. Ten athletes enter a 10 km race. The number of weeks they spent training and their race times are shown in the scatter graph. If another athlete trains for 10 weeks, how many minutes can they expect to take to complete the race?
Don't include the units in your answer.

■ 47

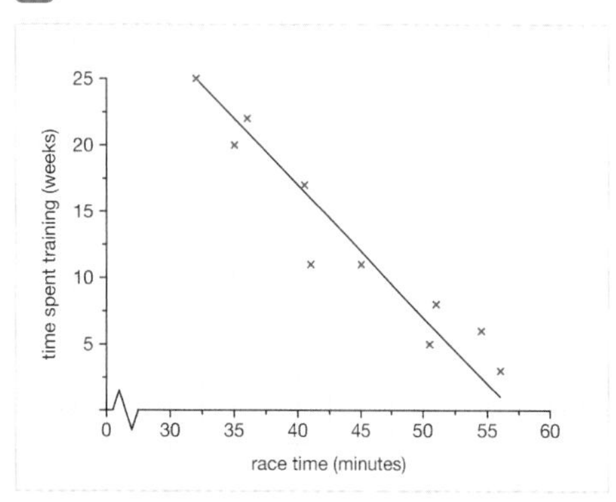

14. An ice cream van records the midday temperature and the number of ice creams sold each day. If 68 ice creams are sold one day, what is a reasonable estimate for the midday temperature?
Do not include the unit in your answer.

■ 22

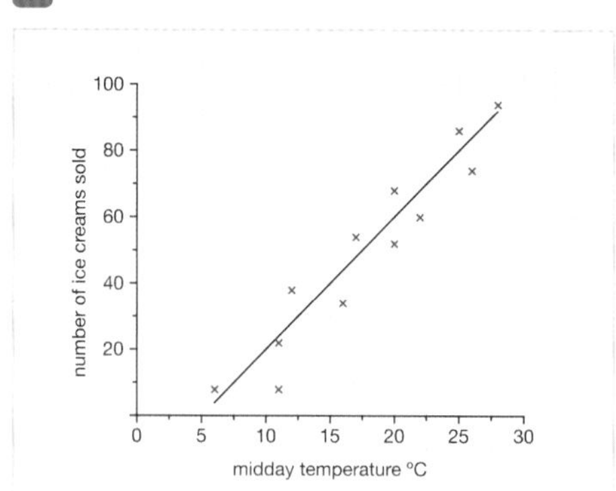

15. Ten athletes enter a 10 km race. The number of weeks they spent training and their race times are shown in the scatter graph. Another athlete completes the race in 50 minutes. What is a reasonable estimate for the number of weeks they spent training for the race?

■ 7

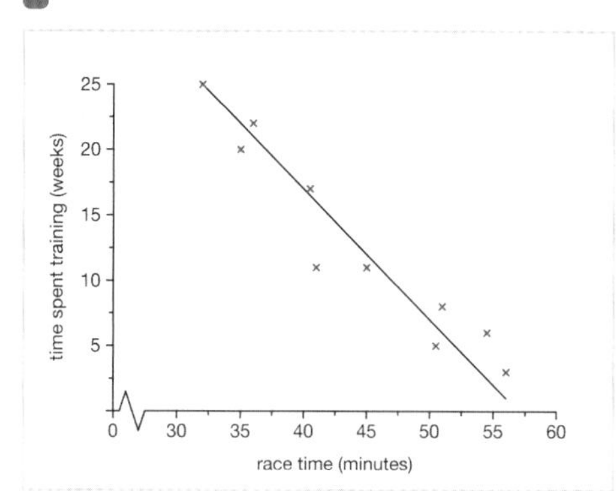

16. An ice cream van records the midday temperature and the number of ice creams sold each day. If tomorrow's midday temperature is forecast to be 16 °C, how many ice creams can they expect to sell?

■ 44

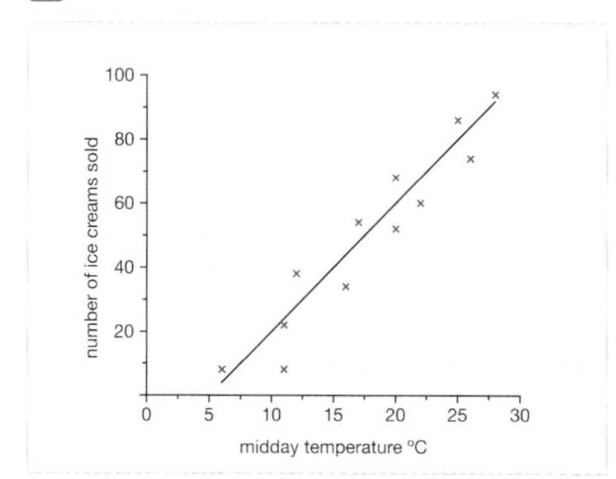

Level 2 continued

17. The scatter graph shows information about the distance of ten flats from London and their weekly rental cost. Estimate how many kilometres from London a flat that costs £500 per week to rent is likely to be.
Give your answer to 1 d.p. and don't include the units.

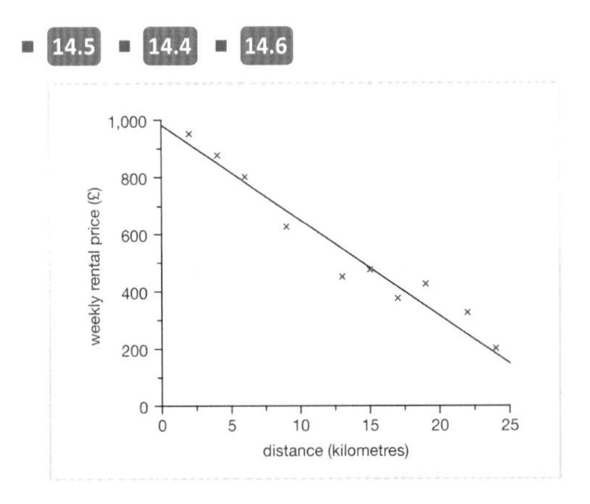

18. Which is a good line of best fit?

1/4

■ (i) ■ (ii) ■ (iii) ■ (iv)

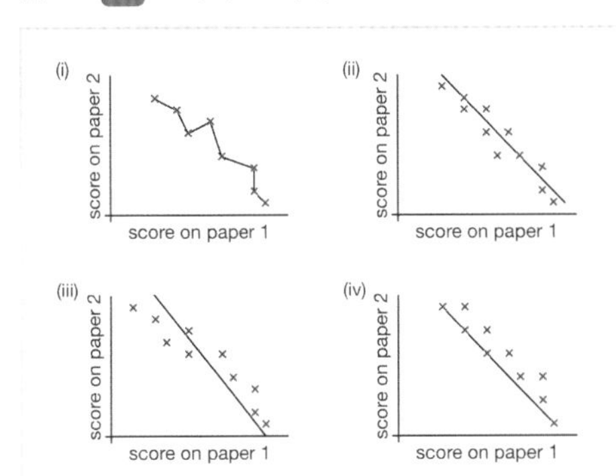

19. Ten athletes enter a 10 km race. The number of weeks they spent training and their race times are shown in the scatter graph. Another athlete completes the race in 35 minutes. What is a reasonable estimate for the number of weeks they spent training for the race?

 ■ 20

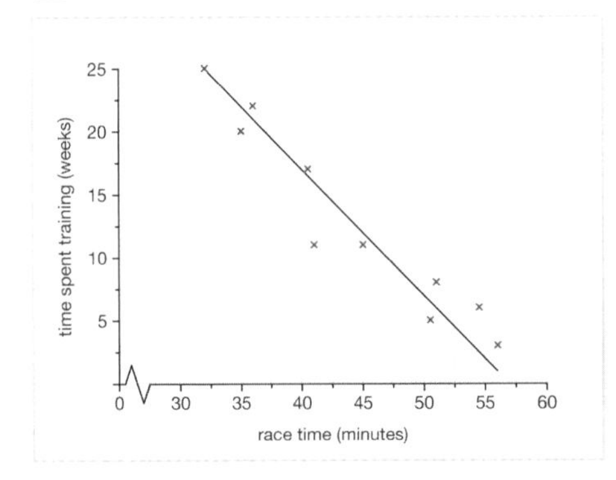

20. The scatter graph shows the marks that a maths class received for two test papers. A student scored 20 marks on the non-calculator paper. Use a line of best fit to estimate their score on the calculator paper.

1/5

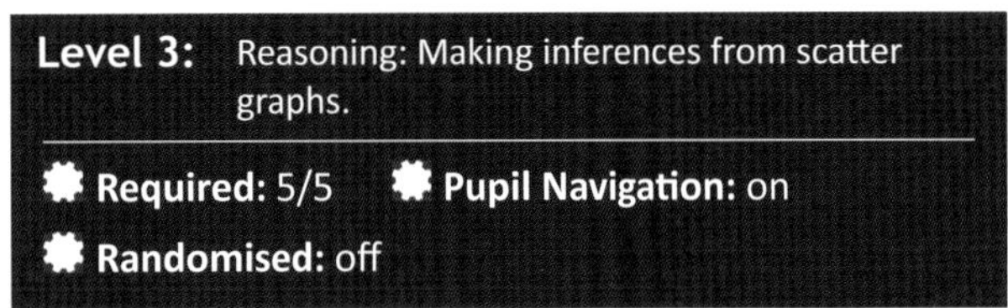

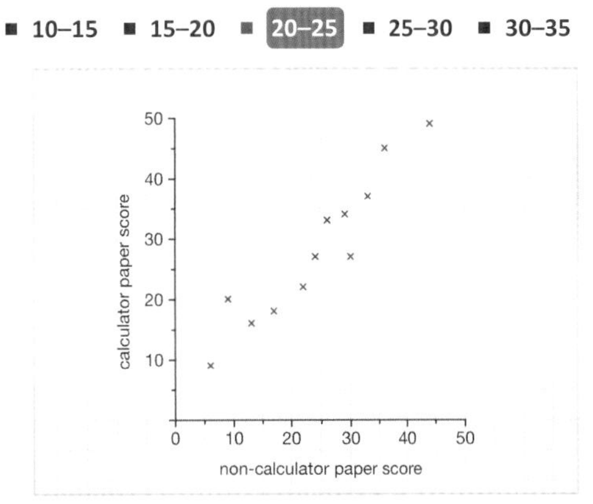

Level 3: Reasoning: Making inferences from scatter graphs.

Required: 5/5 **Pupil Navigation:** on

Randomised: off

21. The scatter graph shows information about the distance of ten flats from London and their weekly rental cost. Why is it not a good idea to use the graph to estimate the weekly rental cost of a flat 50 kilometres from London?

- Open question, no set answer

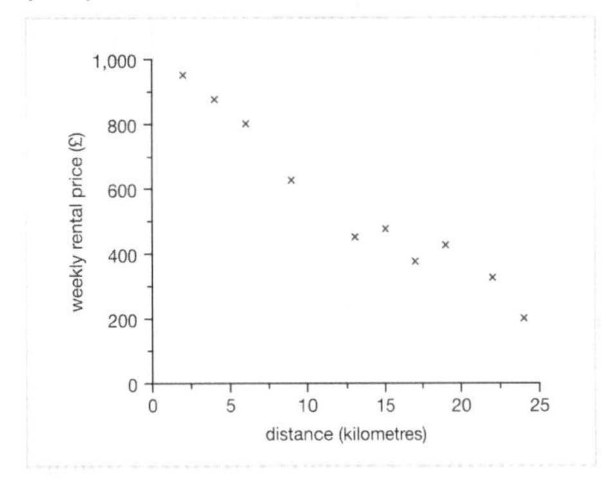

Level 3 *continued*

22. The first graph shows the English and art exam scores for a class. The second graph shows the maths and English scores for the same class. What is the correlation between their maths and art scores?

1/3

- positive
- none
- **negative**

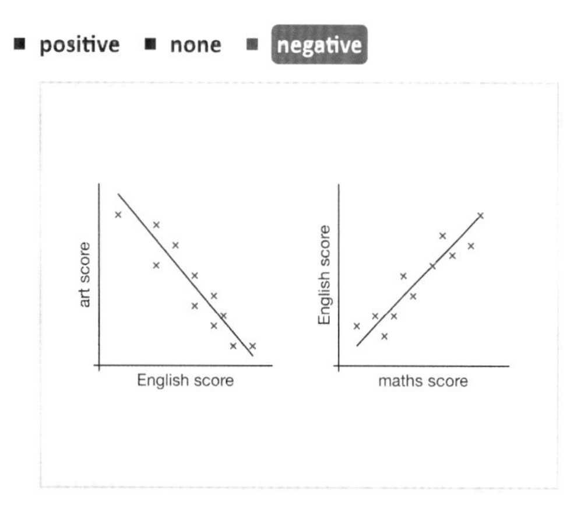

23. The graph shows the second-hand sale price for different ages of a particular make of car. Describe the relationship between the age of the car and the sale price.

- Open question, no set answer

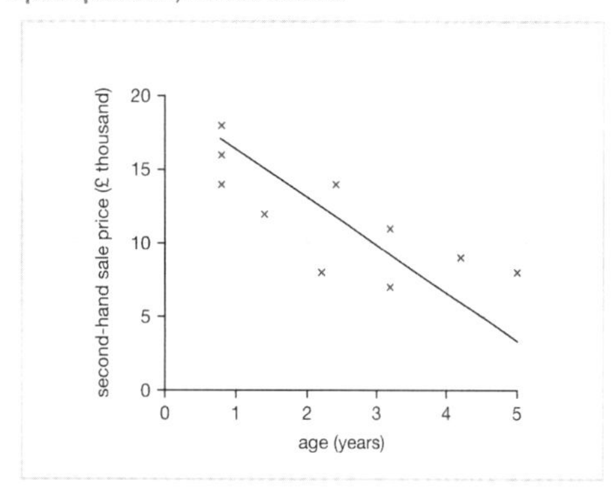

24. The scatter graph shows the time students spent revising and their marks in a test. One of the student's score is an outlier. What is their score?

- **46**
- 14

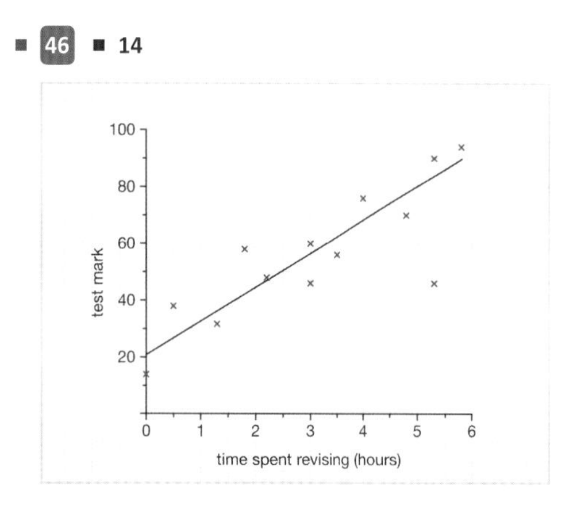

25. A car manufacturer has two models of a car; one is petrol and one is diesel. The graphs show the second-hand sale price of both models. For which model would you be most confident in predicting the sale price of a 3-year-old car? Explain your answer.

- Open question, no set answer

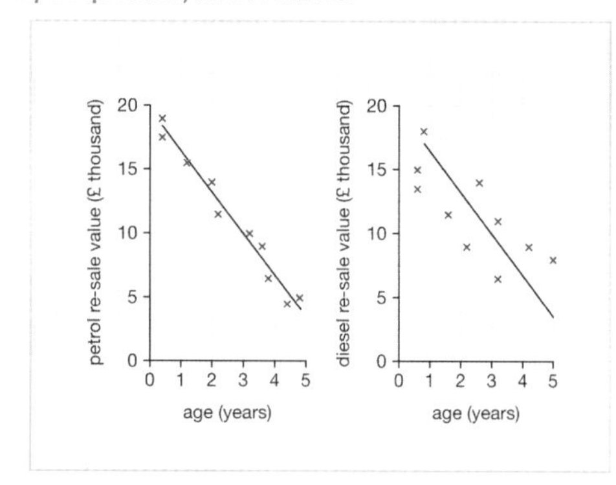

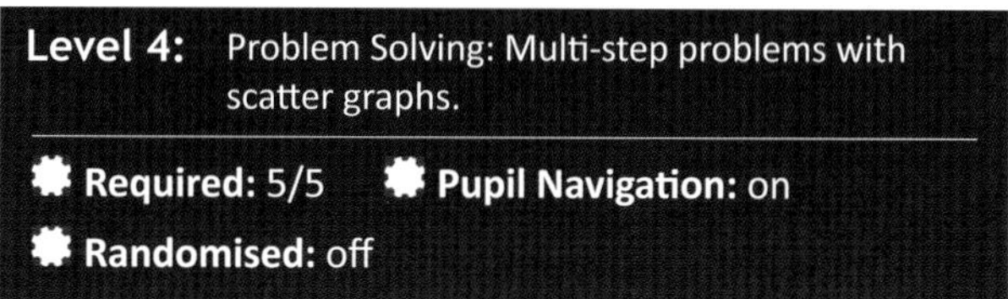

Level 4: Problem Solving: Multi-step problems with scatter graphs.

Required: 5/5 **Pupil Navigation:** on

Randomised: off

26. An ice cream van records the midday temperature and the number of ice creams sold each day. Calculate the mean midday temperature.
Give your answer to one decimal place. Don't include the units.

- **17.8**
- 18

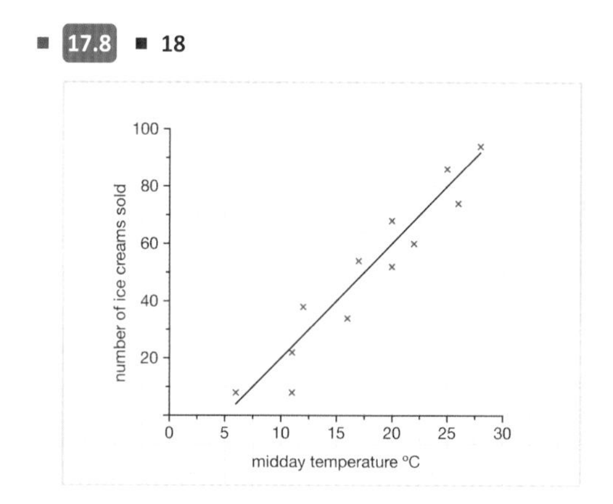

Level 4 *continued*

27. a b c The scatter graph shows information about the distance of 10 flats from London and their weekly rental cost. If a flat is selected at random, what is the probability it is greater than 15 kilometres from London?
Give your answer as a fraction in its simplest form.

- 4/10
- 2/5
- 1/2
- 5/10

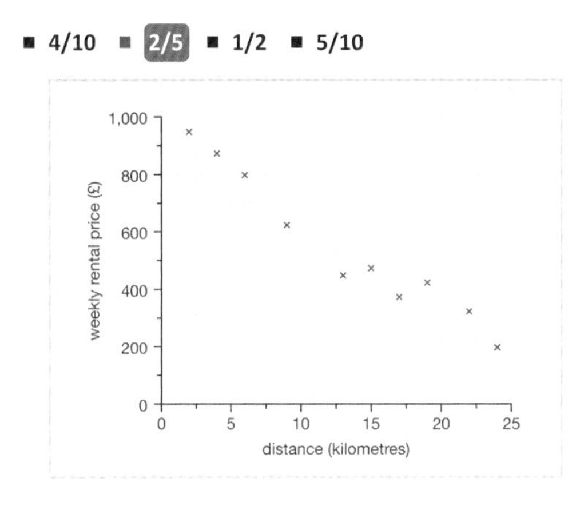

28. 1 2 3 Ten athletes enter a 10 km race. The number of weeks they spent training and their race times are shown in the scatter graph. What is their median race time?

- 43

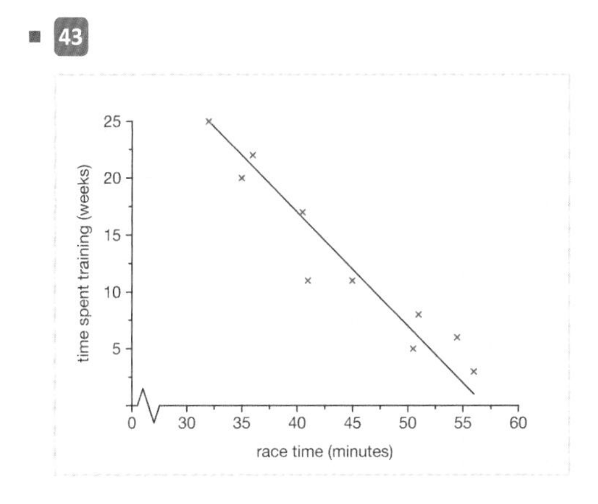

29. a b c The scatter graph shows the average price of a loaf of bread from 1975 to 2015. Use the line of best fit to estimate how much a loaf will cost in 2055.
Give your answer to the nearest pence and include the units.

- 83
- £0.83
- 83p
- 82
- 81p
- £0.81
- £0.82
- 81
- 0.83
- 0.81
- 0.82

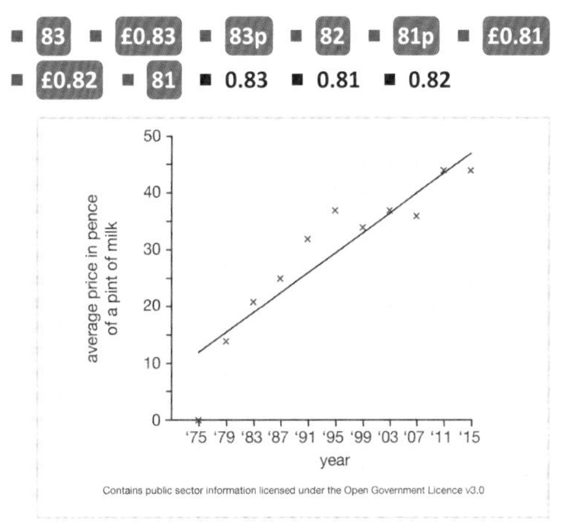

30. 1 2 3 The scatter graph shows the marks that a maths class received for two test papers. What was the mean mark on the calculator paper?
Give your answer rounded to one decimal place.

- 28.1
- 28
- 28.08

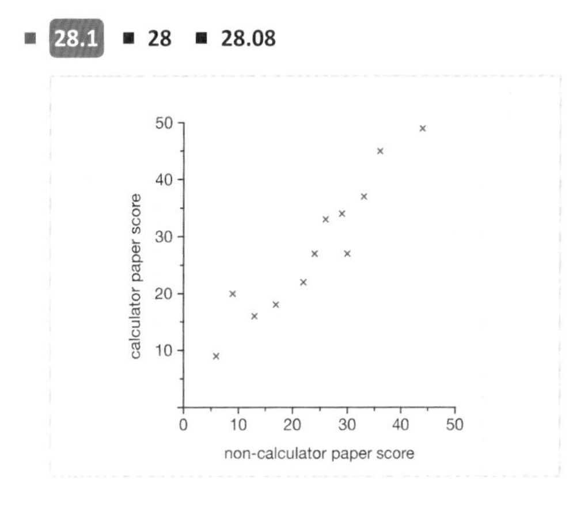

Estimate the Mean from a Grouped Frequency Table

Objective: I can estimate the mean from a grouped frequency table.

Quick Search Ref: 10441

Level 1: Understanding: Reading a grouped frequency table and estimating mean.

Required: 6/8 **Pupil Navigation:** on **Randomised:** off

1. Select one advantage and one disadvantage of using a grouped frequency table instead of a frequency table.

2/4

- **more concise display of information**
- more accurate information
- less concise display of information
- **less accurate information**

2. Hallie records how long it takes some of her friends to get to school.
What is the frequency for the $15 < t \leq 20$ class interval?

- **2**
- 3

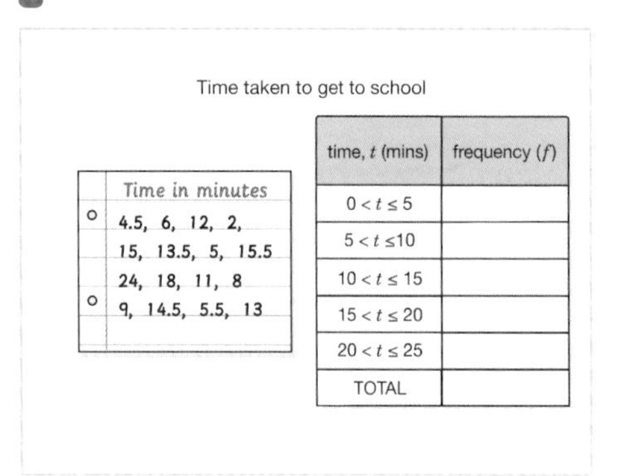

Time taken to get to school

time, t (mins)	frequency (f)
$0 < t \leq 5$	
$5 < t \leq 10$	
$10 < t \leq 15$	
$15 < t \leq 20$	
$20 < t \leq 25$	
TOTAL	

3. Hallie records how long it takes some of her friends to get to school. According to the table, how many people were surveyed?

-  **16**

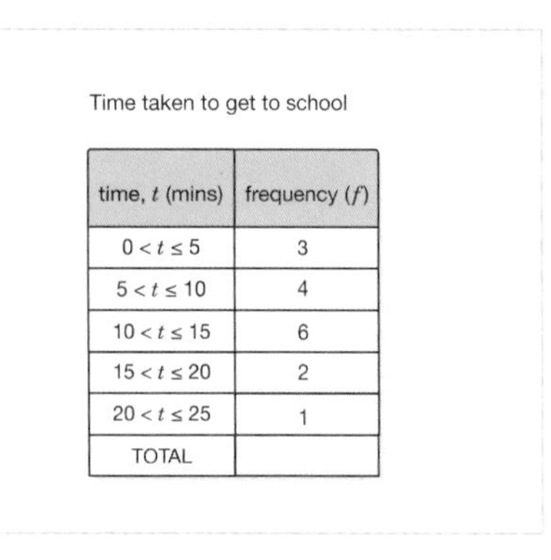

Time taken to get to school

time, t (mins)	frequency (f)
$0 < t \leq 5$	3
$5 < t \leq 10$	4
$10 < t \leq 15$	6
$15 < t \leq 20$	2
$20 < t \leq 25$	1
TOTAL	

4. Arrange the steps in order to estimate the mean of a grouped frequency table.

- **Find the midpoint of each class interval (x).**
- **Multiply the frequency (f) by the midpoint for each class interval (x) to find the fx value.**
- **Calculate the sum of f and the sum of fx.**
- **Divide the sum of fx by the sum of f.**

Time taken to get to school

time, t (mins)	frequency (f)	midpoint (x)	fx
$0 < t \leq 5$	3		
$5 < t \leq 10$	4		
$10 < t \leq 15$	6		
$15 < t \leq 20$	2		
$20 < t \leq 25$	1		
TOTAL			

5. Hallie records how long it takes some of her friends to get to school. What is the fx value for the $15 < t \leq 20$ class interval?

- **35**

Time taken to get to school

time, t (mins)	frequency (f)	midpoint (x)	fx
$0 < t \leq 5$	3	2.5	7.5
$5 < t \leq 10$	4	7.5	30
$10 < t \leq 15$	6	12.5	75
$15 < t \leq 20$	2	17.5	?
$20 < t \leq 25$	1	22.5	22.5
TOTAL	16		

Level 1 *continued*

6. Hallie records how long it takes some of her friends to get to school.
1 2 3 What is the estimate for the mean for this data set?
Give your answer as a decimal and don't include the units.

■ 10.625

Time taken to get to school

time, t (mins)	frequency (f)	midpoint (x)	fx
$0 < t \le 5$	3	2.5	7.5
$5 < t \le 10$	4	7.5	30
$10 < t \le 15$	6	12.5	75
$15 < t \le 20$	2	17.5	35
$20 < t \le 25$	1	22.5	22.5
TOTAL	16		

7. Megan records how long some of her friends spend watching television in a week.
1 2 3 Calculate the midpoint for the $2 < h \le 4$ class interval.

■ 3

Time spent watching TV in a week

time, t (hours)	frequency (f)	midpoint (x)	fx
$0 < t \le 2$	1		
$2 < t \le 4$	4	?	
$4 < t \le 6$	9		
$6 < t \le 8$	12		
$8 < t \le 10$	4		
TOTAL	30		

8. Megan records how long some of her friends spend watching television in a week.
1 2 3 Estimate the mean of this data set to the nearest whole hour.
Don't include the units in your answer.

■ 6 ■ 5.93

Time spent watching TV in a week

time, t (hours)	frequency (f)	midpoint (x)	fx
$0 < t \le 2$	1	1	1
$2 < t \le 4$	4	3	12
$4 < t \le 6$	9	5	45
$6 < t \le 8$	12	7	84
$8 < t \le 10$	4	9	36
TOTAL	30		178

Level 2: Fluency: Estimating the mean from a grouped frequency table with missing data.

Required: 5/8 **Pupil Navigation:** on
Randomised: off

9. The table shows information about the number of hours that 40 children exercise
1 2 3 every week. Work out an estimate for the mean time spent exercising to the nearest whole hour.
Don't include the units in your answer.

■ 4 ■ 3.65 ■ 29

Time spent exercising each week

time, t (hours)	frequency (f)	midpoint (x)	fx
$0 < t \le 2$	14		
$2 < t \le 4$	8		
$4 < t \le 6$	11		
$6 < t \le 8$	5		
$8 < t \le 10$	2		
TOTAL	40		

10. Joseph asked each of his friends how many minutes it took them to get to school.
1 2 3 Estimate the mean time taken to the nearest whole minute.
Don't include the units in your answer.

■ 14.23 ■ 14 ■ 74

Time taken to travel to school

time, t (mins)	frequency (f)	midpoint (x)	fx
$0 < t \le 5$	2		
$5 < t \le 10$	4		
$10 < t \le 15$	8		
$15 < t \le 20$	7		
$20 < t \le 25$	5		
TOTAL	26		

Level 2 *continued*

11. Isabella measured the heights of the 20 children in her class. Estimate the mean height of a child to the nearest whole centimetre (cm).
Include the units cm (centimetres) in your answer.

■ 156 cm ■ 164 ■ 164 centimetres ■ 625 centimetres ■ 156 centimetres ■ 156.25 cm ■ 156.25 centimetres ■ 156 ■ 625 cm ■ 625 ■ 156.25 ■ 164 cm

Height of children in Year 8

height, h (cm)	frequency (f)	midpoint (x)	fx
$145 < h \leq 150$	7		
$150 < h \leq 155$	4		
$155 < h \leq 160$	0		
$160 < h \leq 165$	5		
$165 < h \leq 170$	4		
TOTAL			

12. Marie collected some shells and measured the mass of each one. Work out an estimate for the mean mass of the shells.
Include the units g (grams) in your answer.

■ 15 grams ■ 15 g ■ 75 grams ■ 15 ■ 75 ■ 75 g

Mass of different shells

mass, m	frequency (f)	midpoint (x)	fx
$0 < m \leq 10$	9		
$10 < m \leq 20$	8		
$20 < m \leq 30$	7		
$30 < m \leq 40$	1		
$40 < m \leq 50$	0		
TOTAL			

13. 90 students took a maths exam. Estimate the mean score for the grouped frequency table.

■ 21.561 ■ 22 ■ 388.1 ■ 388

Scores in a maths test

score	frequency (f)	midpoint (x)	fx
0–5	3		
6–11	0		
12–17	9		
18–23	41		
24–29	37		
TOTAL	90		

14. Estimate the mean height of the flowers in Joey's garden.
Give your answer to 1 decimal place and include the units cm (centimetres).

■ 10.5 cm ■ 10.5 centimetres ■ 10.5 ■ 50.4 ■ 50.4 centimetres ■ 50.4 cm

Height of flowers in Joey's Garden

height, h (cm)	frequency (f)	midpoint (x)	fx
$6 < h \leq 9$	7		
$9 < h \leq 12$	12		
$12 < h \leq 15$	4		
$15 < h \leq 18$	0		
$18 < h \leq 21$	1		
TOTAL			

15. Lucy works in a school office. She records the length of time spent on each phone call. Estimate the mean length of a phone call to the nearest minute.
Don't include the units in your answer.

■ 3 ■ 3.05 ■ 46

Time spent on a phone call

time, t (mins)	frequency (f)	midpoint (x)	fx
$0 < t \leq 3$	41		
$3 < t \leq 6$	12		
$6 < t \leq 9$	2		
$9 < t \leq 12$	5		
TOTAL			

16. Edward collects comic books. Calculate an estimate for the mean price of one of Edward's comic books.
Include the £ (pound) sign in your answer.

■ £5.26 ■ 5.26

Price of comic books

price, p (£)	frequency (f)	midpoint (x)	fx
$0 < p \leq 2$	1		
$2 < p \leq 4$	8		
$4 < p \leq 6$	9		
$6 < p \leq 8$	1		
$8 < p \leq 10$	0		
$10 < p \leq 12$	4		
TOTAL			

Level 3: Reasoning: Estimating the mean from a grouped frequency table.

Required: 5/5 **Pupil Navigation:** on
Randomised: off

17. The mean for a grouped frequency table is only an estimate. Explain why.

- Open question, no set answer

18. Sophie estimates the mean length of a pencil crayon to be 3.8 cm. Explain how you know she is incorrect.

- Open question, no set answer

Length of pencil crayons

length, l (cm)	frequency (f)
$0 < l \le 2$	0
$2 < l \le 4$	0
$4 < l \le 6$	3
$6 < l \le 8$	12
$8 < l \le 10$	9
$10 < l \le 12$	8
$12 < l \le 14$	7

19. Toni calculates the mean estimate as 300. What mistake has she made?

- Open question, no set answer

Mass of model toys

mass, m (g)	frequency (f)	midpoint (x)	fx
$0 < m \le 15$	4	7.5	30
$15 < m \le 30$	7	22.5	157.5
$30 < m \le 45$	0	37.5	0
$45 < m \le 60$	8	52.5	420
$60 < m \le 75$	3	67.5	202.5
$75 < m \le 90$	12	82.5	990
TOTAL	34		1,800

20. Ashraf is calculating the mean for a set of test scores. He says, "I shouldn't include the score from the $0 < s \le 10$ class interval". Is Ashraf correct? Explain your answer.

- Open question, no set answer

Test scores (s)

score	frequency (f)
$0 < s \le 10$	1
$10 < s \le 20$	0
$20 < s \le 30$	0
$30 < s \le 40$	0
$40 < s \le 50$	7
$50 < s \le 60$	19
$60 < s \le 70$	13
$70 < s \le 80$	10
TOTAL	50

21. Jenny and Sonny record the same data in frequency tables but use different class intervals. Whose mean would you expect to be the most accurate? Explain your answer.

- Open question, no set answer

Jenny

age, a (years)	frequency (f)
$0 < a \le 2$	1
$2 < a \le 4$	3
$4 < a \le 6$	5
$6 < a \le 8$	9
$8 < a \le 10$	2
$10 < a \le 12$	5
TOTAL	25

Sonny

age, a (years)	frequency (f)
$1 < a \le 4$	4
$4 < a \le 7$	8
$7 < a \le 10$	8
$10 < a \le 15$	5
TOTAL	25

Level 4: Problem Solving: Estimating the mean from a grouped frequency table.

Required: 5/5 **Pupil Navigation:** on
Randomised: off

22. Jon records the heights of some of his classmates in centimetres (cm). Using the data shown, calculate the difference between the actual mean and the estimated mean.
Include the units cm (centimetres) in your answer.

- 1.025 centimetres
- 1.025 cm
- 1.025

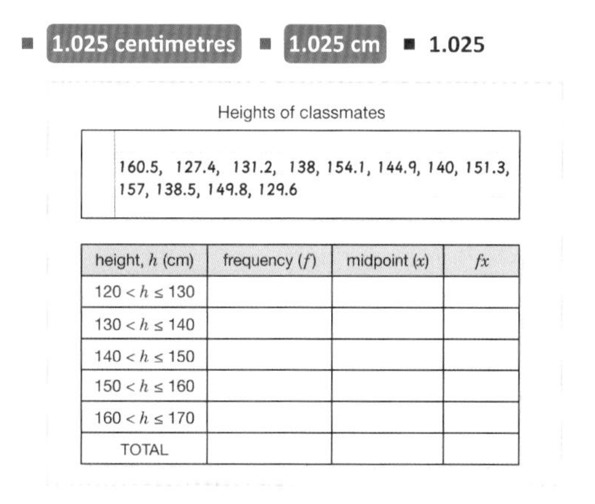

Heights of classmates

160.5, 127.4, 131.2, 138, 154.1, 144.9, 140, 151.3, 157, 138.5, 149.8, 129.6

height, h (cm)	frequency (f)	midpoint (x)	fx
$120 < h \le 130$			
$130 < h \le 140$			
$140 < h \le 150$			
$150 < h \le 160$			
$160 < h \le 170$			
TOTAL			

23. How many children took between 6 and 8 minutes to complete the race?

- 8

Time taken to complete a race

time, t (mins)	frequency (f)	midpoint (x)	fx
$0 < t \le 2$	0	1	
$2 < t \le 4$	2	3	
$4 < t \le 6$	9	5	
$6 < t \le 8$	?	7	
$8 < t \le 10$	5	9	
$10 < t \le 12$	6	11	
TOTAL			218

Level 4 *continued*

24. Using the data shown, calculate the difference between the highest possible mean and the estimated mean.
Don't include the units in your answer.

- 1

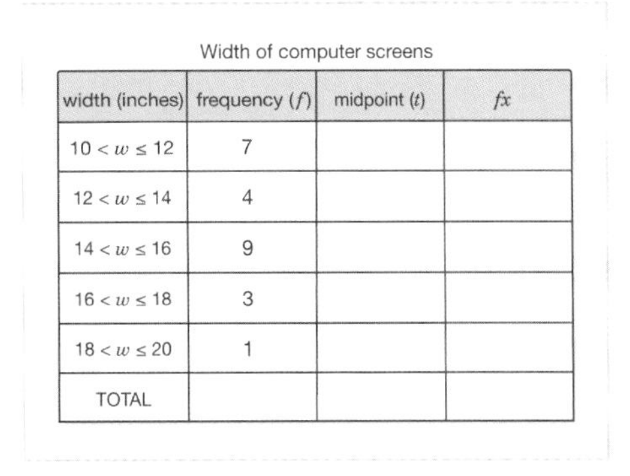

Width of computer screens

width (inches)	frequency (f)	midpoint (t)	fx
$10 < w \leq 12$	7		
$12 < w \leq 14$	4		
$14 < w \leq 16$	9		
$16 < w \leq 18$	3		
$18 < w \leq 20$	1		
TOTAL			

25. Rachel calculated the mean mass of an apple using the data shown in the table. Karl used the same data but used class intervals of 10 grams starting at $70 < m \leq 80$. Calculate the difference between their two answers.
Give your answer to 1 decimal place and include the units g (grams) in your answer.

- 0.3 g
- 0.3 grams
- 0.3

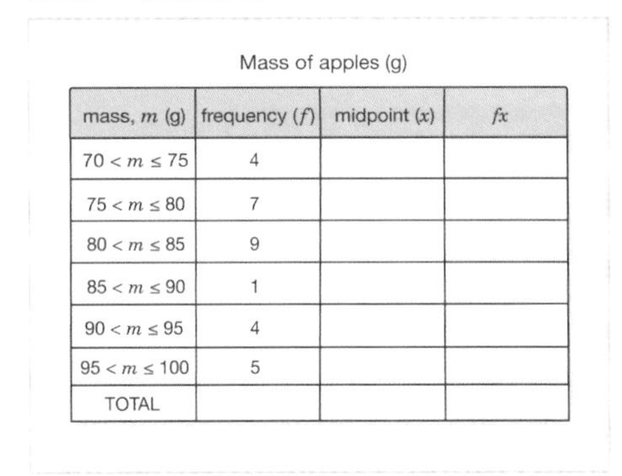

Mass of apples (g)

mass, m (g)	frequency (f)	midpoint (x)	fx
$70 < m \leq 75$	4		
$75 < m \leq 80$	7		
$80 < m \leq 85$	9		
$85 < m \leq 90$	1		
$90 < m \leq 95$	4		
$95 < m \leq 100$	5		
TOTAL			

26. The mean estimate for the age of a child in the group is 5.8. If there are y children aged between 4 and 8, what is the value of y?

- 9

Ages of a group of children

age, a (years)	frequency (f)	midpoint (x)	fx
$0 < a \leq 4$	6	2	12
$4 < a \leq 8$	y	6	$6y$
$8 < a \leq 12$	5	10	50
TOTAL			

Find the Mean from an Ungrouped Frequency Table

Objective: I can find the mean from an ungrouped frequency table.

Quick Search Ref: 10453

Level 1: Understanding: Interpreting a frequency table and finding the mean.

Required: 7/10 **Pupil Navigation:** on **Randomised:** off

1. Brooke records the shoe sizes of ten of her friends:
7, 4, 5, 4, 2, 5, 8, 6, 6, 3
What is the mean shoe size of Brooke's friends?

- **5**
- 50

2. Brooke records the shoe sizes of ten of her friends:
7, 4, 5, 4, 2, 5, 8, 6, 6, 3
What is the missing number from the frequency table?

- **2**
- 10

shoe size	frequency
2	1
3	1
4	?
5	2
6	2
7	1
8	1

3. Brooke records the shoe sizes of ten of her friends. Her results are shown in the frequency table. What is the missing value?

- **12**
- 8

shoe size	frequency	total of the shoe sizes
2	1	2
3	1	3
4	2	8
5	2	10
6	2	?
7	1	7
8	1	8

4. Ethan asks his friends how many goals they have scored for the school football team in the last month. His results are displayed in the frequency table. What is the total number of goals scored by everyone in the last month?

- **45**
- 15
- 18

goals scored	frequency	total goals
0	3	
1	4	
2	1	
3	3	
4	5	
5	2	
		?

5. Brooke records the shoe sizes of ten of her friends. Her results are shown in the frequency table. What is the mean shoe size of Brooke's friends?

- 10
- 7.1
- **5**
- 50
- 35

shoe size	frequency	total of the shoe sizes
2	1	2
3	1	3
4	2	8
5	2	10
6	2	12
7	1	7
8	1	8
	10	50

Level 1 *continued*

6. Arjun says that the mean number of goals scored by his friends is 8. Which table would make Arjun correct?

1/2

■ A ■ B

A

number of goals	frequency	total goals
7	2	14
8	5	40
9	0	0
10	1	10
	8	64

B

number of goals	frequency	total goals
1	1	1
2	5	10
3	0	0
4	2	8
	8	19

7. Rizwana asks some students how many times they had been to the cinema in the last month. Her results are shown in the frequency table. What is the mean number of times that students had been to the cinema in the last month?

■ 1 ■ 42 ■ 8.4 ■ 10

cinema visits	frequency	total number of cinema visits
0	22	0
1	8	8
2	5	10
3	4	12
4	3	12
	42	42

8. Amy asked her friends how many pens they each had in their pencil case. The results are shown in the frequency table. What is the missing value in the table?

■ 20 ■ 9

number of pens	frequency	total number of pens
1	5	5
2	4	8
3	2	6
4	5	?
5	3	15
6	1	6

9. Spencer asked a group of students how many days they had eaten carrots that week. The results are shown in the frequency table. What is the mean number of days that the students had eaten carrots?

■ 35 ■ 2 ■ 14 ■ 10 ■ 70

number of days	frequency	total number of days
0	4	0
1	8	8
2	13	26
3	4	12
4	6	24
	35	70

10. Gabriel asked his friends from swimming club how many days per week they went to the pool. His results are shown in the frequency table. What is the mean number of days that Gabriel's friends went to the pool?

■ 4 ■ 12 ■ 44 ■ 14.7 ■ 11

number of days at the pool	frequency	total number of days at the pool
3	4	12
4	3	12
5	4	20
	11	44

Level 2: Fluency: Finding the mean from an ungrouped frequency table including missing values.

Required: 7/10 **Pupil Navigation:** off
Randomised: off

11. 1 2 3 Paulo records the number of letters that his school receives each day for 39 days. His results are shown in the frequency table. What is the mean number of letters that Paulo's school receives each day?
Give your answer to one decimal place.

- **9.3**
- **9.2**
- **361**
- **72.2**
- **9**

number of letters	frequency	
7	6	
8	5	
9	9	
10	11	
11	8	

12. 1 2 3 Juanita counted the number of raisins in 37 packets. The results are shown in the frequency table. What is the mean number of raisins in each packet?
Give your answer to one decimal place.

- **216**
- **23.4**
- **864**
- **23**
- **23.3**

number of raisins	frequency
22	8
23	17
24	3
25	9

13. 1 2 3 Larry counts the number of potatoes in each sack of potatoes. The results are shown in the frequency table. What is the mean number of potatoes in each sack?
Give your answer to one decimal place.

- **54.2**
- **54**
- **352.3**
- **1409**
- **54.1**

number of potatoes	frequency
53	9
54	7
55	6
56	4

14. 1 2 3 Guang records the number of matchsticks in several boxes. The results are shown in the frequency table. What is the mean number of matchsticks in each box?
Give your answer to one decimal place.

- **107.6**
- **618.8**
- **108**
- **2475**

number of matchsticks	frequency
106	3
107	8
108	7
109	5

15. 1 2 3 Louise records the weight in grams of several copies of the same book. The results are shown in the frequency table. What is the mean weight of each book?
Give your answer to one decimal place.

- **5718.3**
- **1009.1**
- **1009**
- **17155**

weight of book (g)	frequency	
1,008	4	
1,009	7	
1,010	6	

Level 2 *continued*

16. Agatha records the peak temperature in degrees Celsius of the school yard over 14 days. The results are shown in the frequency table. What is the mean peak temperature over the 14 days?
Give your answer to one decimal place and don't include the units.

- 17
- 17.1
- 240
- 60

temperature °C	frequency
16	4
17	6
18	
19	2
	14

17. Lucas records his mum's maximum speed on the journey to school every day for a period of 30 days. The results are shown in the frequency table. What is the mean speed over the 30 days?
Give your answer to one decimal place and don't include the units.

- 38.4
- 1153
- 230.6
- 38

speed (mph)	frequency	
36		
37	3	
38	6	
39	14	
40	4	

18. At the school sports day, Sadie records the time it takes 60 runners in the 100 metre sprint and rounds their time to the nearest second. The results are shown in the frequency table. What is the mean time taken to run 100 metres?
Give your answer to 2 two decimal places and do not include the units in your answer.

- 14.25
- 855
- 142.50
- 142.5
- 14.3
- 14

time (seconds)	frequency
12	7
13	
14	18
15	19
16	6
17	2
	60

19. Anton records the speed of 100 cars on the motorway. The results are shown in the frequency table. What is the mean speed of the cars in miles per hour?
Give your answer to one decimal place and do not include the units.

- 58.3
- 58
- 5826
- 1165.2

speed (mph)	frequency	
56	20	
57		
58	14	
59	18	
60	32	

Level 2 *continued*

20. Ruby records the number of paperclips in several boxes. The results are shown in the frequency table. What is the mean number of matchsticks in each box? Give your answer to one decimal place.

- **50.7**
- 811
- 202.8
- 51

number of paperclips	frequency
49	2
50	4
51	7
52	3

Level 3: Reasoning: Misconceptions within finding the mean from a frequency table.

Required: 5/5 **Pupil Navigation:** on
Randomised: off

21. The total number of goals scored by everybody in a hockey competition is 30. How many people scored two goals?

- **6**
- 12

number of goals	frequency
0	2
1	3
2	?
3	1
4	3

22. 13 students in a class have a total of 32 goals. Find the missing value in the frequency table.

- **7**
- 5

number of siblings	frequency
0	2
1	3
2	1
3	4
4	2
?	1

23. Arthur records the number of text messages he receives over 12 hours. He says that the total number of text messages he received is 10. Is Arthur correct? Explain your answer.

- Open question, no set answer

number of texts	frequency
0	2
1	3
2	1
3	4
4	2

24. Salman says that the mean number of goals scored is 5. Is Salman correct? Explain your answer.

- Open question, no set answer

number of goals	frequency	total number of goals
0	2	0 × 2 = 0
1	4	1 × 4 = 4
2	1	2 × 1 = 2
3	1	3 × 1 = 3
4	4	4 × 4 = 16
		25

25 ÷ 5 = 5 goals

25. The frequency table shows the number of pets owned by 24 students. The mean number of pets owned is 2.5. If a student who owns 10 pets is added to the the table, what is the mean number of pets owned now?
Give your answer to one decimal place.

- **2.8**
- 2.9

number of pets	frequency
0	2
1	4
2	4
3	8
4	6

Level 4: Problem solving: Solve problems involving finding the mean from ungrouped frequency tables.

Required: 5/5 **Pupil Navigation:** off
Randomised: off

26. The total number of points scored by everybody in a competition was 36. How many people took part in the competition?

- **22**
- 10

number of points	frequency
0	4
1	5
2	
3	1
4	2
	?

27. The mean number of fish caught in an angling competition is 2.9. Find the missing value in the frequency table.

- **8**
- 58
- 50

number of fish caught	frequency
0	0
1	4
2	5
3	4
4	6
?	1

28. The number of points scored by a group of basketball players is shown in the frequency table. Six more players who each scored the same number of points are added to the table. The mean number of points changes to 29.5.
How many points did each of the six players score?

- **30**

number of points	frequency
27	2
28	3
29	8
30	6
31	5

29. Using the information below, calculate the mean number of penalties that the players scored.
Give your answer to two decimal places.

In a penalty shoot-out competition, each player gets four shots.
Each goal is awarded one point, so each player can get a maximum of four points.
30 players took part in the competition.
7 players scored the maximum score.
Half of the players scored more than 2 goals.
10 players scored half of their penalties.
3 players didn't score any goals.

- **2.47**
- 2
- 2.5
- 2.46

goals scored	frequency	
0		
1		
2		
3		
4		

Level 4 *continued*

30. The frequency table shows the number of fish caught by each person in a group of anglers. The mean number of fish caught is 2.5. Find the missing value in the frequency table.

number of fish caught	frequency
0	0
1	4
2	0
3	6
4	?

Find the Mode, Median and Range from a Grouped Frequency Table

Objective: I can find the mode, median and range from a grouped or ungrouped frequency table.

Quick Search Ref: 10444

Level 1: Understanding: Finding the mode, median and range from ungrouped frequency tables and understand class intervals.

Required: 7/10 Pupil Navigation: on Randomised: off

1. Arrange the following terms in the same order as their definitions.

- Grouped data
- Frequency
- Class interval
- Modal class interval

Data that can be ordered and sorted into groups called classes.
The number of times a particular event or item appears in a set of data.
The range of values for each class.
The class interval with the highest frequency.

2. The frequency table shows the heights of ten sunflowers. If an extra sunflower with a height of 60 centimetres (cm) is added to the table, which class interval would this sunflower be grouped in?

1/5

- $0 < h \leq 20$
- $20 < h \leq 40$
- **$40 < h \leq 60$**
- $60 < h \leq 80$
- $80 < h \leq 100$

height, h of sunflower (cm)	frequency
$0 < h \leq 20$	3
$20 < h \leq 40$	1
$40 < h \leq 60$	2
$60 < h \leq 80$	2
$80 < h \leq 100$	2

3. 23 girls compete in a 100 metre sprint and their times are recorded. Which of the following strategies will find the position of the girl with the **median** time?

1/3

- Divide the number of girls by 2 and then add 1.
- **Add 1 to the total number of girls and divide this number by 2.**
- Halve the number of girls and then subtract 0.5.

4. 24 football players scored goals for the school team last year. The frequency table gives information about the number of goals they scored.
Select the **modal** number of goals scored.

1/6

- 2
- 3, 6 and 7
- 7
- 5
- **2 and 5**
- 1

goals scored	frequency
1	3
2	8
3	1
4	2
5	8
6	1
7	1

5. 73 students are asked about how many jobs they completed for their parents last week. The results are shown in the frequency table. Give the **median** number of jobs completed.

- **4**
- 3.5
- 11.5

jobs completed	frequency
1	8
2	9
3	18
4	16
5	14
6	8

Level 1 *continued*

6. a b c 24 football players scored goals for the school team last year. The table gives information about the number of goals they scored. What is the **range** of the number of goals scored?

- 8-1 ■ 1-8 ■ 6 ■ 7-1 ■ 7 ■ 1-7

goals scored	frequency
1	3
2	8
3	1
4	2
5	8
6	1
7	1

7. 1 2 3 A group of students are asked about how many jobs they completed for their parents last week. The results are shown in the frequency table.
Give the **modal** number of jobs completed.

■ 3 ■ 6

jobs completed	frequency
1	8
2	9
3	18
4	16
5	14
6	8

8. 1 2 3 24 football players scored goals for the school team last year. The table gives information about the number of goals they scored. What is the **median** number of goals scored?

■ 3.5 ■ 4 ■ 3

goals scored	frequency
1	3
2	8
3	1
4	2
5	8
6	1
7	1

9. 1 2 3 A group of students are asked about how many jobs they completed for their parents last week. The results are shown in the frequency table.
Give the **range** of the number of jobs completed.

■ 5 ■ 10

jobs completed	frequency
1	8
2	9
3	18
4	16
5	14
6	8

10. a b c A group of people are asked about how many days per week they used the local bus service. The results are shown in the frequency table.
Give the **median** number of days that the local bus service was used.

■ 2 ■ 3 ■ 4

number of days	frequency
1	14
2	0
3	3
4	3
5	3
6	1
7	4

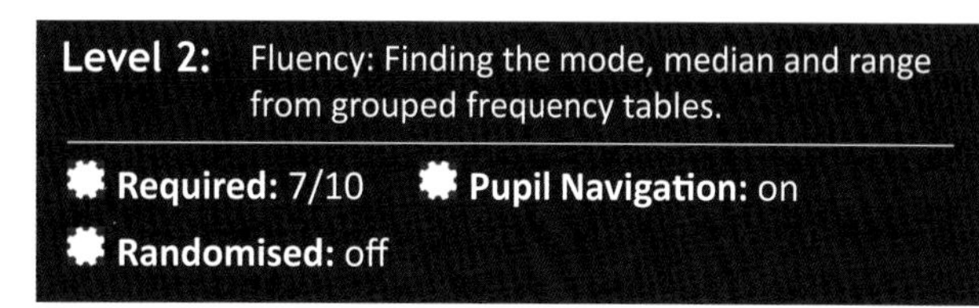
Level 2: Fluency: Finding the mode, median and range from grouped frequency tables.

Required: 7/10 **Pupil Navigation:** on
Randomised: off

11. A garage keeps records of all the costs of repairs to its vehicles. The table gives information about the repair costs which were less than £300 in one week.
Find the class interval containing the median cost.

1/5

- $0 < c \leq 50$
- $50 < c \leq 100$
- $100 < c \leq 150$
- **$150 < c \leq 200$**
- $200 < c \leq 300$

cost, c (£)	frequency
$0 < c \leq 50$	5
$50 < c \leq 100$	7
$100 < c \leq 150$	7
$150 < c \leq 200$	11
$200 < c \leq 300$	10

12. A garage keeps records of all the costs of repairs to its vehicles. The table gives information about the repair costs which were less than £300 in one week.
Calculate the maximum possible range of the costs of the repairs.

- 6
- **300**
- 5-11
- 0-300
- 225

cost, c (£)	frequency
$0 < c \leq 50$	5
$50 < c \leq 100$	7
$100 < c \leq 150$	7
$150 < c \leq 200$	11
$200 < c \leq 300$	10

13. A garage keeps records of all the costs of repairs to its vehicles. The table gives information about the repair costs which were less than £300 in one week. What is the modal class interval?

1/5

- $0 < c \leq 50$
- $50 < c \leq 100$
- $100 < c \leq 150$
- **$150 < c \leq 200$**
- $200 < c \leq 300$

cost, c (£)	frequency
$0 < c \leq 50$	5
$50 < c \leq 100$	7
$100 < c \leq 150$	7
$150 < c \leq 200$	11
$200 < c \leq 300$	10

14. A group of students compete in a cross country race and their times are recorded in a frequency table. Find the class interval containing the median time.

1/5

- $0 < t \leq 20$
- $20 < t \leq 25$
- $25 < t \leq 30$
- **$30 < t \leq 35$**
- $35 < t \leq 50$

time, t (mins)	frequency
$0 < t \leq 20$	1
$20 < t \leq 25$	4
$25 < t \leq 30$	11
$30 < t \leq 35$	14
$35 < t \leq 50$	3

15. A group of students compete in a cross country race and their times are recorded in a frequency table. Calculate the maximum possible range of the times of the students.

- **50**
- 32.5
- 30
- 35
- 13

time, t (mins)	frequency
$0 < t \leq 20$	1
$20 < t \leq 25$	4
$25 < t \leq 30$	11
$30 < t \leq 35$	14
$35 < t \leq 50$	3

Level 2 *continued*

16. A group of students compete in a cross country race and their times are recorded in a frequency table. What is the modal class interval?

1/5

- $0 < t \le 20$
- $20 < t \le 25$
- $25 < t \le 30$
- $30 < t \le 35$
- $35 < t \le 50$

time, t (mins)	frequency
$0 < t \le 20$	1
$20 < t \le 25$	4
$25 < t \le 30$	11
$30 < t \le 35$	14
$35 < t \le 50$	3

17. A group of Year 11 students are asked about how much time they spend completing homework each day after school. Find the class interval containing the median time to complete homework.

1/5

- $0 < t \le 15$
- $15 < t \le 30$
- $30 < t \le 45$
- $45 < t \le 60$
- $60 < t \le 120$

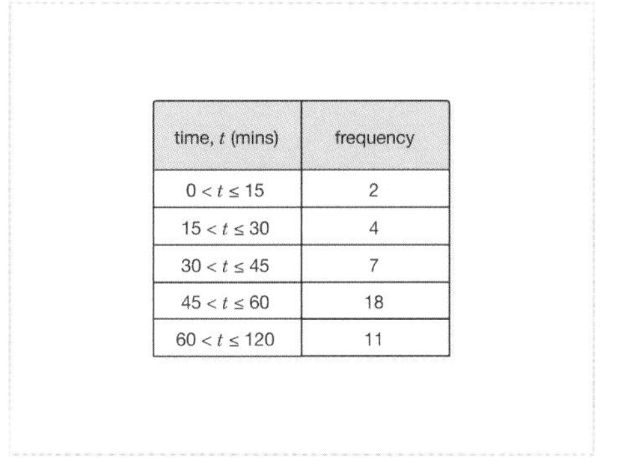

time, t (mins)	frequency
$0 < t \le 15$	2
$15 < t \le 30$	4
$30 < t \le 45$	7
$45 < t \le 60$	18
$60 < t \le 120$	11

18. A group of Year 11 students were asked about how much time they spend completing homework each day after school. What is the modal class interval?

1/5

- $0 < t \le 15$
- $15 < t \le 30$
- $30 < t \le 45$
- $45 < t \le 60$
- $60 < t \le 120$

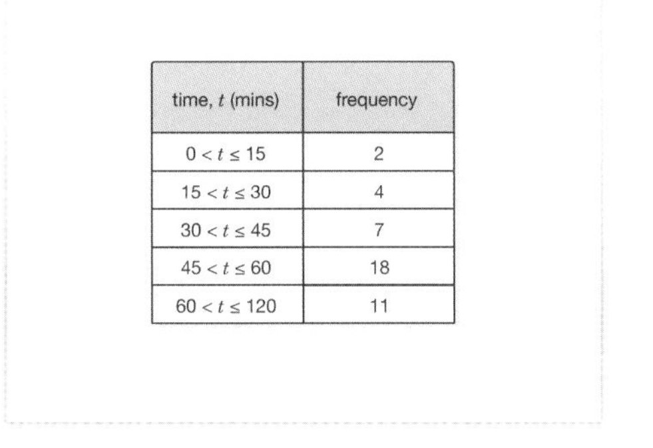

time, t (mins)	frequency
$0 < t \le 15$	2
$15 < t \le 30$	4
$30 < t \le 45$	7
$45 < t \le 60$	18
$60 < t \le 120$	11

19. A group of Year 11 students are asked about how much time they spend completing homework each day after school. Calculate the maximum possible range of the times of the students.

- 120
- 16
- 2-18
- 0-120

time, t (mins)	frequency
$0 < t \le 15$	2
$15 < t \le 30$	4
$30 < t \le 45$	7
$45 < t \le 60$	18
$60 < t \le 120$	11

20. A group of students are asked about how long they spend walking their dog each week. Find the class interval containing the median time.

1/5

- $0 < t \le 30$
- $30 < t \le 60$
- $60 < t \le 90$
- $90 < t \le 120$
- $120 < t \le 360$

time, t (mins)	frequency
$0 < t \le 30$	1
$30 < t \le 60$	1
$60 < t \le 90$	14
$90 < t \le 120$	0
$120 < t \le 360$	16

Level 3: Reasoning: Find the mode, median and range from a grouped frequency table.

Required: 5/5 **Pupil Navigation:** on
Randomised: off

21. a b c A group of people are asked how much time they spend cleaning each day. The results are shown in the frequency table. Jennifer says that the median class interval is $20 < t \le 30$. Is she correct? Explain your answer.

- Open question, no set answer

time, t (mins)	frequency
$0 < t \le 10$	2
$10 < t \le 20$	7
$20 < t \le 30$	4
$30 < t \le 40$	11
$40 < t \le 50$	3

22. a b c A group of people are asked how much time they spend cleaning each day. The results are shown in the frequency table. Nathan says that the median class interval is 35. Is he correct? Explain your answer.

- Open question, no set answer

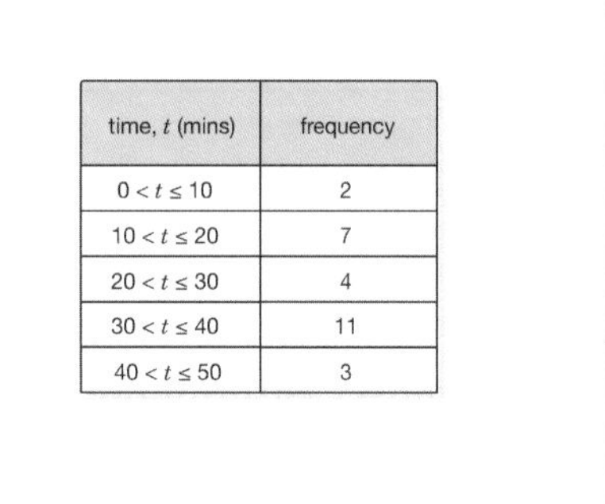

time, t (mins)	frequency
$0 < t \le 10$	2
$10 < t \le 20$	7
$20 < t \le 30$	4
$30 < t \le 40$	11
$40 < t \le 50$	3

23. 1 2 3 A group of students each grow a sunflower as part of their homework and the height of the sunflower is recorded. The median height lies in the class interval $60 < h \le 80$. How many students had a sunflower with a height which was in the class interval $40 < h \le 60$?

■ 4

height, h (cm)	frequency
$0 < h \le 20$	3
$20 < h \le 40$	5
$40 < h \le 60$	?
$60 < h \le 80$	1
$80 < h \le 100$	12

24. a b c The grouped frequency table shows the weights of 25 mice. Brontie estimates that the range of weights of the mice is 80 g. Why is her value only an estimate?

- Open question, no set answer

weight, w (g)	frequency
$20 < w \le 30$	9
$30 < w \le 70$	1
$70 < w \le 80$	6
$80 < w \le 100$	7

Level 3 *continued*

25. A group of people are asked how much time they spend cleaning each day. The results are shown in the frequency table.
The median time lies in the class interval 30 < t ≤ 40. One extra person, not included in the table, spends 70 minutes cleaning.
Derek says that the class interval which contains the median will now change.
Is Derek correct? Explain your answer.

- Open question, no set answer

time, t (mins)	frequency
$0 < t \le 10$	2
$10 < t \le 20$	7
$20 < t \le 30$	4
$30 < t \le 40$	11
$40 < t \le 50$	3

Level 4: Problem solving: Find the mode, median and range from a grouped frequency table.

Required: 5/5 **Pupil Navigation:** on
Randomised: off

26. Brendan has answered four maths questions correctly but spilt ink over his homework before answering question 5. What is the correct answer to question 5?

■ 7

time (t seconds)	frequency
$0 < t \le 5$	[illegible]
$5 < t \le 15$	[illegible]
$15 < t \le 25$	[illegible]
$25 < t \le 40$	[illegible]
$40 < t \le 50$	[illegible]

1. How many pupils were there? 15
2. How many times are in the class interval 15 < t ≤ 25? 6
3. What is the range of times? 40
4. What is the modal class interval? 5 < t ≤ 15
5. How many students are in the modal class interval? ___

27. Nyree has lost her homework which was based on a grouped frequency table of the times it took her friends to run a race. She writes down the facts she can remember.

How many friends had a time in the class interval 75 < t ≤ 100?

■ 5

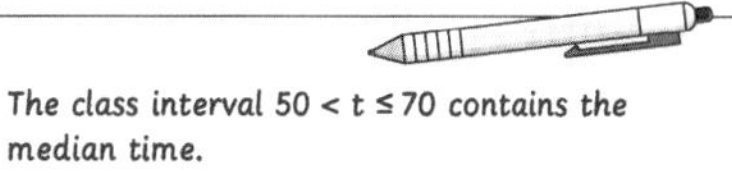

The class interval 50 < t ≤ 70 contains the median time.
The modal class is 75 < t ≤ 100.
Nobody took over 100 seconds.
The class interval 25 < t ≤ 50 had a frequency of 4.
The frequency table has four class intervals of equal width.
There were 11 friends.

28. Jasper has answered one of his frequency table questions incorrectly. What is the correct answer to the question he got wrong?

■ 8

time (t seconds)	frequency
$50 < t \le 65$	7
$65 < t \le 75$	8
$75 < t \le 80$	3
$80 < t \le 90$	9
$90 < t \le 110$	2

1. What is the maximum range of times? 60 seconds
2. How wide is the modal class interval? 10 seconds
3. How many students are in the class interval which contains the median? 3

29. Darren recorded how long 26 students could juggle three tennis balls. Use the information shown to calculate how many students juggled for a length of time that lies in the modal class.

■ 9

The modal class interval is $30 < t \le 40$.
There are five class intervals in total, each with an equal width.
No student juggled for more than 50 seconds.
Eight students took more than 20 seconds but less than or equal to 30 seconds.
Seven students took more than 10 seconds but less than or equal to 20 seconds.
One student juggled for 8.3 seconds and another student juggled for 42.0 seconds.

Level 4 *continued*

30. Erica spilled ink all over the frequency column of her homework before she answered question 5. What is the correct answer to question 5?

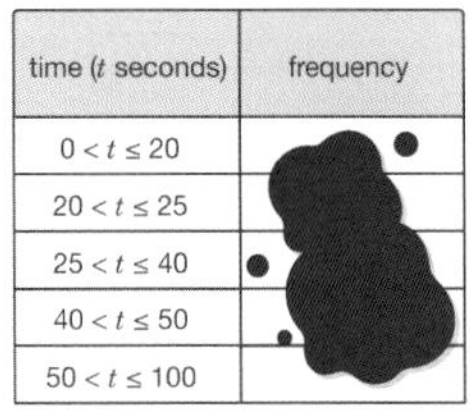

time (t seconds)	frequency
$0 < t \le 20$	
$20 < t \le 25$	
$25 < t \le 40$	
$40 < t \le 50$	
$50 < t \le 100$	

1. What is the width of the modal class interval? 5
2. What is the range of times? 30 seconds
3. How many students are in the class interval $25 < t \le 40$? 14
4. How many students are there in total? 30
5. How many students are in the class interval $40 < t \le 50$? ____

Interpret Frequency Polygons

Objective: I can plot, read and interpret frequency polygons.

Quick Search Ref: 10442

Level 1: Understanding: Plotting and reading frequency polygons.

Required: 7/10 **Pupil Navigation:** on **Randomised:** off

1. Select three properties of a frequency polygon.

3/6

- Frequency is plotted on the horizontal axis.
- Points are plotted at the end of each class interval for grouped data.
- Data points are joined with straight lines.
- Frequency is plotted on the vertical axis.
- Data points are joined with a curved line.
- Points are plotted at the midpoint of each class interval for grouped data.

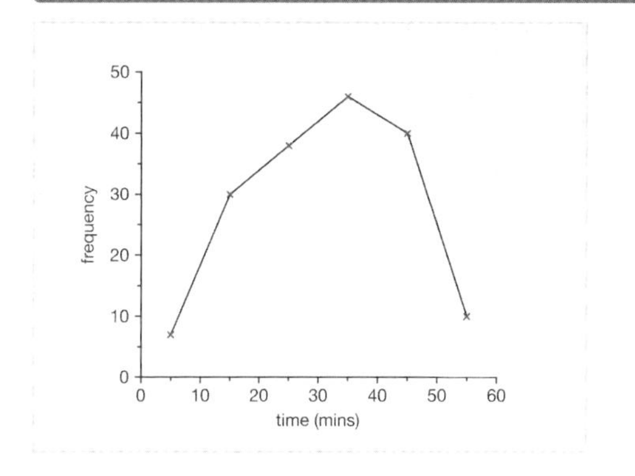

2. Which of the following data would you show using a frequency polygon?

1/3

- The sale price of five different cars.
- The temperature at a weather station during one day.
- The race times of a year group.

3. The frequency table shows the amount of time Year 9 students spend on maths homework each week. Select the **three** values that would be in the class interval $10 < t \leq 20$

3/6

- 9 minutes
- 10 minutes
- 15 minutes
- 17 minutes
- 20 minutes
- 21 minutes

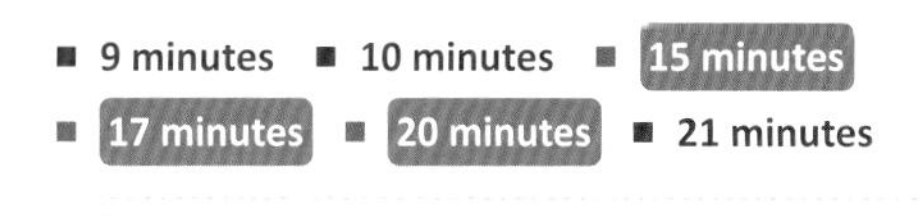

time, t (mins)	frequency
$0 < t \leq 10$	
$10 < t \leq 20$	
$20 < t \leq 30$	
$30 < t \leq 40$	
$40 < t \leq 50$	
$50 < t \leq 60$	

4. The number of siblings that students in a class have is shown in a frequency polygon. How many students have two siblings?

- 16

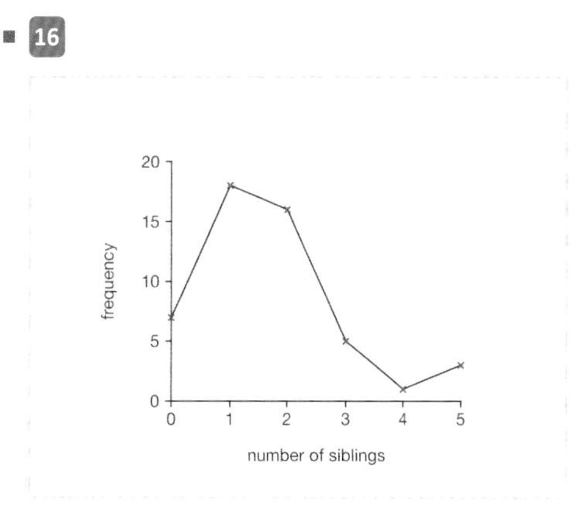

5. The frequency table shows the amount of time Year 9 students spend on maths homework each week. Where should the data point representing $10 \leq t < 20$ minutes be plotted?

1/6

- A
- B
- C
- D
- E
- F

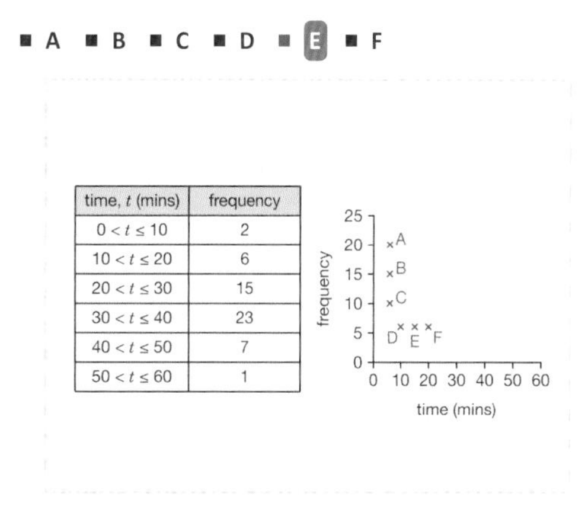

time, t (mins)	frequency
$0 < t \leq 10$	2
$10 < t \leq 20$	6
$20 < t \leq 30$	15
$30 < t \leq 40$	23
$40 < t \leq 50$	7
$50 < t \leq 60$	1

6. A frequency polygon of grouped data is drawn for the length of time students spend on mobile phone calls each week. Which class interval has a frequency of 40?

1/5

- $45 \leq t < 55$
- 40
- $40 \leq t < 50$
- 45
- $35 \leq t < 45$

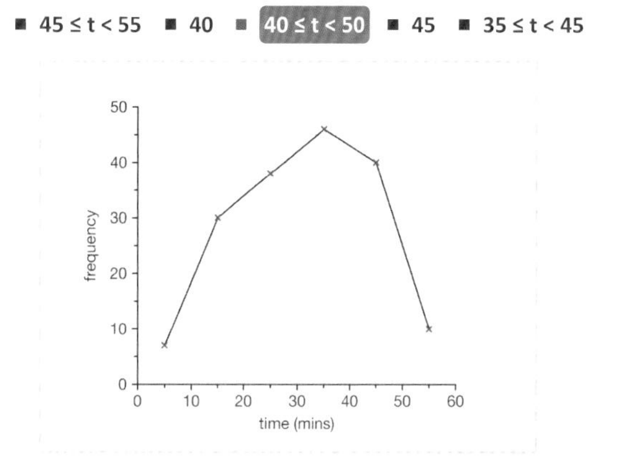

Level 1 *continued*

7. The frequency table shows the marks students received in a test. Where should the data point representing the number of students scoring 1–5 marks be plotted?

1/6

■ A ■ **B** ■ C ■ D ■ E ■ F

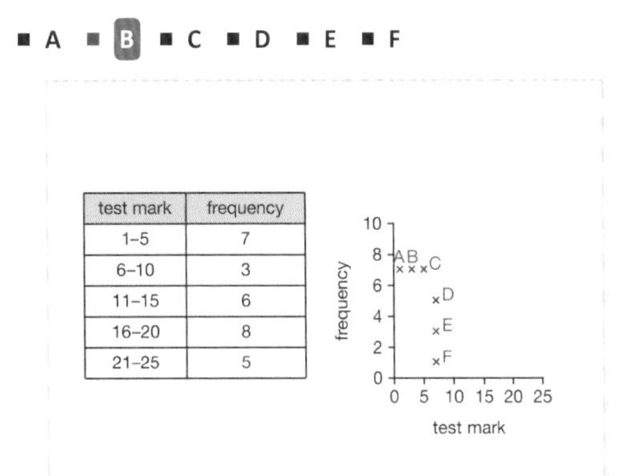

test mark	frequency
1–5	7
6–10	3
11–15	6
16–20	8
21–25	5

8. The frequency polygon shows the number of siblings students in a class have. How many students have four siblings?

■ **1**

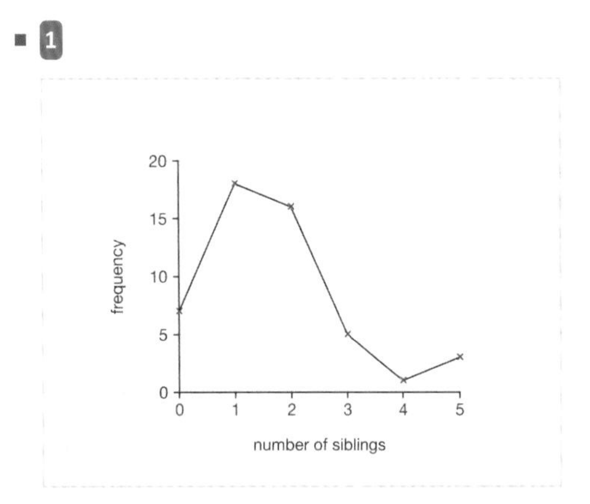

9. A frequency polygon of grouped data is drawn for the heights of Year 9 students. Which class interval has a frequency of 14?

1/5

■ **140 < h ≤ 160** ■ 14 ■ 130 < h ≤ 150 ■ 150
■ 150 < h ≤ 130

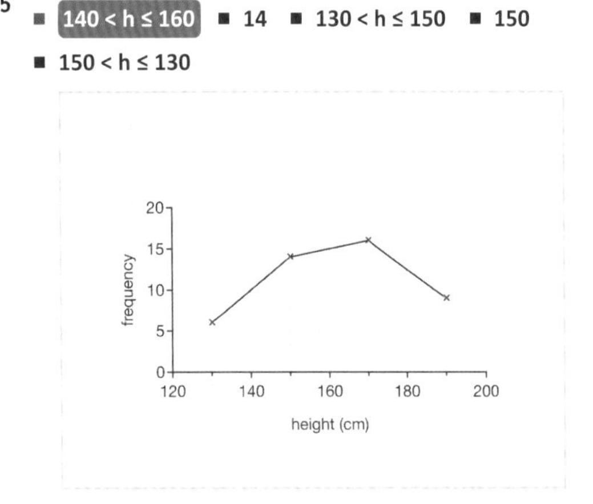

10. The frequency table shows the number of letters in each word in a sentence. Where should the data point representing 4–6 letters be plotted?

1/6

■ A ■ B ■ C ■ D ■ **E** ■ F

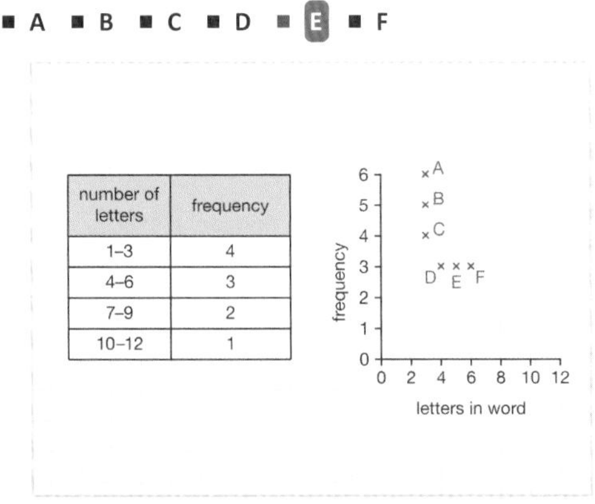

number of letters	frequency
1–3	4
4–6	3
7–9	2
10–12	1

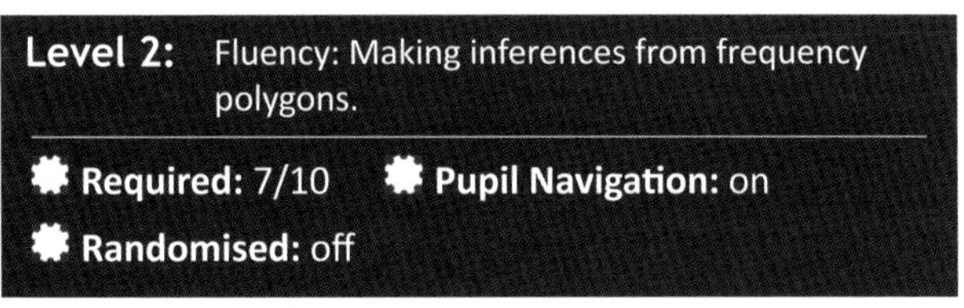

Level 2: Fluency: Making inferences from frequency polygons.

Required: 7/10 **Pupil Navigation:** on

Randomised: off

11. The number of siblings that students in a class have is shown in a frequency polygon. What is the mode number of siblings?

■ **1** ■ 18 ■ 5

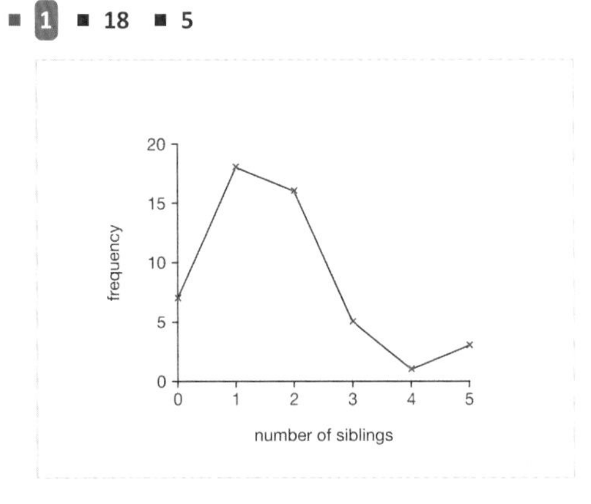

12. The frequency polygon shows the shoe size of students in a class rounded to the nearest integer. What is the range of shoe sizes?

1/6

■ 12 ■ 9–3 ■ **6** ■ 3–9 ■ 12–2 ■ 9

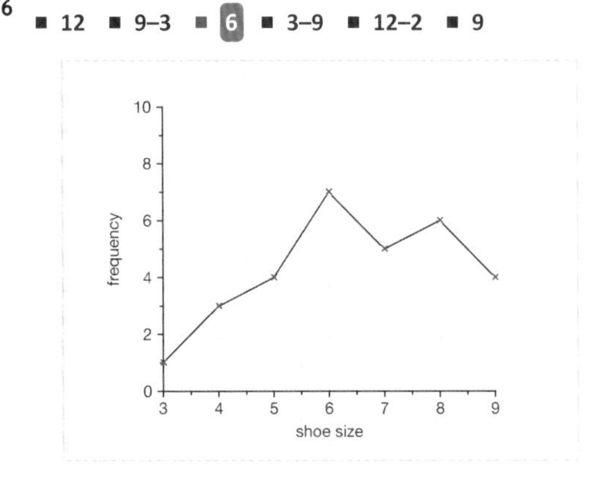

Level 2 *continued*

13. The number of siblings that students in a class have is shown in a frequency polygon. How many students are in the class?

■ 50 ■ 6

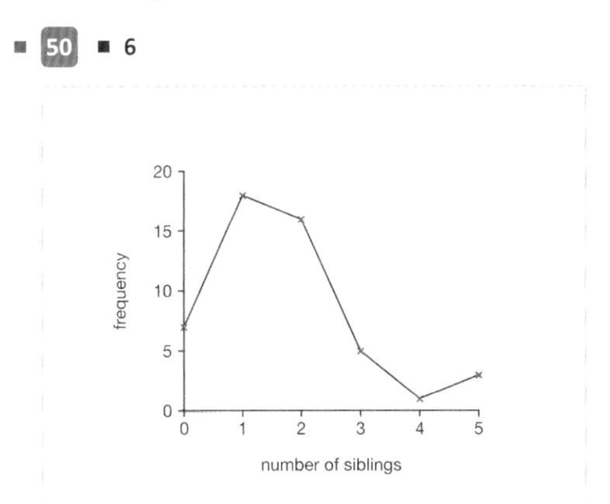

14. A frequency polygon of grouped data is drawn for the length of time students spend on mobile phone calls each week. What is the modal class interval?

1/6

■ 40

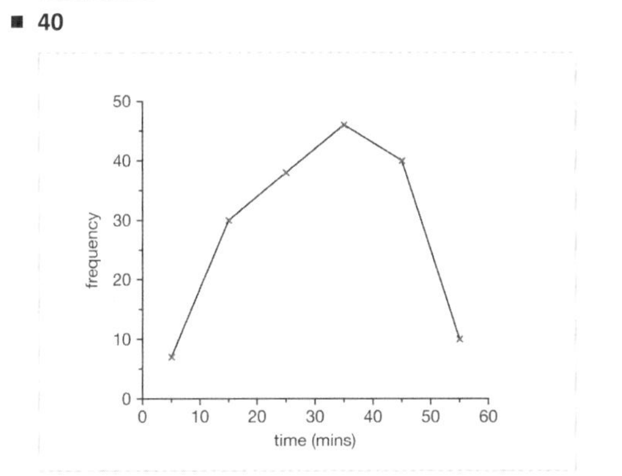

15. The frequency polygons of grouped data show the amount of time spent each week on maths homework by Year 9 and Year 11 students. How many more students spend between 0 and 30 minutes on homework in Year 9 than Year 11?

■ 37

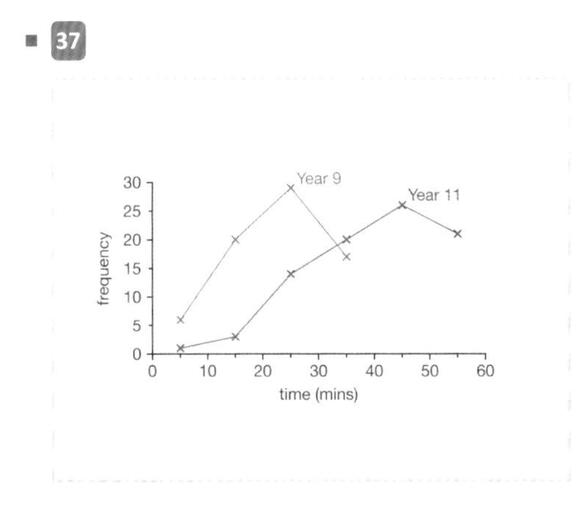

16. A frequency polygon of grouped data is drawn for the heights of Year 9 students. What is the maximum possible range of heights?
Include the units cm (centimetres) in your answer.

■ 70 cm ■ 80

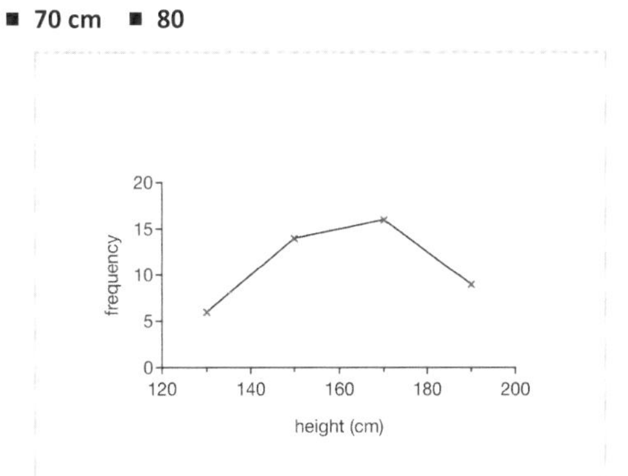

17. A frequency polygon of grouped data is drawn for the length of time students spend on mobile phone calls each week. How many students spend less than 20 minutes per week on phone calls?

■ 37 ■ 30

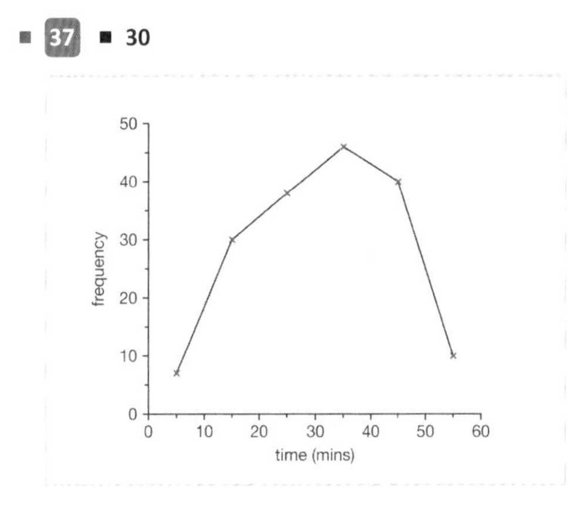

18. A frequency polygon of grouped data is drawn for the heights of Year 9 students. What is the modal class interval?

1/6

■ 160 ■ 170 ≤ h < 190 ■ 170 ■ 160 ≤ h < 180
■ 180 ■ 150 ≤ h < 170

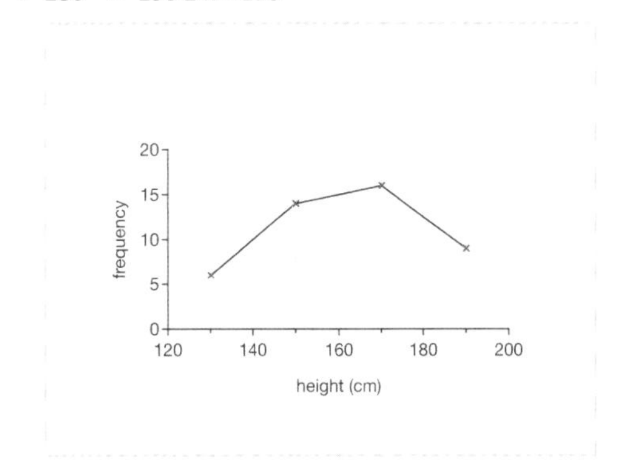

Level 2 *continued*

19. A frequency polygon of grouped data is drawn for the length of time students spend on mobile phone calls each week. What is the maximum possible range of time spent on phone calls?
Give your answer in minutes and don't include the units.

- **60**
- **50**

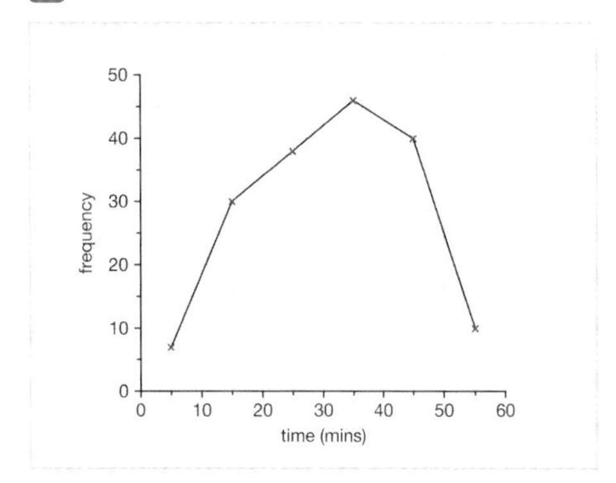

20. A frequency polygon of grouped data is drawn for the heights of Year 9 students. How many students are shorter than 180 centimetres?

- **36**
- **16**

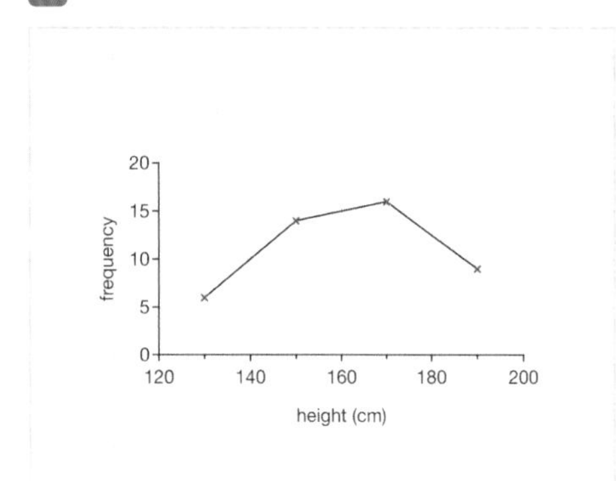

Level 3: Reasoning: Misconceptions and comparing frequency polygons.

Required: 5/5 **Pupil Navigation:** on
Randomised: off

21. Harvey is plotting the times that a football team concede goals during matches. What mistake has Harvey made? Explain your answer.

- Open question, no set answer

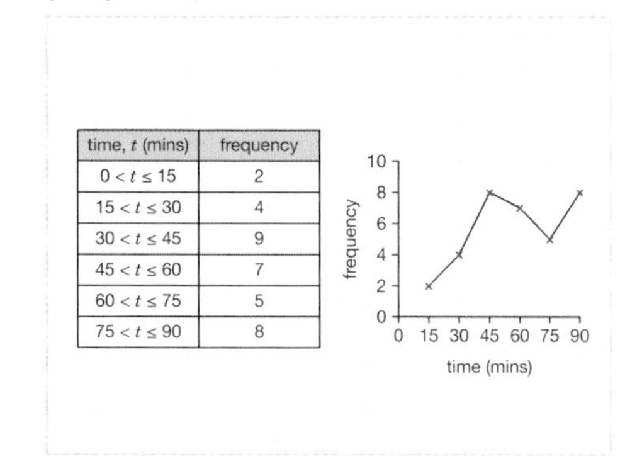

time, t (mins)	frequency
$0 < t \leq 15$	2
$15 < t \leq 30$	4
$30 < t \leq 45$	9
$45 < t \leq 60$	7
$60 < t \leq 75$	5
$75 < t \leq 90$	8

22. The frequency polygons of grouped data show the amount of time spent each week on maths homework by Year 9 and Year 11 students. Select the two statements which correctly compare the frequency polygons.

2/4

- **The range of time spent on homework is larger for Year 9 than Year 11.**
- **The mode of time spent on homework is smaller for Year 9 than Year 11.**
- **The mode of time spent on homework is larger for Year 9 than Year 11.**
- **The range of time spent on homework is smaller for Year 9 than Year 11.**

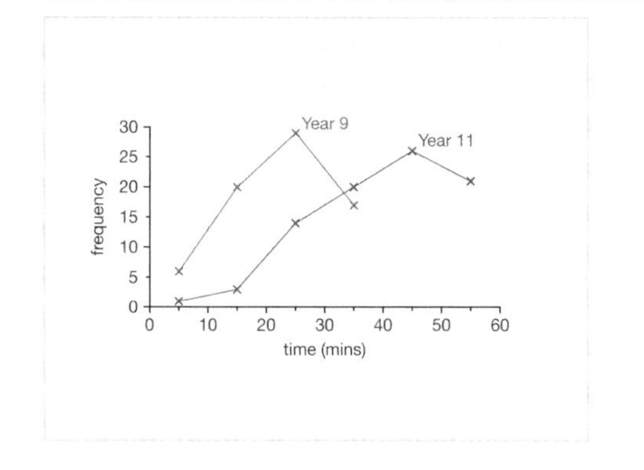

Level 3 continued

23. a b c A frequency polygon of grouped data is drawn for the length of time students spend on mobile phone calls each week. What is a reasonable estimate for the number of students who spend between 40 and 45 minutes each week on phone calls? Explain your answer.

- Open question, no set answer

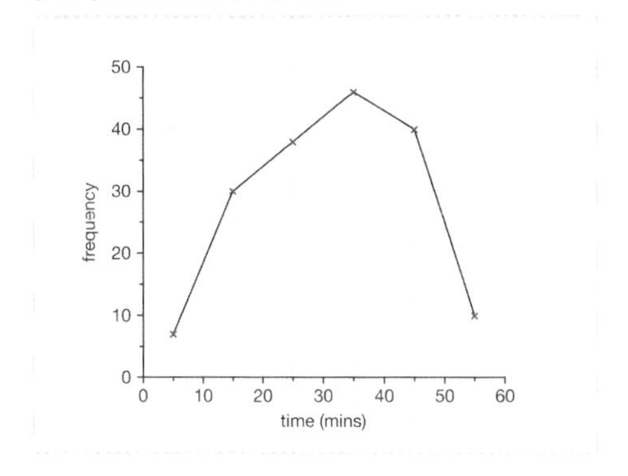

24. Which of the following graphs is a frequency polygon?

1/4

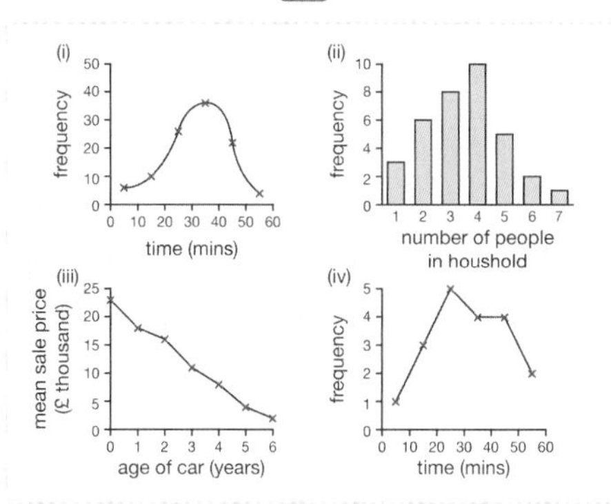

25. a b c The frequency polygon shows the times taken by runners to complete a 5 km race. Priya says the mode time is 35 minutes. Is Priya correct? Explain your answer.

- Open question, no set answer

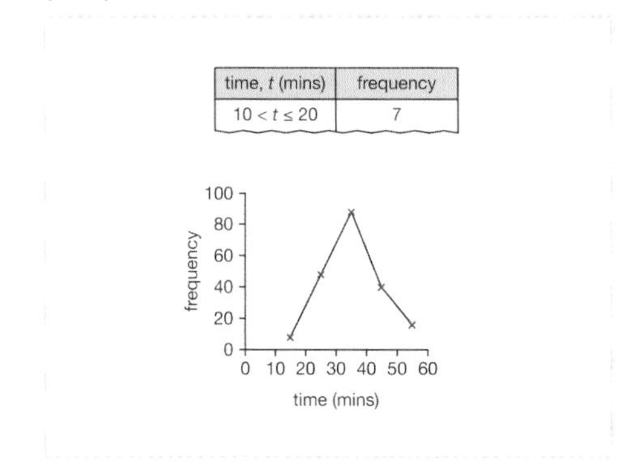

time, t (mins)	frequency
$10 < t \le 20$	7

Level 4: Problem Solving: Multi-step problems with frequency polygons.

Required: 5/5 **Pupil Navigation:** on **Randomised:** off

26. 1 2 3 The frequency polygon shows the shoe size of students in a class rounded to the nearest integer. What is the median shoe size?

- 6.5
- 6

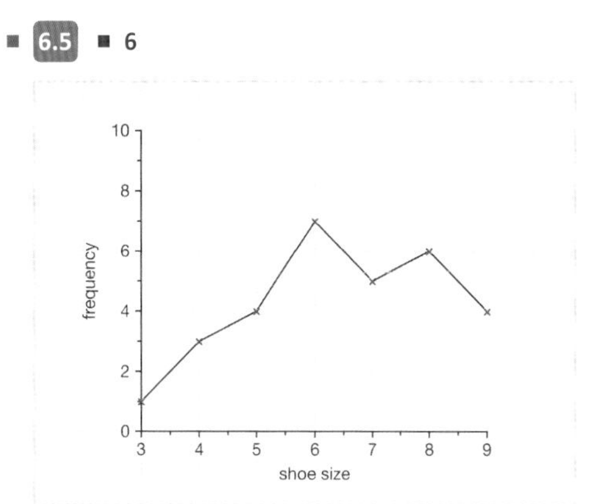

27. a b c The frequency polygon shows the number of siblings students in a class have. What percentage of the students have more than two siblings?
Include the % (percent) sign in your answer.

- 18
- 18%
- 9%
- 42%
- 9
- 42

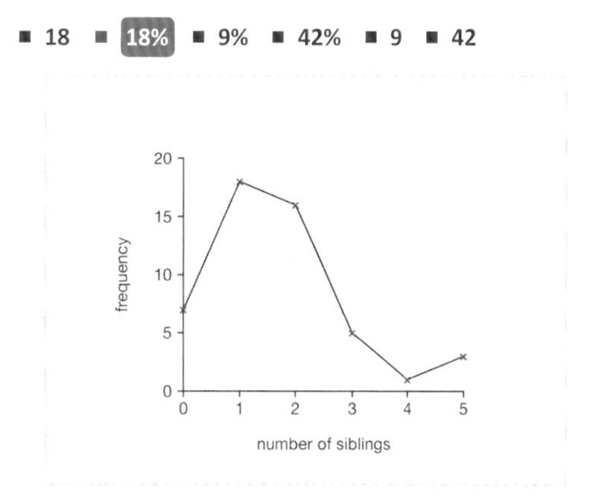

28. 1 2 3 The frequency polygon shows the shoe size of students in a class rounded to the nearest integer. What is the mean average shoe size?
Give your answer to one decimal place.

- 6.5

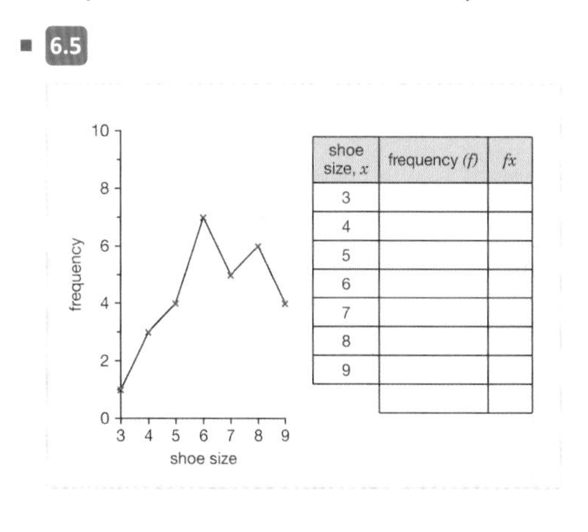

shoe size, x	frequency (f)	fx
3		
4		
5		
6		
7		
8		
9		

Level 4 *continued*

29. a b c A frequency polygon of grouped data is drawn for the heights of Year 9 students. If a student is selected at random, what is the probability that they are shorter than 160 cm?
Give your answer as a fraction in its simplest form.

- 4/9
- **20/45**

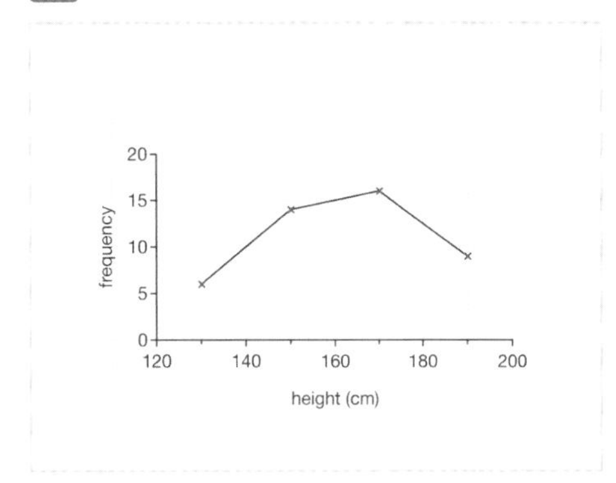

30. 1 2 3 The frequency polygon shows the number of siblings students in a class have. What is the mean number of siblings students have?
Give your answer to two decimal places.

- 1.68

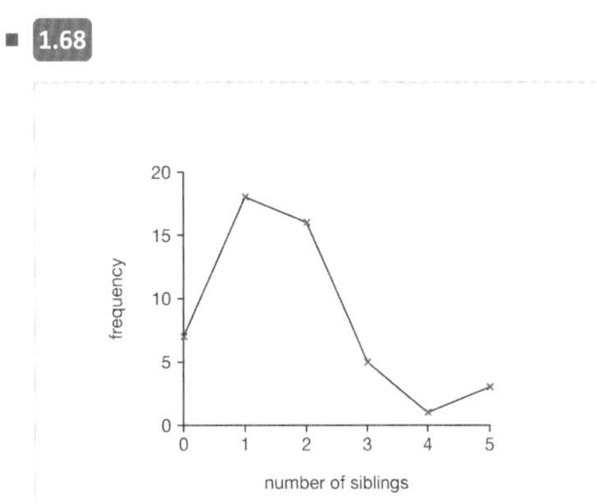

Statistics Topic Review

Objective: I can answer questions about statistics.

Quick Search Ref: 10729

Level 1: Understanding

Required: 10/10 **Pupil Navigation:** on **Randomised:** off

1. In a pie chart, what angle would represent the number of games lost by the football team?
Don't include the ° (degrees) symbol in your answer.

- **72**

Football team results

result	number of games	fraction of total	angle in pie chart
won	7	$\frac{7}{20}$	126°
drew	1		
lost	4		?°
total	20	1	360°

2. Here is a stem and leaf diagram of the ages of people who work in an office. What age is the oldest person in the office?

- 5
- **55**
- 9

stem	leaf
1	8 9 9
2	1 3 6 7 7 7 9
3	0 2 2 5 8
4	1 6 7
5	3 5

key: 2 | 3 means 23 years

3. The number of siblings that students in a class have is shown in a frequency polygon. How many students have two siblings?

- **16**

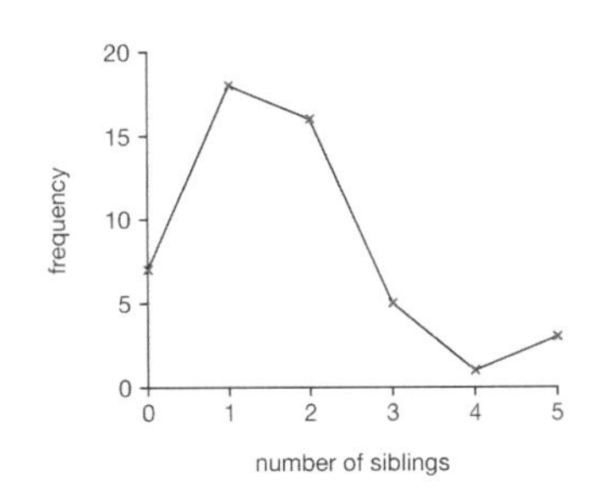

4. The scatter graph shows information about the distance of ten flats from London and their weekly rental cost. What type of correlation does the graph show?

1/3

- **negative**
- none
- positive

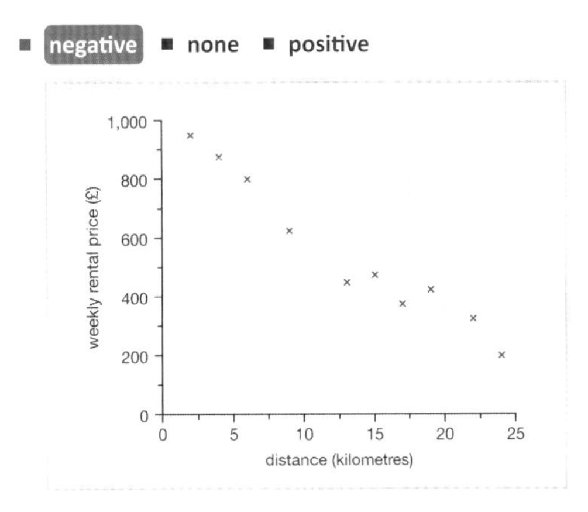

5. A group of students are asked about how many jobs they completed for their parents last week. The results are shown in the frequency table.
Give the **range** of the number of jobs completed.

- **5**
- 10

jobs completed	frequency
1	8
2	9
3	18
4	16
5	14
6	8

Level 1 ***continued***

6. Amy asked her friends how many pens they each had in their pencil case. The results are shown in the frequency table. What is the missing value in the table?

■ **20** ■ 9

number of pens	frequency	total number of pens
1	5	5
2	4	8
3	2	6
4	5	?
5	3	15
6	1	6

7. A nurse records the temperature of patients on her ward in a stem and leaf diagram. What is the lowest temperature of all the patients?
Don't include the units in your answer.

■ 361 ■ **36.1** ■ 1

stem	leaf
36	1 4 6 6 8
37	0 2 3 5 7 9
38	1

key: 37 | 5 means 37.5 °C

8. The frequency polygon shows the number of siblings students in a class have. How many students have four siblings?

■ 

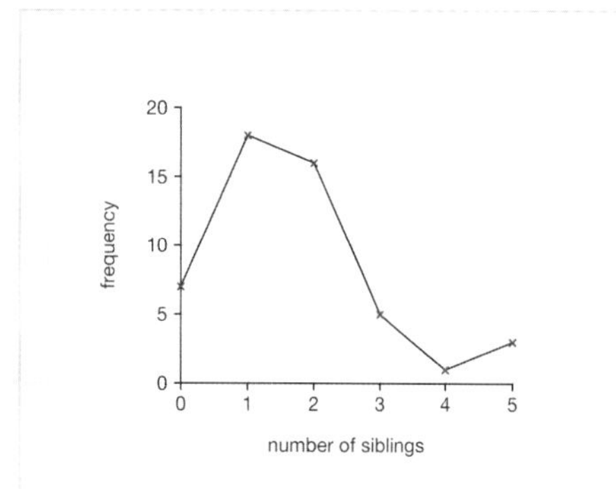

9. 24 football players scored goals for the school team last year. The frequency table gives information about the number of goals they scored.
1/6
Select the **modal** number of goals scored.

■ 2 ■ 3, 6 and 7 ■ 7 ■ 5 ■ **2 and 5** ■ 1

goals scored	frequency
1	3
2	8
3	1
4	2
5	8
6	1
7	1

10. The scatter graph shows the marks that a maths class received for two test papers. Joel scored 9 marks on the calculator paper. What was his score on the non-calculator paper?

■ 

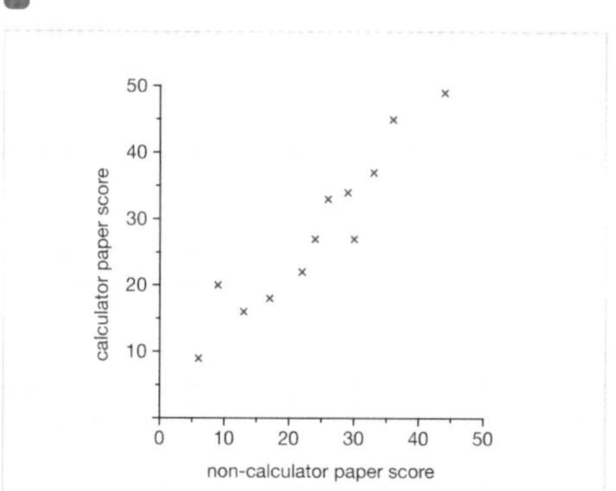

Level 2: Fluency

Required: 10/10 **Pupil Navigation:** on

Randomised: off

11. What angle in a pie chart would represent the number of Year 11s in drama club?
Don't include the ° (degrees) symbol in your answer.

■ **126**

Year groups of students in drama club

year group	fraction/ percentage of total	angle in pie chart
7	$\frac{3}{20}$	
8	$\frac{1}{5}$	
9	5%	
10	$\frac{1}{4}$	
11	35%	
total	1	360°

Level 2 *continued*

12. 1 2 3 Gabriel asked his friends from swimming club how many days per week they went to the pool. His results are shown in the frequency table. What is the mean number of days that Gabriel's friends went to the pool?

- **4**
- 44
- 11
- 12
- 14.7

number of days at the pool	frequency	total number of days at the pool
3	4	12
4	3	12
5	4	20
	11	44

13. 1 2 3 The back-to-back stem and leaf diagram shows the times boys and girls took to run 400 metres. What is the **median** of the times for the boys?
Give your answer in seconds and don't include the units.

- **93**

boys		girls
9 3	7	7
8 7 2 0	8	3 7 9
9 8 4 4 3 0	9	0 2 5 8
7 6 4	10	1 4 5 7 7 9
	11	3

key: 8 | 4 means 84 seconds

14. 1 2 3 Anton records the speed of 100 cars on the motorway. The results are shown in the frequency table. What is the mean speed of the cars in miles per hour?
Give your answer to one decimal place and do not include the units.

- **58.3**
- 5826
- 58
- 1165.2

speed (mph)	frequency	
56	20	
57		
58	14	
59	18	
60	32	

15. 1 2 3 Megan records how long some of her friends spend watching television in a week. Estimate the mean of this data set to the nearest whole hour.
Don't include the units in your answer.

- **6**
- 5.93

Time spent watching TV in a week

time, t (hours)	frequency (f)	midpoint (x)	fx
$0 < t \leq 2$	1	1	1
$2 < t \leq 4$	4	3	12
$4 < t \leq 6$	9	5	45
$6 < t \leq 8$	12	7	84
$8 < t \leq 10$	4	9	36
TOTAL	30		178

16. 1 2 3 Ten athletes enter a 10 km race. The number of weeks they spent training and their race times are shown in the scatter graph. Another athlete completes the race in 35 minutes. What is a reasonable estimate for the number of weeks they spent training for the race?

- **22**
- 20

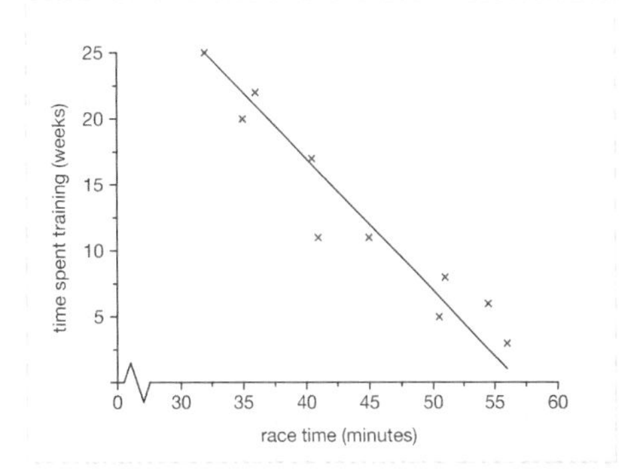

Level 2 *continued*

17. a b c Edward collects comic books. Calculate an estimate for the mean price of one of Edward's comic books.
Include the £ (pound) sign in your answer.

- **£5.26**
- **5.26**

Price of comic books

price, p (£)	frequency (f)	midpoint (x)	fx
$0 < p \le 2$	1		
$2 < p \le 4$	8		
$4 < p \le 6$	9		
$6 < p \le 8$	1		
$8 < p \le 10$	0		
$10 < p \le 12$	4		
TOTAL			

18. a b c Ruby records the number of paperclips in several boxes. The results are shown in the frequency table. What is the mean number of matchsticks in each box? Give your answer to one decimal place.

- **50.7**
- **202.8**
- **811**
- **51**

number of paperclips	frequency
49	2
50	4
51	7
52	3

19. 1 2 3 A frequency polygon of grouped data is drawn for the heights of Year 9 students. How many students are shorter than 180 centimetres?

- **36**
- **16**

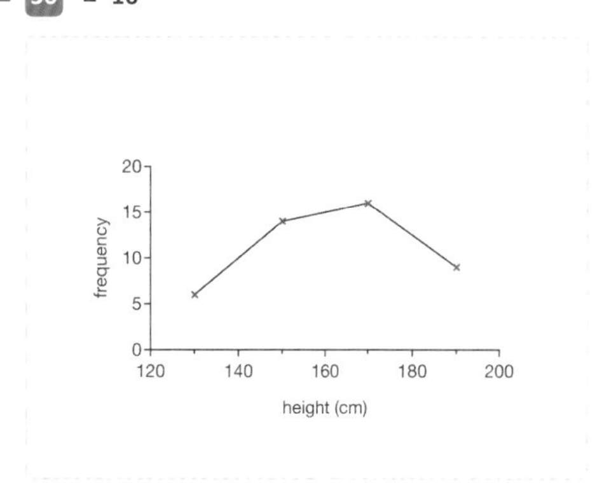

20. 1 2 3 Lucy works in a school office. She records the length of time spent on each phone call. Estimate the mean length of a phone call to the nearest minute.
Don't include the units in your answer.

- **3**
- **46**
- **3.05**

Time spent on a phone call

time, t (mins)	frequency (f)	midpoint (x)	fx
$0 < t \le 3$	41		
$3 < t \le 6$	12		
$6 < t \le 9$	2		
$9 < t \le 12$	5		
TOTAL			

Level 3: Reasoning

Required: 5/5 **Pupil Navigation:** on
Randomised: off

21. a b c The frequency polygon shows the times taken by runners to complete a 5 km race. Priya says the mode time is 35 minutes. Is Priya correct? Explain your answer.

- Open question, no set answer

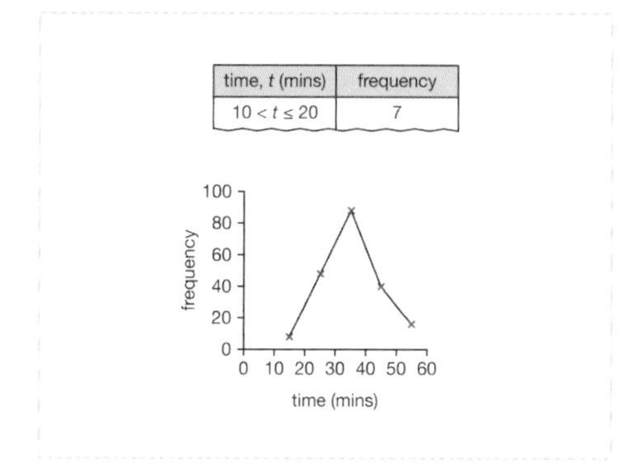

Level 3 *continued*

22. The frequency table shows the number of pets owned by 24 students. The mean number of pets owned is 2.5. If a student who owns 10 pets is added to the the table, what is the mean number of pets owned now?

1 2 3

Give your answer to one decimal place.

- **2.8**
- **2.9**

number of pets	frequency
0	2
1	4
2	4
3	8
4	6

23. A group of students each grow a sunflower as part of their homework and the height of the sunflower is recorded. The median height lies in the class interval $60 < h \le 80$. How many students had a sunflower with a height which was in the class interval $40 < h \le 60$?

1 2 3

- **4**

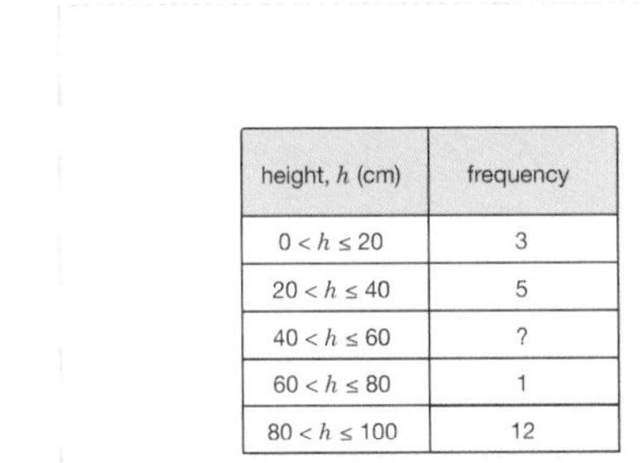

height, h (cm)	frequency
$0 < h \le 20$	3
$20 < h \le 40$	5
$40 < h \le 60$	?
$60 < h \le 80$	1
$80 < h \le 100$	12

24. The back-to-back stem and leaf diagram shows the times boys and girls took to run 400 metres. Select the statements which correctly compare their times.

2/4

- **The mode time is slower for the boys than for the girls.**
- **The range of times is smaller for the boys than for the girls.**
- **The range of times is larger for the boys than for the girls.**
- **The mode time is faster for the boys than for the girls.**

boys		girls
9 3	7	7
8 7 2 0	8	3 7 9
9 8 4 4 3 0	9	0 2 5 8
7 6 4	10	1 4 5 7 7 9
	11	3

key: 8 | 4 means 84 seconds

25. The mean for a grouped frequency table is only an estimate. Explain why.

a b c

- Open question, no set answer

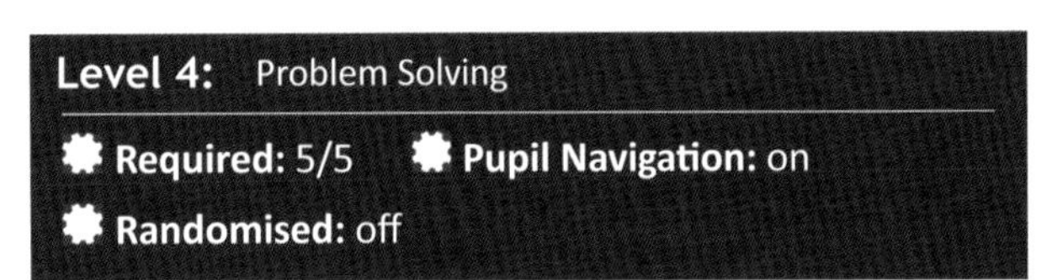

Level 4: Problem Solving

Required: 5/5 **Pupil Navigation:** on

Randomised: off

26. The frequency polygon shows the shoe size of students in a class rounded to the nearest integer. What is the median shoe size?

1 2 3

-  **6.5**
- **6**

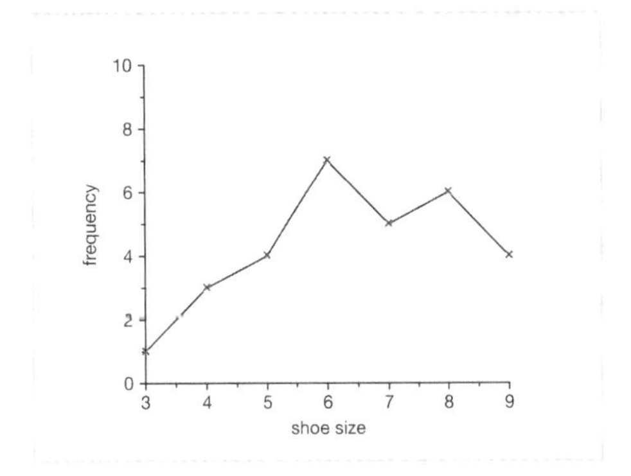

Level 4 *continued*

27. The total number of points scored by everybody in a competition was 36. How many people took part in the competition?

number of points	frequency
0	4
1	5
2	
3	1
4	2
	?

28. A biologist makes a stem and leaf diagram of the weights of salmon in a river. The biologist can only carry 25 kg of fish. What is the greatest number of fish that they can carry?

0	8 9
1	2 5 8 8 9
2	0 1 2 3 5 7
3	0 2 4

key: 2 | 4 means 2.4 kg

29. Erica spilled ink all over the frequency column of her homework before she answered question 5. What is the correct answer to question 5?

time (t seconds)	frequency
$0 < t \le 20$	
$20 < t \le 25$	
$25 < t \le 40$	
$40 < t \le 50$	
$50 < t \le 100$	

1. What is the width of the modal class interval? 5
2. What is the range of times? 30 seconds
3. How many students are in the class interval $25 < t \le 40$? 14
4. How many students are there in total? 30
5. How many students are in the class interval $40 < t \le 50$? ____

30. Using the data shown, calculate the difference between the highest possible mean and the estimated mean.
Don't include the units in your answer.

Width of computer screens

width (inches)	frequency (f)	midpoint (t)	fx
$10 < w \le 12$	7		
$12 < w \le 14$	4		
$14 < w \le 16$	9		
$16 < w \le 18$	3		
$18 < w \le 20$	1		
TOTAL			

Mathematics Y9

Probability

Probability Scale
Probability of Single Events
Probability of Combined Events
Sets and Venn Diagrams

Understand and Use the Probability Scale

Objective: I can understand and use probability vocabulary and the probability scale.

Quick Search Ref: 10416

Level 1: Understand: Using probability vocabulary with the probability scale.

 **Required:** 7/10 **Pupil Navigation:** on **Randomised:** off

1. Probability is . . .

1/3

- when two ratios are equal.
- how much of one thing there is compared to another thing.
- **how likely something is to happen.**

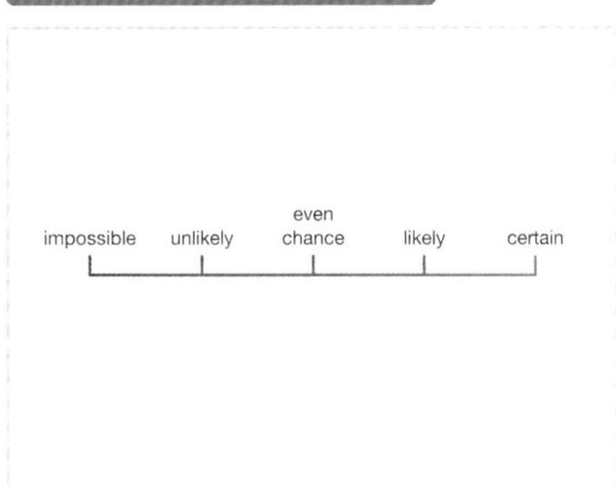

2. Which arrow represents 'impossible' on the probability scale?

1/3

- **A**
- B
- C

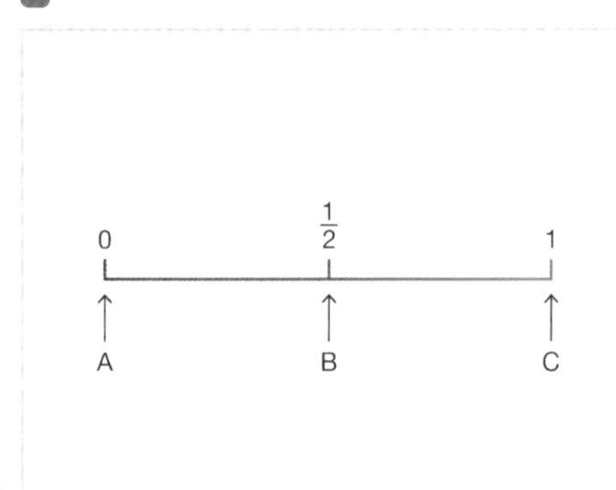

3. What does arrow A represent on the probability scale?

1/5

- unlikely
- likely
- impossible
- **certain**
- even chance

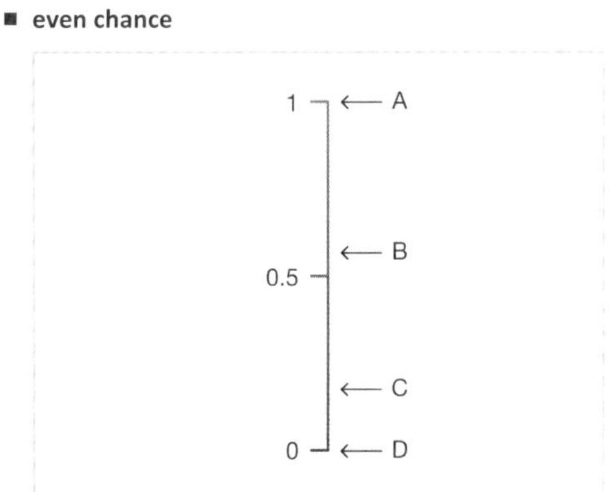

4. Which arrow best represents 'likely' on the probability scale?

1/5

- V
- W
- X
- **Y**
- Z

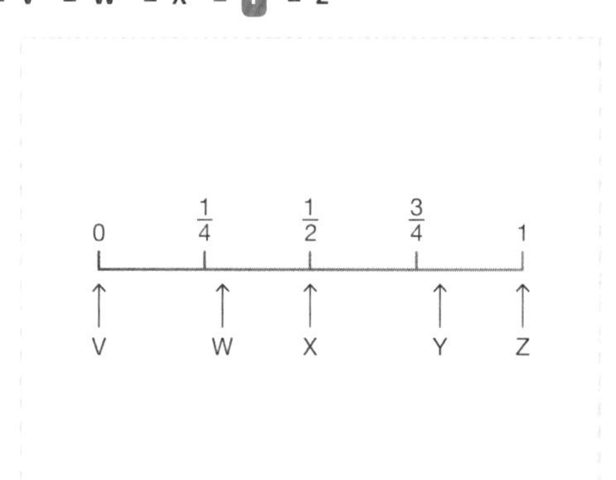

5. Which arrow best represents 'unlikely' on the probability scale?

1/5

- S
- **T**
- U
- V
- W

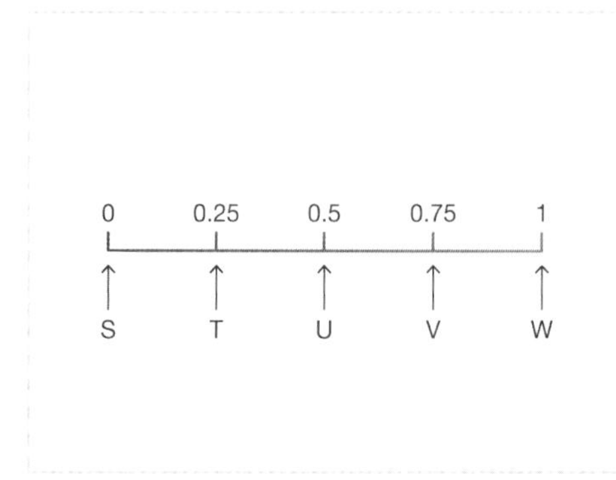

6. Which arrow represents 'even chance' on the probability scale?

1/5

- D
- E
- **F**
- G
- H

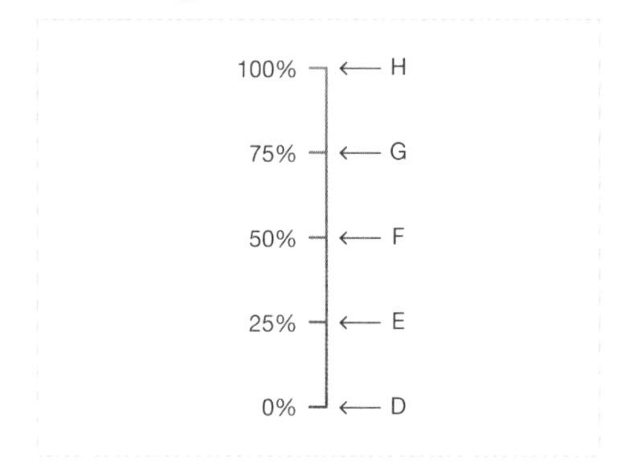

Level 1 *continued*

7. Arrange the options in ascending order of likelihood (least likely first).

event	probability
A	$\frac{3}{10}$
B	1
C	0
D	$\frac{3}{5}$
E	$\frac{7}{10}$

0 $\frac{1}{2}$ 1

8. Which arrow represents 'certain' on the probability scale?

1/5

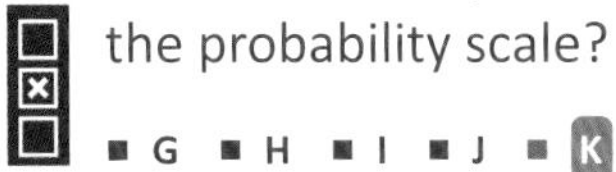

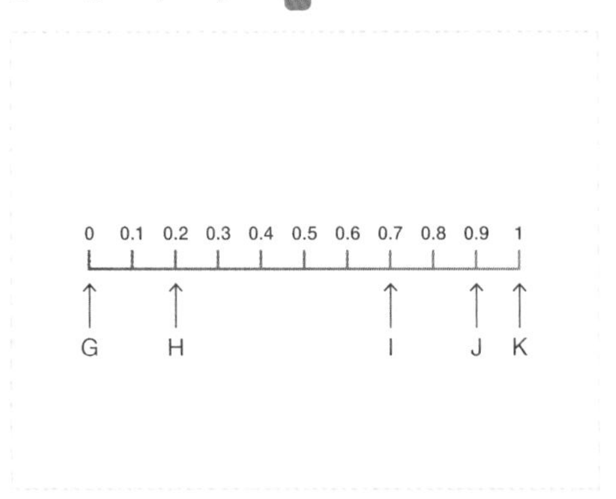

9. Which arrow represents 'even chance' on the probability scale?

1/5

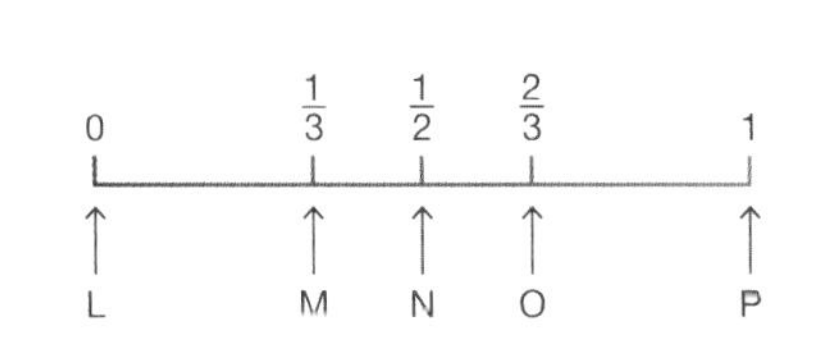

10. What does the arrow represent on the probability scale?

1/5

- certain
- unlikely
- impossible
- likely
- even chance

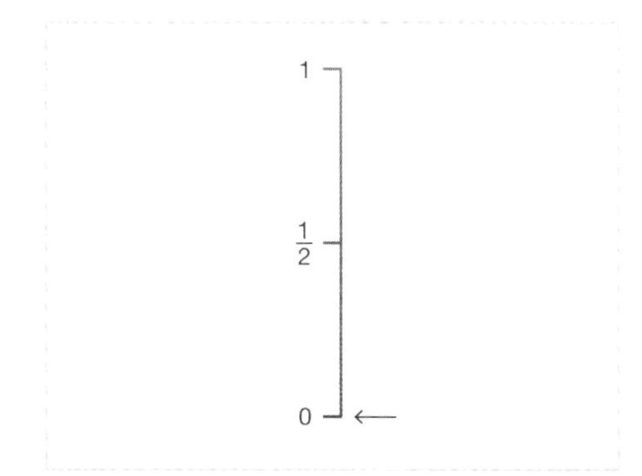

Level 2: Fluency: Estimating the probability of events and placing them on the probability scale.

Required: 8/10 **Pupil Navigation:** on

Randomised: off

11. Which event could be represented by the arrow shown?

1/4

- You will grow a second nose.
- It will snow next week.
- The sun will rise tomorrow.
- The next child born is a boy.

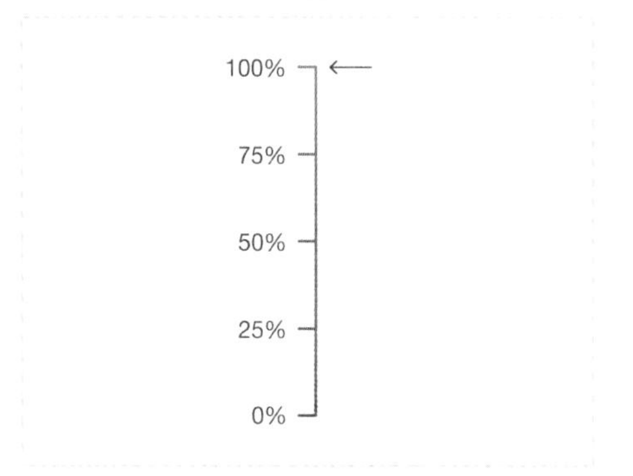

12. Which arrow best represents the probability of you selecting at random a tile with a vowel on it?

1/4

- (i)
- (ii)
- (iv)
- (v)

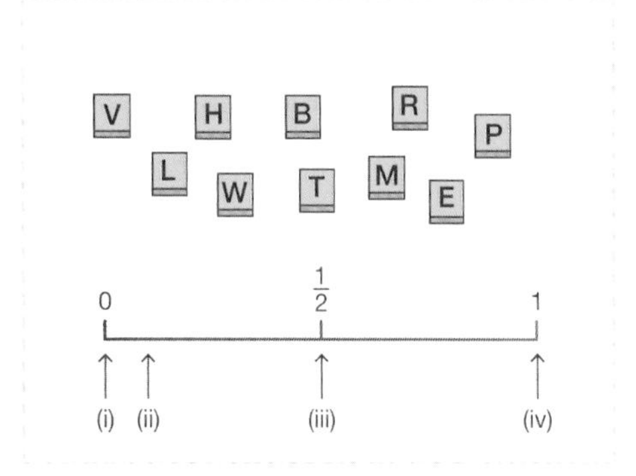

Level 2 *continued*

13. A fair 6-sided dice has sides marked 2, 2, 2, 2, 5 and 5. What is the likelihood of rolling a 2?

1/5

- likely
- impossible
- certain
- even chance
- unlikely

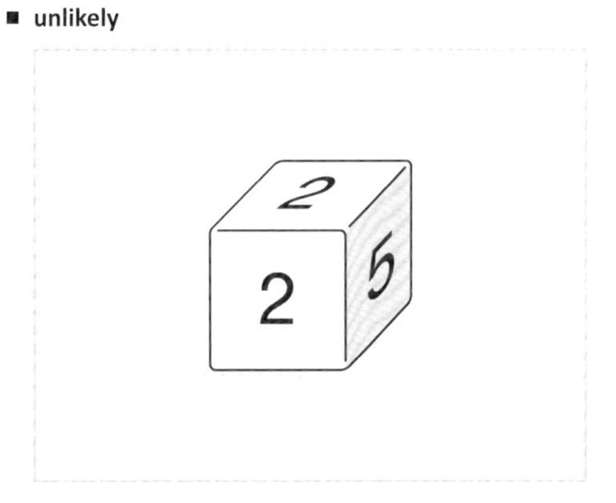

14. Which arrow best represents the probability of you being older than your biological mother?

1/4

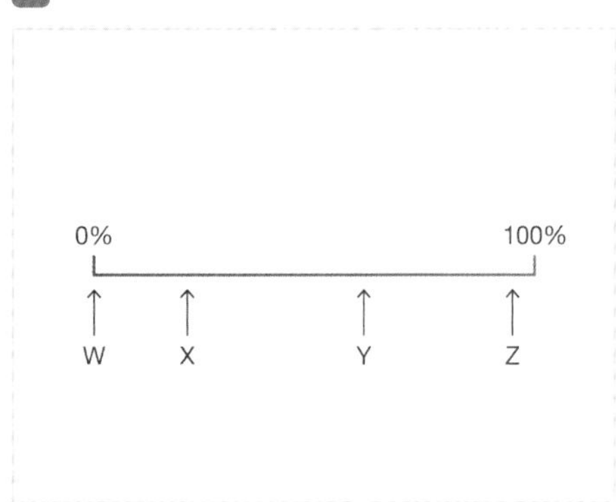

15. Eliza chooses a counter from a bag containing only red and orange counters. What is the likelihood of her choosing a blue counter?
Give your answer as a percentage.

- 0%
- 0

0% 50% 100%

16. If Mark flips a fair coin, what is the likelihood it will land on tails?
Give your answer as a fraction.

- 1/2

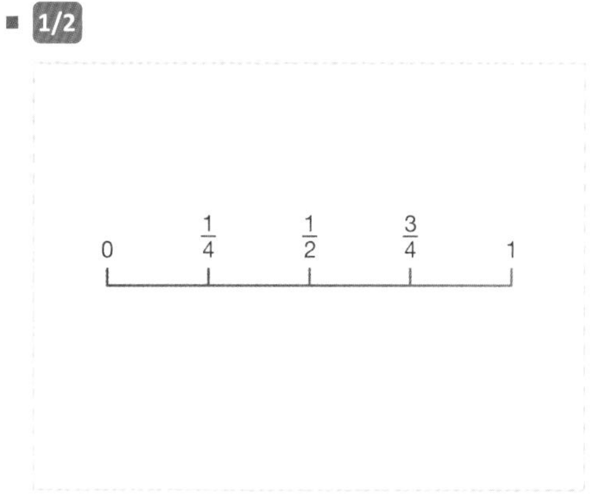

17. If you pull a random counter from the bag, what number is it most likely to have on it?

18. Rosie picks a piece of fruit from a bag of apples. What is the probability she picks an apple?
Give your answer as a percentage.

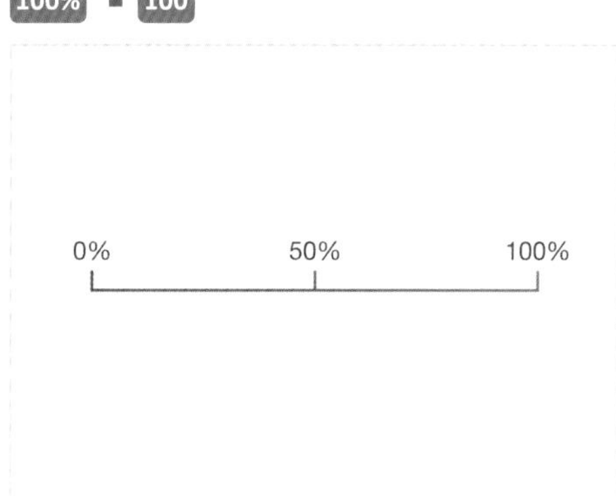

Level 2 *continued*

19. Which event could be represented by the arrow shown?

1/4

- rolling a 4 on a regular 6-sided dice
- a fair coin landing on heads
- choosing the King of Hearts from a deck of cards
- having your birthday on the 35th of April

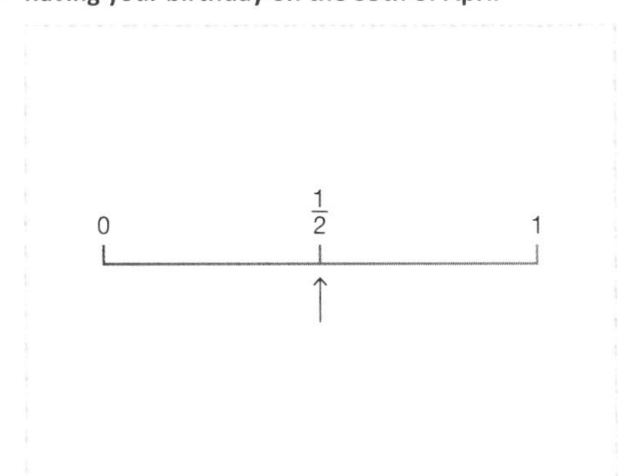

20. If Bella takes a random piece of fruit from those shown, which piece of fruit is she least likely to pick?

1/4

- banana
- orange
- apple
- peach

Level 3: Reasoning: Probability can't be more than one, ordering events according to probability.

Required: 5/5 **Pupil Navigation:** on

Randomised: off

21. Bob says the probability of his football team winning is greater than 1. Can Bob be correct? Explain your answer.

- Open question, no set answer

22. Which of these lottery tickets is most likely to win?

1/6

- None will win
- (a)
- (b)
- (c)
- (d)
- All have the same likelihood of winning

23. Jamil says, "The probability of throwing heads on a fair coin is 0.5, so the probability of throwing 3 heads in a row is 1.5". Is Jamil correct? Explain your answer.

- Open question, no set answer

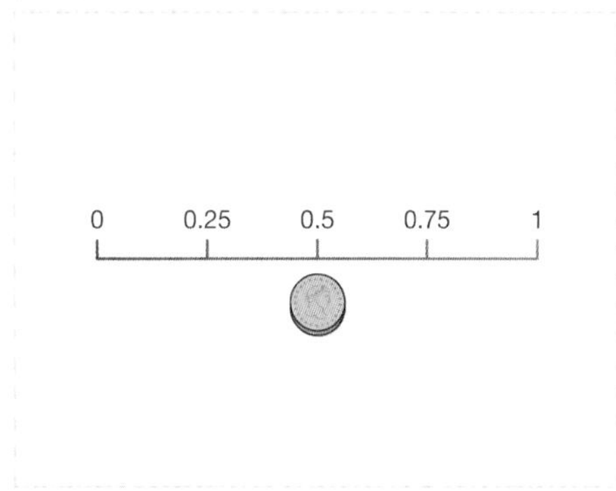

24. Arrange the events in descending order according to likelihood (most likely first).

- An object thrown into the air will fall back down.
- A positive integer selected at random will be odd.
- Aliens will land on Earth.
- You will get younger tomorrow.

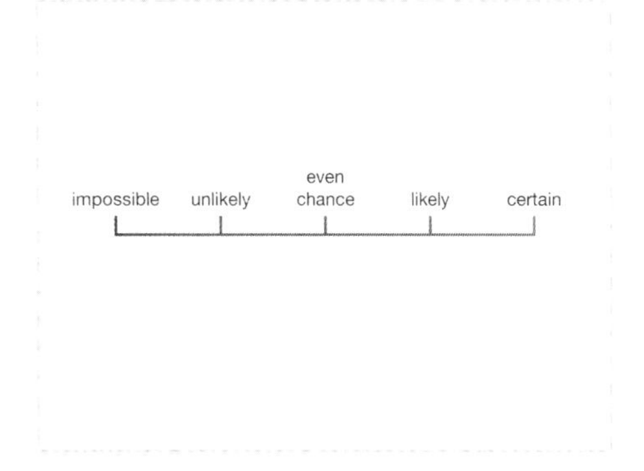

Level 3 *continued*

25. Jenny is playing a game with a standard, fair six-sided dice. The game costs £1 per roll and if she rolls a 6, she wins £5. Is Jenny likely to win more money than she spends? Explain your answer.

- Open question, no set answer

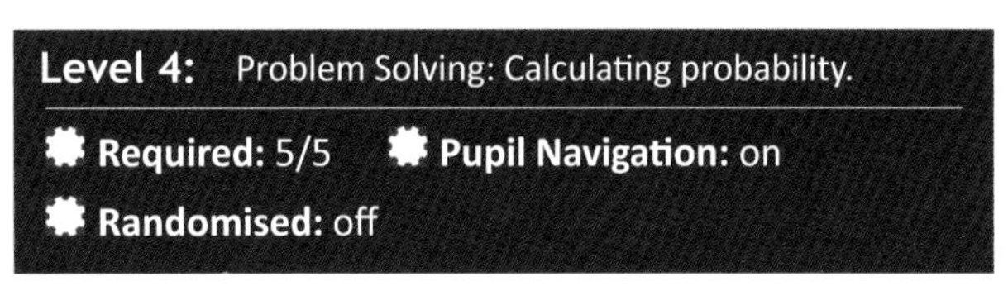

26. What is the probability of randomly picking a triangle from the bag?
Give your answer as a fraction.

- 1/5

27. What is the probability of the spinner stopping on a shaded area?
Give your answer as a decimal.

- 0.25

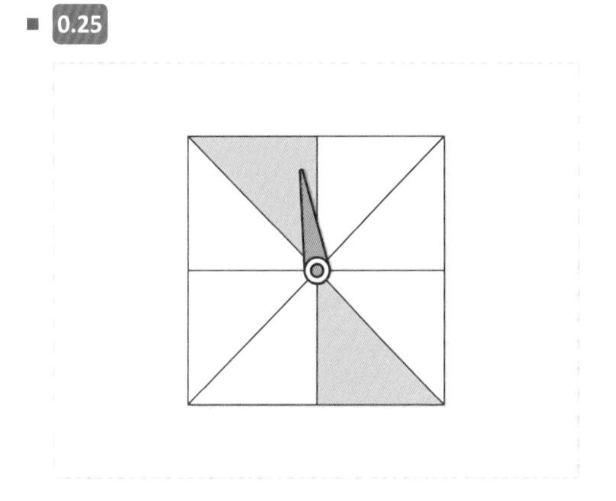

28. There are four types of sweets in a bag: raspberry, lime, orange and cherry. Evie is asked to pick one at random. There is a 40% chance that she will pick raspberry, a 12% chance that she will pick lime and a 27% chance that she will pick orange. What is the probability that she will pick cherry?
Give your answer as a percentage.

- 21%
- 21

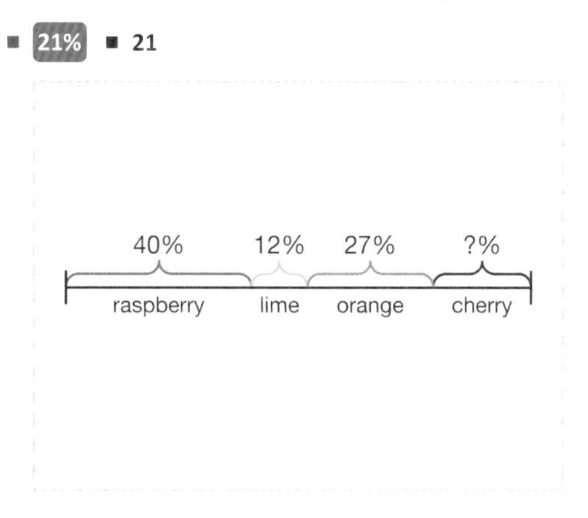

29. For this group of children, what is the probability that two children were born on the same day of the week?
Give your answer as a decimal.

- 1

30. Fiaz is rolling a fair, regular 6-sided dice. Arrange the events in ascending order of likelihood (least likely first).

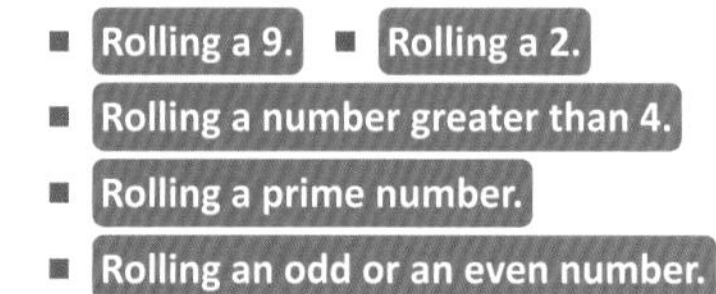

Understand That All Probabilities Add up to 1

Objective: I can use the fact that probabilities add to 1 to calculate the probabilities of events not happening and of outcomes within exhaustive events.

Quick Search Ref: 10423

Level 1: Understanding: Calculating the probability of events not happening.

Required: 7/10 **Pupil Navigation:** on **Randomised:** off

1. Probability is . . .

1/3

- when two ratios are equal.
- how much of one thing there is compared to another thing.
- **how likely something is to happen.**

2. The probability of an event . . .

1/2

- can be more than 1.
- **cannot be more than 1.**

3. A set of events are exhaustive if . . .

1/3

- it is possible for none of them to occur.
- **at least one of the events must occur.**
- the events can't happen at the same time.

4. If the probability of it raining tomorrow is 40%, what is the probability of it not raining?
Give your answer as a percentage including the % (percent) sign.

- **60%**
- 60

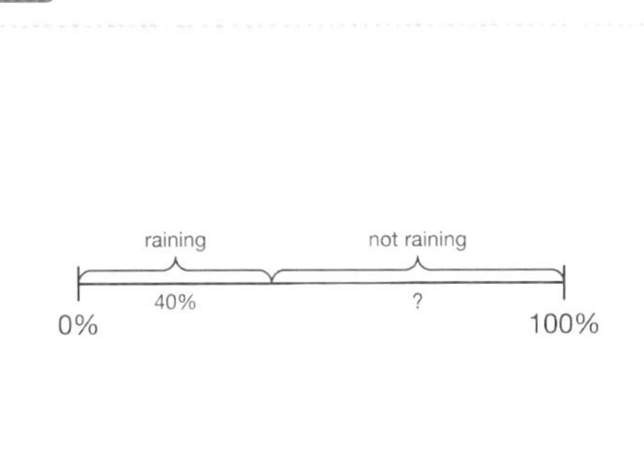

5. If the probability of being chosen for the school chess team is 2/5, what is the probability of not being chosen?
Give your answer as a fraction.

- **3/5**

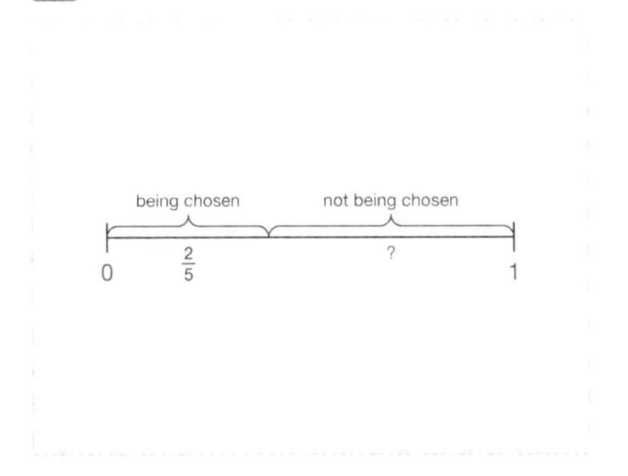

6. A bag contains only blue and red counters. If the probability of picking a red counter is 0.4, what is the probability of picking a blue counter?
Give your answer as a decimal.

- **0.6**

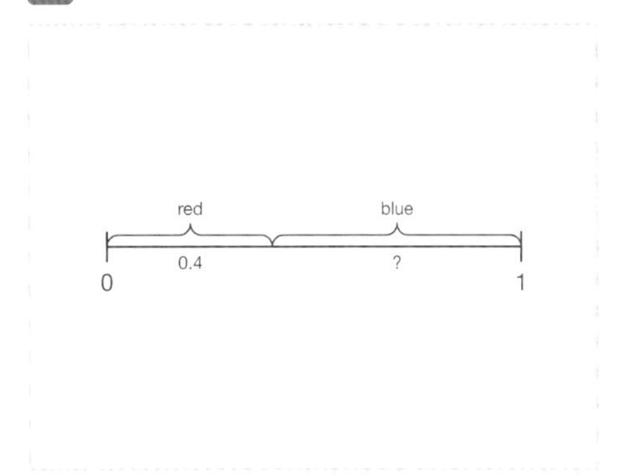

7. The probability of picking a boy from a group of children is 1/3. What is the probability of picking a girl?
Give your answer as a fraction.

- **2/3**

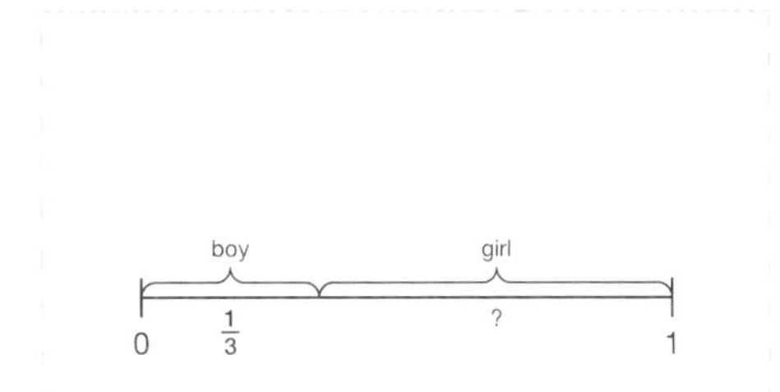

Level 1 *continued*

8. If the probability of it snowing tomorrow is 1/8, what is the probability of it not snowing?
Give your answer as a fraction.

- 7/8

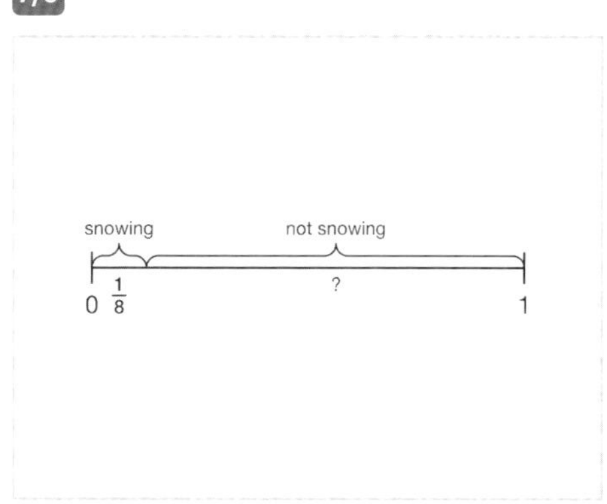

9. A biased coin has a 0.59 probability of landing on heads. What is the probability it won't land on heads?
Give your answer as a decimal.

- 0.41

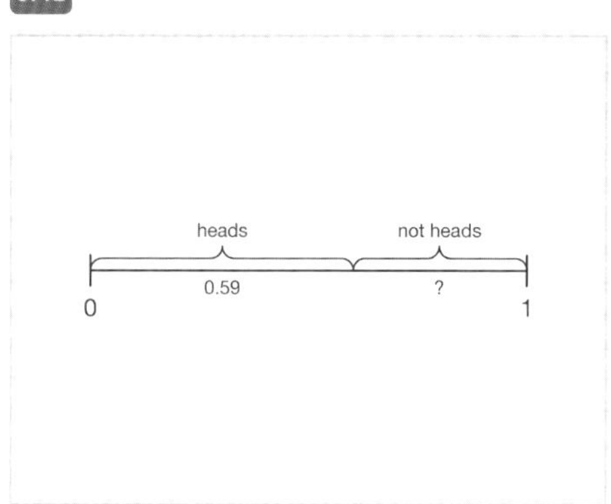

10. At break time, children are given an apple or an orange. If there is a 52% probability they will get an apple, what is the probability they will get an orange?
Give your answer as a percentage including the % (percent) sign.

- 48%
- 48

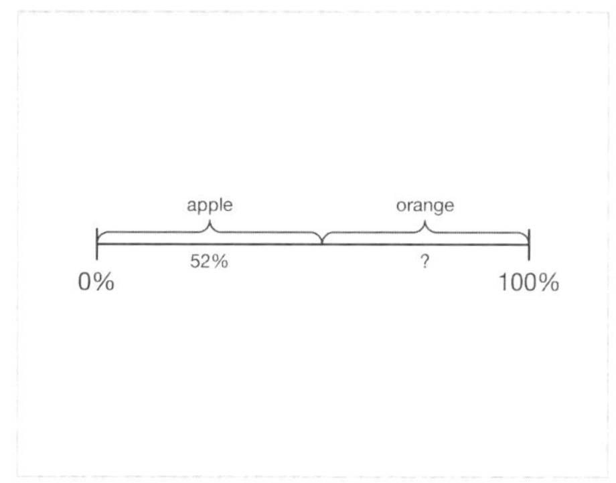

Level 2: Fluency: Calculating the probability of exhaustive events.

Required: 7/10 **Pupil Navigation:** on
Randomised: off

11. What is the probability of not rolling a 3 on a fair, standard six-sided dice?
Give your answer as a fraction.

- 5/6
- 1/6

12. There are only red, green and blue counters in a bag. If the probability of choosing a a red counter is 20% and the probability of choosing a green counter is 25%, what is the probability of choosing a blue counter?
Give your answer as a percentage including the % (percent) sign.

- 55
- 55%
- 45
- 45%

13. There are yellow, white, black, red and orange coloured marbles in a bag. If each colour has the same chance of being picked, what is the probability of not choosing a red marble?
Give your answer as a fraction.

- 4/5
- 1/5

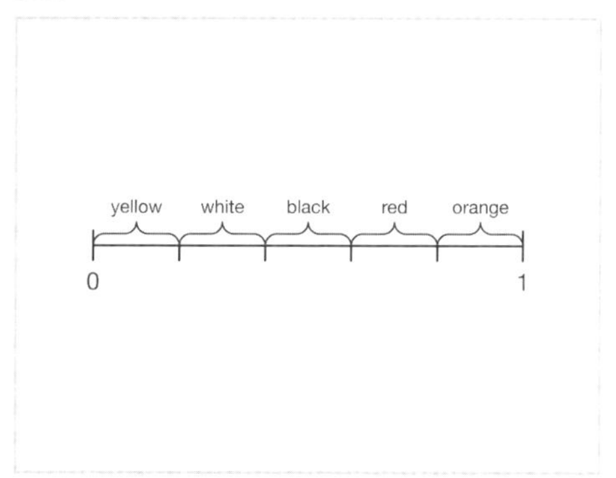

14. Children are being split into three groups: A, B and C. There is a 0.52 chance of being in group A and an equally likely chance of being in group B as in group C. What is the probability of being in group C?
Give your answer as a decimal.

- 0.24
- 0.48

Level 2 *continued*

15. The probability of each student being chosen for a race is shown in the table. What is the probability that Jim will be picked?
Give your answer as a fraction in its simplest form.

- **1/6** (correct)
- 5/6
- 5/30

Lucy	Jim	Chloe	Nick
$\frac{1}{5}$	?	$\frac{3}{10}$	$\frac{1}{3}$

16. The probability of each student getting the highest mark in the class is shown in the table. If Charlie is twice as likely as Grace to get the highest mark, what is the probability that Charlie will get the highest mark?
Give your answer as a decimal.

- **0.3** (correct)
- 0.45
- 0.15

Luke	Grace	Michelle	Charlie
0.32	a	0.23	$2a$

17. What is the probability Billy won't pick a 9?
Give your answer as a percentage including the % (percent) sign.

- 30%
- **70%** (correct)
- 70
- 30

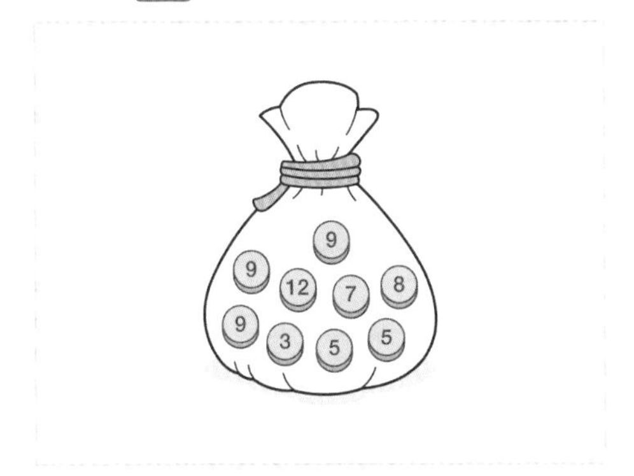

18. What is the probability of not picking a raspberry sweet?
Give your answer as a fraction.

- **4/7** (correct)
- 3/7

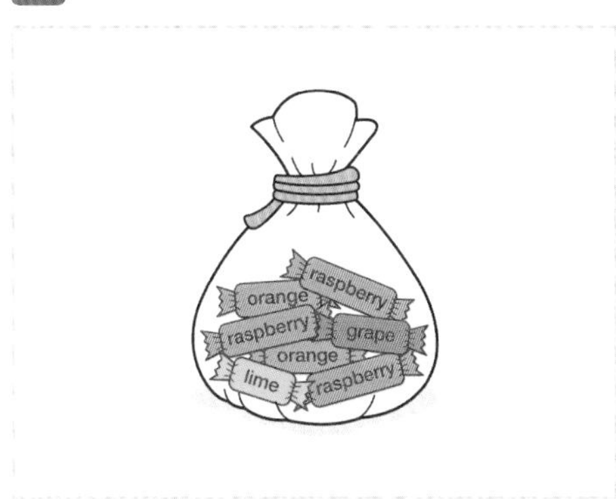

19. Lilly chooses a number from 20 to 29 (inclusive). What is the probability of her not choosing a prime number?
Give your answer as a decimal.

- **0.8** (correct)
- 0.2

20. The probability of each student being chosen for the main part in the school play is shown in the table. Jenny and Becca have an equal chance of being picked. What is the probability of Becca getting the part?
Give your answer as a decimal.

- **0.1** (correct)
- 0.2

Jenny	?
Romana	0.3
Elsie	0.45
Becca	?
Isla	0.05

Level 3: Reasoning: With probability questions.

Required: 5/5 **Pupil Navigation:** on
Randomised: off

21. Select the two numbers that cannot be a probability.

2/5

- 0
- **-1** (correct)
- 23%
- 2/7
- **1.2** (correct)

Level 3 *continued*

22. abc When rolling a fair, standard six-sided dice, are the following events exhaustive? Explain your answer.

- rolling an even number
- rolling a multiple of 5
- rolling a prime number

- Open question, no set answer

23. abc Rahib says, "The probability of picking a circle is 0.6 and the probability of picking a square is 0.5". Is Rahib correct? Explain your answer.

- Open question, no set answer

24. ↑↓ Four bags contain different counters as shown. Arrange the statements about the contents of the bags into the same order as the bags they describe.

- The probability of picking a 2 is 0.
- The probability of not picking a 1 is 50%.
- The probability of not picking a 2 is 7/8.
- The probability of not picking a 2 is 0.2.

25. abc Jasper says, "There is a greater chance of spinning red on spinner 2 than on spinner 1". Explain why Jasper is incorrect.

- Open question, no set answer

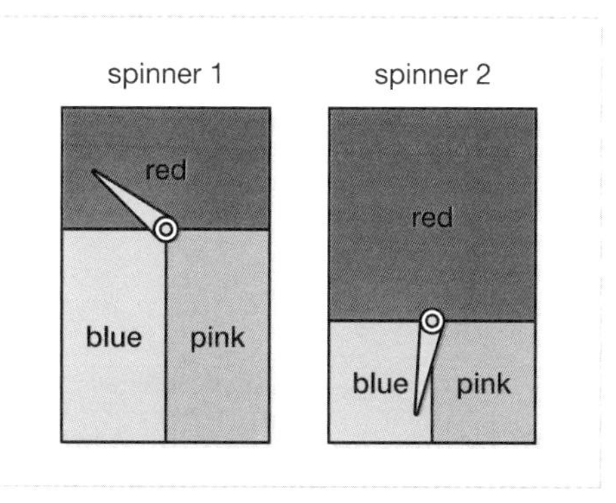

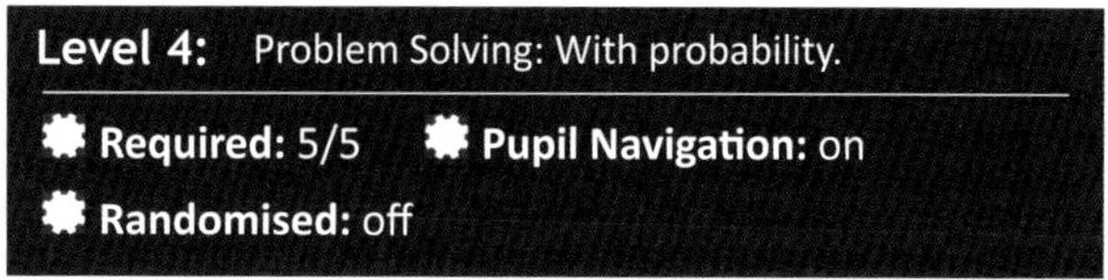
Level 4: Problem Solving: With probability.

Required: 5/5 **Pupil Navigation:** on

Randomised: off

26. 123 A bag contains red, purple, green and orange counters only. The probability of picking a counter of each colour is shown in the table. What is the probability of picking a green counter?
Give your answer as a decimal.

- 0.1
- 0.35
- 0.9

colour	red	purple	green	orange
probability	$\frac{1}{4}$	0.6	?	5%

27. abc There are only red, green and yellow marbles in a bag. The probability of picking a red marble is x, green is $2x$ and yellow is $3x$. What is the value of x as a fraction?

- 1/6

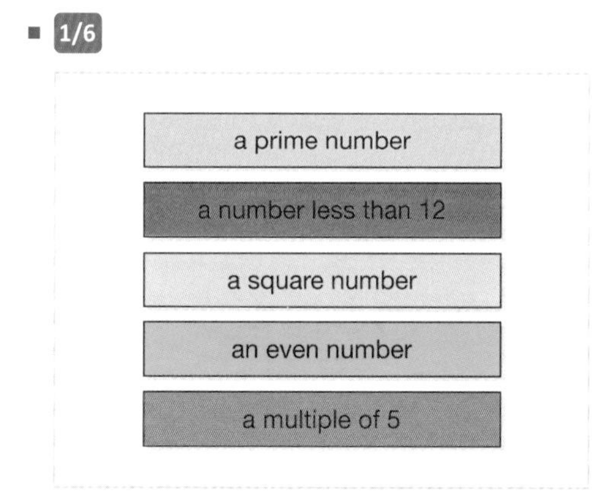

Level 4 *continued*

28. abc A spinner has equally-sized red, yellow, purple, green and blue sections. If the pointer is spun twice, what is the probability that it lands on red on both spins?

- **1/25**
- 1/5
- 2/5

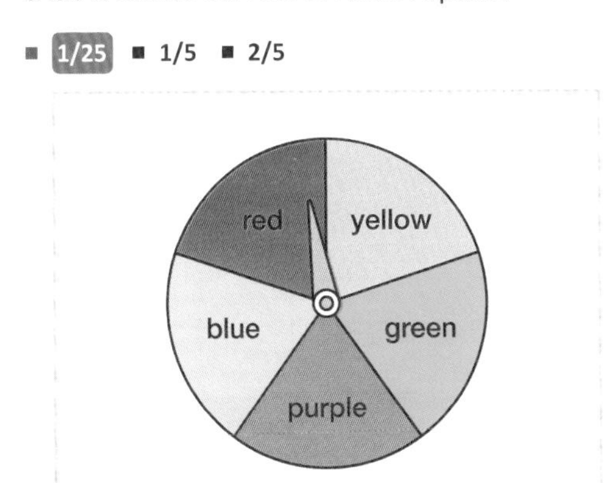

29. 123 A bag contains 15 pink counters and 11 yellow counters. More counters are added in the ratio of 1 pink for every 2 yellow. If the probability of picking a pink counter is now ½, how many counters were added?

- 19
- **12**
- 38

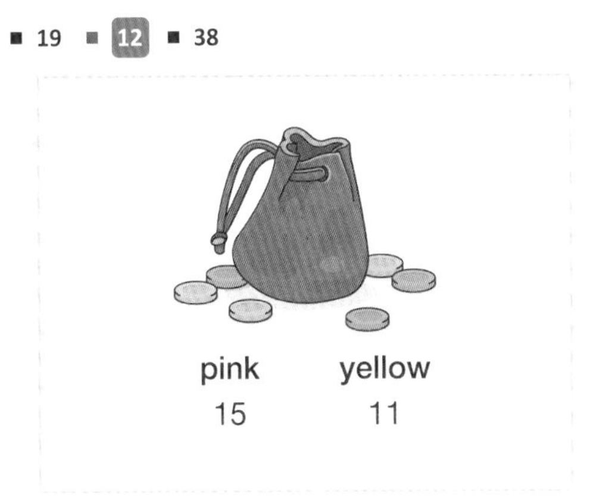

30. abc What is the difference between the probabilities of spinner A and spinner B landing on 4?
Give your answer as a fraction in its simplest form.

- **1/5**
- 0.2
- 4/20

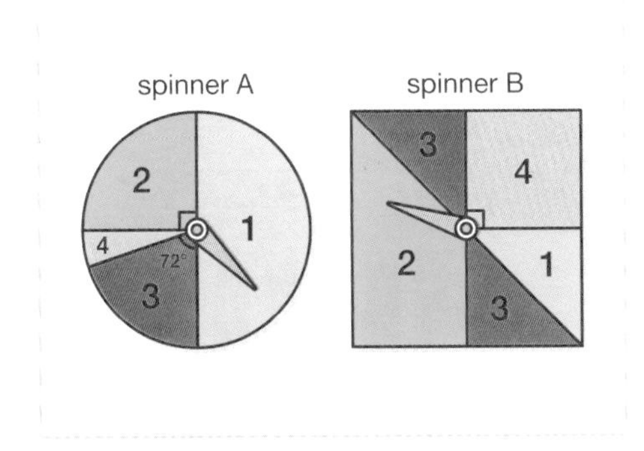

Find Probabilities for Single Events

Objective: I can calculate the probability of a single event.

Quick Search Ref: 10419

Level 1: Understanding: Probability of single events.

Required: 7/10 **Pupil Navigation:** on **Randomised:** off

1. Probability is . . .

1/3

- how much of one thing there is compared to another thing.
- how likely an event is to happen.
- when two ratios are equal.

2. Mutually exclusive events . . .

1/3

- always happen at the same time.
- sometimes happen at the same time.
- never happen at the same time.

3. What is the probability of rolling a 5 on a fair, standard six-sided dice?
Give your answer as a fraction.

- 1/6 ■ 1 ■ 5/6

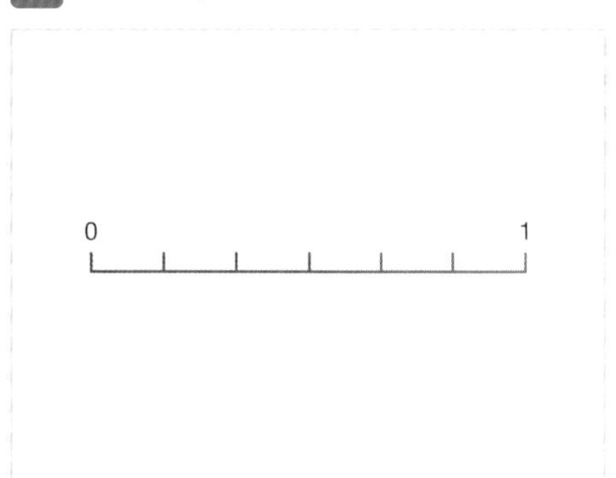

4. When a coin is flipped, what is the probability of the coin landing on tails?
Give your answer as a decimal.

- 0.5

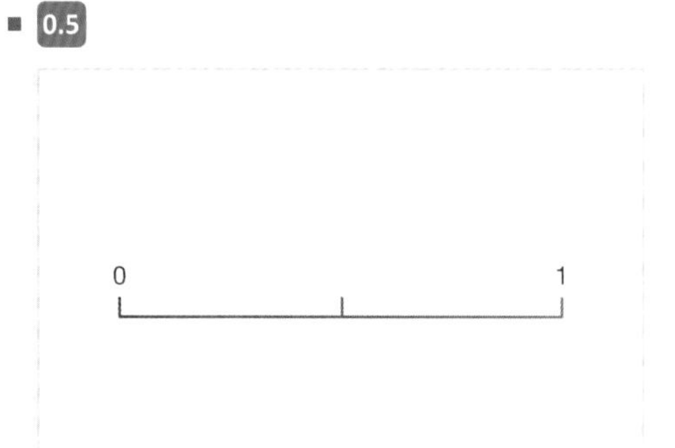

5. What is the probability of the spinner landing on red?
Give your answer as a percentage including the % (percent) sign.

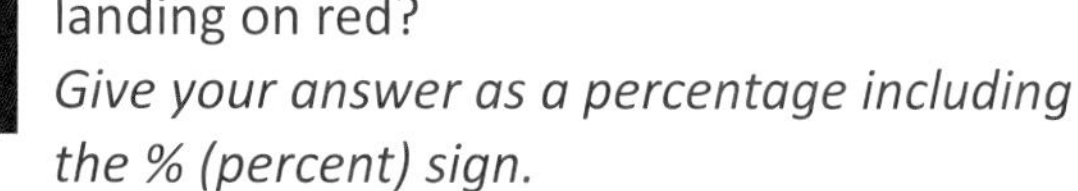

- 25% ■ 25

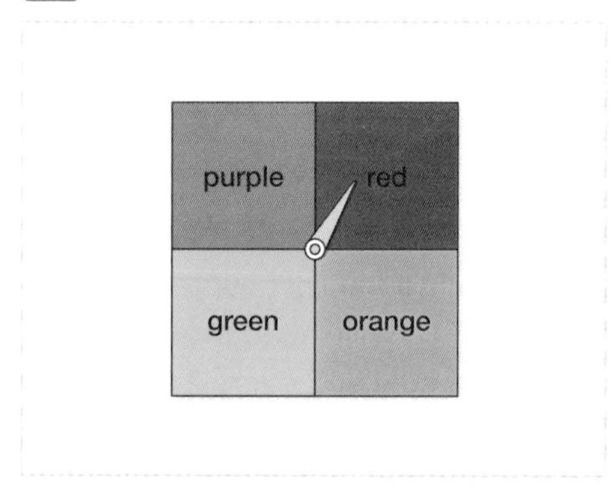

6. Cards numbered 1 to 9 are placed in a bag. What is the probability of choosing the card numbered 7?
Give your answer as a fraction.

- 1/9 ■ 1 ■ 1/8

7. What is the probability of choosing the strawberry ice pop?
Give your answer as a decimal.

- 0.2 ■ 0.5 ■ 0.25

Level 1 *continued*

8. Millie has 8 cards numbered 1 to 8. What is the probability of someone choosing the card with 2 on it?
Give your answer as a fraction.

- **1/8**
- **1**

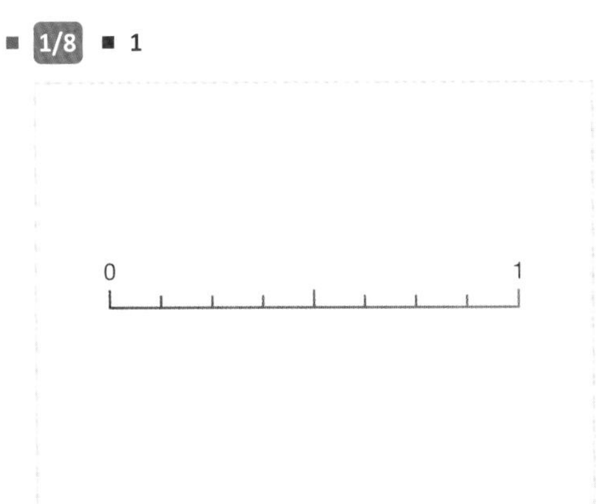

9. What is the probability of choosing the yellow marble?
Give your answer as a percentage including the % (percent) sign.

- **50%**
- **50**

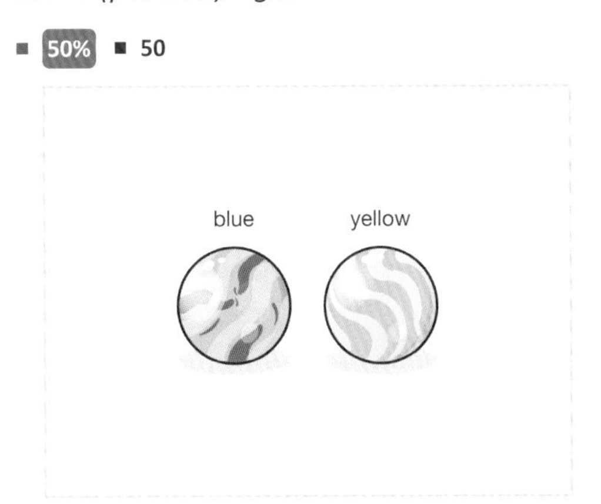

10. What is the probability of picking the cube?
Give your answer as a decimal.

- **0.25**

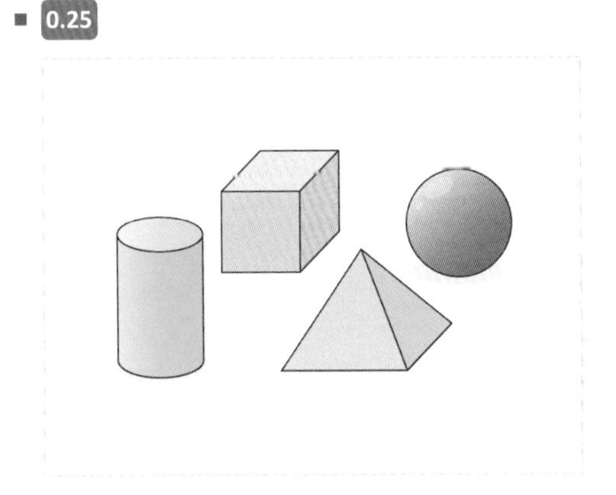

Level 2: Fluency: Probability with multiple outcomes.

Required: 7/10 **Pupil Navigation:** on
Randomised: off

11. What is the probability of rolling an even number with a fair 6-sided dice?
Give your answer as a fraction in its simplest form.

- **3**
- **1/2**
- **3/6**

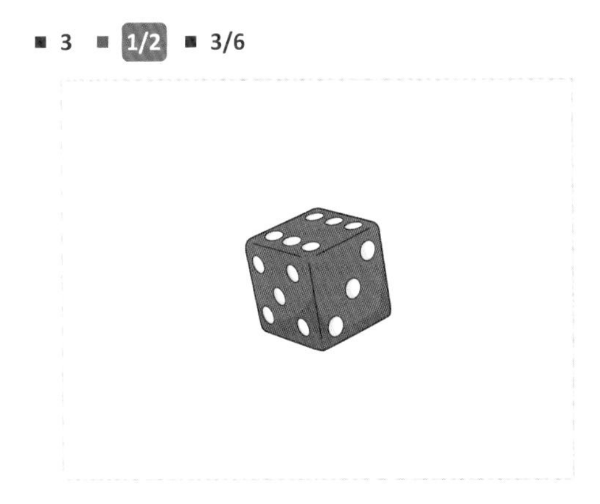

12. Thom is thinking of a number between 1 and 30. What is the probability he is thinking of a prime number?
Give your answer as a fraction in its simplest form.

- **10**
- **1/3**
- **10/30**

1	2	3	4	5
6	7	8	9	10
11	12	13	14	15
16	17	18	19	20
21	22	23	24	25
26	27	28	29	30

13. There are 52 cards in a deck of playing cards. If four of them are queens, what is the probability of picking a queen?
Give your answer as a fraction in its simplest form.

- **1/13**
- **4/52**
- **2/26**

Level 2 *continued*

14. Ethan is going to take a shape from the bag. Select the statement that is true.

1/4

- Ethan's shape will definitely be a square.
- **The probability of Ethan picking a triangle is 1/5.**
- There is an even chance of picking a triangle or star.
- The difference between the probabilities of choosing a circle and a star is 2.

15. Jill is choosing fruit for her lunch. There is a 32% chance she will choose an apple and a 10% chance she will choose a banana. What is the probability she will choose an apple or a banana?
Give your answer as a percentage including the % (percent) sign.

- 22%
- 42
- **42%**
- 320%
- 22
- 320

16. On a 5-coloured spinner, there is a probability of 1/4 of spinning red and a probability of 1/3 of spinning orange. What is the probability of spinning red or orange?
Give your answer as a fraction in its simplest form.

- 2/7
- **7/12**
- 1/12

17. What is the probability that the spinner lands on blue or orange?
Give your answer as a decimal.

- **0.5**
- 1/2

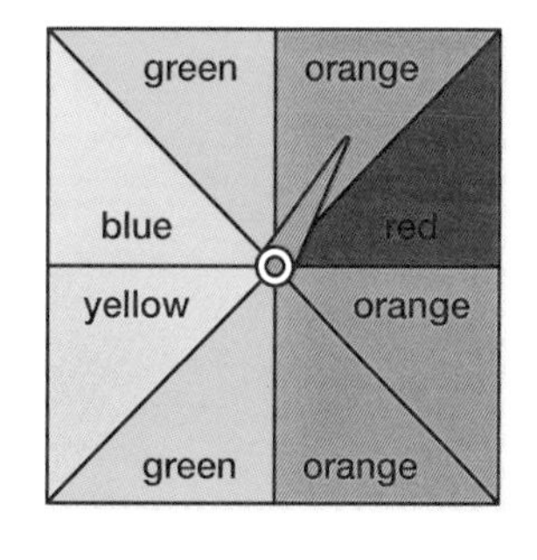

18. Harvey is thinking of a number from 10 to 19 (inclusive). What is the probability he is thinking of a number that is a multiple of 3?
Give your answer as a percentage including the % (percent) sign.

- **30%**
- 30

19. Sammy is choosing a sports club to join. There is a probability of 0.43 that she will choose basketball and a probability of 0.38 that she will choose tennis. What is the probability she will choose basketball or tennis?
Give your answer as a decimal.

- **0.81**
- 0.05

20. What is the probability that the spinner lands on red or green?
Give your answer as a fraction in its simplest form.

- **3/8**

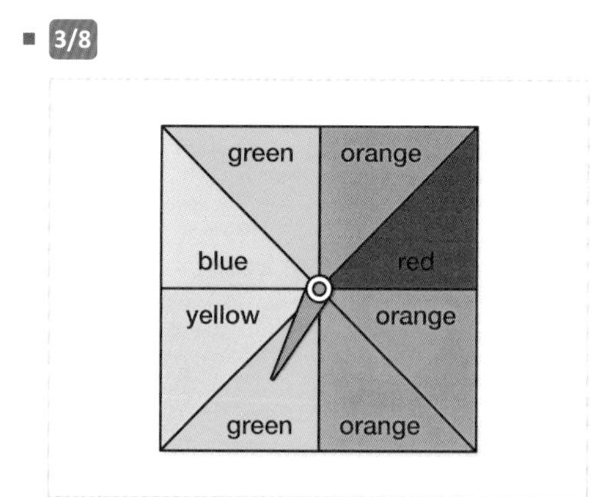

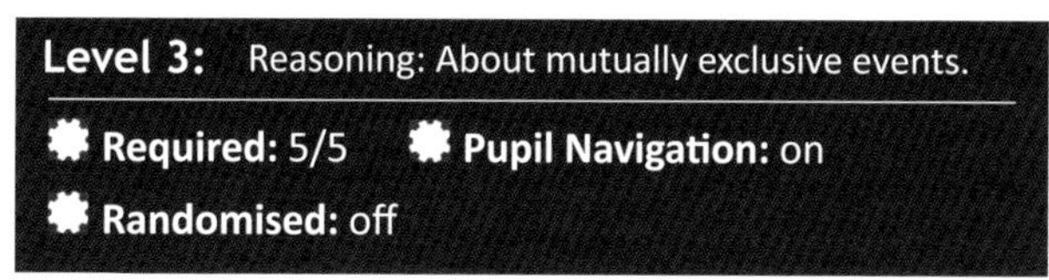

21. Jack rolls two fair, standard six-sided dice. He says the probability of rolling a total of one is 1/12. Is Jack correct? Explain your answer.

- Open question, no set answer

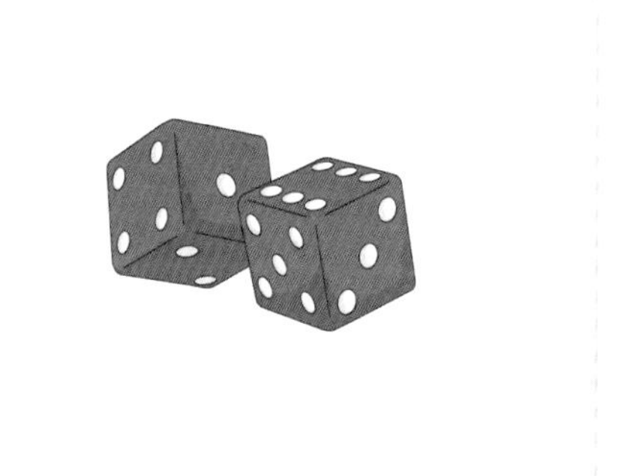

Level 3 continued

22. Melanie threw a coin 50 times and it landed heads up 9 times. What is the probability of the coin landing heads up on the next throw?
Give your answer as a fraction in its simplest form.

- 1/2

23. Each money pot contains only 1, 5 and 10 pence coins. Which pot are you most likely to pick a 10 pence coin from? Explain your answer.

- Open question, no set answer

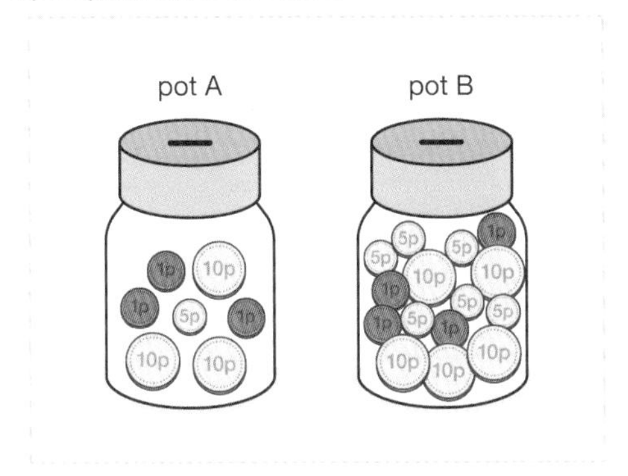

24. If two events are mutually exclusive, what is the probability of them both happening at the same time?

- 0

25. Jay says, "In a football match, we can only win, draw or lose, so the probability that my team will draw is 1/3". Explain why Jay is incorrect.

- Open question, no set answer

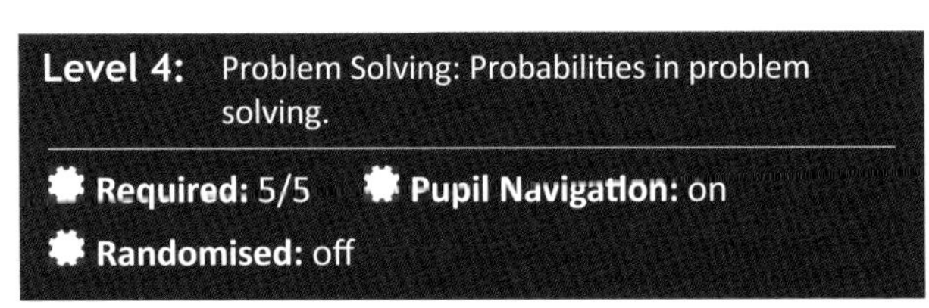

Level 4: Problem Solving: Probabilities in problem solving.

Required: 5/5 **Pupil Navigation:** on
Randomised: off

26. When throwing two fair coins, what is the probability of one landing on heads and one landing on tails?
Give your answer as a fraction in its simplest form.

- 1/2
- 1/3

27. Select the two events that are mutually exclusive for picking a number between 1 and 30.

1/5

For choosing a number between 1 and 99, which two options that show mutually exclusive events?

- picking a prime number
- picking a number less than 12
- picking a square number
- picking an even number
- picking a multiple of 5

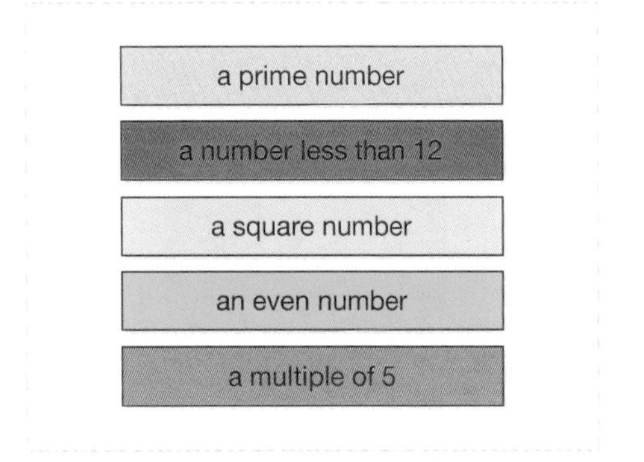

28. 64 people each pay £2 to play a game. The probability that a person wins is 1/8. If the winners each receive £5, how much profit should the game make?
Include the £ sign in your answer.

- 128
- £88
- £88.00
- £40.00
- 88.00
- 128.00
- 40.00
- £128.00
- £128
- 40
- £40
- 88

29. Amir throws a counter onto a play mat. The counter has an equal chance of landing on eeach square. What is the difference in probability between the counter landing on the road and landing on the pond?
Give your answer as a fraction in its simplest form.

- 11/64
- 4/64
- 1/16
- 2/32
- 7/64

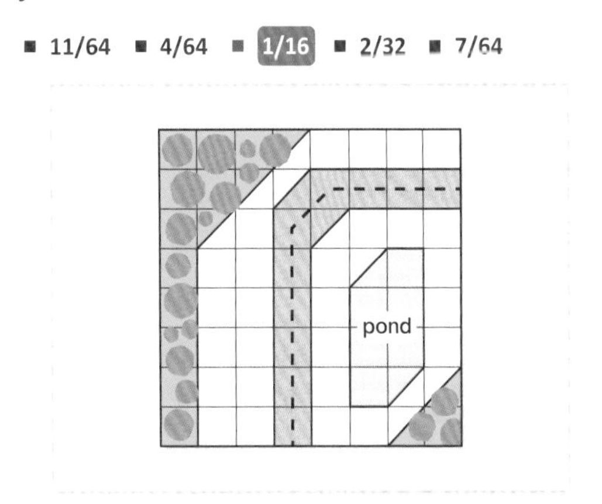

Level 4 *continued*

30. In this regular octagonal spinner, what number goes in the shaded section to make all the following statements true?

1 2 3

- There is an even chance of spinning a prime number and an even number.
- There is a 1/4 chance of rolling the missing number.
- There is a 50% chance of rolling a number greater than 5.

■ 7 ■ 3 ■ 9

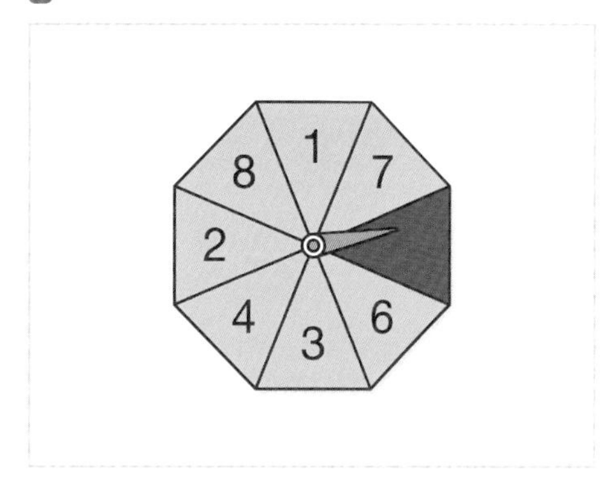

Use Sample Spaces to Calculate Probabilities of Single and Combined Events

Objective: I can calculate probabilities for single and combined events.

Quick Search Ref: 10424

Level 1: Understanding: Listing outcomes and enumerating sample spaces.

Required: 7/10 **Pupil Navigation:** on **Randomised:** off

1. How many possible outcomes are there for flipping a coin?

- **2** (correct)
- 3

2. How many possible outcomes are there for rolling a standard six-sided dice once?

- **6** (correct)

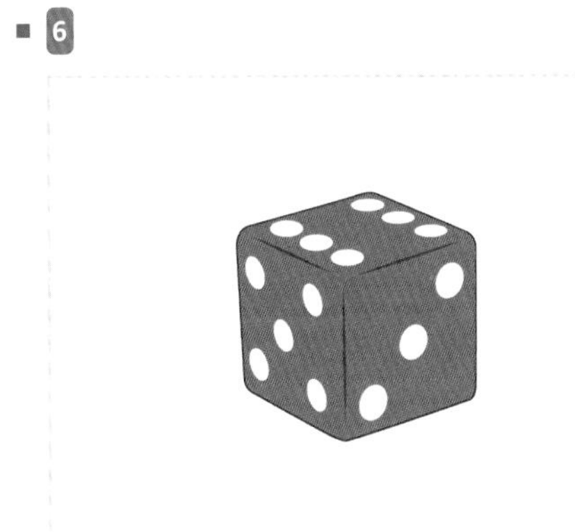

3. How many possible outcomes are there for flipping a 2p and 10p coin once each?

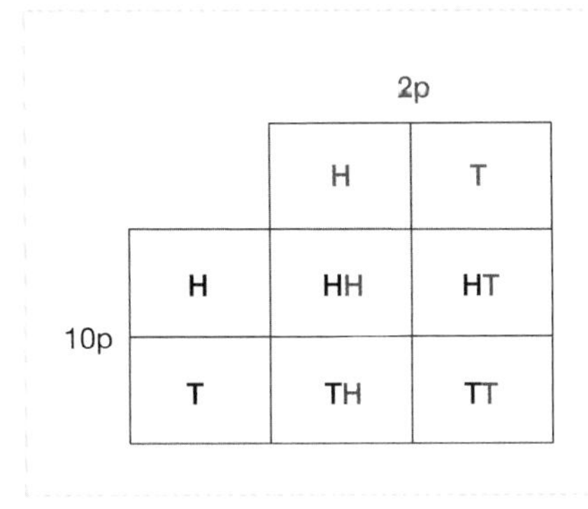

		2p	
		H	T
10p	H	HH	HT
	T	TH	TT

4. How many combinations of outcomes are there for rolling a standard 6-sided dice once and flipping a coin once?

- **12** (correct)
- 21
- 20

	1	2	3	4	5	6
H	H1	H2	H3	H4	H5	H6
T	T1	T2	T3	T4	T5	T6

5. A cafe offers a choice of two starters and three main courses. How many combinations of one starter and one main course can a customer choose?

- 12
- **6** (correct)
- 11

		main		
		fish and chips	pie and mash	vegetable curry
starter	soup	SF	SP	SV
	melon	MF	MP	MV

6. A sandwich shop offers a choice of three different breads and five different fillings. How many different sandwiches can you order if each has one type of bread and one filling?

- 23
- **15** (correct)
- 8
- 24

Level 1 *continued*

7. How many possible outcomes are there for spinning the spinner once and rolling the standard six-sided dice once?

1 2 3

- 10
- **24**
- 35
- 34

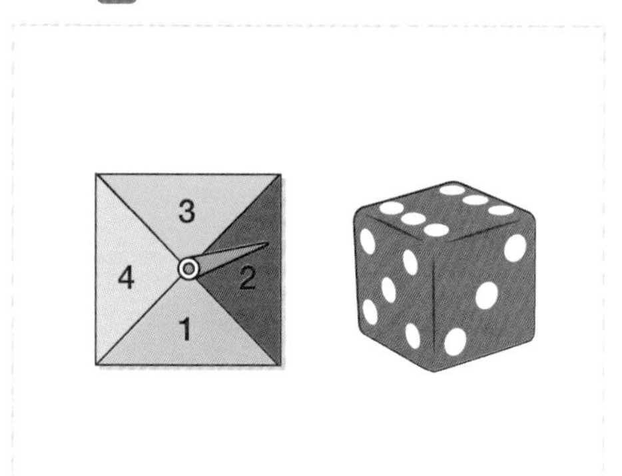

8. A school canteen offers a choice of 4 main meals and 3 desserts. How many combinations of one main meal and one dessert can a student choose?

1 2 3

- 20
- **12**
- 19

main

dessert	curry	pizza	stew	lasagne
fruit				
yoghurt				
sponge				

9. A fruit machine has two reels and four items on each reel. How many possible ways can the two reels land?

1 2 3

- 8
- **16**
- 25
- 24

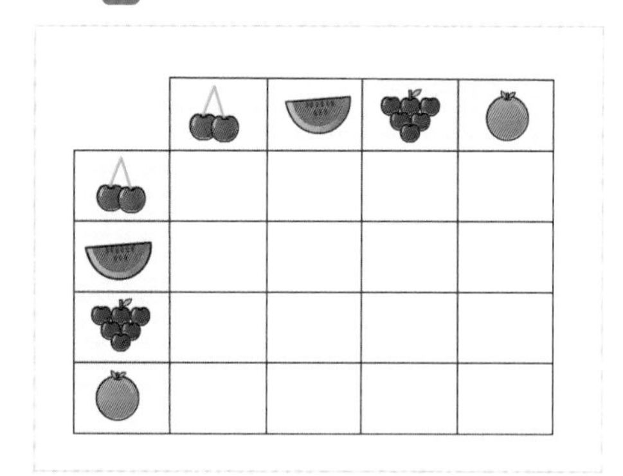

10. A sandwich shop offers a choice of five different breads and eight different fillings. How many different sandwiches can you order if each has one type of bread and one filling?

1 2 3

- 53
- **40**
- 54
- 13

Level 2: Fluency: Calculating probability from a sample space.

Required: 7/10 **Pupil Navigation:** on
Randomised: off

11. When flipping a 2p coin and a 10p coin, what is the probability of both coins landing on heads?
Give your answer as a fraction.

a b c

- **1/4**
- 1/3

2p

10p	H	T
H	HH	HT
T	TH	TT

12. Arthur flips a coin and rolls a standard six-sided dice. What is the probability that the coin lands on heads and the dice lands on 6?
Give your answer as a fraction.

a b c

- **1/12**

	1	2	3	4	5	6
H	H1	H2	H3	H4	H5	H6
T	T1	T2	T3	T4	T5	T6

13. Meg rolls two standard six-sided dice. What is the probability she gets a total of 12?
Give your answer as a fraction.

a b c

- **1/36**

blue die

red die	1	2	3	4	5	6
1	2	3	4	5	6	7
2	3	4	5	6	7	8
3	4	5	6	7	8	9
4	5	6	7	8	9	10
5	6	7	8	9	10	11
6	7	8	9	10	11	12

Level 2 *continued*

14. Gary has five letter tiles. He selects a tile, replaces it and then selects a second tile. What is the probability that Gary selects the letter M twice?
Give your answer as a fraction.

- 1/25
- 2/5

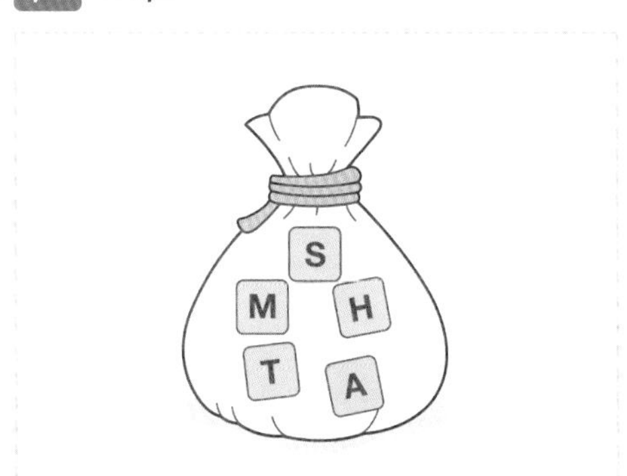

15. Caden rolls two standard six-sided dice. What is the probability he gets a total less than 5?
Give your answer as a fraction in its simplest form.

- 3/18
- 1/6
- 10/36
- 1/36
- 6/36
- 5/18
- 2/12

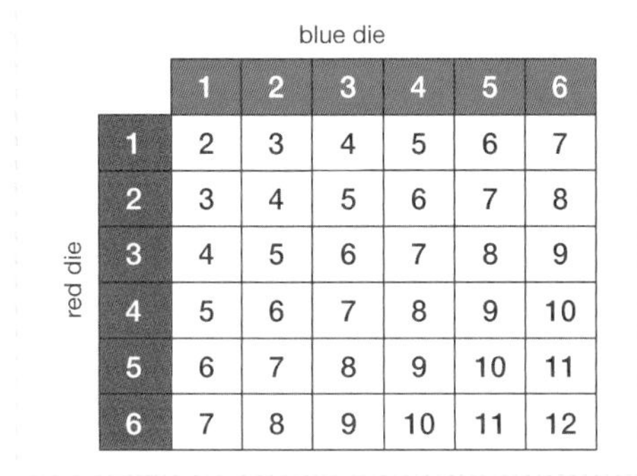

blue die (columns) / red die (rows)

	1	2	3	4	5	6
1	2	3	4	5	6	7
2	3	4	5	6	7	8
3	4	5	6	7	8	9
4	5	6	7	8	9	10
5	6	7	8	9	10	11
6	7	8	9	10	11	12

16. A bag contains a 5p coin and two 10p coins. Another bag contains two 5p coins and three 10p coins. If a coin is selected at random from each bag, what is the probability that they total 20p?
Give your answer as a fraction in its simplest form.

- 2/5
- 6/15

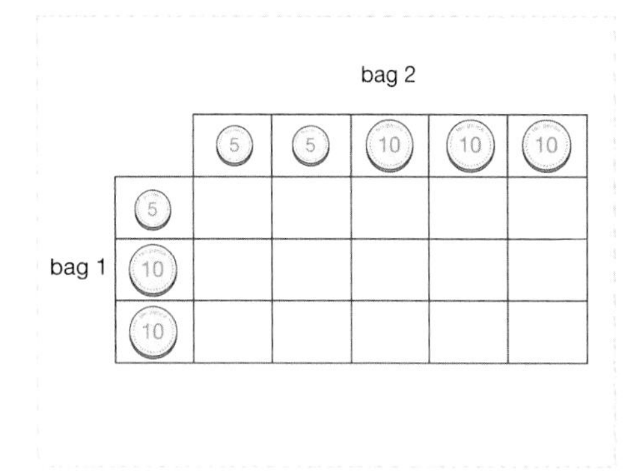

17. Five cards are numbered 1 to 5. Ruben chooses a card, replaces it and then selects a second card. What is the probability he selects two prime numbers?
Give your answer as a fraction.

- 1/25
- 9/25
- 16/25

second card (columns) / first card (rows)

	1	2	3	4	5
1					
2					
3					
4					
5					

18. Beth spins the two spinners shown. What is the probability she gets a total of 9?
Give your answer as a fraction.

- 1/20

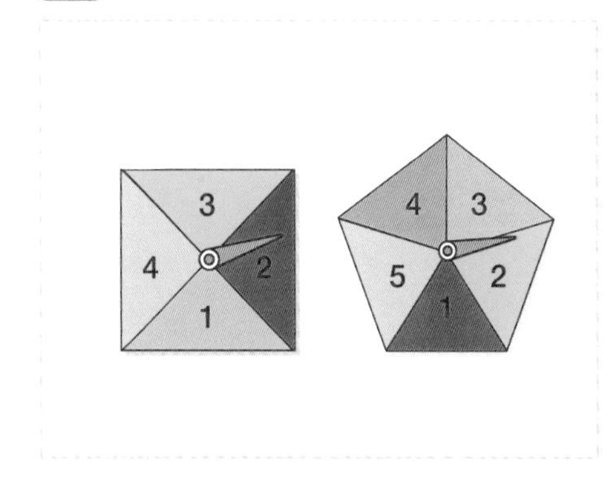

Level 2 *continued*

19. Leah flips a coin and rolls a standard six-sided dice. What is the probability that the coin lands on heads and the dice lands on an even number?
Give your answer as a fraction in its simplest form.

- 1/12 ■ **1/4** ■ 3/12

	1	2	3	4	5	6
H	H1	H2	H3	H4	H5	H6
T	T1	T2	T3	T4	T5	T6

20. Kobi has six letter tiles that together spell out the word 'chance'. Kobi selects a tile, replaces it and then selects a second tile. What is the probability he selects two vowels?
Give your answer as a fraction in its simplest form,

- 2/18 ■ **1/9** ■ 4/36 ■ 1/36

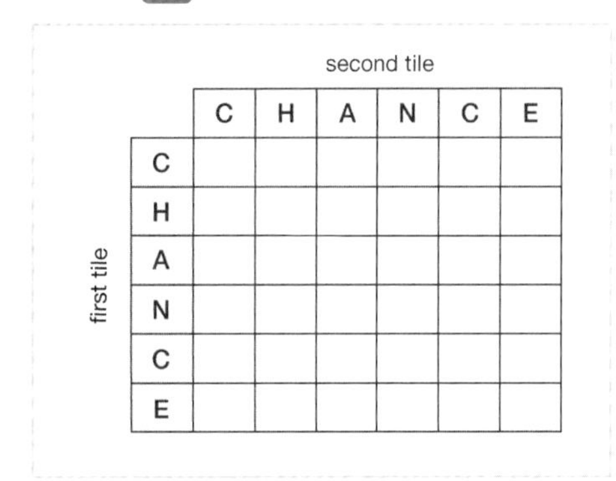

second tile

first tile	C	H	A	N	C	E
C						
H						
A						
N						
C						
E						

Level 3: Reasoning: Misconceptions and reasoning with sample spaces.

Required: 5/5 **Pupil Navigation:** on
Randomised: off

21. James is spinning the two spinners shown. James says the smallest possible total is 1 and the largest total is 8, so the probability of getting a total of 5 is 1/8. Is James correct? Explain your answer.

- Open question, no set answer

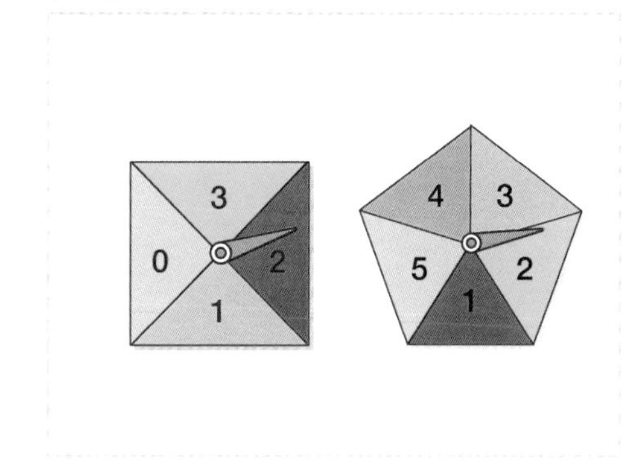

22. Callie rolls two standard six-sided dice. What is the most likely total?

- **7**

	1	2	3	4	5	6
1						
2						
3						
4						
5						
6						

23. Jayden charges £1 for his game of chance. Players roll a standard six-sided dice, and if they get a 6, Jayden gives them their money back plus a £5 prize.
Is this a fair game? Explain your answer.

- Open question, no set answer

Level 3 *continued*

24. The probability of picking a triangle from bag A is 2/5. The probability of picking a triangle from bag B is 1/4. Chloe says the probability of picking a triangle from both bags is 1/10. Is Chloe correct? Explain your answer.

- Open question, no set answer

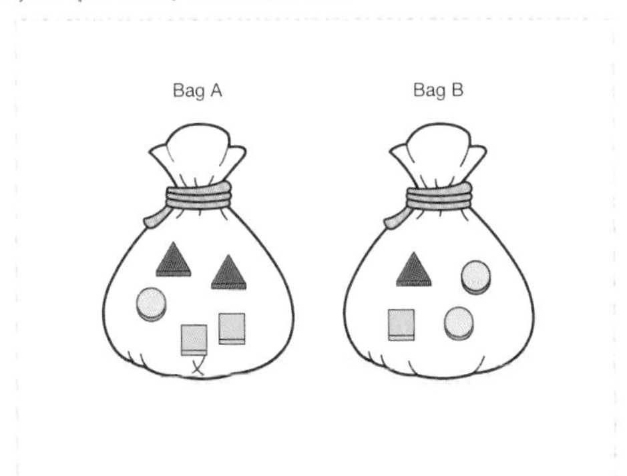

25. Zenya flips a 1p coin, a 5p coin and a 10p coin. What is the probability that all three coins land on heads?

1/4 ■ **1/8** ■ 1/6 ■ 1/3 ■ 3/6

Level 4: Problem solving: Multi-step problems with sample spaces.

Required: 5/5 **Pupil Navigation:** on
Randomised: off

26. Fazel rolls two standard six-sided dice and multiplies together the numbers they land on. What is the probability his product is even?
Give your answer as a fraction in its simplest form.

■ **3/4** ■ 1/2 ■ 27/36

	1	2	3	4	5	6
1						
2						
3						
4						
5						
6						

27. Rafael rolls two standard six-sided dice and finds the positive between them. What is the most likely difference?

■ **1**

	1	2	3	4	5	6
1						
2						
3						
4						
5						
6						

28. Five cards are numbered 1 to 5 and placed in a box. Halima chooses a card, keeps it and then selects a second card. What is the probability that her two cards total 5?
Give your answer as a fraction in its simplest form,

■ 4/20 ■ **1/5** ■ 4/25

second card

first card	1	2	3	4	5
1					
2					
3					
4					
5					

29. Fleur charges £1 for her game of chance. Players roll two dice and if they get a total of 12 they win £25. If they get a total of 7, they roll the dice again for free.
If 180 people play the game, how much profit can she expect to make?
Include the £ sign in your answer.

■ **£30.00** ■ **£30** ■ £55.00 ■ 30 ■ 30.00 ■ 55.00 ■ 55 ■ £55

	1	2	3	4	5	6
1						
2						
3						
4						
5						
6						

Level 4 *continued*

30. Wayne spins a four-sided spinner numbered 1 to 4 and a five-sided spinner numbered 1 to 5. He multiplies the numbers they land on together. What is the mean average score Wayne can expect to get?
Give your answer as a decimal.

■ 

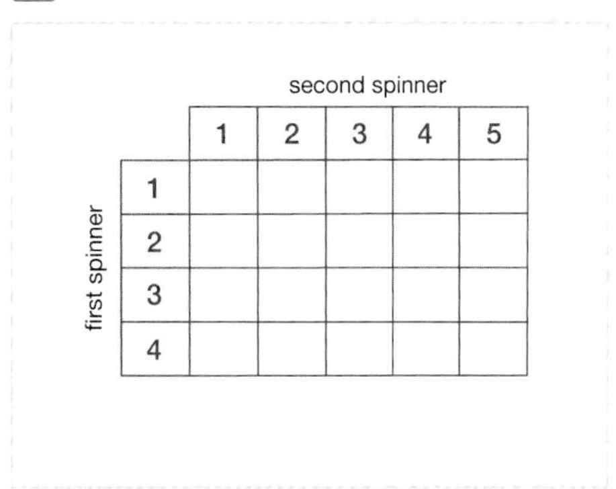

Interpreting Venn Diagrams and Multiple Sets

Objective: I can identify the elements of sets and unions/intersections of sets using Venn diagrams.

Quick Search Ref: 10410

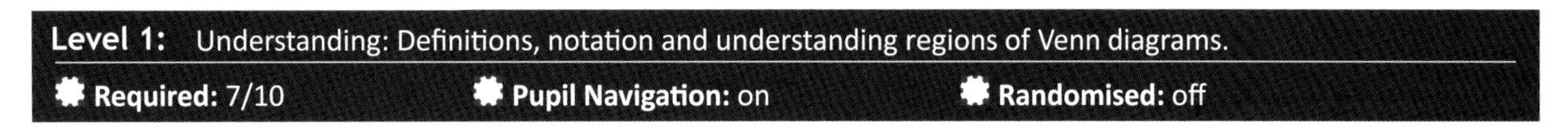

Level 1: Understanding: Definitions, notation and understanding regions of Venn diagrams.

Required: 7/10 **Pupil Navigation:** on **Randomised:** off

1. Arrange the following terms into the same order as their definitions.

- A union B
- A intersection B
- compliment of A
- empty set
- universal set

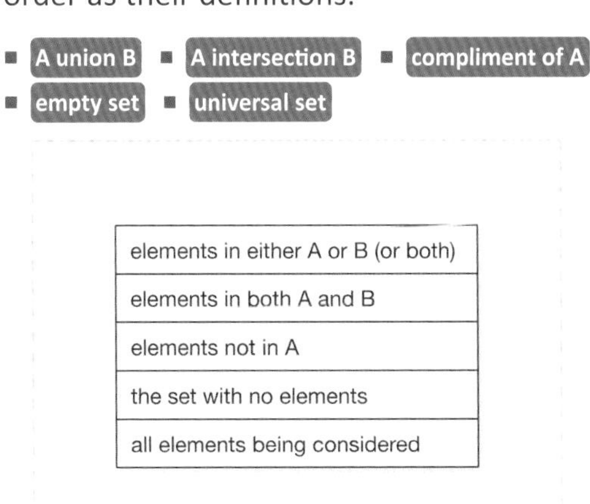

elements in either A or B (or both)
elements in both A and B
elements not in A
the set with no elements
all elements being considered

2. Arrange the following symbols and notations into the same order as their definitions.

- A ∪ B
- A ∩ B
- A'
- ∅
- ξ

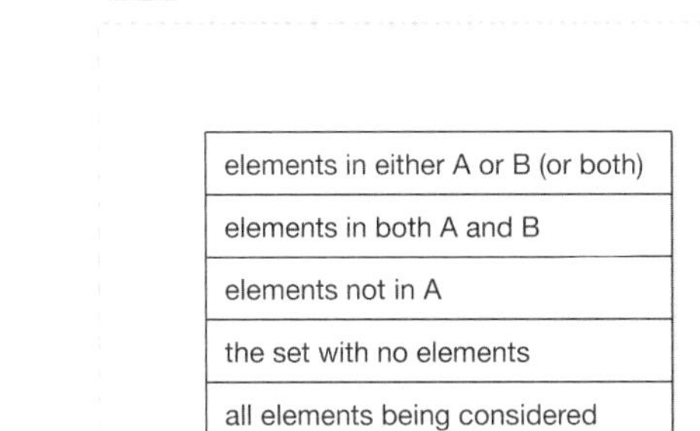

elements in either A or B (or both)
elements in both A and B
elements not in A
the set with no elements
all elements being considered

3. In which diagram does the shaded region represent set T?

1/4

- (i)
- (ii)
- (iii)
- (iv)

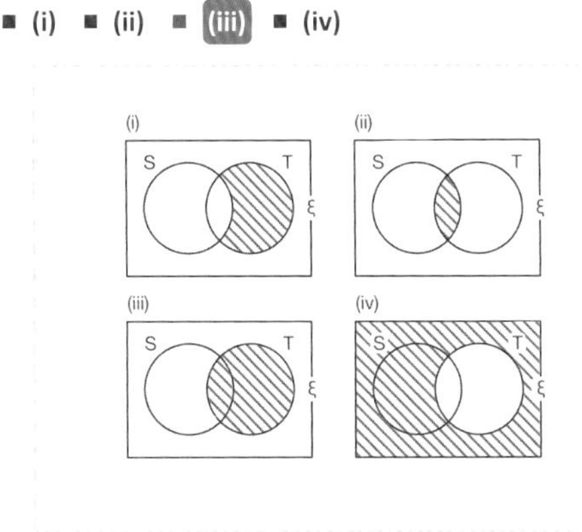

4. Which region is represented by the diagram?

1/4

- E'
- E ∩ P
- P
- E ∪ P

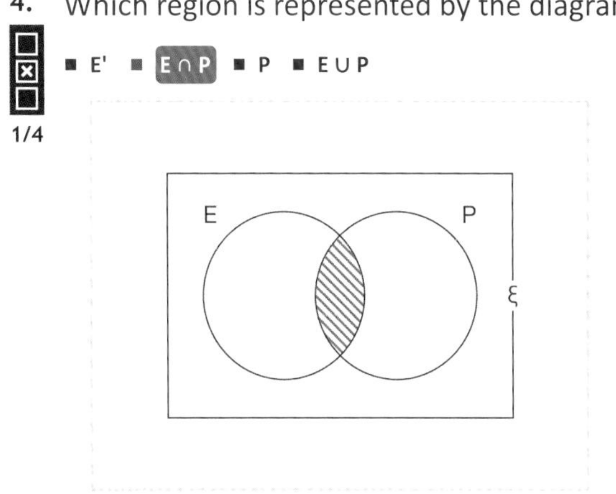

5. Which regions would you need to shade to represent R ∪ S?

3/5

- (i)
- (ii)
- (iii)
- (iv)
- (v)

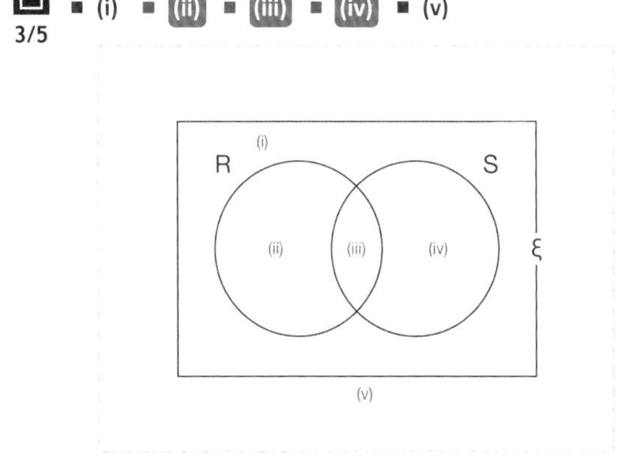

6. Which region is represented by the diagram?

1/7

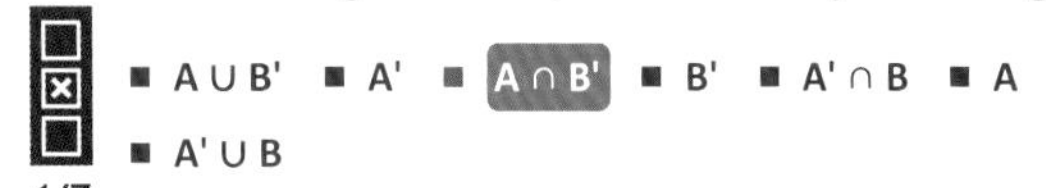

- A ∪ B'
- A'
- A ∩ B'
- B'
- A' ∩ B
- A
- A' ∪ B

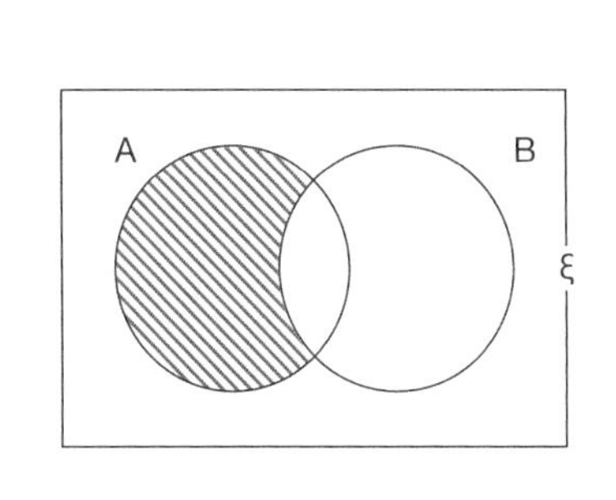

Level 1 *continued*

7. Which regions would you need to shade to represent D'?

2/5

- **(i)** (highlighted)
- **(ii)** (highlighted)
- (iii)
- (iv)
- (v)

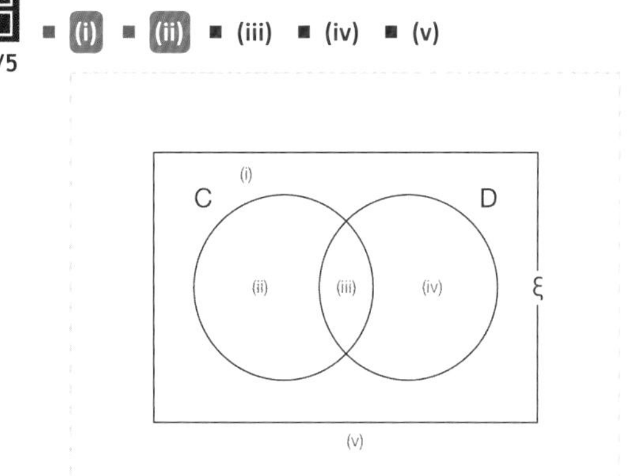

8. In which diagram does the shaded region represent F ∪ S'?

1/4

- (i)
- **(ii)** (highlighted)
- (iii)
- (iv)

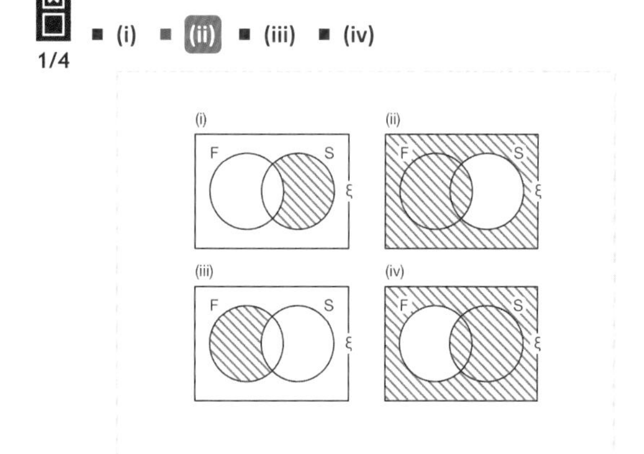

9. Which region is represented by the diagram?

1/5

- A' ∩ B
- **A' ∩ B'** (highlighted)
- A' ∪ B'
- A ∪ B'
- A ∩ B'

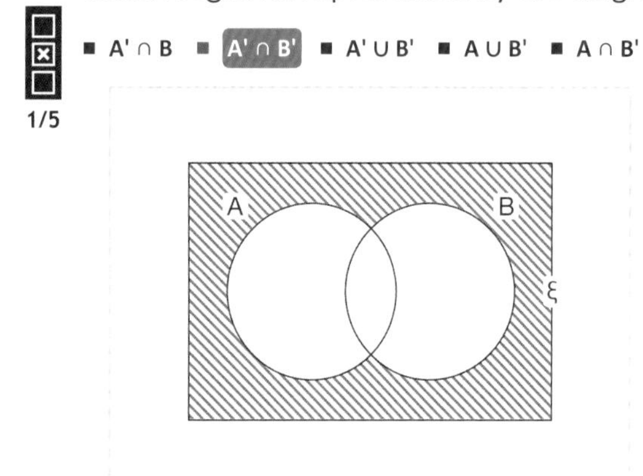

10. Which region would you need to shade to represent E ∩ M'?

1/5

- (i)
- **(ii)** (highlighted)
- (iii)
- (iv)
- (v)

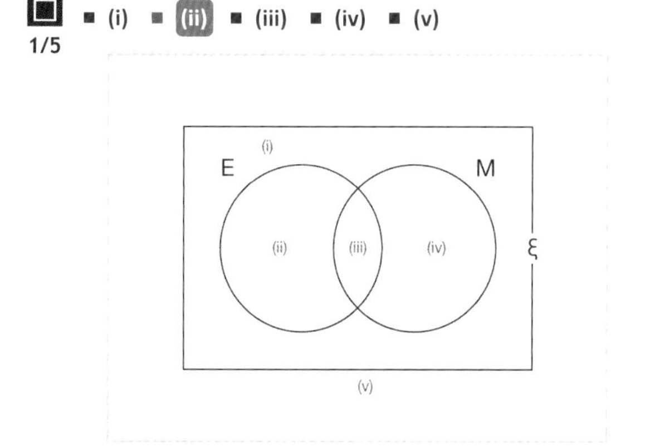

Level 2: Fluency: Enumerating multiple sets supported by Venn diagrams.

Required: 7/10 **Pupil Navigation:** on

Randomised: off

11. ξ = {integers from 1 to 12 inclusive}
F = {factors of 12}
P = {prime numbers}
Which set represents F ∪ P?

1/5

- {8, 9, 10}
- {2, 3, 5, 7, 11}
- **{1, 2, 3, 4, 5, 6, 7, 11, 12}** (highlighted)
- {2, 3}
- {1, 2, 3, 4, 6, 12}

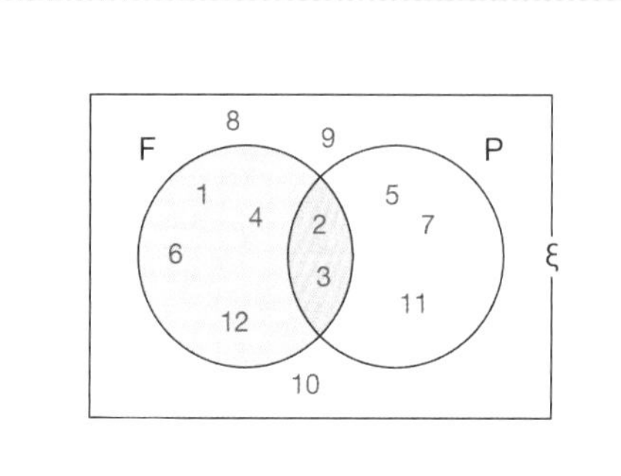

12. ξ = {letters in the alphabet}
H = {letters in the word 'homework'}
I = {letters in the word 'improve'}
Which elements are members of H ∩ I?

4/7

- **o** (highlighted)
- **m** (highlighted)
- p
- **e** (highlighted)
- h
- i
- **r** (highlighted)

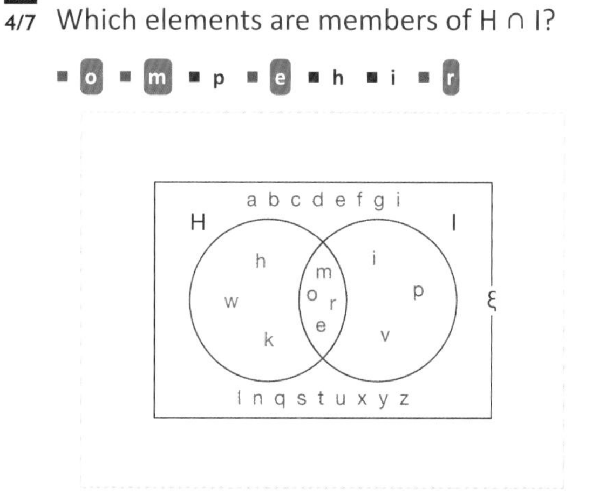

Level 2 *continued*

13.   ξ = {integers from 1 to 15 inclusive}
F = {factors of 15}
S = {square numbers}
1/5 Which set represents F ∪ S?

- {1, 3, 4, 5, 9, 15}
- {1, 4, 9}
- {2, 6, 7, 8, 10, 11, 12, 13, 14}
- {1}
- {1, 3, 5, 15}

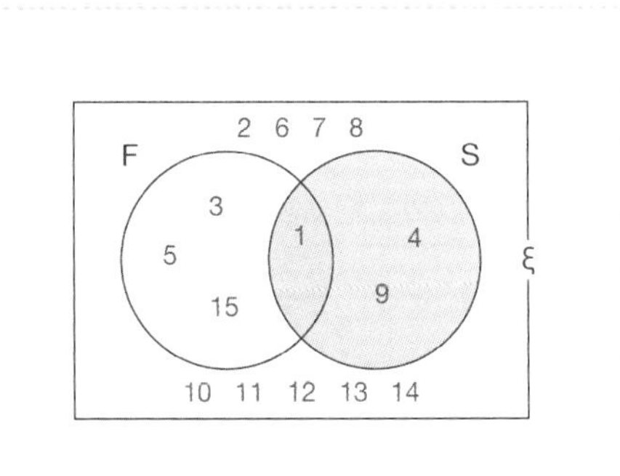

14.  ξ = {integers from 1 to 12 inclusive written in words}
E = {words that contain the letter e}
T = {words that contain the letter t}
How many elements are in the set E ∩ T?

- 4
- 10

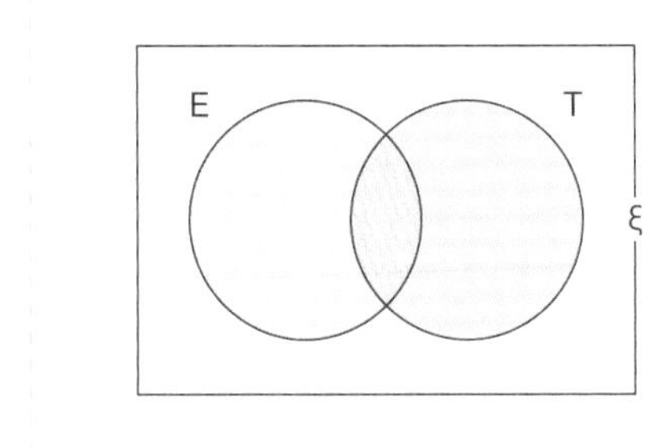

15.  ξ = {quadrilaterals}
E = {shapes with two pairs of sides with equal length}
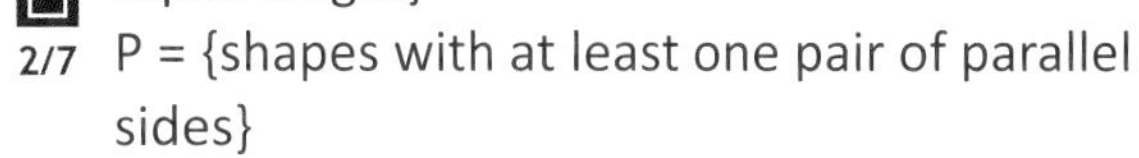
2/7 P = {shapes with at least one pair of parallel sides}
Which elements are members of E ∩ P'?

- rhombus
- trapezium
- arrowhead
- irregular quadrilateral
- parallelogram
- kite
- isosceles trapezium

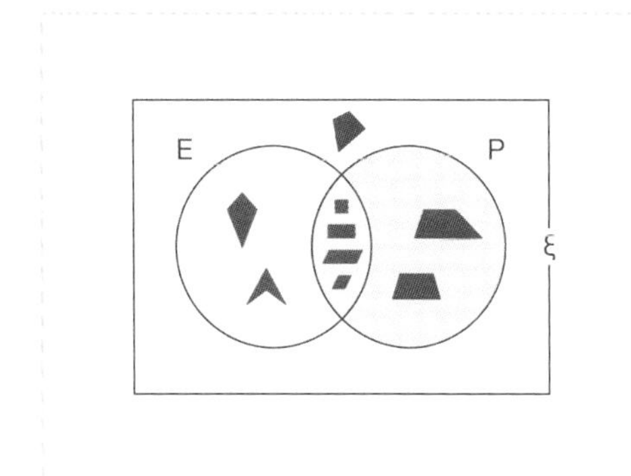

16.  ξ = {integers from 1 to 10 inclusive}
F = {factors of 10}

P = {prime numbers}
How many elements are in F ∪ P'?

- 8

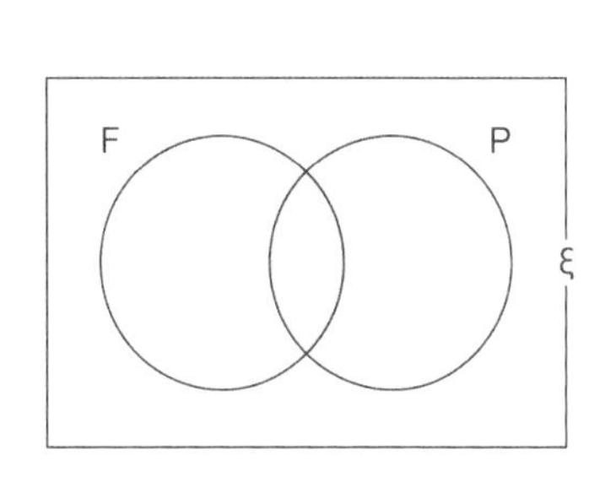

17.  ξ = {the first 15 letters of the alphabet}
C = {letters in the word 'Chile'}
E = {letters in the word 'England'}
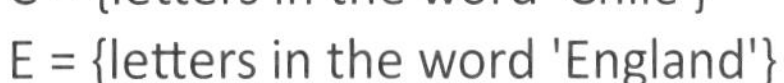
Arrange the sets into the same order as the descriptions of the regions.

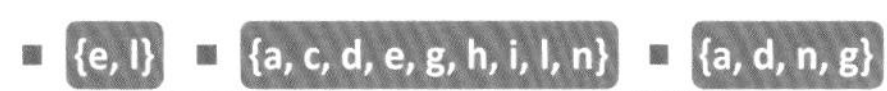

- {e, l}
- {a, c, d, e, g, h, i, l, n}
- {a, d, n, g}
- {b, f, j, k, m, o}
- {b, c, f, h, i, j, k, m, o}

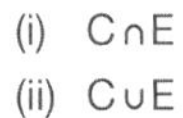

(i) C∩E
(ii) C∪E
(iii) E∩C'
(iv) E'∩C'
(v) E'

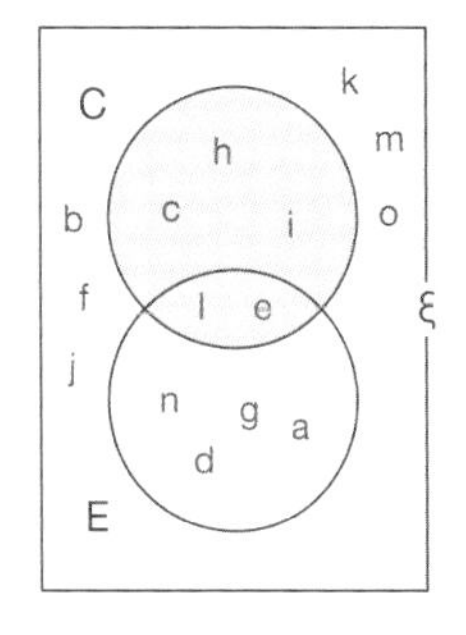

Level 2 *continued*

18. ξ = {integers from 1 to 15 inclusive}
M = {multiples of 2}
P = {prime numbers}
How many elements are in P'?

- 9

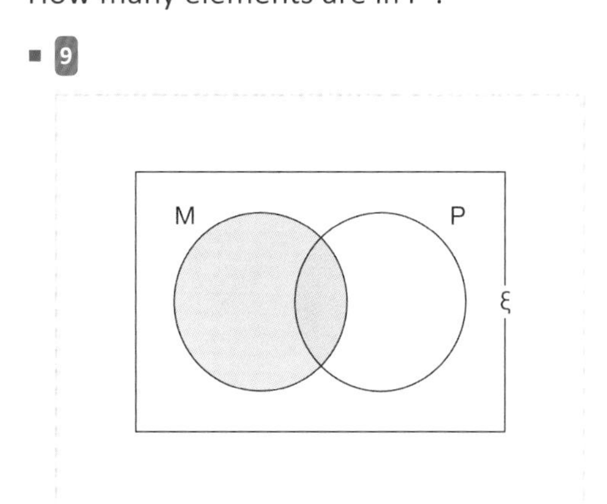

19. ξ = {letters in the alphabet}
H = {letters in the word 'holiday'}
S = {letters in the word 'school'}
4/7 Which elements are members of H ∩ S'?

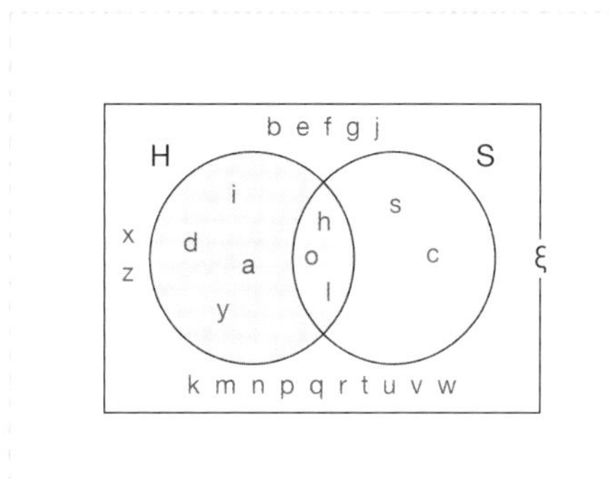

20. ξ = {integers from 1 to 15 inclusive}
F = {factors of 42}
P = {prime numbers}
Arrange the sets according to the regions described.

- {2, 3, 7}
- {1, 2, 3, 5, 6, 7, 11, 13, 14}
- {4, 8, 9, 10,12, 15}
- {4, 5, 8, 9, 10, 11, 12, 13, 15}
- {1, 6, 14}

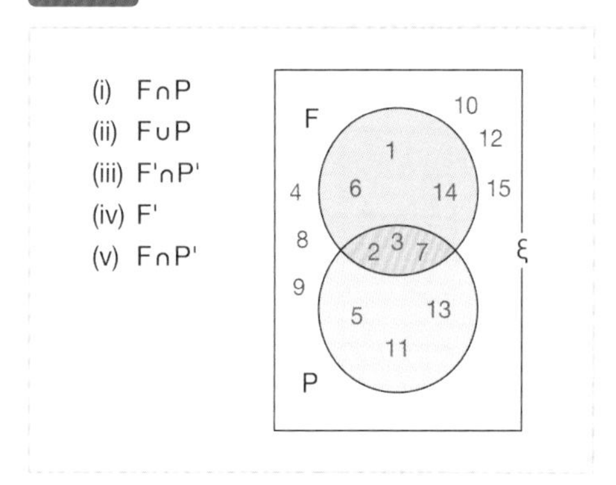

Level 3: Reasoning: Drawing conclusions from Venn diagrams.

Required: 5/5 **Pupil Navigation:** on
Randomised: off

21. ξ = {quadrilaterals}
E = {shapes with at least two pairs of sides with equal length}
P = {shapes with at least one pair of parallel sides}
Is kite ∈ (E ∩ P)? Explain your answer.

- Open question, no set answer

22. The Venn diagram shows the numbers of students who study French and Spanish in one year group at school. If a student is selected at random, what is the probability that they study both French and Spanish?
Give your answer as a fraction in its simplest form.

- 5/24
- 25/120

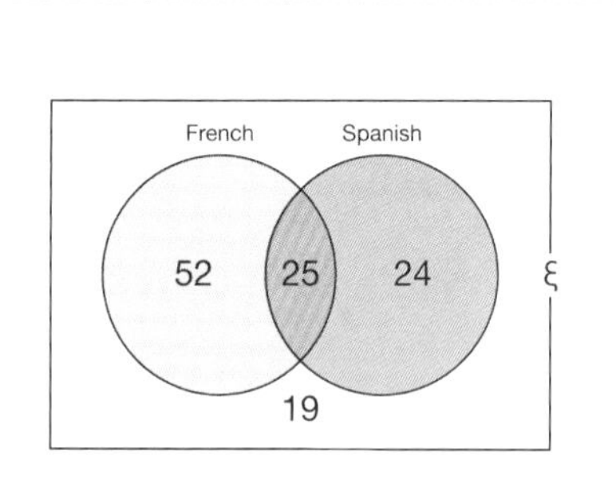

23. Zenya is drawing a Venn diagram to show elements of two sets A and B. Use her diagram to decide which statement must be
1/4 true.

- A ∪ B = ∅
- A ∩ B = ξ
- A ∪ B = ξ
- A ∩ B = ∅

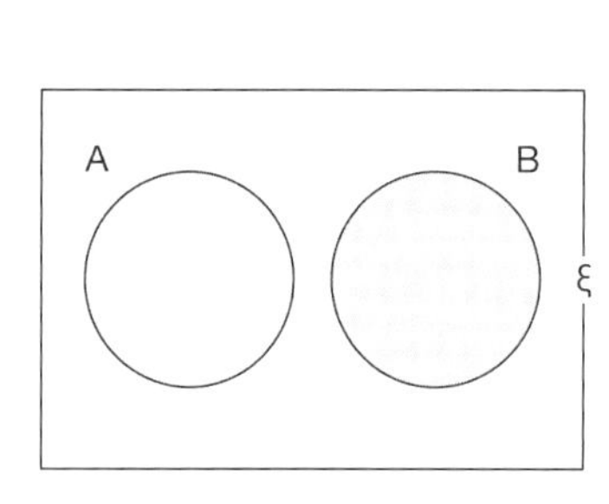

Level 3 *continued*

24. The diagram shows which activities 100 students did in their free time last week.
How many students read, played sport and listened to music?

- **13**

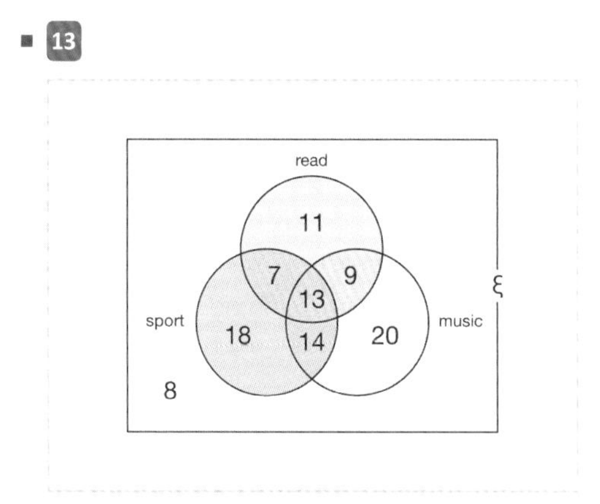

25. Fabio says, "The number of elements in A plus the number of elements in A' equals the number of elements in ξ".
Is Fabio correct? Explain your answer.

- Open question, no set answer

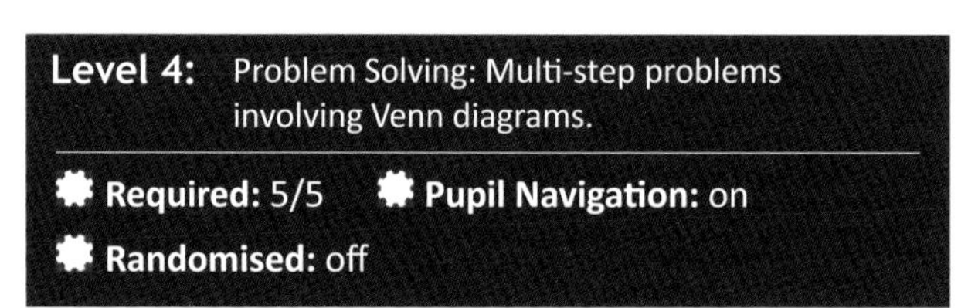

26. Select the two regions that are equivalent to each other.

2/5

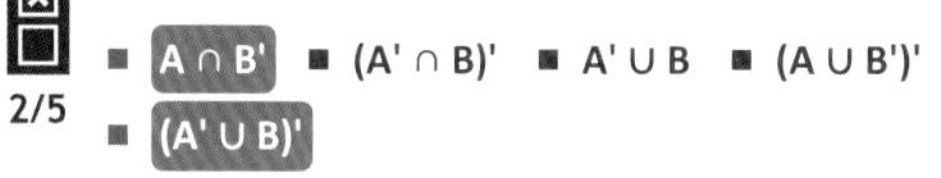

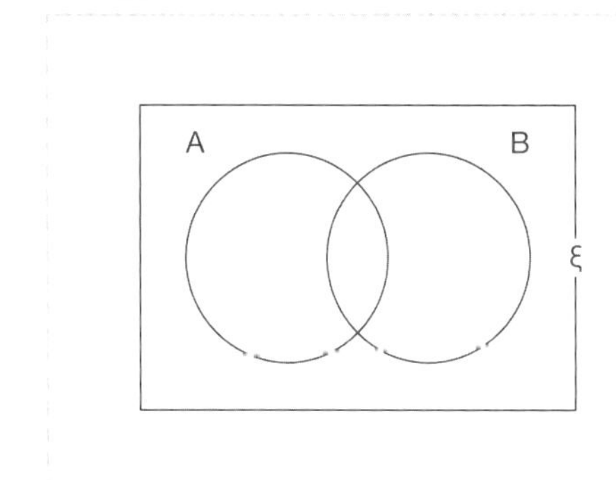

27. In a year group of 120 students, 83 students study geography, but not art, 31 study geography and art, and 18 don't study either geography or art.
How many students study art?

- **50**
- 19

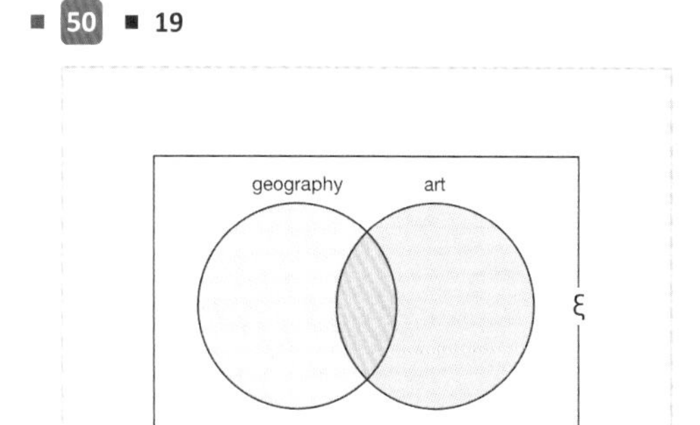

28. Which regions would you need to shade to represent $(A \cup B) \cap C$?

3/7

- **(i)**
- (ii)
- **(iii)**
- **(iv)**
- (v)
- (vi)
- (vii)

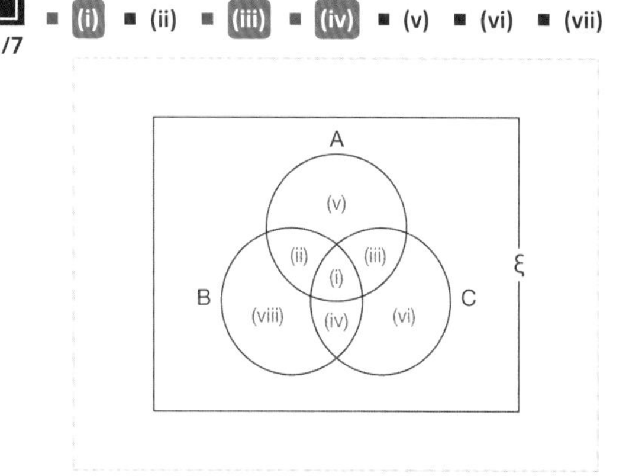

29. In a class of 30 students, 15 students play football, 7 boys don't play football and 6 girls do play football.
If a girl is selected at random, what is the probability that they play football?
Give your answer as a fraction in its simplest form.

- 6/14
- **3/7**
- 6/30
- 1/5

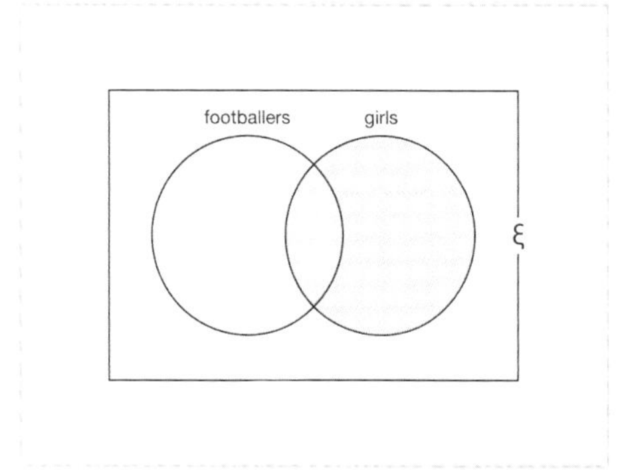

Level 4 *continued*

30. A year group of 120 students went on an adventure holiday and completed various activities.
15 only did bowling, and 42 did bowling altogether.
60 students did canoeing.
35 students did canoeing and archery.
28 students did bowling and canoeing.
12 students did all three activities and 7 did none of them.
How many students only did archery?

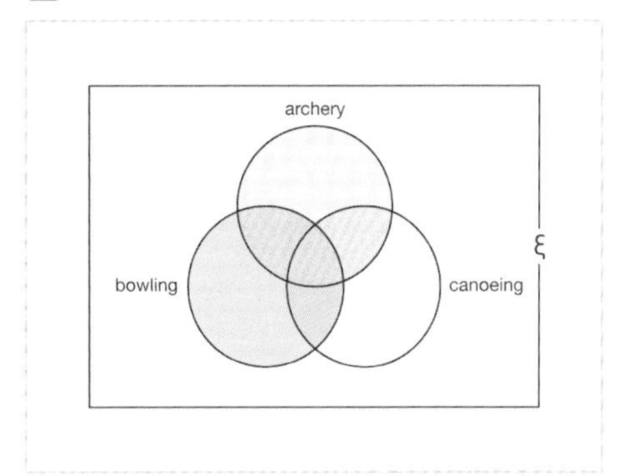

Understanding Sets and Set Notation

Objective: I can interpret and use set notation.

Quick Search Ref: 10409

Level 1: Understanding: Definitions, notation and writing sets.

Required: 7/8 **Pupil Navigation:** on **Randomised:** off

1. Arrange the following terms into the same order as their definitions.

- set
- element
- universal set
- empty set

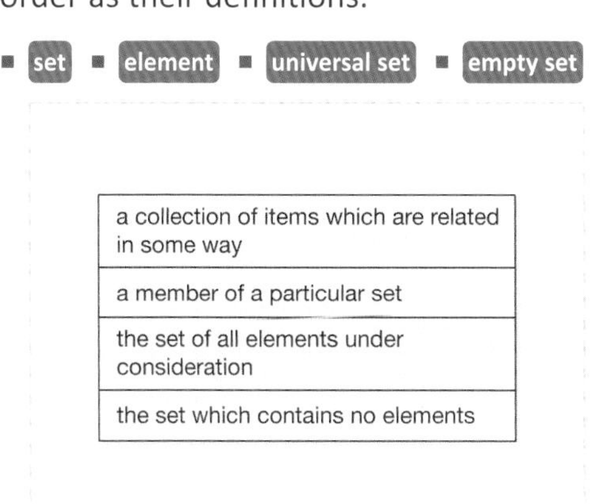

2. Arrange the following symbols in the same order as the definitions.

- ξ
- ∅
- ∈
- ∉

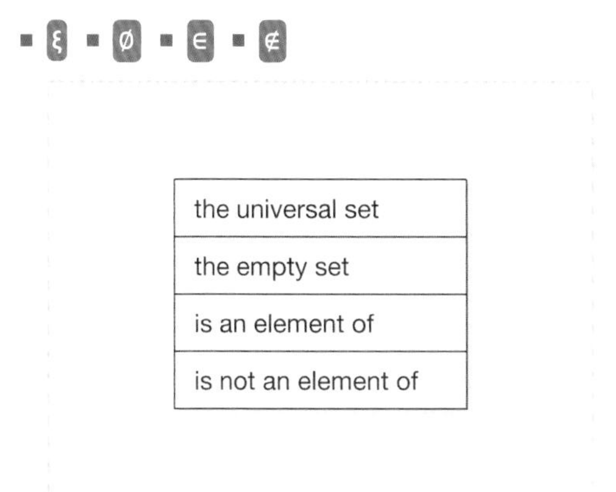

3. Which set matches this description?
{the first 4 multiples of 5}

1/5

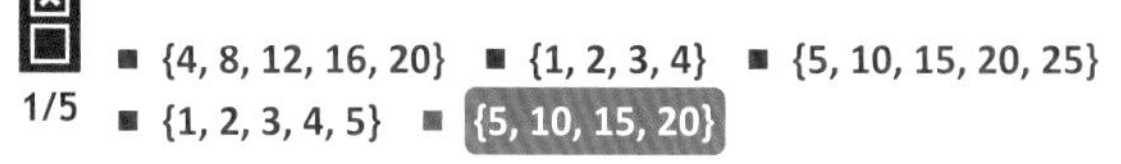

- {4, 8, 12, 16, 20}
- {1, 2, 3, 4}
- {5, 10, 15, 20, 25}
- {1, 2, 3, 4, 5}
- {5, 10, 15, 20}

4. Which of these numbers are elements of this set?
{the factors of 9}

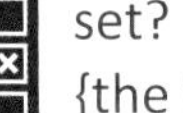

3/7

- 27
- 3
- 9
- 12
- 18
- 1
- 6

5. Which description fully describes this set?
{square, rhombus}

1/5

- {quadrilaterals}
- {quadrilaterals with 2 pairs of equal angles}
- {quadrilaterals with 4 equal sides}
- {quadrilaterals with 2 pairs of parallel sides}
- {quadrilaterals with 4 equal angles}

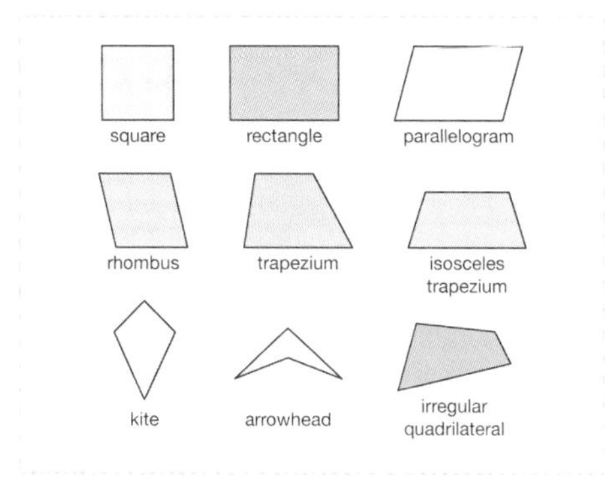

6. Which set matches this description?
{the letters in the word tomato}

1/4

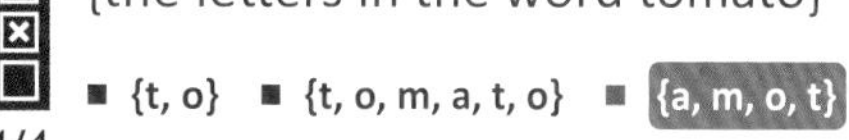

- {t, o}
- {t, o, m, a, t, o}
- {a, m, o, t}
- {a, m, o, o, t, t}

7. Enter the set that is represented by the following:
{the first 2 square numbers}
Separate the elements with a comma.

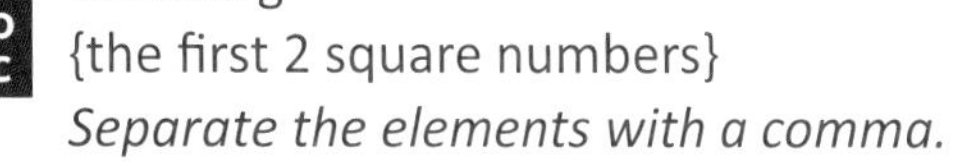

- {1, 4}
- 1, 4
- {4, 1}
- 4, 1

Level 1 *continued*

8. Which numbers are elements of the following set?
{prime numbers between 20 and 30}

2/7

- 31
- 21
- **29**
- 27
- 19
- **23**
- 17

20, 21, 22, 23, 24, 25, 26, 27, 28, 29, 30

Level 2: Fluency: The universal set and empty set.

Required: 6/8 **Pupil Navigation:** on

Randomised: off

9. If the universal set ξ = {positive integers 1 to 20}, what is the set M = {multiples of 6}?

1/5

- {12, 18, 24}
- {6, 36, 216}
- {1, 2, 3, 6}
- **{6, 12, 18}**
- {1, 2, 3}

ξ = { positive integers 1 to 20 }

M = { multiples of 6 }

M = { ? }

10. If the universal set ξ = {positive integers 10 to 20}, which numbers are elements of the set P = {prime numbers}?

4/7

- **19**
- 15
- **17**
- 5
- **11**
- 7
- **13**

ξ = { positive integers 10 to 20 }

P = { prime numbers }

P = { ? }

11. The universal set ξ = {odd numbers} and F = {factors of 70}. How many values of x are there such that $x \in F$?

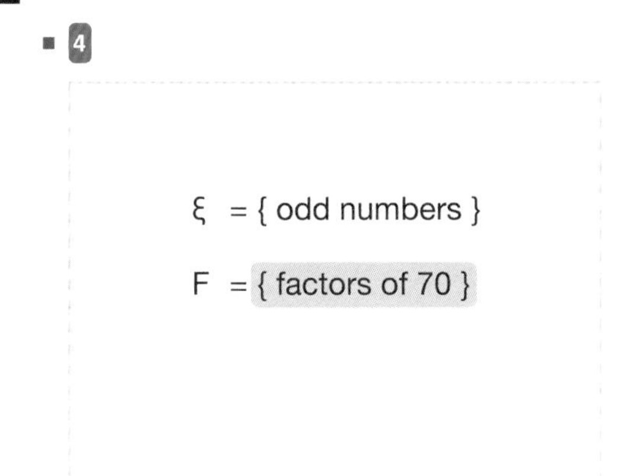

12. If the universal set ξ = {even numbers}, which set is the set F = {factors of 45}?

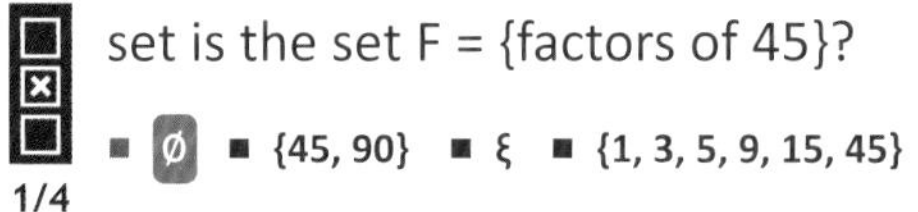

1/4

- **∅**
- {45, 90}
- ξ
- {1, 3, 5, 9, 15, 45}

ξ = { even numbers }

F = { factors of 45 }

F = { ? }

13. If the universal set ξ = {the first six months of the year}, select the elements of the set U = {words containing the letter u}.

3/6

- August
- May
- **February**
- July
- **January**
- **June**

ξ = { the first six months of the year }

U = { words containing the letter u }

U = { ? }

Level 2 *continued*

14. If the universal set ξ = {square numbers between 1 and 100}, type in the set that is represented by the following:
S = {numbers that end with the digit 6}
Separate the elements with a comma.

- {16, 36}
- 16, 36
- {36, 16}
- 36, 16

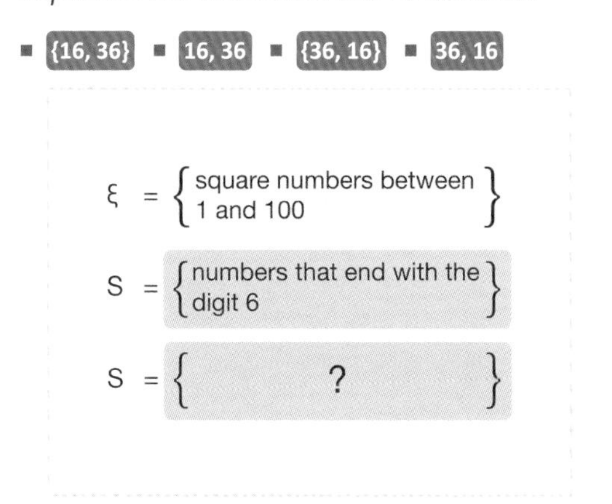

15. The universal set ξ = {integers 1 to 40} and S = {multiples of 7}. How many values of x are there such that $x \in S$?

- 5

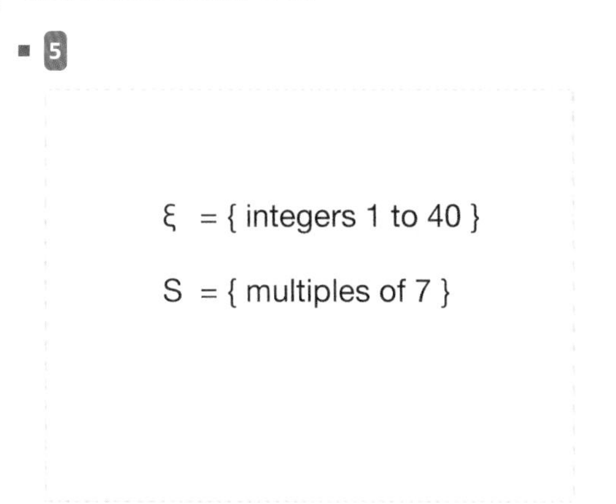

16. If the universal set ξ = {vowels}, enter the set that is represented by the following:
G = {letters in the word 'geometry'}.
Separate the elements with a comma.

- {o, e}
- e, o
- {e, o}
- o, e

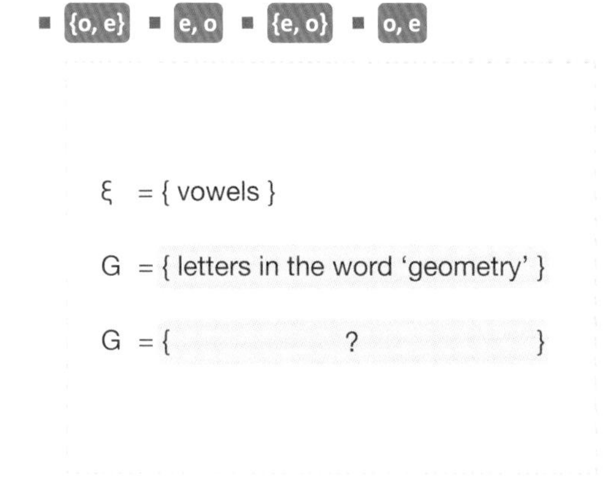

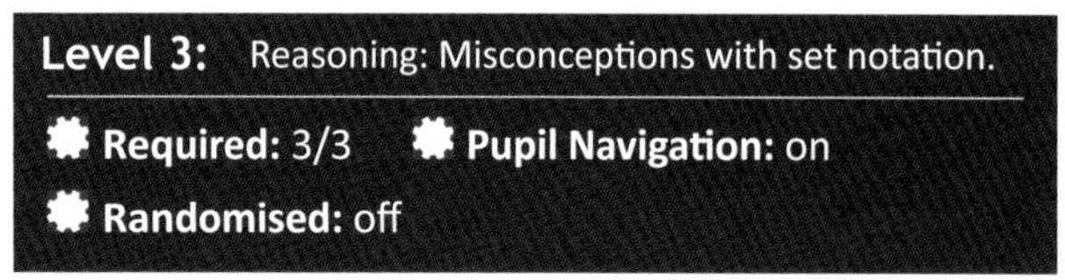

Level 3: Reasoning: Misconceptions with set notation.

Required: 3/3 **Pupil Navigation:** on
Randomised: off

17. Owen is completing his homework on set notation. Which question has Owen got wrong? Explain your answer.

- Open question, no set answer

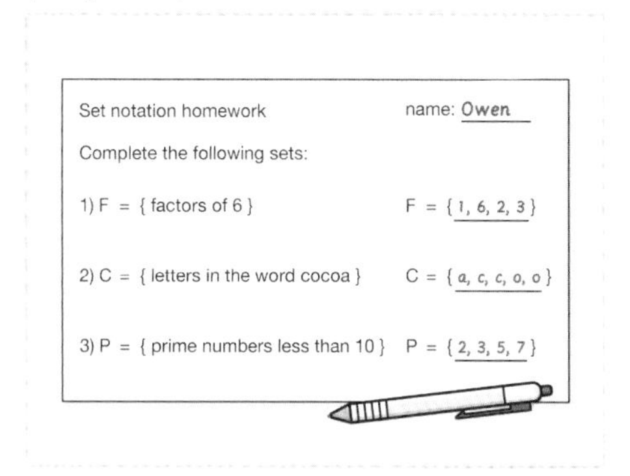

Set notation homework name: Owen

Complete the following sets:

1) F = { factors of 6 } F = { 1, 6, 2, 3 }

2) C = { letters in the word cocoa } C = { a, c, c, o, o }

3) P = { prime numbers less than 10 } P = { 2, 3, 5, 7 }

18. The universal set ξ = {positive integers from 1 to 20 inclusive} and the set S = {the set of square numbers}. How many values of x are there such that $x \notin S$?

- 16
- 4

19. Aman says that there is a mistake in the text book.
Is Aman correct? Explain your answer.

- Open question, no set answer

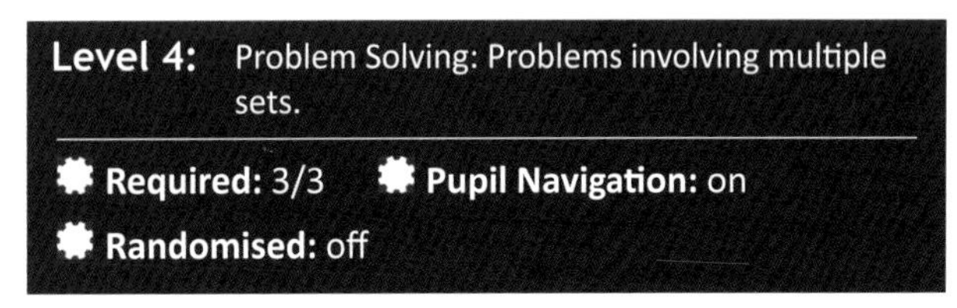

20. P = {the set of prime numbers} and S = {the factors of 42}.
Select the numbers that are elements of both P and S.

3/7

- 1 - **2** - **3** - 4 - 5 - 6 - **7**

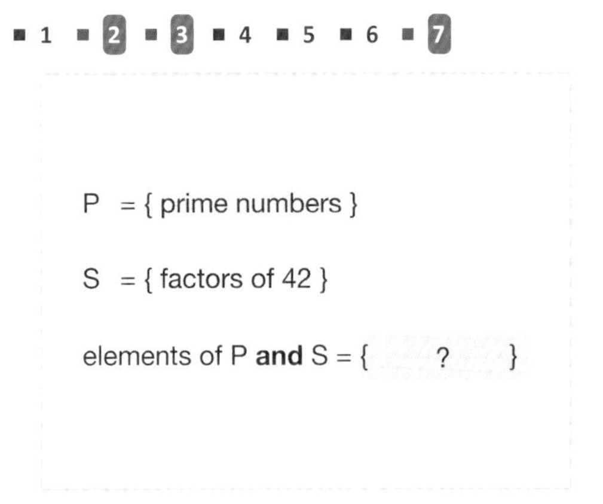

21. Three sets are defined as follows:

- T = {multiples of 3}
- F = {multiples of 4}
- S = {square numbers}

What is the smallest number that is an element of T, F and S?

- **36**

22. If B = {letters in the word banana} and O = {letters in the word orangutan}, how many elements are in B or O or both?

B = { letters in the word banana }

O = { letters in the word orangutan }

Probability and Venn Diagrams Topic Review

Objective: I can answer questions about probability and Venn diagrams.

Quick Search Ref: 10754

Level 1: Understanding

Required: 10/10 **Pupil Navigation:** on **Randomised:** off

1. Which arrow represents 'even chance' on the probability scale?

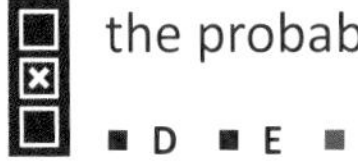

■ D ■ E ■ F ■ G ■ H

1/5

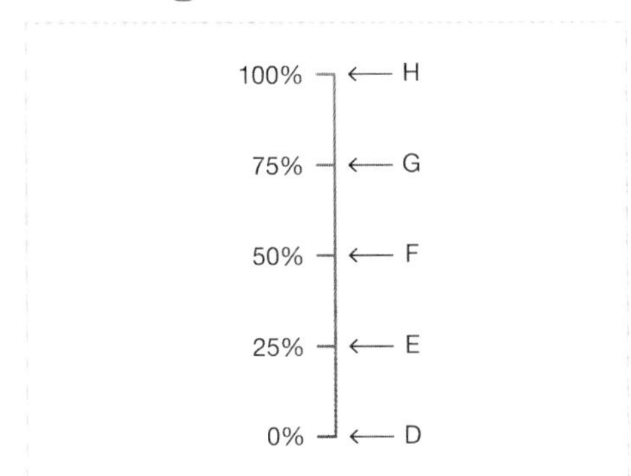

2. What is the probability of picking the cube?
Give your answer as a decimal.

■ 0.25

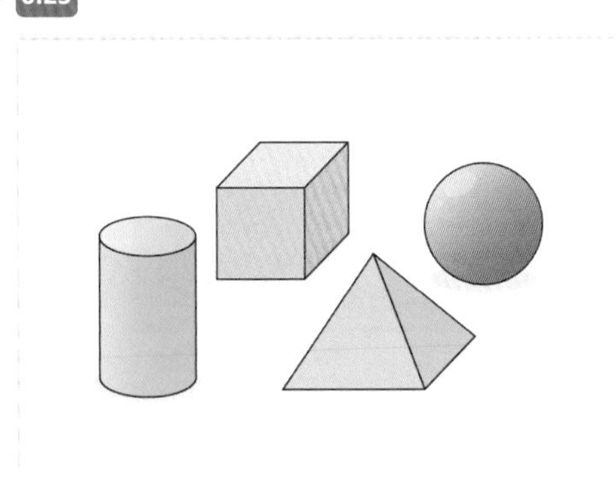

3. What is the probability of not picking a raspberry sweet?
Give your answer as a fraction.

■ 4/7 ■ 3/7

4. If Mark flips a fair coin, what is the likelihood it will land on tails?
Give your answer as a fraction.

■ 1/2

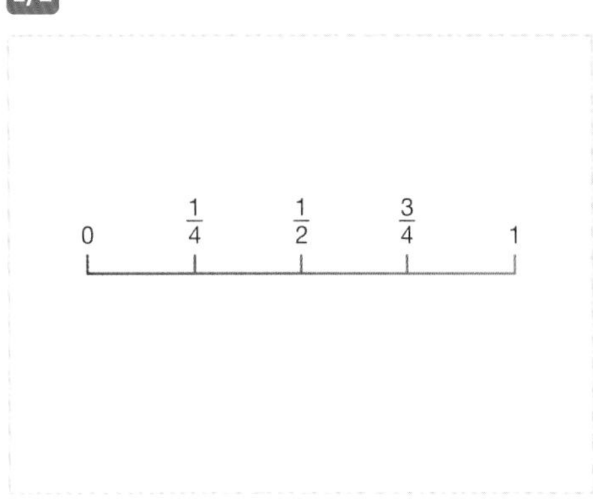

5. A school canteen offers a choice of 4 main meals and 3 desserts. How many combinations of one main meal and one dessert can a student choose?

■ 12 ■ 20 ■ 19

main

dessert	curry	pizza	stew	lasagne
fruit				
yoghurt				
sponge				

6. Which set matches this description?
{the letters in the word tomato}

■ {t, o} ■ {t, o, m, a, t, o} ■ {a, m, o, t}

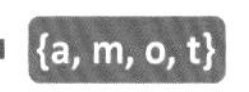

1/4

■ {a, m, o, o, t, t}

Level 1 *continued*

7. Which regions would you need to shade to represent D'?

2/5

■ (i) ■ (ii) ■ (iii) ■ (iv) ■ (v)

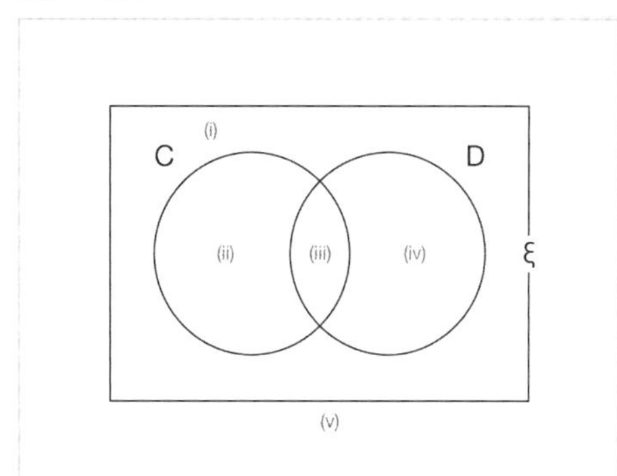

8. What is the probability of rolling an even number with a fair 6-sided dice?
Give your answer as a fraction in its simplest form.

■ 1/2 ■ 3 ■ 3/6

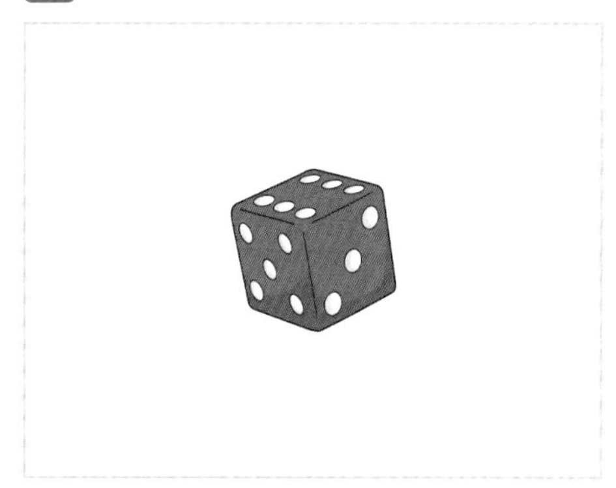

9. If the probability of it snowing tomorrow is 1/8, what is the probability of it not snowing?
Give your answer as a fraction.

■ 7/8

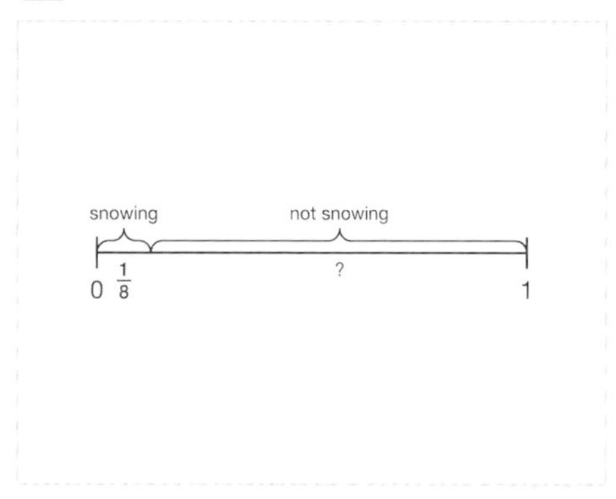

10. Which region is represented by the diagram?

1/4

■ E' ■ E ∩ P ■ P ■ E ∪ P

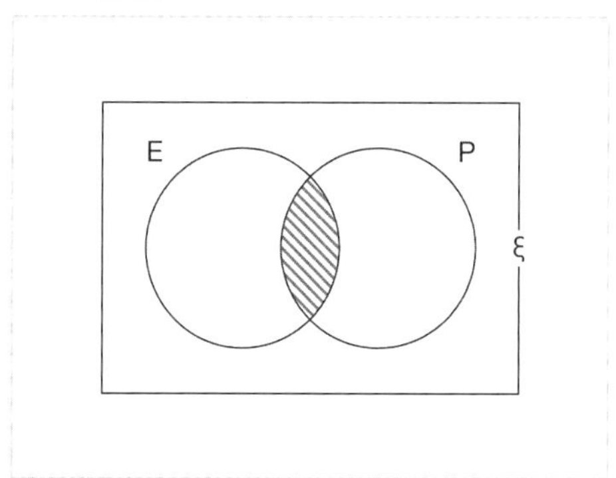

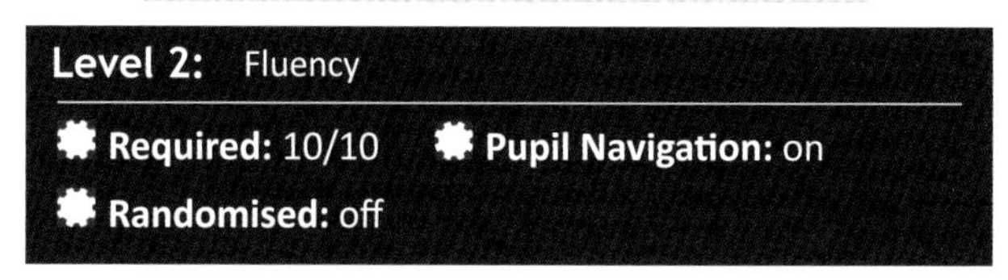

Level 2: Fluency

Required: 10/10 **Pupil Navigation:** on

Randomised: off

11. What is the probability that the spinner lands on blue or orange?
Give your answer as a decimal.

■ 0.5 ■ 1/2

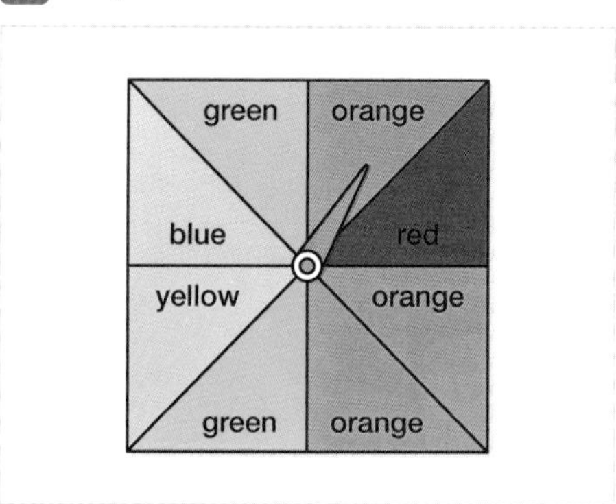

12. What is the probability of not rolling a 3 on a fair, standard six-sided dice?
Give your answer as a fraction.

■ 5/6 ■ 1/6

Level 2 *continued*

13. Kobi has six letter tiles that together spell out the word 'chance'. Kobi selects a tile, replaces it and then selects a second tile. What is the probability he selects two vowels?
Give your answer as a fraction in its simplest form,

- 1/9 ■ 1/36 ■ 2/18 ■ 4/36

second tile / first tile

	C	H	A	N	C	E
C						
H						
A						
N						
C						
E						

14. ξ = {integers from 1 to 12 inclusive}
F = {factors of 12}
P = {prime numbers}
1/5 Which set represents F ∪ P?

- {8, 9, 10} ■ {2, 3, 5, 7, 11} ■ {1, 2, 3, 4, 5, 6, 7, 11, 12}
- {2, 3} ■ {1, 2, 3, 4, 6, 12}

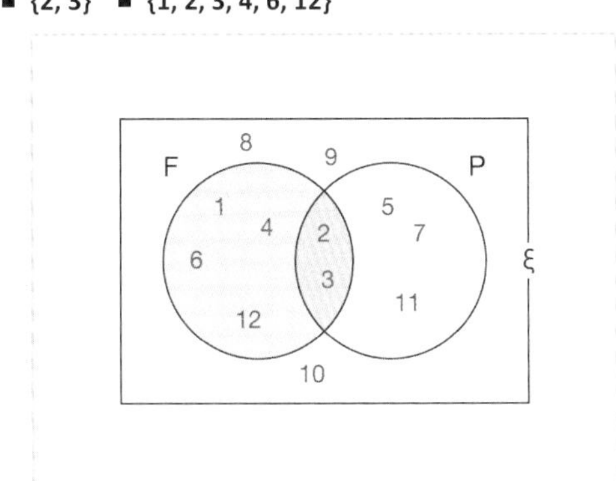

15. ξ = {letters in the alphabet}

H = {letters in the word 'holiday'}
S = {letters in the word 'school'}
4/7 Which elements are members of H ∩ S'?

- c ■ a ■ z ■ y ■ i ■ s ■ d

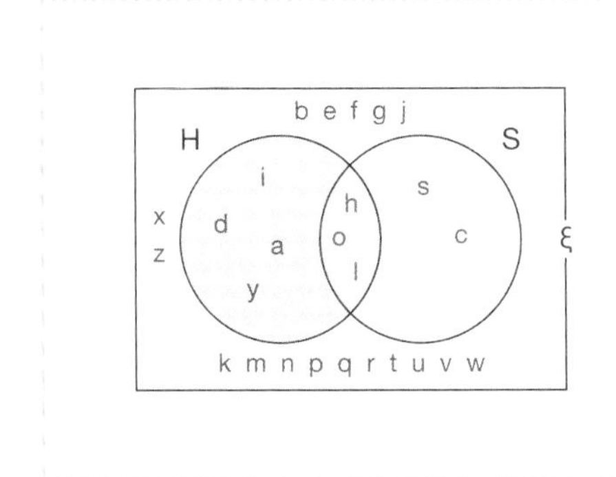

16. Ethan is going to take a shape from the bag. Select the statement that is true.

1/4

- Ethan's shape will definitely be a square.
- The probability of Ethan picking a triangle is 1/5.
- There is an even chance of picking a triangle or star.
- The difference between the probabilities of choosing a circle and a star is 2.

17. A bag contains a 5p coin and two 10p coins. Another bag contains two 5p coins and three 10p coins. If a coin is selected at random from each bag, what is the probability that they total 20p?
Give your answer as a fraction in its simplest form.

- 2/5 ■ 6/15

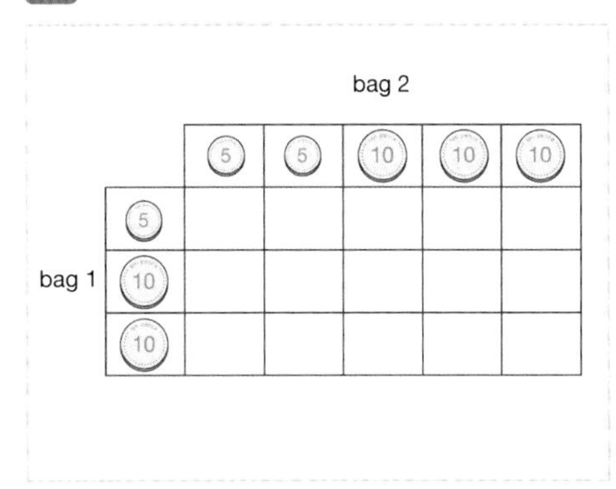

18. The probability of each student being chosen for the main part in the school play is shown in the table. Jenny and Becca have an equal chance of being picked. What is the probability of Becca getting the part?
Give your answer as a decimal.

- 0.1 ■ 0.2

Jenny	?
Romana	0.3
Elsie	0.45
Becca	?
Isla	0.05

Level 2 *continued*

19. Which description fully describes this set? {square, rhombus}

1/5

- {quadrilaterals}
- {quadrilaterals with 2 pairs of equal angles}
- **{quadrilaterals with 4 equal sides}**
- {quadrilaterals with 2 pairs of parallel sides}
- {quadrilaterals with 4 equal angles}

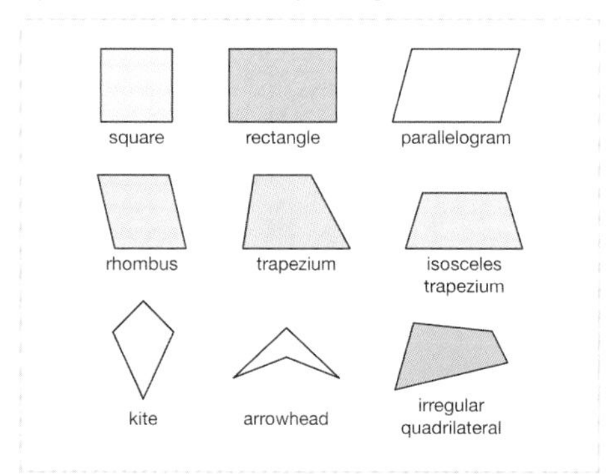

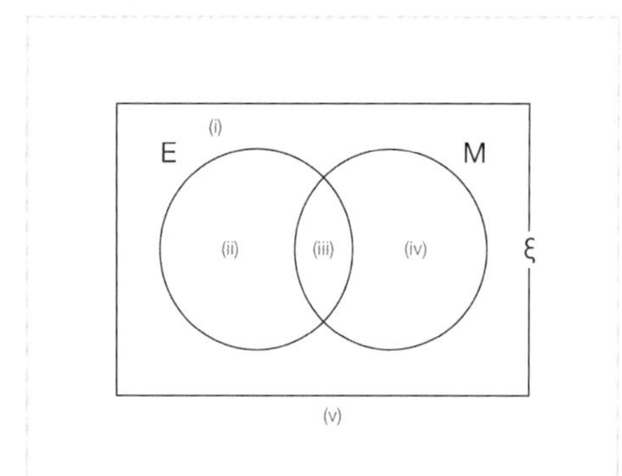

20. Which region would you need to shade to represent E ∩ M'?

1/5

- (i)
- **(ii)**
- (iii)
- (iv)
- (v)

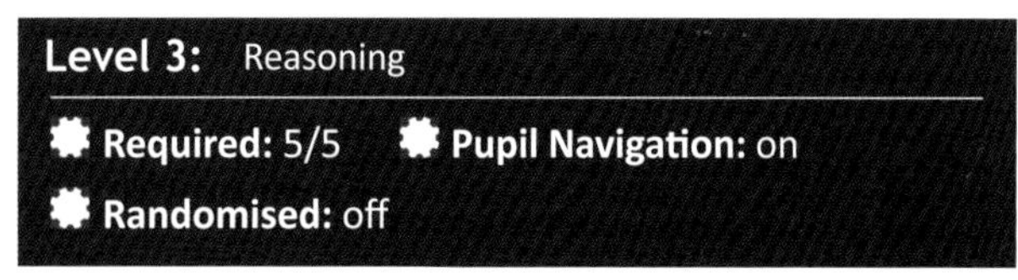

Level 3: Reasoning

Required: 5/5 **Pupil Navigation:** on

Randomised: off

21. When rolling a fair, standard six-sided dice, are the following events exhaustive? Explain your answer.

- rolling an even number
- rolling a multiple of 5
- rolling a prime number

- Open question, no set answer

22. Melanie threw a coin 50 times and it landed heads up 9 times. What is the probability of the coin landing heads up on the next throw?

Give your answer as a fraction in its simplest form.

- **1/2**

23. Owen is completing his homework on set notation. Which question has Owen got wrong? Explain your answer.

- Open question, no set answer

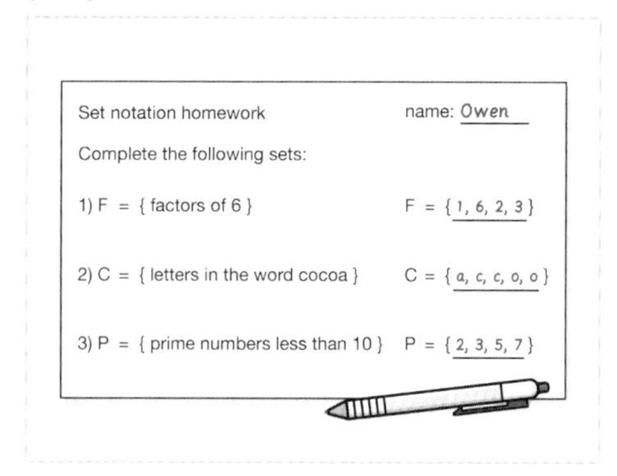

Set notation homework name: Owen

Complete the following sets:

1) F = { factors of 6 } F = { 1, 6, 2, 3 }

2) C = { letters in the word cocoa } C = { a, c, c, o, o }

3) P = { prime numbers less than 10 } P = { 2, 3, 5, 7 }

24. Callie rolls two standard six-sided dice. What is the most likely total?

- **7**

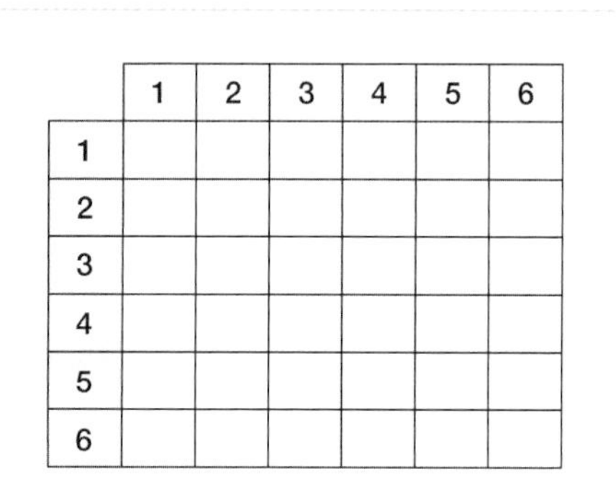

	1	2	3	4	5	6
1						
2						
3						
4						
5						
6						

25. Zenya is drawing a Venn diagram to show elements of two sets A and B. Use her diagram to decide which statement must be true.

1/4

- A ∪ B = ∅
- A ∩ B = ξ
- A ∪ B = ξ
- **A ∩ B = ∅**

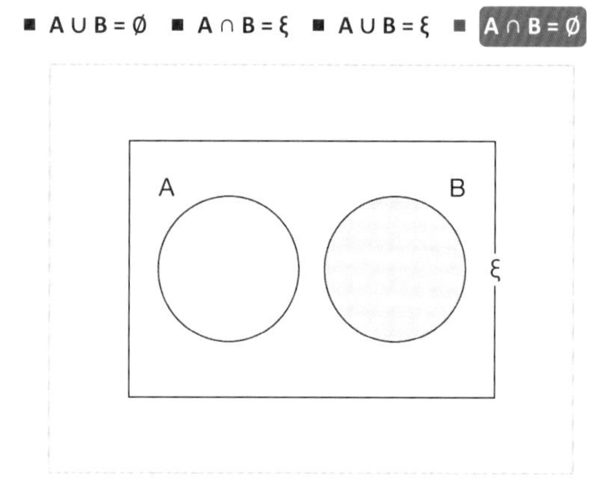

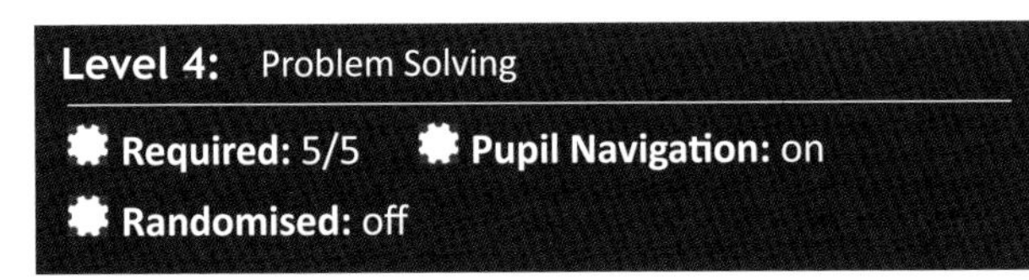

26. Fazel rolls two standard six-sided dice and multiplies together the numbers they land on. What is the probability his product is even?
Give your answer as a fraction in its simplest form.

- **3/4** ▪ 27/36 ▪ 1/2

	1	2	3	4	5	6
1						
2						
3						
4						
5						
6						

27. In a class of 30 students, 15 students play football, 7 boys don't play football and 6 girls do play football.
If a girl is selected at random, what is the probability that they play football?
Give your answer as a fraction in its simplest form.

- **3/7** ▪ 1/5 ▪ 6/14 ▪ 6/30

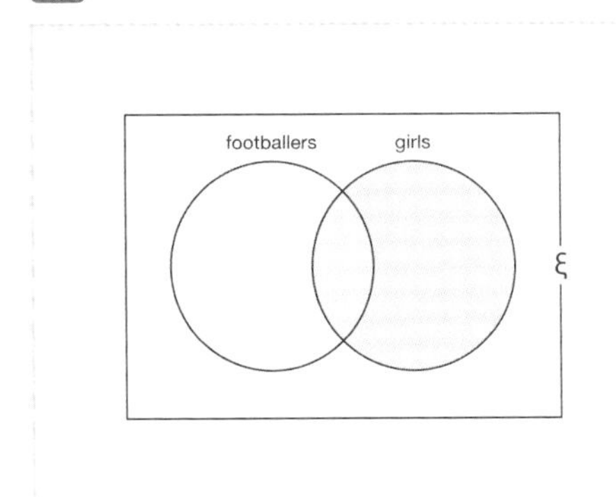

28. Three sets are defined as follows:

- T = {multiples of 3}
- F = {multiples of 4}
- S = {square numbers}

What is the smallest number that is an element of T, F and S?

29. Five cards are numbered 1 to 5 and placed in a box. Halima chooses a card, keeps it and then selects a second card. What is the probability that her two cards total 5?
Give your answer as a fraction in its simplest form,

- **1/5** ▪ 4/20 ▪ 4/25

second card

first card	1	2	3	4	5
1					
2					
3					
4					
5					

30. A year group of 120 students went on an adventure holiday and completed various activities.
15 only did bowling, and 42 did bowling altogether.
60 students did canoeing.
35 students did canoeing and archery.
28 students did bowling and canoeing.
12 students did all three activities and 7 did none of them.
How many students only did archery?

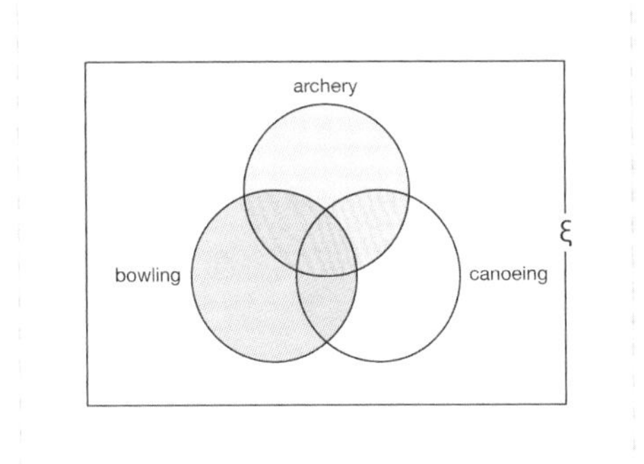

LbQ Super Deal

Class set of tablets and charging cabinet

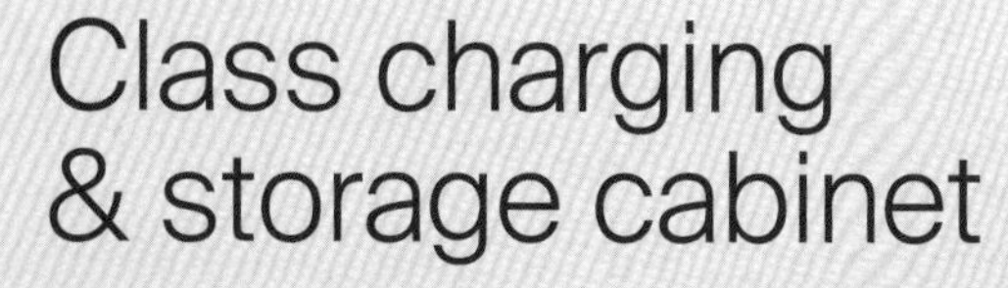

+ 32 x
Pupil 7" HD tablets with protective cover

+ 1 x
Teacher 10" HD tablet with protective cover

*Special Offer Price

£1,100 per year on 3 years compliant operating lease

Subject to a £150 initial documentation fee

LbQ Question Set subscription required to be eligible

Min 1 LbQ subscription per set £200/year or £500/3 years

Learning by Questions App pre-loaded

Includes 3 years advanced replacement warranty on tablets - damage not covered

Prices exclude VAT and delivery

Option to renew equipment or purchase at end of agreement

Available in United Kingdom and Republic of Ireland only

Product Code TC001

£1,100* per year for 3 years including warranty

Place your orders with our sales partner LEB who will organise the paperwork for you:

Email: **orders@lbq.org**
Tel: 01254 688060

Specifications

Charging & Storage Cabinet

- 32-bay up to 10" tablet charging cabinet
- 2 easy access sliding shelves – 16 bays on each shelf
- 4 efficient quiet fans for ventilation
- Locking Doors (4 keys)
- 4 castors / 2 x handling bars Overload, leakage and lightning surge protection
- CE / ROHS / FCC compliancy
- Warranty 3 years

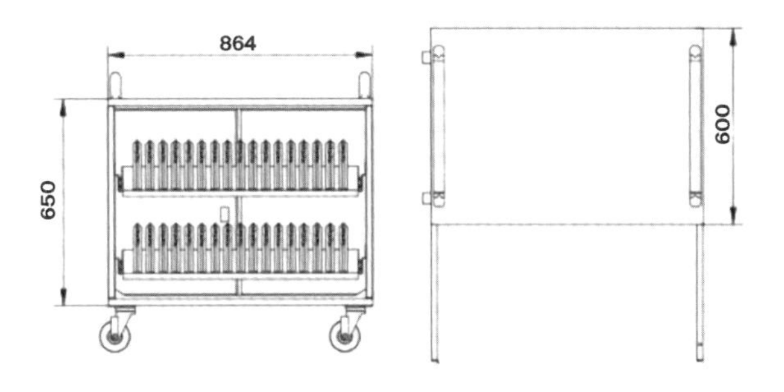

Student and Teacher tablets configuration

All tablets with LbQ Tablet Tasks App pre-loaded and installed in cabinet including charging cables and mains adapters for quick and easy deployment when onsite in classroom.

	7" Android Tablet with Protective Cover	10" Android Tablet with Protective Cover
Display		
1920 * 1200 IPS	✓	✓
16:10 Display ratio	✓	✓
Capacitive 5-touch	Capacitive 5-touch	Capacitive 10-touch
System		
Cortex 64bit Quad Core 1.5GHz CPU	✓	✓
2GB of RAM	✓	✓
16Gb of storage	✓	✓
Android 7.0	✓	✓
Front 2M and rear 5M Camera	✓	✓
Input / Output Ports		
1 x Micro SD Slot	✓	✓
1 x Micro USB (PC / device / charger)	✓	✓
Micro-HDMI output	✓	✓
1 x Earphone, 1 x Speaker, 1 x Mic	1 x Earphone, 1 x Speaker, 1 x Mic	1 x Earphone, 2 x Speaker, 1 x Mic
Communication		
Wifi – 802.11a/b/g/n	✓	✓
GPS module	✓	✓
Bluetooth	✓	✓
Power		
5V 2A	✓	✓
Battery	3000mAh battery	7000mAh battery
Physical		
Colour: Metal black	✓	✓
Weight:	230g	560g
Dimensions:	192 x 112 x 9mm (approx.)	263 x 164 x 9mm (approx.)
Warranty		
3 years Advanced Replacement for faulty tablets - does not cover damage CE / ROHS / FCC compliant	✓	✓